TRANSPORT AND DIFFUSION IN TURBULENT FIELDS

TRANSPORT AND DIFFUSION IN TURBULENT FIELDS

Modeling and Measurement Techniques

35th Oholo Conference
Israel Institute for Biological Research

Edited by

HADASSAH KAPLAN, NATHAN DINAR
AVI LACSER and YEHUDAH ALEXANDER

Reprinted from *Boundary-Layer Meteorology*, Volume 62, 1993

SPRINGER SCIENCE+BUSINESS MEDIA, B.V.

Library of Congress Cataloging-in-Publication Data

Transport and diffusion in turbulent fields : modeling and measurement techniques / edited by Hadassah Kaplan, Nathan Dinar, Avi Lacser and Yehudah Alexander.
p. cm.
Contains lectures presented at the 35th Oholo Conference, held in Eilat, Israel, Oct. 28-Nov. 1, 1991.
ISBN 978-0-7923-2035-7 ISBN 978-94-011-2749-3 (eBook)
DOI 10.1007/978-94-011-2749-3
1. Atmospheric turbulence--Congresses. 2. Atmospheric diffusion--Congresses. 3. Atmospheric physics--Congresses. 4. Transport theory--Congresses. I. Kaplan, Hadassah. II. Dinar, Nathan. III. Oholo Conference (35th : 1990 : Elat, Israel)
QC880.4.D44T7 1992
551.55--dc20 92-35071

ISBN 978-0-7923-2035-7

Printed on acid-free paper

TABLE OF CONTENTS

PART II: ANALYSIS OF CONCENTRATION FLUCTUATIONS

PART III: EXPERIMENTAL TECHNIQUES AND METHODS OF DATA ACQUISITION

PREFACE

This volume contains the invited and contributed lectures that were presented at the 35th Oholo Conference, held in Eilat, Israel, during October 28–November 1, 1991. This meeting, in common with the 28th Oholo Conference (1984) in Zichron Yaakov, was organized by the Environmental Sciences Division of the Israel Institute for Biological Research. The present conference dealt with theoretical, experimental, and practical aspects of "Transport and Diffusion in Turbulent Fields".

The scientists attending the meeting came from over a dozen different countries, and met to exchange ideas and share results of research in the following broad range of topics:

Basic studies and novel methods in modeling atmospheric flows – theory and experiments involving eddy flow; turbulent mixing, transport and diffusion in various media, ranging from microscale to macroscale.

Analysis of concentration fluctuations – theoretical developments; observations in different atmospheric conditions.

Experimental techniques and methods of data acquisition – use of various sensors; laboratory experiments and observations in the atmosphere and in the ocean.

Finally, on the last morning a round table discussion – punctuated by lively audience participation – was devoted to the topic of: *Interactions and feedback between theoretical models and experimental results.*

The advances in our understanding of turbulent diffusion and transport over the past few decades have been quite impressive; it is our hope that the fruitful discussions held in Eilat will lead to cooperation and still further progress in this wide ranging field.

THE EDITORS

PERMANENT COMMITTEE

SCIENTIFIC ORGANIZING COMMITTEE
(OHOLO Conference)

ACKNOWLEDGEMENTS

The Organizing Committee of the 35th. OHOLO Conference gratefully acknowledges the generous support of the following organizations:

U.S. ARMY RESEARCH AND DEVELOPMENT, EUROPEAN BRANCH, LONDON, U.K.
ISRAEL ACADEMY OF SCIENCES AND HUMANITIES, JERUSALEM, ISRAEL.
JOSEPH MEYERHOFF FUND INC., BALTIMORE, MD., U.S.A.
MINISTRY OF SCIENCE & TECHNOLOGY, JERUSALEM, ISRAEL.
MINISTRY OF ENVIRONMENT, JERUSALEM, ISRAEL.
ISRAEL ELECTRIC CORP LTD., HAIFA, ISRAEL.
ATOMIC ENERGY COMMISSION, TEL-AVIV, ISRAEL.
ISRAEL CHEMICALS LTD., TEL-AVIV, ISRAEL.
MINISTRY OF TOURISM, JERUSALEM, ISRAEL.

WELCOMING ADDRESS

I want to welcome you to Israel, to sunny Eilat, and to the 35th OHOLO conference on 'Transport and Diffusion in Turbulent Fields'.

IIBR is primarily oriented towards applied research and development, mainly in the fields of biology, chemistry, ecology and public health. The OHOLO conferences have been organized by the Institute for 35 years. In 1984 we organized a conference on 'The Boundary Layer Structure – Modeling and Application to Air Pollution and Wind Energy'. That conference took place in Zichron Yaakov, a somewhat sleepy town and I think that among our guests from abroad only Dr. Hunt and Dr. Hanna participated then.

In 1993 we are planning to hold together with IGAC (International Global Atmospheric-Biospheric Chemistry Project) a conference on Atmospheric Chemistry and I invite you all to participate.

As you all know the conference was scheduled for April 1991 but Sadam didn't get our invitation or misinterpreted our call for contributions and we had to postpone the conference.

I guess that the relatively small number of participants today is because of the political situation in the region but we believe that science and politics shouldn't be mixed, especially when dealing with ways of better understanding our environment.

I wish you all an interesting conference and hope for fruitful discussions leading perhaps to joint projects.

I wish you again a pleasant stay in Israel.

DR. A. LACSER
Head Environ. Res. Div.

LIST OF PARTICIPANTS

ALEXANDER, Y.	Israel Institute for Biological Research, Ness-Ziona, Israel.
ALMOG, H.	Israel Electric Corp., Haifa, Israel.
ALPERT, P.	Tel-Aviv University, Tel-Aviv, Israel.
ARELLANO V-G, J.	University of Utrecht, Utrecht, The Netherlands.
AZOR, M.	Israel Institute for Biological Research, Ness-Ziona, Israel.
BANGE, P.	NV KEMA, Arnhem, The Netherlands.
BARKAN, Y.	Israel Institute for Biological Research, Ness-Ziona, Israel.
BAUER, T.	U.S. Army, USA.
BAUMGARTEN, Y.	Israel Institute for Biological Research, Ness-Ziona, Israel.
BRANOVER, H.	Ben-Gurion University, Beer Sheva, Israel.
BRIGGS, G.A.	USEPA, Research Triangle Park, NC., U.S.A.
CHATWIN, P.	University of Sheffield, Sheffield, U.K.
DAHLQUIST, H.	Swedish Air Force, Stockholm, Sweden.
DINAR, N.	Israel Institute for Biological Research, Ness-Ziona, Israel.
DUYNKERKE, P.G.	Univ. of Utrecht, Utrecht, The Netherlands.
EGERT, S.	Israel Institute for Biological Research, Ness-Ziona, Israel.
FELIKS, Y.	Israel Institute for Biological Research, Ness-Ziona, Israel.
GAVZE, E.	Israel Institute for Biological Research, Ness-Ziona, Israel.
GLAZER E.	Israel.
GOLSDTEIN, J.	Tel-Aviv University, Tel-Aviv, Israel.
GRABER, M.	Ministry of the Environment, Jerusalem, Israel.
GUTMAN, A.	Kinneret Limnological Lab., Tiberias, Israel.
HADAD, A.	Technion, Haifa, Israel.
HANNA, S.R.	Sigma Res. Corp. Westford, MA, U.S.A.
HUNT, J.C.R	University of Cambridge, Cambridge, U.K.
JORGENSEN, H.E.	Riso Nat. Lab., Roskilde, Denmark.
KALLOS, G.	Univ. of Athens, Athens, Greece.
KAPLAN, H.	Israel Institute for Biological Research, Ness-Ziona, Israel.
KLEIMAN, M.	Israel Institute for Biological Research, Ness-Ziona, Israel.
KLUG, W.	Inst. of Met., Darmstadt, Germany.
LACSER, A.	Israel Institute for Biological Research, Ness-Ziona, Israel.
LEMPERT, Y.	Israel Institute for Biological Research, Ness-Ziona, Israel.
LIEMAN, R.	Tel-Aviv University, Tel-Aviv, Israel.
LOFSTROM, P.	Nat. Env. Res. Inst., Roskilde, Denmark.
MAHRER, I.	Hebrew Univ., Rehovot, Israel.

Boundary-Layer Meteorology **62**: xiii–xiv, 1993.

MALINA, J.	Soreq Nuclear Res. Center, Yavne, Israel.
ODUYEMI, K.	Dundee Inst. of Techn., Dundee, Scotland.
PIRINGER, M.	Central Inst. of Met., Hohewarte, Vienna, Austria.
POREH, M.	Technion, Haifa, Israel.
RANTALAINEN, L.	Imatron Voima OY, Vantaa, Finland.
RIDE, D.J.	CBDE Porton Down, Salisbury, U.K.
SADEH, W.Z.	Colorado State University, Colorado, U.S.A.
SAWFORD, B.	CSIRO, Mordialloc, Victoria, Australia.
SCHMIDT, J.	IABG, Germany.
SETTER, I.	Israel Meteorological Service, Beit Dagan, Israel.
SHTEINMAN, B.	Kinneret Limnological Lab., Tiberias, Israel.
SHTEINMAN, W.	Tiberias, Israel.
SIVAN, Y.	Israel Institute for Biological Research, Ness-Ziona, Israel.
SKIBIN, D.	Nuclear Research Center, Beer-Sheva, Israel.
STIASSNIE M.	Technion, Haifa, Israel.
STULL, R.B.	Univ. of Wisconsin, Madison, U.S.A.
TERCIER, P.	Station Aerologique, Payerne, Switzerland.
THOMAS, P.	Kernforschungszentrum Karlsruhe, Germany.
TOKAR, Y.	Tel-Aviv University, Tel-Aviv, Israel.
UMBERTO, G.	Italy.
VAN DOP, H.	World Met. Org., Geneve, Switzerland.
VOGT, S.	Kernforschungszentrum Karlsruhe, Germany.
WEBMAN F.	IBM Haifa, Israel.
WILSON, J.D.	University of Alberta, Edmonton, Canada.

PART I

BASIC STUDIES AND NOVEL METHODS IN MODELING ATMOSPHERIC FLOWS

UNCERTAINTIES IN AIR QUALITY MODEL PREDICTIONS

S. R. HANNA
Sigma Research Corporation, 234 Littleton Road, Suite 2E
Westford, Massachusetts 01886, U.S.A.

(Received October 1991)

Abstract. As a result of several air quality model evaluation exercises involving a large number of source scenarios and types of models, it is becoming clear that the magnitudes of the uncertainties in model predictions are similar from one application to another. When considering continuous point sources and receptors at distances of about 0.1 km to 1 km downwind, the uncertainties in ground-level concentration predictions lead to typical mean biases of about ±20 to 40% and typical relative root-mean-square errors of about 60 to 80%. In fact, in two otherwise identical model applications at two independent sites, it is not unusual for the same model to overpredict by 50% at one site and underpredict by 50% at the second site. It is concluded that this fundamental level of model uncertainty is likely to exist due to data input errors and stochastic fluctuations, no matter how sophisticated a model becomes. The tracer studies that lead to these conclusions and have been considered in this study include: (1) tests of the Offshore and Coastal Dispersion (OCD) model at four coastal sites; (2) tests of the Hybrid Plume Dispersion Model (HPDM) at five power plants; (3) tests of a similarity model for near-surface point source releases at four sites; and (4) tests of 14 hazardous gas models at eight sites including six sets of experiments where dense gases were released.

1. Introduction

There is not a shortage of air quality models. There are currently available dozens of alternate forms of the Gaussian plume model, several three-dimensional numerical codes, a plethora of similarity and statistical models, and even a few so-called state-of-the-art models based on recent advances in boundary-layer theories. But the uncertainties of these models are not well-known, because of the lack of comprehensive studies in which various models are evaluated with independent data sets.

Some authors have attempted to estimate the uncertainties in air quality models applied to continuous point sources and receptors on the order of 1 km downwind and there is growing body of evidence to support Pasquill's (1974) contention that the uncertainties in typical applications are about a factor of two, and that the uncertainties in high-grade research programs are about ±20%. These numbers apply to concentration averaging times on the order of 10 minutes and to maximum plume centerline concentrations at a given distance (i.e., the observed and predicted maxima may be at different receptor position). Fox (1984), Smith (1984), and Irwin et al. (1987) have set the stage for the "modern era" in air quality model evaluations by outlining the components of model uncertainty, suggesting methods for estimating this uncertainty, and providing some examples of comparisons with databases. They point out that, no matter how good the science is in a model, there will always be uncertainties due to data input errors and due to stochastic (turbulence) processes. Fox (1984) states that there are preliminary formulas available for estimating the magnitude of model uncertainties due to these components, which together define what may be called the "minimum achievable model uncertainty". However, these formulas have not reached the operational stage where they can be applied with confidence to a wide range of scenarios.

The objective of this paper is to attempt to quantify the "minimum achievable model uncertainty" based on a large number of independent exercises in which the predictions of air quality models are compared with observations during field programs. It is expected that if enough of these exercises are analyzed, then a consistent pattern may appear. The author has been involved in four model evaluation programs over the past two years (Hanna and Paine, 1989; DiCristofaro and Hanna, 1989; Hanna, Strimaitis and Chang, 1991; Hanna, Chang and Strimaitis, 1990) and believes that it is useful to summarize the results in order to demonstrate that a consistent pattern is indeed emerging.

The results that are given below cover 21 models and 21 field sites. Because it is not possible to describe each model and each field site within a reasonable number of pages, the reader is referred to papers and reports for these details. Instead, we will emphasize general characteristics of the models and the datasets and the typical uncertainties of the models.

2. Measures of Model Uncertainty

In most air quality model evaluation programs, the variable that is analyzed is the normalized concentration, C/Q, averaged over a period of several minutes to one hour, where Q is the source emission rate. Recognizing that it is very difficult for a model to match the exact orientation of a plume (due to uncertainties in wind directions), the concentration C is often defined as the maximum observed at a given time anywhere along a given downwind monitoring arc. For example, Figure 1 present the positions of the tracer gas (SF_6) monitors for one set of field experiments at the Kincaid field site (from Hanna and Paine, 1989). In this case, there are five monitoring arcs. For each one-hour sampling period, five pairs of observed and predicted concentrations could be selected for analysis. It is possible that the observed maximum would fall on the northern edge of a given arc and the predicted maximum would fall on the southern edge.

As another possibility, the effect of downwind distance could be removed and the concentration, C, to be evaluated could be defined as the maximum anywhere on the monitoring network. This definition may be appropriate for the monitoring network for Summit County, Ohio, pictured in Figure 2. There are 12 monitors interspersed among nine major point sources, plus a large number of smaller point sources in the area round Akron, Ohio. In the model evaluation exercises in Section 3, both optional definitions of C/Q mentioned above are used (i.e., maximum C/Q at given downwind distances or maximum C/Q at any location on the monitoring network).

The statistical evaluation methods used in the study are those described by

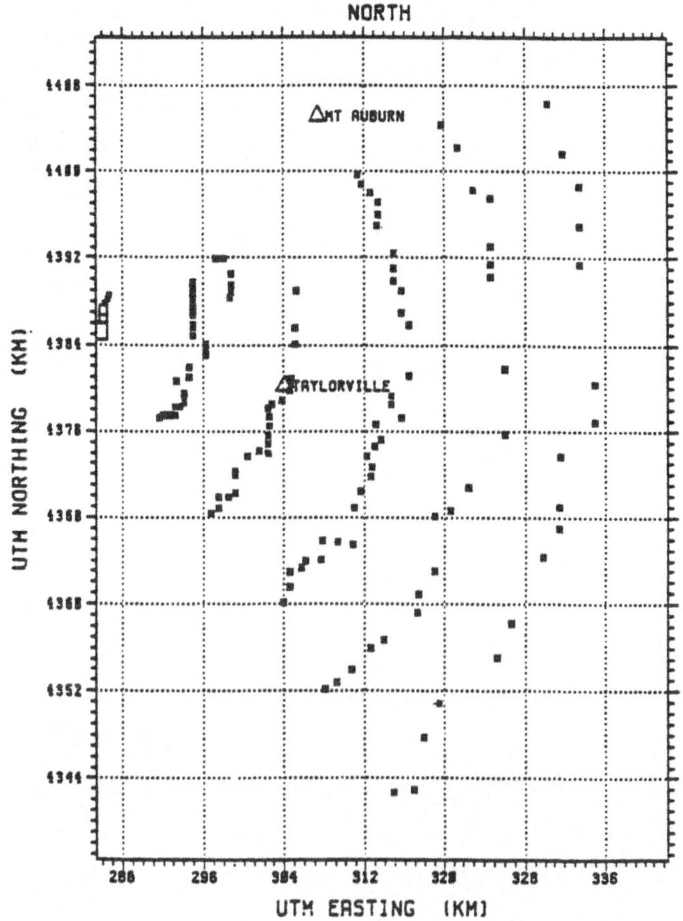

Fig. 1. The SF_6 tracer sampling array at Kincaid for neutral conditions and westerly winds (from Hanna and Paine, 1989).

Hanna (1989), who has applied various versions of the software to several air quality model scenarios. These procedures are similar to those proposed by Cox and Tikvart (1991). The software package can calculate the dimensionless model performance measures known as the fractional bias (FB), geometric mean bias (MG), normalized mean square error (NMSE), geometric mean variance (VG), correlation coefficient (R) and fraction within a factor of two (FAC2), which are defined below:

$$FB = (\bar{X}_o - \bar{X}_p)/(0.5(\bar{X}_o + \bar{X}_p)) \tag{1}$$

$$MG = exp(\overline{lnX_o} - \overline{lnX_p}) \tag{2}$$

$$NMSE = \overline{(X_o - X_p)^2}/(\bar{X}_o\bar{X}_p) \tag{3}$$

$$VG = exp[\overline{(lnX_o - lnX_p)^2}] \tag{4}$$

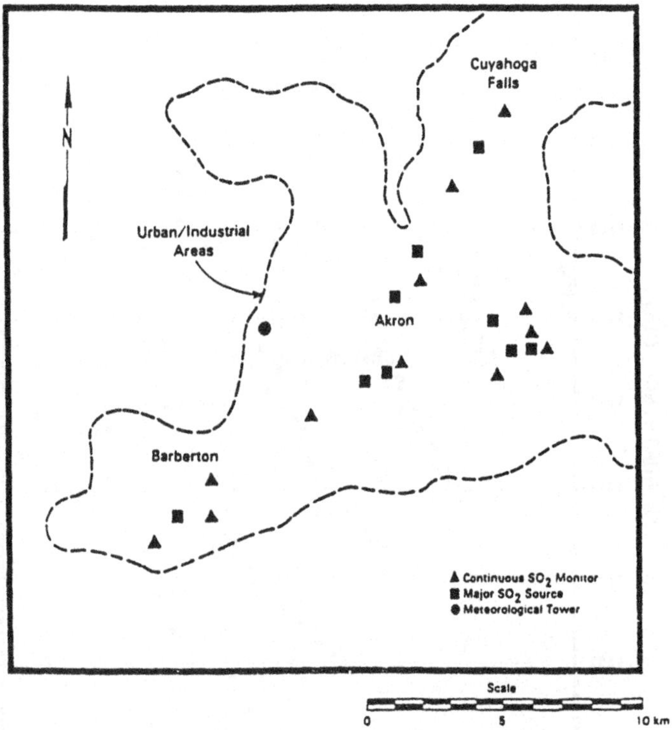

Fig. 2. Summit County Ohio, SO_2, monitoring network

$$R = \overline{(X_o - \bar{X}_o)(X_p - \bar{X}_p)}/\sigma_{X_p}\sigma_{X_o} \tag{5}$$

$$FAC2 = \textit{fraction of data for which } 0.5 \leq X_p/X_o \leq 2. \tag{6}$$

where X_o is an observed quantity, and X_p is the corresponding modeled quantity. The geometric mean variance, VG, and the geometric mean bias, MG, are recommended when there is a wide range in values of X_p/X_o in the date set, and are substitutes for the normalized mean square error, NMSE, and fractional bias, FB, respectively. The software also calculates 95% confidence limits on these performance measures using the bootstrap resampling method.

Residual plots are used as a method for evaluating the scientific credibility of a model. "Residual" is defined in this application as the ratio of the predicted to the observed concentration (note that the logarithm of this ratio equals the difference between the logarithm of the two concentrations). In other applications, the residual could be defined as the arithmetic difference between the observed and predicted concentrations. Values of the residual can be plotted versus variables such as wind speed or stability. The residual of a good model (1) should not exhibit any trend with variables such as wind speed and stability class, and (2) should not exhibit large deviations from unity (implying a perfect match between the model and the observed).

The residuals are grouped for plotting by means of "box plots." Grouping is usually necessary if there is a large number of data points in the set of field experiments being analyzed. The cumulative distribution function (cdf) of the residuals within each group is represented by the 2nd, 16th, 50th, 84th, and 98th percentiles. These five significant points in the cdf are then plotted in a "box" pattern. As mentioned above, the residual boxes should not exhibit any systematic dependence on primary variables. It is also desirable that the residual boxes should be compact and should not deviate too much from unity.

3. Results of Analyses of Model Predictions at Field Sites

The results of four separate model evaluation programs are summarized below.

3.1. Offshore and Coastal Dispersion (OCD) Model Evaluation

The Offshore and Coastal Dispersion (OCD) model has been developed (Hanna et al., 1984; DiCristofaro and Hanna, 1989) to simulate the effect of offshore emissions from point, area, or line sources on the air quality of coastal regions. Modifications were made to an existing EPA model (MPTER) to incorporate over-water plume transport and dispersion as well as changes that occur as the plume crosses the shoreline. The model, which is thoroughly discussed in the references, parameterizes the boundary layer over the water based on observations of the air-water temperature difference and the wind speed.

It was fortunate that several tracer experiments were conducted over coastal areas in order to develop a database for better understanding of coastal dispersion and for evaluation of models.These experiments took place at four independent sites (Ventura, CA; Pismo Beach, CA; Cameron, LA; and Carpinteria, CA). Separate experiments were conducted during two different seasons at the first three sites, and three types of independent experiments were conducted at the Carpinteria site. The characteristics of these nine experiments are given in Table I, and the locations of the sampling monitors, the meteorological towers, and the release site at Cameron, LA, are given in Figure 3 as an example of a typical site map. All pertinent data are listed in appendixes in the report by DiCristofaro and Hanna (1989).

Two versions of the OCD model were evaluated, and the performance of the model for several optional combinations of input data was assessed. Table II gives some of the results for the latest version of the model, OCD-4, for the case in which all available overwater meteorological observations (including lateral turbulence) are used as input to the model. It is seen that a total of 110 hours of tracer data are available over all the field studies.

Table II lists three performance measures: the maximum normalized C/Q during each experiment, the fractional bias (defined in equation (1)), and the normalized mean square error (defined in equation (3)). The results at the Ventura site are typical: the OCD model underpredicts the observed maximum C/Q by about 30% in the set of fall data, but overpredicts the observed maximum C/Q by about 20% in

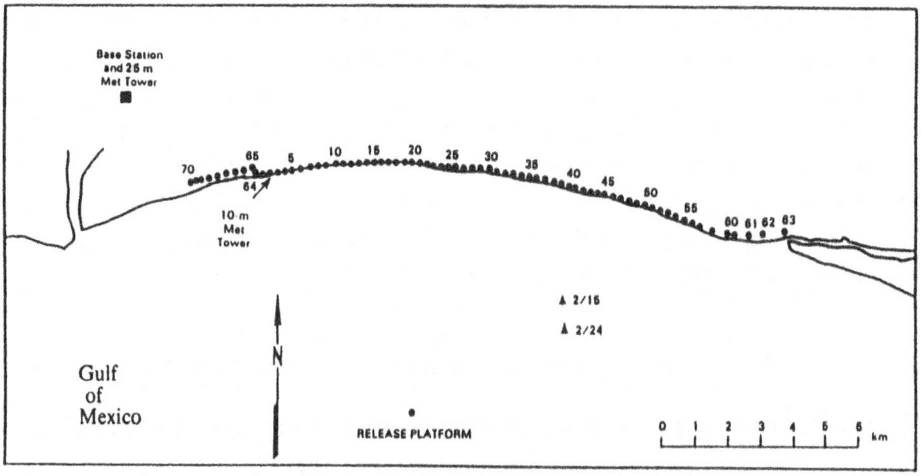

Fig. 3. Tracer sampling sites used in the Cameron, LA, experiment (sites have been renumbered for model evaluation). The tracer release points for 2/15 and 2/24 are indicated by triangles.

the set of winter data. The figures for fractional bias also indicate a swing from an overprediction of the mean by 35% in the fall and an underprediction by 22% in the winter. In fact, looking over the nine experiments, it is seen that there are five cases of overpredictions of the maximum C/Q, and four cases of underpredictions, with the median absolute magnitude of these uncertainties equal to 47%. For the mean values, the fractional bias has a median absolute magnitude of 0.22, indicating a typical uncertainty in the mean prediction of ±22% at any site. The normalized mean square error (NMSE) has a median value of 0.52 and a range from 0.17 to 1.38. The relative root mean square error (RMSE), or the square root of NMSE, has a median of about 70% for these nine sites.

The practical implication of these numbers is that it is natural for the mean bias of OCD model to vary by ±20% from site to site. Also, the relative RMSE is expected to be in the range from about 40% to 100%.

3.2. HPDM Model for Power Plant Stacks

The Hybrid Plume Dispersion Model (HPDM) was developed by Hanna and Paine (1989) for estimating transport and dispersion of air pollutants emitted by power plant stacks.

It was evaluated using tracer data from field experiments at the Kincaid, IL, and Bull Run, TN, power plants (Figure 1 provides an example of the monitoring network at Kincaid). Stack heights are about 200 m at those sites. Emphasis was on maximum hourly-averaged normalized concentration, C/Q, observed and predicted anywhere on the monitoring network. Following that comparison, Hanna and Chang (1991) modified the HPDM model so that it would be more valid over urban areas and evaluated the revised model using field data from Indianapolis. More recently, we have been evaluating the model using one-year of field data

TABLE I
Characteristics of nine coastal experiments used to evaluate the OCD model

Site	Hours, this Analysis	Source	Turbulence Obs.	Wind Velocity	Vert. T. Profile	Monitors
Ventura Fall	9	Boat, 13 m, 5–7 km offshore	Boat σ_θ	Boat	Aircraft	Arcl: 0.5 km onshore; Arc 2: 6–8 km onshore
Ventura Winter	8	Boat, 13 m, 5–7 km offshore	Boat σ_θ	Boat	Aircraft	Same as fall
Pismo Beach Summer	16	Boat, 13 m, 6–8 km offshore	Boat σ_θ	Boat	Aircraft	Arcl: Shoreline; Arc2: 6–8 km onshore
Pismo Beach Winter	15	Boat, 13 m, 6–8 km offshore	Boat σ_θ	Boat	Aircraft	Same as summer
Cameron Summer	9	Platform, 13 m, 7 km offshore	Shoreline, σ_θ, σ_w	Platform	Aircraft	Shoreline arc
Cameron Winter	17	Platform, 13 m, 7 km offshore*	Shoreline, σ_θ, σ_w	Platform	Aircraft	Shoreline arc
Carpinteria SF_6 Complex Terrain	18	Boat, 20–30 m, 0.3–0.7 offshore	Tethersonde	Tethersonde	Tethersonde	Arcl: Shoreline; Arc2: 1 km onshore
Carpinteria Freon,Complex terrain	10	Boat, 20–70 m, 0.3–0.7 offshore	Tethersonde,σ_θ	Tethersonde	Tethersonde	Arcl: Shoreline
Carpinteria Fumigation	9	Boat, 70–100 m, 0.3–0.7 offshore	Tethersonde,σ_θ	Tethersonde	Tethersonde	Arcl: Shoreline; Arc2: 1 km onshore

* At Cameron, the releases on 2/15 and 2/24 were from a boat located 4km offshore. It was evaluated using tracer data from field experiments at the Kincaid,IL site.

TABLE II
Summary of OCD-4 model performance for nine independent experiments

Dataset	Hours, this analysis	Maximum C/Q ($\mu s/m^3$)		FB=	NMSE=
		Observed	Predicted	$\frac{2(\overline{C_0/Q}-\overline{C_p/Q})}{(\overline{C_0/Q}+\overline{C_p/Q})}$	$\frac{\overline{(C_o/Q-C_p/Q)^2}}{(\overline{C_o/Q})(\overline{C_p/Q})}$
Ventura, Fall	9	2.8	1.9	-0.35	0.36
Ventura, Winter	8	3.0	3.7	0.22	0.17
Pismo Beach, Summer	16	7.8	15.0	-0.32	0.81
Pismo Beach, Winter	15	9.2	13.7	0.02	0.89
Cameron, Summer	9	3.0	2.4	-0.46	0.52
Cameron, Winter	17	37.0	31.4	-0.21	0.36
Carpinteria SF_6 Compl. Terrain	18	109.0	231.0	-0.22	1.38
Carpinteria Freon Terrain	9	25.0	26.7	-0.32	0.43
Carpinteria Fumigation	9	15.2	9.8	0.16	0.97

from the Baldwin, IL power plant and from an urban study in Akron, OH.

The HPDM model is similar to a standard Gaussian model in many respects, but makes better use of observed boundary-layer data. Also, it employs improved algorithms for dispersion in convective conditions and for plume interactions with the capping inversion at the top of the boundary layer. Nevertheless, it exhibits quite a bit of uncertainty when compared hour-by-hour with observations, as seen by the scatter plot in Figure 4. Even though the maxima observed and predicted C/Q agree within $\pm 10\%$, the correlation among all the points is not significantly different from zero.

Table III contains the characteristics of the datasets at the five sites and the model evaluation results, following the same format as Table II. Looking at the maximum hourly-averaged concentrations, we see that the predictions are within $\pm 10\%$ of the observation at four of the sites. However, there can be as much as a 30 to 40% over or underprediction. The median absolute magnitude of the fractional bias is 0.24, indicating that the relative bias in the mean is typically $\pm 24\%$. The normalized mean square error values fall into two groups-one group with a relative small value ($\sim$ 1) for tracer studies and urban areas, and a second group with NMSE $\sim$ 3 to 10 for hourly SO_2 data collected over periods of several months or more. The latter datasets tend to be dominated by low near-background values. Another reason for the relatively large scatter is the fact that the characteristics of the ambient boundary layer are simply not well known at all hours of the day at plume elevations of several hundred meters.

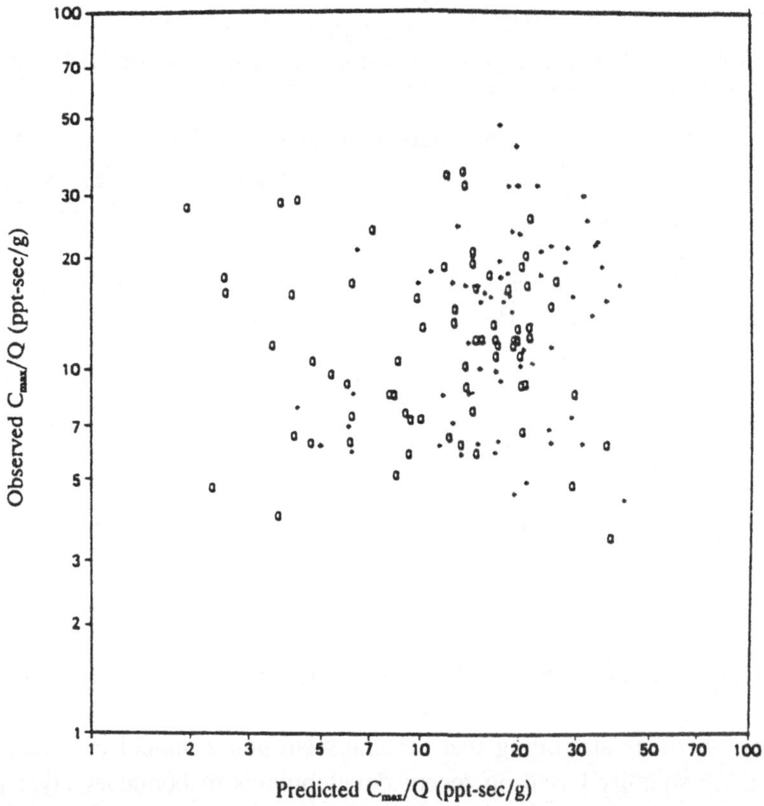

Fig. 4. Observed and predicted normalized hourly maximum ground-level concentration C_{max}/Q for Kincaid (circles) and Bull Run (crosses),(from Hanna and Paine, 1989).

3.3. Similarity Model Applied to Continuous Ground-Level Tracer Gas Releases

The Gaussian plume model for continuous point source releases is used in most EPA regulatory models. This model and its dispersion coefficients (σ_y and σ_z) were originally tested with high-quality research grade field data from the 1950's and 1960's, including data from the well-known Prairie Grass, Ocean Breeze, Dry Gulch, and Green Glow experiments. In fact these data have been used to calibrate the σ_y and σ_z coefficients. In view of the simplistic nature of these field experiments and the accuracy of the data, it is felt that the models should demonstrate optimum performance with these data.

These data have been analysed by many scientists. We obtained most of the data from Mr. Bruce Kunkel of the Air Force Geophysics Laboratory, who had been using them in the development of the new AFTOX model. The full databases are given in the report by Hanna and Chang (1991), and a brief summary of their characteristics is given in Table IV. It is seen that there are almost 250 different runs, for a variety of stability classes, and at downwind distances ranging from 50 to 25600 m. In general, the source emissions of tracer gas were continuous over a

TABLE III
Evaluation of HPDM at five sites (Note: C/Q is used at the first three sites, and C at the last two sites)

Site	Source	Tracer	Hours	Maximum C/Q ($\frac{10^{-9}s}{m^3}$) Observed	Maximum C/Q ($\frac{10^{-9}s}{m^3}$) Predicted	FB= $\frac{2(\overline{C_0/Q}-\overline{C_p/Q})}{(\overline{C_0/Q}+\overline{C_p/Q})}$	NMSE= $\frac{\overline{(C_o/Q-C_p/Q)^2}}{\overline{(C_o/Q)}\,\overline{(C_p/Q)}}$
Kincaid	187m stack	SF_6	175	208	202	0.06	0.90
		SO_2	2880	269	264	0.41	2.82
Bull Run	243m stack	SF_6	158	338	383	0.33	1.78
Indianapolis	83m stack	SF_6	89	1250	1300	0.09	0.39
Baldwin	3 183m stacks	SO_2	Full year	2090	1890	-0.33	10.00
Akron	Many stacks	SO_2	Full year	1120	1575	-0.24	1.15

period of 20 minutes or more, and the elevation of the sources was on the order of 1 m.

Many scientists are urging that the Gaussian plume model be replaced by a model more soundly based on recent developments in boundary-layer physics. Consequently, the model that is evaluated in this section is a similarity model which scales distances by the Monin-Obukhov length, L, and scales velocities by the friction velocity, u_*. The ability of these scaling relations to order the data is illustrated in Figure 5, for stable runs at the Prairie Grass site. However, it is interesting to note that the scatter in the data points about the line from the similarity formula is about a factor of 3 or 4 at any given x/L.

Table V contains the summary statistics for the similarity model applied to the five field databases. The analysis is restricted to observations and predictions at the closest downwind monitoring arc in each database, since the models are expected to be most accurate at close distances. The relative or fractional bias results are similar to those in the previous two sets of field data, with a median absolute fractional bias of 0.26, and a variation from -0.30 +0.26. The median normalized mean square error is 0.56, which is typical of tracer studies where the plume always impacts the monitoring network. The median value of the relative RMSE is therefore about 75% for these data. The fraction of predictions within a factor of two of the observations is also shown on the table, and has a median value of 0.71.

As mentioned in Section 2, it is important that the model residuals, expressed as C_p/C_o, not show any trend with independent variables such as wind speed, stability, or downwind distance. Figure 6 contains plots of residuals for five optional models at Prairie Grass, as a function of downwind distance ranging from 50 to 800 m. The

TABLE IV
Characteristics of field databases used to evaluate the similarity model

Database	Number of Runs Analyzed	Distances of Monitoring Arcs	Stabilities
Prairie grass	44	50, 100, 200, 400, 800m	All
Ocean Breeze	72	1200, 2400, 4800 m	Mostly Neutral/Unstable
Dry Gulch B	55	2301, 5665 m	Mostly Neutral/Unstable
Dry Gulch D	51	853, 1500, 4715 m	Mostly Neutral/Unstable
Green Glow	27	200, 800, 1600, 3200, 12800, 25600 m	Mostly Stable

TABLE V
Summary of similarity model performance evaluation, where only the closest monitoring arc is used from each experiment.

Database	Fractional bias $FB = \frac{\overline{C_o}-\overline{C_p}}{0.5(\overline{C_o}+\overline{C_p})}$	NMSE $= \frac{\overline{(C_o-C_p)^2}}{\overline{C_o}\ \overline{C_p}}$	Fraction Within Factor of 2
Prairie Grass	0.26	0.15	0.90
Ocean Breeze	–0.19	0.50	0.66
Dry Gulch B	–0.30	1.23	0.69
Dry Gulch D	–0.28	0.58	0.71
Green Glow	0.04	0.18	0.91

"Similarity-C" model is the one whose performance measures are given in Table V, and the OB/DG, Regression-C, GPM, and VSDM are other optional models. It is seen that all models except the similarity model show a clear downward trend in C_p/C_o, with a median C_p/C_o of about 2 at x = 50 m and a median C_p/C_o of about 0.5 at x = 800 m. In contrast, the similarity model has a median C_p/C_o of nearly 1.0 at all downwind distances.

3.4. 14 Hazardous Gas Models Applied to 8 Databases

A comprehensive model evaluation project was recently completed (Hanna, Strimaitis and Chang, 1991) in which the predictions of 14 hazardous gas models were evaluated using data from independent field experiments at eight sites. The field data are summarized in Table VI. Most of the releases lasted no more than a few minutes and the gas/aerosol that was released in six of the field experiments (all except Hanford and Prairie Grass) behaved as a dense gas in the initial stages

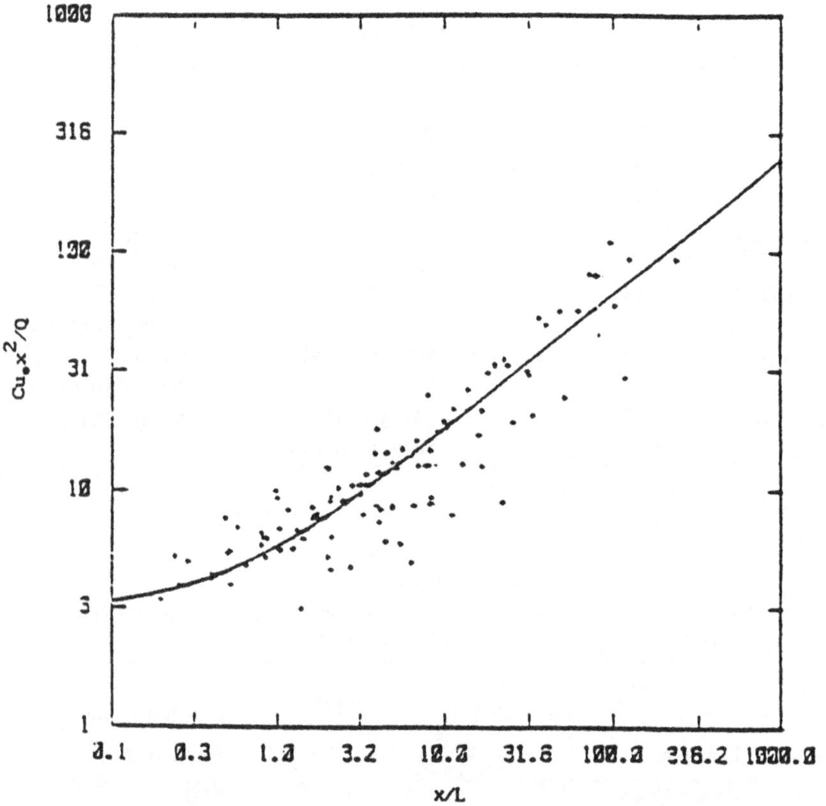

Fig. 5. Plot of Prairie Grass stable data, where concentrations are scaled by $Q/(u_*x^2)$ and distances are scaled by L. The similarity formula $Cu_*^2x/Q = 3.01(1 + 2.20x/L_0)^{0.57}$ is drawn as a curved line.

of transport and dispersion. It is seen that there are generally fewer trials in the hazardous gas studies than in the studies discussed earlier, due to the fact that is relatively difficult (and hence more expensive) to conduct experiments in which the tracer gas is toxic or flammable.

Fourteen dispersion models are evaluated, including six publicly-available computer models (AFTOX, DEGADIS, HEGADAS, INPUFF, OB/DG, and SLAB) and six proprietary computer models (AIRTOX, CHARM, FOCUS, GASTAR, PHAST, and TRACE). A simple Gaussian plume formula and a set of nomograms (Britter and McQuaid, 1988) are also evaluated for comparative purposes. Several of the models explicity treat the effects of dense gases (AIRTOX, CHARM, DEGADIS, FOCUS, HEGADAS, GASTAR, PHAST, TRACE, and Britter and McQuaid). The characteristics of the models are discussed in detail in the reference.

Because of the large numbers of models and the different types of datasets, the model performance exercise was simplified by grouping the field datasets into five groups with similar characteristics. Furthermore, the logarithm of the observed and predicted concentration was used as a primary variable because of the several

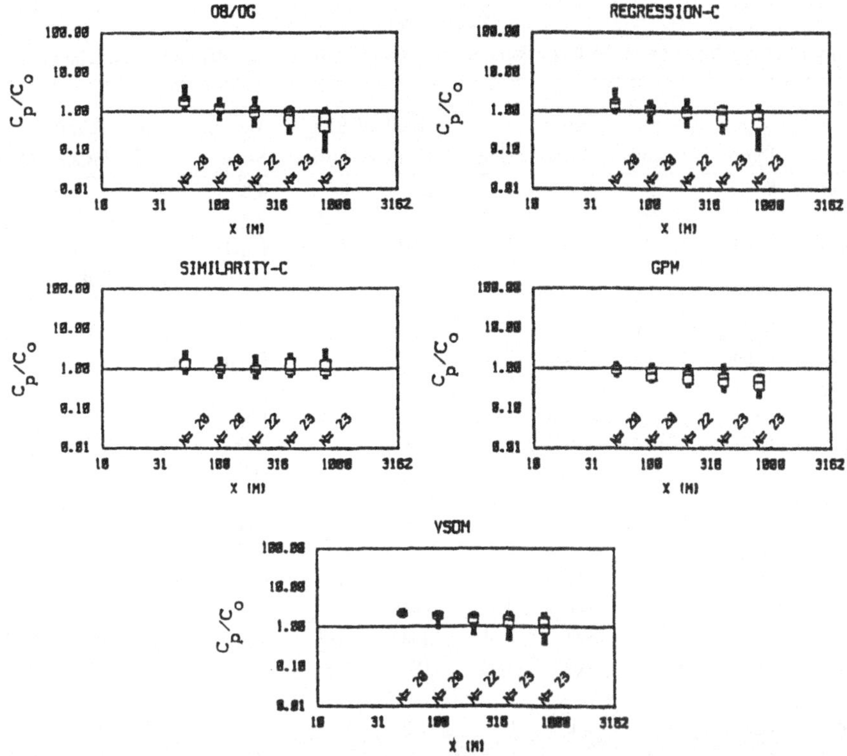

Fig. 6. Ratios of predicted to observed concentrations (C/Q) for the Prairie Grass data set, as a function of downwind distance, for five models (OB/DG, REGRESSION-C, SIMILARITY-C, GPM, and VSDM) for stable conditions. N is the number of data points at each distance. The midline of each box plot is the median, and the other lines represent ± one and two standard deviations.

order of magnitude range of the observed concentrations over all the experiments and all downwind monitoring arc distances. The mean bias is expressed by the geometric mean, MG=exp($\overline{lnC_o} - \overline{lnC_p}$), and the mean-square-error is expressed by the geometric variance VG=exp($\overline{(lnC_o - lnC_p)^2}$). A perfect model will have MG=1 and VG=1. The vertical dotted lines represent "factor of two" agreement. An example of the results is given in Figure 7 for the group of datasets characterized by continuous releases of dense gases and short averaging times. Values of MG and VG for each model are plotted. It is seen that the points for ten of the models are clustered in one zone of the figure, while the points for four of the models are relative outliers. The ten better models all have better than "factor of two" agreement in the mean bias.

The behavior of the residual plots for two models (SLAB and TRACE) for the Thorney Island Field Data are shown in Figure 8. The SLAB model is seen to have little trend in its residuals, while the TRACE model shows a downward trend in C_p/C_o with distance, an upward trend with wind speed, and a downward trend with PG stability class. These results suggest that the physical assumptions in the

TABLE VI

Summary of characteristics of the hazardous gas datasets (from Hanna, Strimaitis and Chang, 1991)

	Burro	Coyote	Desert Tortoise	Goldfish	Hanford Kr^{85} Continu.	Instant.	Maplin Sands	Prairie Grass	Thorney Island Instant.	Cont.
Number of Trials	8	3	4	3	5	6	4,8	44	9	2
Material	LNG	LNG	NH_3	HF	Kr^{85}	Kr^{85}	LNG, LPG	SO_2	Freon and N_2	Freon and N_2
Type of Release	Boiling Liquid	Boiling Liquid	2-Phase Jet	2-Phase Jet	Gas	Gas	Boiling Liquid	Gas Jet	Gas	Gas
Total Mass (kg)	10700-17300	6500-12700	10000-36800	3500-3800	11-24*	10*	LNG: 2000-6600, LPG: 100-380	23-63	3150-8700	4800
Duration (s)	79-190	65-98	126-381	125-360	598-1191	(Inst.)	60-360	600	(Inst.)	460
Surface	Water	Water	Soil	Soil	Soil	Soil	Water	Soil	Soil	Soil
Roughness (m)	.0002	.0002	.003	.003	.03	.03	.0003	.006	.005-.018	.01
Stability Class	C-E	C-D	D-E	D	C-E	C-F	D	A-F	D-F	E-F
Max. Distance (m)	140-800	300-400	800*	3000	800	800	400-650	800	500-580	472
Min. Averaging Time (s)	1	1	1	66.6-88.3	38.4	4.8	3	(Dos.)	0.06	30
Averaging Time (s)	40-140	50-90	80-300	66.6-88.3	270-845	4.8	3	600	0.06	30

* Curies, rather than kg, are used as a measure of the amount of this radioactive tracer released.

** Concentrations are measured beyond 800 m, but these are not well- instrumented measurement arcs.

TRACE model need some improvement.

There has been space in this paper to present only a few of the results of this extensive study, and many more details are given by Hanna, Strimaitis, and Chang (1991). The general conclusions are given below:

a. A few models can successfully predict concentrations with a mean bias of 20% or less, a relative scatter of 50% or less, and little variability of the residual errors with input parameters.

b. The performance of any model is not related to its cost or complexity. Once the accuracy mentioned in conclusion (a) is achieved, it appears that it is very difficult to demonstrate improved model performance as enhancements in model physics are added.

TABLE VII
Summary of Model Evaluation Results

Data Group	Number of Independent Sets of Data	Fractional Bias Median Magnitude	Fractional Bias Range	Normalized Mean Square Error(Mean)
Offshore Tracer Data	9	0.22	-0.40 to +0.22	0.52
Power Plant Stacks	6	0.24	-0.33 to +0.41	1.78
Ground Level Tracer Data	5	0.26	-0.30 to +0.26	0.56
Hazardous Gas Data	8	0.20(*)		0.50(*)

*For the best models.

c. Relatively simple screening formulas which contain the basic physics are seen to perform adequately in the hazardous gas field experiments.

4. Conclusions and Recommendations

The results of four model evaluation exercises were summarized in Section 3. These results can be further combined into one set of performance measures as listed in Table VII. It is seen that the median absolute magnitude of the fractional bias or the relative bias in the mean predictions is consistently in the range from 0.20 to 0.27 (i.e., 20% to 27%). At any given site, the relative bias could be anywhere from -46% to +41%. Similarly, the median normalized or relative mean square error is consistently about 0.50 (the one exception occurs in some of the power plant data sets, which over a year will contain many zeros and a few large observed concentrations). The relative root-mean-square error (or scatter) is therefore about 0.7 (or 70%).

These results point out the danger in using data from only one field site in an air quality model evaluation exercise. It is not unusual for the same model to overpredict by 40% at one site and underpredict by 40% at another site.

These figures also suggest that there are certain minimum values of relative mean bias and relative scatter which are very difficult to improve upon, even with advanced, state-of-the-art models. Once a given model achieves these values (i.e., relative mean bias of about ±20% and a relative scatter of about 70%), it may not be possible for newer models to demonstrate statistically significant improvements in these performance measures.

The irreducible values of scatter are caused by (1) data errors and (2) stochastic fluctuations. It may be possible to eliminate many of the effects of data errors once accurate instruments can be set up at representative locations. However, the stochastic fluctuations are a natural characteristic of the atmosphere that cannot

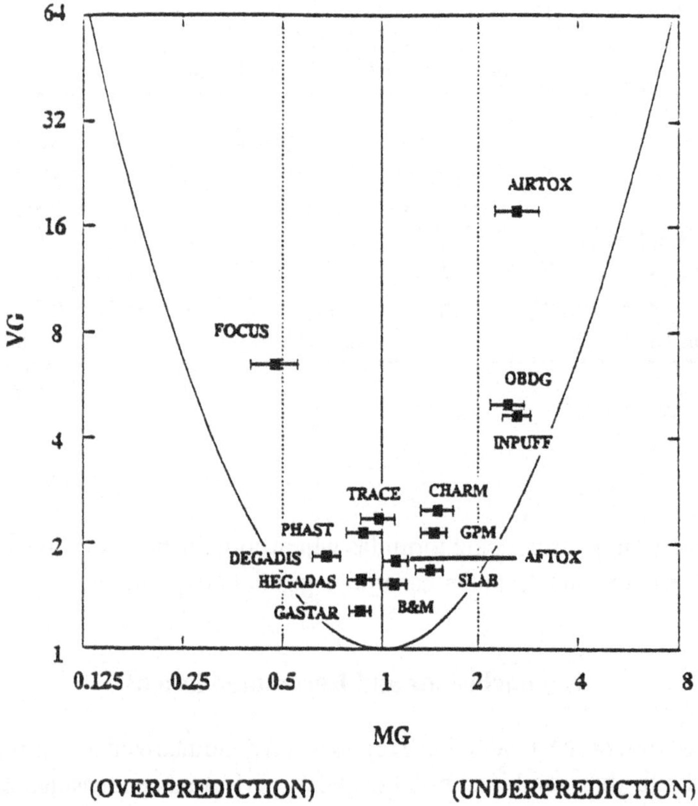

Fig. 7. Geometric Mean, MG, and Geometric Variance, VG, for 14 models for continuous dense gas field data with short averaging times. Horizontal bars represent 95% confidence intervals. Note: $MG = exp(\overline{lnC_o} - \overline{lnC_p}), VG = exp(\overline{lnC_o - lnC_p})^2$.

be eliminated. It is recommended that further research efforts attempt to develop and test predictive equations for these two components of the scatter. Currently, researchers have studied limited aspects of the data error and stochastic components but there is not a general consensus on the best way to estimate these components for a wide variety of conditions.

Acknowledgements

The research reported in this paper is the result of several projects conducted over the past ten years. Major support has come from the Electric Power Research Institute, the U.S. Air Force, the American Petroleum Institute, the U.S. Army, and the Minerals Management Service. Many scientists have supplied field data and dispersion codes for use in these model evaluation studies, and their help is gratefully acknowledged.

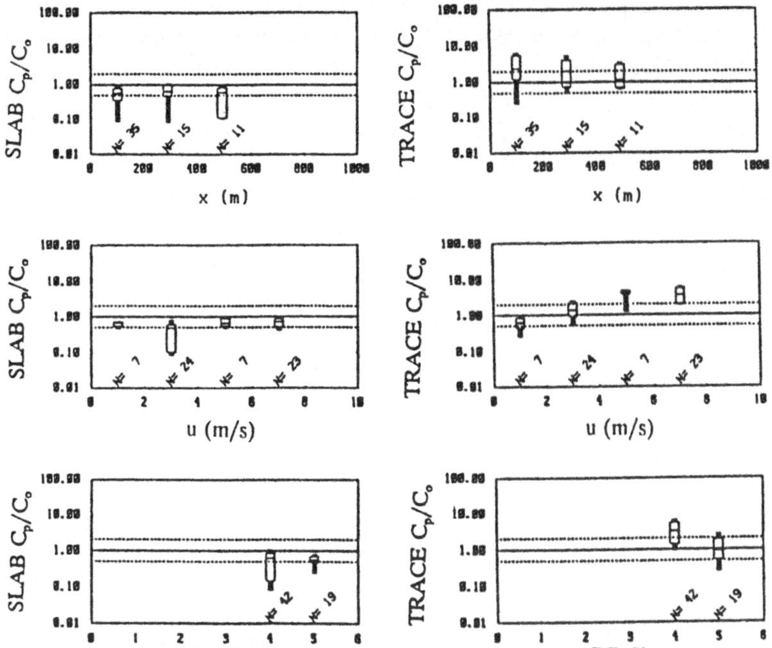

Fig. 8. Residual plots (C_p/C_o) for the SLAB and TRACE models, where the independent variable is downwind distance (Top), wind speed u (Middle) and PG stability class (Bottom) for the Thorney Island instantaneous release field data.

References

Britter, R.E. and McQuaid, J., 1988: *Workbook on the Dispersion of Dense Gases*. HSE Contract Research Report No. 17/1988, Health and Safety Executive, Sheffield, UK.

Cox, W.M. and J.A. Tikvart, 1990:*A statistical procedure for determining the best performing air quality simulation model*. Atmos. Environ., **24A**, 2387-2395.

DiCristofaro, D.C. and S.R. Hanna, 1989:*OCD-The Offshore and Coastal Dispersion Model. Volume I: User's guide*. Minerals Management Service Contract No. 14-120001-30396, 31 Elden Street, Herndon, VA, 22070-4817.

Fox, D.G., 1984: *Uncertainty in air quality modeling*. Bull. Am. Meteorol. Soc., 65,27-36

Hanna, S.R., 1989: *Confidence limits for air quality model evaluations, as estimated by bootstrap and jackknife resampling methods*. Atmos. Environ., **23**, 1385-1398.

Hanna, S.R. and J.C. Chang, 1990: *Modification of the Hybrid Plume Dispersion Model (HPDM) for urban conditions and its evaluation using the Indianopolis data set* EPRI Proj. No. RP-02736-1, EPRI, 3412 Hillview Ave., Palo Alto, CA, 94303.

Hanna, S.R., J.C. Chang and D.G. Strimaitis, 1990: *Uncertainties in source emission rate estimates using dispersion models*. Atmos. Environ., 24A, 2971-2980.

Hanna, S.R. and R.J. Paine, 1989: *Hybrid Plume Dispersion Model (HPDM) development and evaluation*. J. Appl. Meteorol., 28, 206-224.

Hanna, S.R.,L.L. Schulman, R.J. Paine, J.E. Pleim and M. Baer, 1985: *Development and evaluation of the Offshore and Coastal Dispersion (OCD) model*. J. Air Poll. Control Assoc., 35, 1039-1047.

Hanna, S.R., D.G, Strimaitis and J.C. Chang, 1991: *Hazard response modeling uncertainty (A quantitative method). Volume II: Evaluation of commonly-used hazardous gas dispersion models*. AFESC Contract No. FO8635-89-C-0136, H.Q. AFESC/RDVS, Tyndall AFB, FL 32403.

Irwin, J.S., S.T. Rao, W.B. Petersen and D.B. Turner, 1987: *Relating error bounds for maximum concentration estimates to diffusion meteorology uncertaintly*. Atmos. Environ., 21, 1927-1937.

Pasquill, F., 1974: Atmospheric Diffusion. John Wiley, New York, 395-400.
Smith, M.E., 1984:*Review of the attributes and performance of 10 rural diffusion models*. Bull. Am. Meteorol. Soc., 65, 554-558.

REVIEW OF NON-LOCAL MIXING IN TURBULENT ATMOSPHERES: TRANSILIENT TURBULENCE THEORY

ROLAND B. STULL
Dept. of Meteorology, University of Wisconsin, Madison, WI 53706-1695, USA.

(Received October 1991)

Abstract. Some of the larger eddies in a turbulent region can be coherent structures that turbulently advect air parcels across large vertical distances before smaller eddies mix the parcels with the environment. Such a process is nonlocal rather than diffusive. Transilient turbulence theory, named after a Latin word meaning "jump over", provides a framework for considering the ensemble-averaged effect of many eddies of different sizes on the net nonlocal mixing in the vertical. Nonlocal turbulence statistics can then be examined, and nonlocal first-order closure can be formulated.

1. The Nature of Turbulence

1.1. Turbulent Advection

Turbulence is advective, not diffusive. Start with a basic conservation equation, perform the usual perturbation expansion ($S = \bar{S} + s'$) and Reynolds average:

$$\frac{\partial S}{\partial t} \quad + U_j \frac{\partial S}{\partial x_j} = \quad \cdots \quad + \nu_s \frac{\partial^2 S}{\partial x_j^2} \tag{1.1}$$

$$\downarrow \qquad \searrow \qquad\qquad \searrow$$

$$\frac{\partial \bar{S}}{\partial t} \quad + \bar{U}_j \frac{\partial \bar{S}}{\partial x_j} = \quad -\frac{\partial \overline{u_j' S'}}{\partial x_j} \quad + ... + \quad negligible \tag{1.2}$$

Storage	Advection by Mean Wind	Turbulent Flux Diverg.	Molecular Diffusion

where S is a scalar or a momentum component, U_j is velocity component in the j-direction, t is time, x_j is distance in the j-direction, ν is kinematic viscosity, overbars denote ensemble averages, primes denote deviations from that average, and Einstein's summation notation is used (Stull, 1988a). The turbulent flux-divergence term comes from the advection term, while the diffusion term is negligible.

The advective nature of turbulence is apparent in numerical forecast models of successively finer grid resolution. In large-eddy simulation models, the large eddies are resolved by grid-cell-average wind, and by the advection between grid boxes. Effects of medium and smaller eddies are sometimes approximated by diffusion-like parameterizations. Yet if the grid size is made even smaller, as for direct simulation, even these small eddies are resolved by cell-averaged advection (Fig. 1.1).

Boundary-Layer Meteorology **62:** 21–96, 1993.

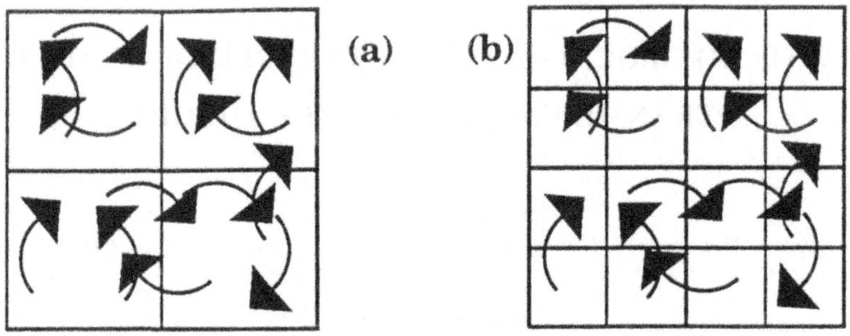

Fig. 1.1. The same turbulent motions that might appear diffusive-like for large grid boxes (a) show their advective nature when the grid resolution is made finer (b).

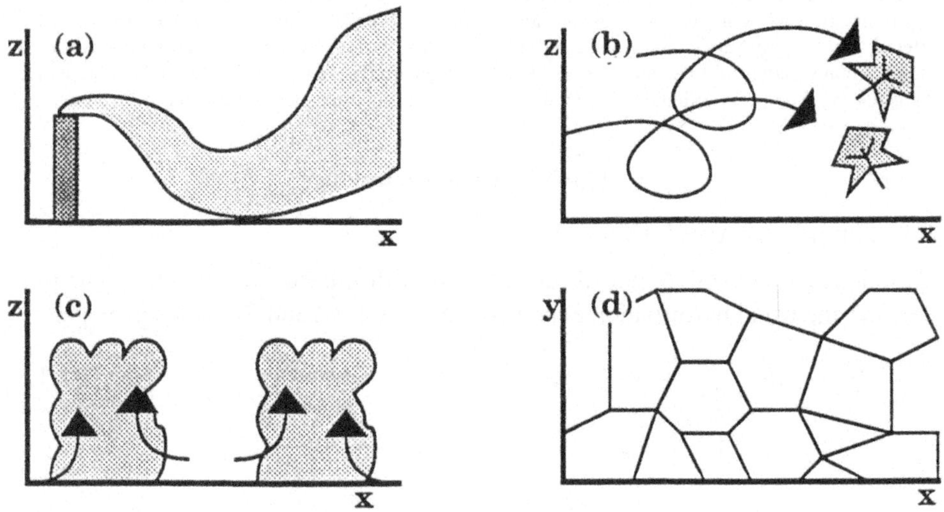

Fig. 1.2. Evidence of large eddy structures: (a) looping smoke plumes; (b) swirls of leaves or snowflakes; (c) aerosol-laden thermals; and (d) cellular cross sections.

Turbulence causes dispersion of tracers. Molecular diffusion causes dispersion of tracers. Although they both cause dispersion, turbulence and diffusion are different physical processes. An appropriate parameterization of the turbulent flux-divergence term must be associated with dispersion, and should be consistent with the advective origin of that term. Care should be taken not to confuse or substitute the word "diffusion" for the word "dispersion".

1.2. Coherent Structures and Large Eddies

We see ample evidence of large organized eddies: thermals rising from the surface, looping smoke plumes, bursts of turbulence in forest canopies, and swirls of leaves or snowflakes around buildings and vehicles (Fig. 1.2). Remote sensors such as dual-Doppler radars detect them as cellular patterns, lidars detect aerosol-laden updrafts, and sodars detect turbulent plumes (Stull, 1984).

These eddies, also known as coherent structures, dominate many atmospheric

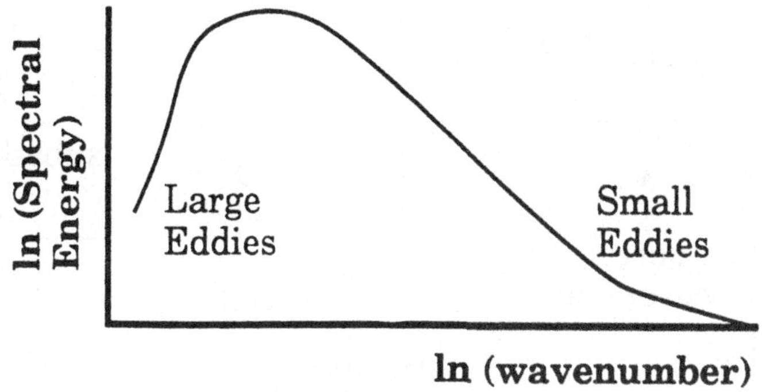

Fig. 1.3. Sketch of the turbulence kinetic energy spectrum.

flows and hold the most turbulent kinetic energy. They are produced directly by, and gain their energy from, external forcings and mean wind shear. They hold more spectral energy than smaller eddies (Fig. 1.3), and make the greatest contributions to vertical turbulent transport and flux. Their contribution to the total flux is also most advective-like.

The largest eddies are usually anisotropic, and have a size comparable to boundary-layer scales and external forcings. Thermals have most of their energy in the vertical, because of buoyancy. Shear-generated eddies have greater energy in the horizontal, because of the vertical shear in the mean horizontal wind. Thermal diameters nearly equal the mixed-layer depth, wake eddies are the size of the obstacle, and shear eddies are the size of the shear domain.

1.3. Nonlocal Static Stability

Static stability indicates whether flow will become or remain turbulent when there is no mean wind shear. Once turbulent, the large coherent structures in the flow advectively communicate this turbulent state across large distances. The resulting turbulent state at any point may thus depend on nonlocal influences (Stull, 1991b).

Local lapse rate is a poor indicator of static stability. Fig. 1.4 shows a typical convective mixed layer over a forest canopy. The turbulent state is determined by nonlocal parcel movement, not by local lapse rate. For this reason, one should not use a stability quantifier such as "unstable" or "neutral" to specify the lapse rate. Instead, one should use the words "superadiabatic", "adiabatic", or "subadiabatic" for lapse rates of $\partial\theta_v/\partial z$ that are negative, zero, or positive, respectively.

To graphically determine static stability, first plot the vertical profile of virtual potential temperature. From each relative maximum in the profile, conceptually lift an air parcel at constant virtual potential temperature until it crosses the sounding. From each relative minimum, similarly lower an air parcel. The domains of these parcel displacements define "statically unstable" regions and turbulence (Fig. 1.4). Locally adiabatic lapse rates in the remaining regions indicate "statically neutral" stability, while locally subadiabatic lapse rates indicate "statically stable" stability, except as described below.

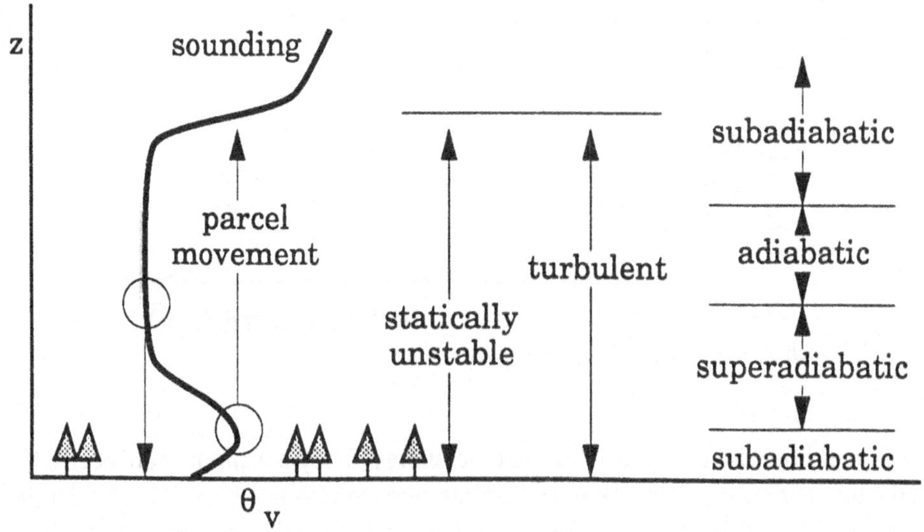

Fig. 1.4. Static stability is based on nonlocal parcel movement, not on local lapse rates.

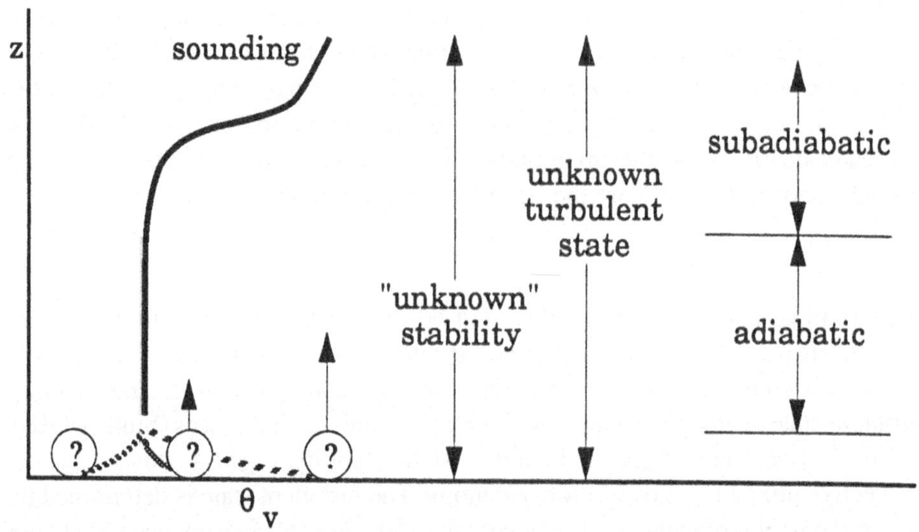

Fig. 1.5. Soundings not ending at a boundary cause uncertainty in the stability classification.

Stability can be determined only within the context of the whole sounding because of the potential for nonlocal influences. If only a sounding segment is known, then portions of the soundings have unknown stability and should be classified as such (Fig. 1.5). Nonlocally unstable portions of a sounding segment are indeed unstable and turbulent, but the remaining regions must be classified as "unknown" regardless of their local lapse rate. The stability of these unknown regions can be resolved from additional information, such as knowledge of external forcings, and observations of turbulence or dispersion.

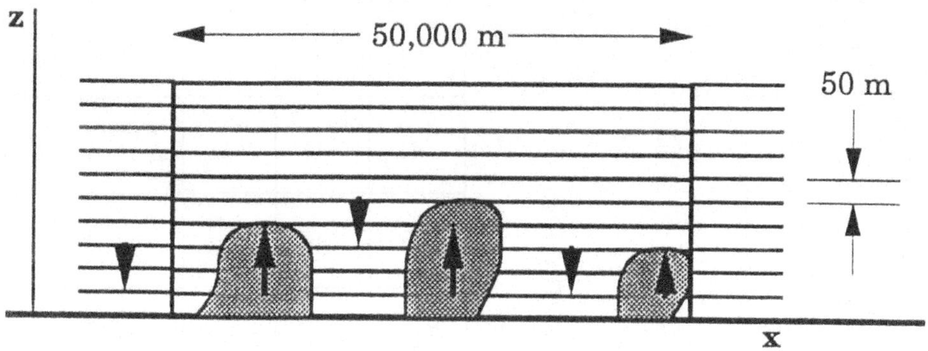

Fig. 2.1. Dimensions of a typical mesoscale-forecast-model grid cell are shown. Turbulent eddies are frequently not horizontally resolved, but their effects are vertically resolved.

2. Transilient Framework

2.1. Horizontal vs. Vertical Grid Resolution

Turbulence statistics describe the resolvable effects of unresolved turbulent motions. Eddies might not be resolved in the horizontal, but their vertical mixing is resolved (Fig. 2.1). This is the case when: (1) the statistic represents an area average over many eddies; (2) the statistic is used in a vertical 1-D model; or (3) the statistic represents a 3-D grid-cell average in a weather forecast model, where horizontal grid resolution (5 to 100 km) is typically greater than the largest eddies but vertical grid resolution (10 to 500 m) is smaller.

A local statistic such as vertical kinematic heat flux, $\overline{w'\theta'}(z, t)$, gives the net heat transport at one specific height and time, caused by many eddies having different sizes and operating at different locations within the averaging domain. Similarly, nonlocal statistics give the net effect between two points in space of many eddies having different sizes and operating at different locations within the averaging domain.

2.2. Discrete View of Nonlocal Mixing

Let $c_{ij}(t, \Delta t)$ represent the fraction of air mixed into a destination grid cell at vertical location index i from a source grid cell at location j during a time interval from t to $t + \Delta t$. For a column of equally-spaced grid cells with layer-thickness Δz, location index i represents a grid point at the center of the cell (Fig. 2.2a), at height $z = (i - 0.5) \cdot \Delta z$. The matrix $\mathbf{C}(t, \Delta t)$ of all such elements c_{ij} is called a "transilient matrix". The name "transilient" comes from a Latin root meaning "jump over", to suggest the nonlocal and advective nature of this statistic (Stull, 1984a).

Elements along the main diagonal of the matrix, c_{ii}, indicate the fraction of air in cell i that is not involved in turbulent mixing between other cells. If such an element has a value of 1, then there is no turbulent mixing to or from that cell.

To help understand the physics in transilient matrices, display them with the

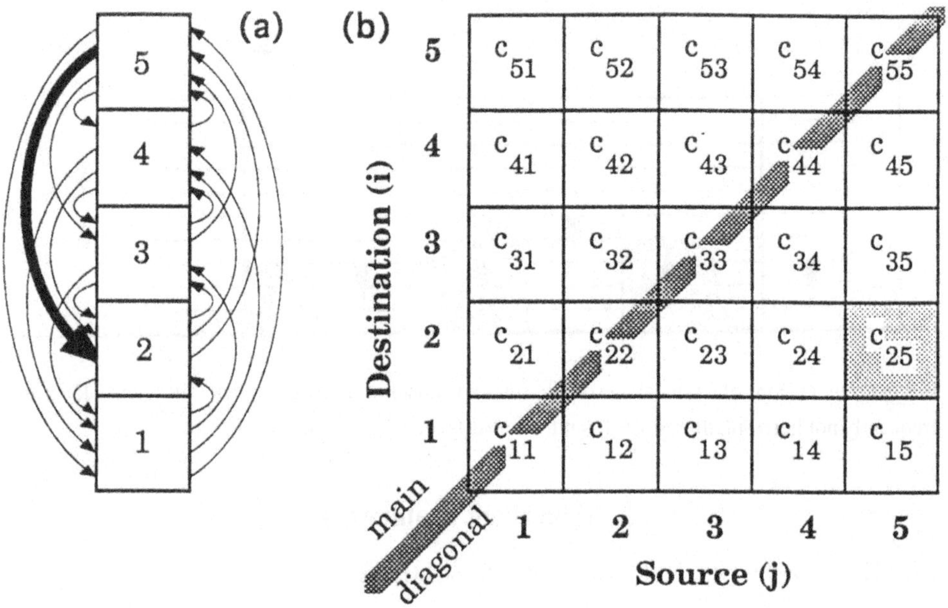

Fig. 2.2. A turbulent field consisting of a superposition of mixing between all possible pairs of N grid cells (a) is described by an NxN transilient matrix (b). Mixing between source cell 5 and destination cell 2 is highlighted in both (a) and (b).

destination index increasing from the bottom to the top of the matrix. Destinations near the bottom of the atmosphere correspond to elements near the bottom of the matrix, and height increases upward. It is not a matrix transpose, but a flip upside-down from the usual mathematical notation (Fig. 2.2).

After the action of turbulence during interval Δt, the state $\bar{S}_i$ of a destination cell, i, is altered. It now contains a mixture of air originating from a variety of sources, where $\bar{S}_j$ describes the state of scalar $\bar{S}$ at source cell j, and c_{ij} gives the relative amount of mixture coming from that source.

For example, the final state of grid cell 2 in the figure above is

$$\bar{S}_2(t+\Delta t) = c_{21} \cdot \bar{S}_1(t) + c_{22} \cdot \bar{S}_2(t) + c_{23}.\bar{S}_3(t) + c_{24} \cdot \bar{S}_4(t) + c_{25} \cdot \bar{S}_5(t).$$

In general, the final state of every cell after interval Δt is

$$\bar{S}_i(t+\Delta t) = \sum_{j=1}^{N} c_{ij}(t,\Delta t)\bar{S}_j(t) \tag{2.1}$$

where N is the number of layers (grid cells) in the physical model. Let **S** be an N-element column vector, and **C** be an $N \times N$ matrix. Equation (2.1) describes a simple matrix multiplication:

$$\mathbf{S}(t+\Delta t) = \mathbf{C}(t,\Delta t)\mathbf{S}(t). \tag{2.2}$$

The arrows in Fig. 2.2a represent net "mixing processes" that occur during time interval Δt ; they do not represent individual eddies. For example, the mixing highlighted in that figure might have been caused by a series of small eddies transporting air parcels from location 5 to 4, and then 4 to 3, and 3 to 2. Alternately, it might have been caused by a large coherent structure advectively transporting air parcels from cell 5 thru cells 4, 3 and 2 enroute to 1 during the time interval, and then by a small eddy transporting them back from 1 to 2.

It is assumed that transported air parcels carry their original characteristics of heat, moisture, tracers, and momentum components. The same transilient matrix can, and should, be applied to all of the state variables at any one time. As time progresses, the transilient matrix changes because the amount of turbulent mixing changes. The transilient matrix describes only the advective portion of turbulence. It does not include other factors such as pressure correlations, which can also alter momentum.

A full transilient matrix holds information on mixing processes ranging from the smallest resolved by the grid spacing to the largest allowed by the domain. It includes the small-eddy effects of K-theory plus the medium and large advective-like coherent structure effects that K-theory misses. It is essentially a Green's-function (Morse and Feshbach, 1953) matrix that yields the distribution of tracer [left side of (2.1)] as a superposition of individual distributions from many point sources [i.e., grid points on the right side of (2.1)].

2.3. Physical Constraints and Conservation

To satisfy conservation of air mass and state, each row and each column of a transilient matrix for uniform grid spacing must sum to 1 (Stull, 1984a):

$$\sum_{j=1}^{N} c_{ij} = 1, \qquad \sum_{i=1}^{N} c_{ij} = 1. \tag{2.3}$$

Each row sums to 1 because 100% of the air at each destination must have come from somewhere. Each column must sum to 1 because all of the air initially within each grid box must go somewhere. This type of matrix is "doubly-stochastic".

If turbulence increases randomness and entropy, then each matrix element must be non-negative. Although it is mathematically possible to have some negative elements and yet satisfy conservation constraints, the resulting matrix is not physically realistic. Such a matrix would cause un-mixing.

Each element must be less than or equal to 1.0. Remembering that each element of the matrix represents the fraction of air arriving at a destination from some source, it is impossible to have more than 100% of the destination air arriving from anywhere. Thus,

$$0 \le c_{ij} \le 1. \tag{2.4}$$

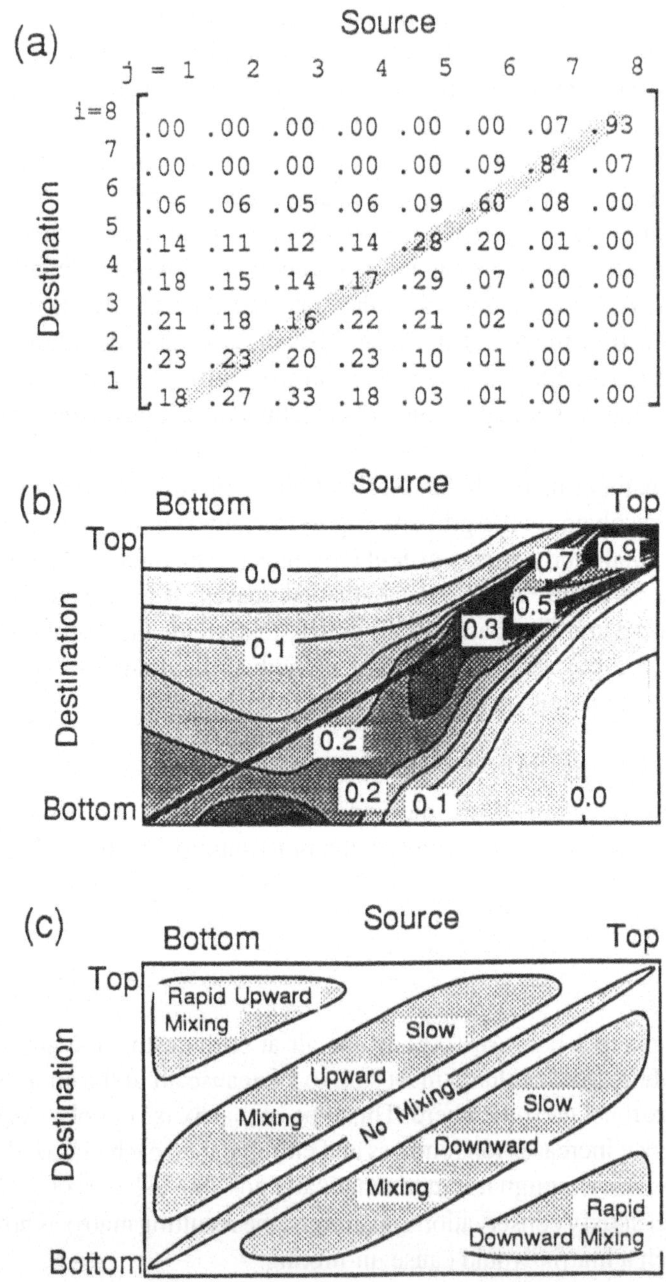

Fig. 2.3. A transilient matrix (a) is contoured (b). Mixing processes can be interpreted from the relative location of an element within the matrix (c), and the amount of air involved in those processes is found from the contoured magnitude at those locations (b). (Ebert *et al.*, 1989)

2.4. Interpretation of Transilient Matrices

Transilient matrices are often displayed as a contour plot of the element values (Ebert *et al.*, 1989) to highlight the underlying physics (Fig. 2.3). The magnitudes of matrix elements tell how much air is involved in a mixing process (Fig. 2.3a & b), and the relative locations of those elements indicate the type of mixing process (Fig. 2.3c).

For example, an element in the lower right corner of the matrix represents a (relatively) high altitude source and a low altitude destination. Mixing between this source and destination occurs during the time interval Δt specified for the matrix; thus, this mixing process is interpreted as rapid downward mixing. Elements closer to the main diagonal move a relatively shorter distance during the same time interval, and represent slower mixing processes.

The structure of well-behaved transilient matrices should be independent of the grid spacing, and should exhibit similarity when properly scaled. In other words, for a given situation such as a convective mixed layer, the contoured transilient matrices should look similar regardless of the $N \times N$ size of the matrix or the N grid resolution of the original model. One would expect that similarity relationships could be developed for transilient matrices when scaled with respect to mixed-layer depth, z_i, convective velocity scale, w_*, friction velocity, u_*, or other relevant variables.

The magnitudes of elements in a transilient matrix are determined by both the physics and the time-step duration. For very long time steps, turbulence can cause significant mixing across large distances and is associated with a full matrix of nearly equal element values. For very short time steps, turbulence has not the time to mix air parcels far from their starting points, leading to a diagonally-dominated matrix with zeros in the upper and lower triangles. To first order, matrices for different time steps are related to each other by

$$\mathbf{C}(t, \Delta t_2) = [\mathbf{C}(t, \Delta t_1)]^a, \qquad \textit{for} \quad \Delta t_2 = a \cdot \Delta t_1, \tag{2.5}$$

assuming turbulence is stationary during those intervals (Stull, 1985). In actuality, factors such as "convective structure memory" (Section 8) cause (2.5) to be slightly inaccurate.

A matrix representing no mixing has 1's along the main diagonal and 0's elsewhere. For a turbulent situation with a nonturbulent subdomain, the matrix has ones along the portion of the main diagonal corresponding to those grid cells that are not turbulent, and values less than one along the main diagonal for the turbulent cells. A simple algorithm to locate resolvable subdomains of turbulence and nonturbulence from a known transilient matrix is:

If $c_{ii} = 1.0$, then grid cell i is nonturbulent,
else cell i is turbulently mixing with other cells.

Idealizations within classical local turbulence theory, such as homogeneity and isotropy, have their nonlocal counterparts (Stull, 1984a). For "homogeneous" turbulence:

$$c_{ij} = c_{i+m,j+m} \tag{2.6}$$

for m = a constant integer. "Isotropic" turbulence in one dimension (up vs down) is given by:

$$c_{i,i+m} = c_{i,i-m}. \tag{2.7}$$

Section 4 gives a greater discussion of anisotropy in one dimension. For "stationary" turbulence:

$$\mathbf{C}(t_1, \Delta t) = \mathbf{C}(t_2, \Delta t), \qquad for \quad t_1 \neq t_2. \tag{2.8}$$

In general, the transilient matrix need not be symmetric, as is shown in Fig. 2.3. If it is symmetric, it satisfies the "exchange hypothesis":

$$c_{ij} = c_{ji}. \tag{2.9}$$

Physically, for any fraction of air that moves from grid point j to i during time step Δt, there is an equal fraction moving from i to j. Hence, there is an equal exchange of air between those two grid points. The exchange idealization is uniquely nonlocal, and does not have a classical counterpart.

The transilient matrix holds a wealth of information about the nature of turbulence, which can easily be extracted and analyzed. Section 4 describes various statistical techniques that can also be used to help interpret the matrices.

2.5. Unequally-Spaced Grid Points

Let m_i represent the mass of air within layer i. The air mass and state conservation relationships of (2.3) can now be rewritten as:

$$\sum_{j=1}^{N} c_{ij} = 1, \qquad \sum_{i=1}^{N} \frac{m_i}{m_j} \cdot c_{ij} = 1 \tag{2.10}$$

respectively. This can be seen by example from Fig. 2.4, where 10 equally-spaced grid layers in Fig. 2.4a are grouped into four layers of unequal size in Fig. 2.4c.

For the special case of perfect exchange (2.9), the transilient matrix elements in the upper and lower triangles are related by:

$$c_{ji} = \frac{m_i}{m_j} c_{ij}. \tag{2.11}$$

Parameterization of the matrix elements for unequal grid spacing is discussed in section 5.6.

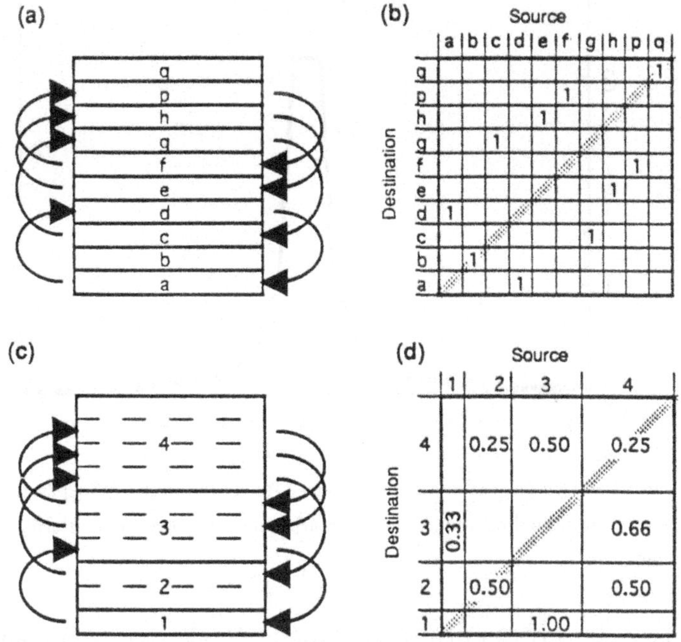

Fig 2.4. (a) Schematic of a column of ten equally-spaced grid points labeled a-q, each representing a unit mass of air contained within the grid boxes shown. For sake of demonstration, assume that air is completely removed from some grid boxes and replaced by air from other grid boxes, as indicated by the arrows. (b) Symmetric transilient matrix corresponding to (a). Zero elements are blank, and each row and column sum to one. (c) Schematic of a column of four unequally-spaced grid points based on groupings of the unit masses from (a). The same amount of mass exchange as (a) is assumed to occur, as indicated by the arrows. (d) Asymmetric transilient matrix corresponding to (c). Each row sums to one, but each column does not. (Raymond and Stull, 1990)

2.6. ANALYTICAL VIEW OF NONLOCAL MIXING

An analytical version of (2.1) is (Stull, 1984, 1986):

$$\frac{d\bar{S}(t,z)}{dt} = \int_{\zeta=a-z}^{b-z} \gamma(t,z,\zeta)\cdot[\bar{S}(t,z+\zeta)-\bar{S}(t,z)]d\zeta \tag{2.12}$$

or

$$\frac{d\bar{S}(t,z)}{dt} = \int_{\xi=a-z}^{b-z} \alpha(t,z,\zeta)\cdot\bar{S}(t,z+\zeta)d\zeta, \tag{2.13}$$

where γ is a transilient rate coefficient having dimensions of (length · time)$^{-1}$, δ^D is the Dirac delta function, ζ is an interval of distance, ζ' is a dummy of integration, a and b are the lower and upper bounds of the domain (where the turbulent flux is zero by definition) and

$$\alpha(t,z,\zeta) = \gamma(t,z,\zeta) - \delta^D(\zeta)\cdot\int\gamma(t,z,\zeta')d\zeta'. \tag{2.14}$$

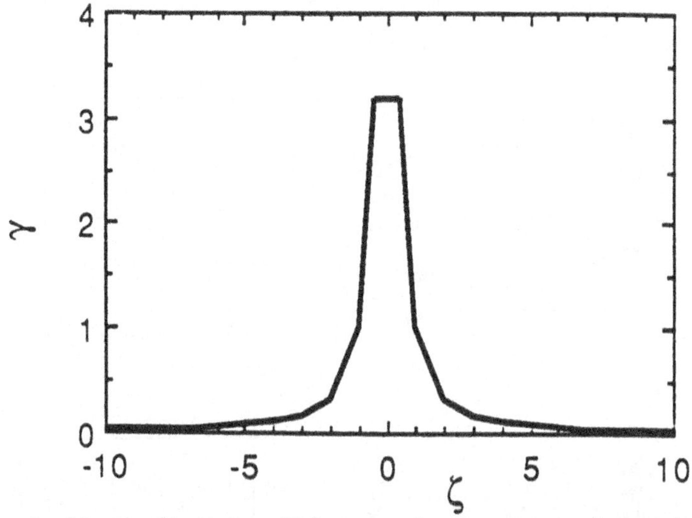

Fig. 2.5. Example of the transilient rate coefficient γ as a function of nonlocal separation distance ζ, for turbulence having an inertial subrange. Units are arbitrary. The spectral similarity is not applied to the smallest values of ζ.

The tendencies in (2.12) and (2.13) represent only the portion of total tendency associated with turbulent mixing. Other factors affecting the total tendency could be advection by mean wind and various production or loss mechanisms.

If $z+\zeta$ is replaced by Z in (2.13), then that equation can be interpreted (Leonard, 1974) as the convolution of $\bar{S}(t, Z)$ with filter function $\alpha(t, Z-z)$. Thus, (2.13) is one of the simplest analytical representations of nonlocal closure.

The relationship between discrete (c_{ij}) and analytical (γ or α) forms for the transilient coefficients is:

$$c_{ij}(t, \Delta t) = \Delta z \cdot \Delta t \cdot \bar{\gamma}(t, z_i, z_i - z_j) + \delta_{ij} \cdot [1 - \Delta z \cdot \Delta t \cdot \sum_k a\bar{\gamma}(t, z_i, z_i - z_k)]$$

where $\bar{\gamma}$ is the vertical average of γ over interval Δz centered at $\zeta = z_i - z_j$.

If we interpret distance interval ζ as a wavelength, then turbulence having an energy spectrum of $E(\kappa)$ gives a transilient rate coefficient of:

$$\gamma(z, \zeta = 1/\kappa) \propto \kappa^2 \cdot [\kappa \cdot E(\kappa)]^{1/2} \tag{2.16}$$

where κ is wavenumber. In the inertial subrange, this becomes

$$\gamma(z, \zeta) \propto \varepsilon^{1/3} \cdot |\zeta|^{5/3}. \tag{2.17}$$

At very small ζ, the inertial subrange is not valid, so (2.17) cannot be extended to $\zeta = 0$. Equations (2.17) and (2.18) are idealizations that apply to stationary turbulence, and thus are not appropriate for most geophysical flows.

The transilient rate coefficient exhibits strong central tendency as a function of ζ. As an example, the function described by (2.17) is plotted in Fig. 2.5. The

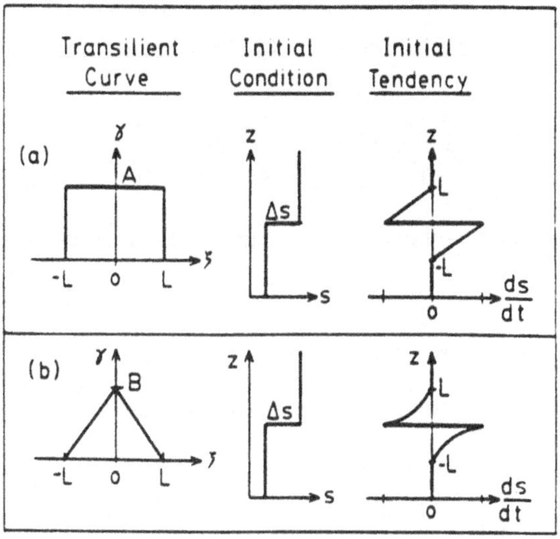

Fig. 2.6. Sketch of the initial tendencies of S caused by nonlocal mixing for idealized γ curves. (a) All eddies equally important within finite range of sizes. (b) Small eddies dominate. (Stull, 1984)

tails of the γ coefficient are small, but significant because they hold the nonlocal information. Neglect or truncation of these tails leads to incorrect forecasts.

One can numerically integrate (2.12) for arbitrary vertical profiles of S, and for arbitrary transilient coefficient functions. For idealized special cases of simple profiles and function shapes, (2.12) can be integrated analytically as shown in Fig. 2.6. From that figure, we see a nonlocal influence on the initial tendency.

2.7. Other Nonlocal Theories

Many investigators have independently reached the conclusion that nonlocal mixing is important, resulting in many similar theories:

ANALYTICAL INTEGRO-DIFFERENTIAL FORMS:

- Direct Interaction Approximation & Two-Point Closure - DIA (Kraichnan, 1959, 1966, 1971a, 1971b, 1976; Roberts, 1961; Leslie, 1973; Chollet and Lesieur, 1981; Stanišić, 1985).
- Spectral Diffusivity Theory - SDT (Berkowicz and Prahm, 1979, 1980, 1984; Prahm *et al.*, 1979; Berkowicz, 1980, 1984; Jenkins, 1985; Imboden, 1981).
- Orthonormal Expansions - ONE (Saffman, 1969; Romanof, 1982, 1987, 1989)
- Nonlocal or Integral Turbulence Closure - ITC (Eringen, 1972; Speziale and Eringen, 1981; Fiedler, 1984; Fiedler and Moeng, 1985; Boudreau and Imboden, 1987; Narasimhan and Saibel, 1989).

DISCRETE (MATRIX) FORMS:

- Convective Circulations - CC (Estoque, 1968; Blackadar, 1979; Zhang and Anthes, 1982).
- Turbulence Adjustment - TA (Klemp and Lilly, 1978; Stull and Hasagawa, 1984).

- Markov and Non-Markov Models - MM (Hostetler and Opitz, 1982; Fortak, 1987)

BOTH ANALYTICAL & DISCRETE FORMS:

- Transilient Turbulence Theory - T3 (Stull & colleagues, see reference list)

Each of these is closely related, if not mathematically identical, to transilient turbulence theory - T3.

2.7.1. *Analytical integro-differential forms*

These are extensions or generalizations of the diffusion equation, which in one dimension is:

$$\frac{\partial \bar{S}(z,t)}{\partial t} = \frac{\partial}{\partial z}[K(z,t)\frac{\partial \bar{S}(z,t)}{\partial z}]. \tag{2.18}$$

$K(z,t)$ is the eddy diffusivity at height z, and is analogous to a molecular diffusivity but of much greater magnitude. Equation (2.18) states that the turbulent flux (the term in square brackets) of S flows down the local vertical gradient of $\bar{S}$, and the local vertical divergence of this flux causes a change in mean concentration $\bar{S}$.

Berkowicz and colleagues suggest that the eddy diffusivity might be a function of spectral wavenumber, κ, yielding a diffusion equation of the form:

$$\frac{\partial \bar{S}(z,t,\kappa)}{\partial t} = \frac{\partial}{\partial z}[K(z,t,\kappa)\frac{\partial \bar{S}(z,t,\kappa)}{\partial z}] \tag{2.19}$$

where $K(z,t,\kappa)$ is the spectral turbulent diffusivity. When integrated over all wavenumbers, this equation can be written in nonlocal form as:

$$\frac{\partial \bar{S}(z,t)}{\partial t} = \frac{\partial}{\partial z}\int[\Xi(z,Z,t)\frac{\partial \bar{S}(Z,t)}{\partial Z}]dZ \tag{2.20}$$

where Z is also a height variable. Berkowicz (1984) and Fiedler (1984) show that (2.20) is mathematically identical to (2.12) of transilient turbulence theory. Troen *et al.* (1980a, b) object to some of the basic assumptions of SDT, and to the non-Gaussian dispersion result at short times, but Prahm and Berkowicz (1980) disagree with those objections.

$\Xi(z,Z,t)$ is the turbulent diffusivity transfer function, which acts like a nonlocal eddy diffusivity. It specifies how the flux at location z is affected by the local $\bar{S}$ gradient at all distant locations Z. The functional shape of $\Xi(z,Z,t)$ vs. $(z-Z)$ is similar to the curve in Fig. 2.5.

Romanof (1982, 1987, 1989) uses orthonormal expansions to develop the most generalized form of the diffusion equation:

$$\frac{\partial \bar{S}(x,t)}{\partial t} = \frac{\partial}{\partial x_i}\int_{t'=-\infty}^{t}\int_{x'=-\infty}^{t}[\Xi(x,x',t,t')\frac{\partial \bar{S}(x't')}{\partial x'_j}]dx'dt' \tag{2.21}$$

where Ξ_{ij} is a generalized diffusion tensor. This equation states that the turbulent flux of tracer S in direction i at point x and time t is proportional to the net effect of gradients of S in all directions j at all (local and nonlocal) locations x' at all previous times t'. Romanof shows how (2.21) relates to SDT and DIA.

Direct Interaction Approximation assumes that turbulent energy at any wavenumber κ is nonlinearly coupled with turbulence at other wavenumbers (Stanišić, 1985). The two-point closure allows coupling between two points in spectral space. The net effect of such coupling is a nonlocal eddy diffusivity in physical space.

Nonlocal or integral closure utilizes equations similar to (2.12) or (2.13). These have been used to describe atmospheric turbulence and top-down bottom-up convection, flow in pipes, lubricant turbulence in bearing journals, and bioturbation caused by animals digging burrows in ocean sediments.

Almost all of the analytical forms achieve closure by making some a-priori assumption about the nature or spectrum of turbulence. Examples include homogeneous and isotropic turbulence, or turbulence having spectra with an inertial subrange described by Kolmogorov similarity theory. Such approaches work well for theoretical studies of idealized situations, but are of limited usefulness for nonstationary, inhomogeneous geophysical flows.

2.7.2. Discrete or Matrix Forms.

Discrete or matrix forms are well-suited for numerical modeling. Estoque and Blackadar recognized that buoyant air parcels can rise from the surface layer to a range of destination grid points within the mixed layer. This results in a transilient matrix with all elements zero except those in the first column and those near the main diagonal. Such a convective circulation boundary-layer parameterization was implemented in a mesoscale weather forecast model by Zhang and Anthes.

Turbulent adjustment is analogous to convective adjustment, whereby a column of grid points within a model is allowed to mix in the vertical to eliminate flow instabilities. For convective adjustment, static instability is eliminated. For turbulent adjustment, dynamic instability (based on the Richardson number) is eliminated. Turbulent adjustment into a stable flow state is possible only if nonlocal mixing is allowed. By making the amount of nonlocal mixing proportional to the amount of nonlocal instability, a responsive-type of parameterization is achieved (see section 5 for details).

The transilient matrix is similar to a transition matrix of a first-order Markov process. It describes the future state of a column of grid points in terms of only one present state. Although a Markov approach has the potential for physically realistic nonlocal effects, many parameterizations of the transition matrix to date have included only Gaussian diffusion and mean advection.

Transilient turbulence theory in discrete form (2.1) has been used with a variety of responsive closure parameterizations that work in heterogeneous nonstationary geophysical flows such as boundary layers.

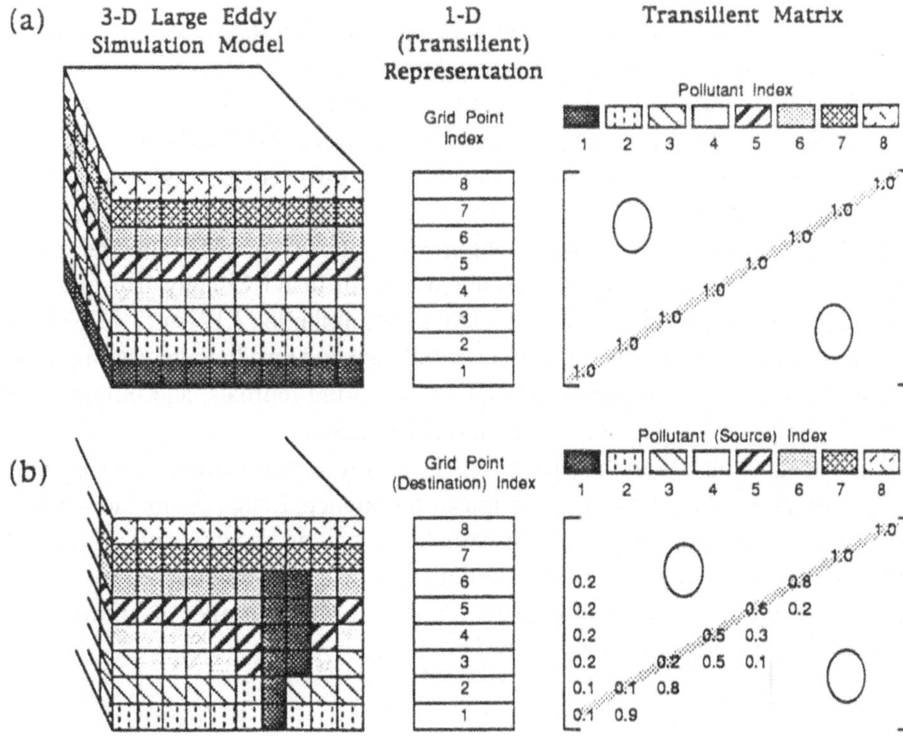

Fig. 3.1. Method used to measure transilient matrices by (a) injecting layers of tracers into a LES. After some finite time interval (b), turbulence has redistributed the tracers. The relative fractions of tracers at each level is a direct measure of the transilient matrix (Ebert *et al.*, 1989).

3. Les Measurements

3.1. Large-Eddy Simulation (LES) Techniques

To "measure" transilient matrices by LES, include prognostic equations for N different tracers in addition to the existing five or more prognostic equations for momentum, heat, and moisture, where N is the number of layers in the model. For example, a 24-layer LES model requires a minimum of $24 + 5 = 29$ equations to be solved at every grid point at every time step. Needless to say, LES model run-time increases considerably.

First, the LES is run with zero tracer concentration to allow the simulated turbulence to reach quasi-steady state. Define time $t = t_o = 0$ at the end of this initialization stage.

Pause the simulation at this time while tracers are 'injected"into the model. Tracers are injected as initially uniform layers, using different tracers for each different height, as sketched in Fig. 3.1a (Ebert *et al.*, 1989). The initial fraction of tracers at each height is indicated by the matrix on the right side of Fig. 3.1a.

Next, continue running the model and measure the relative concentrations of tracers averaged over each layer at a number of successive times, t. An idealized example of tracer transport and the corresponding tracer matrix for a convective boundary layer is sketched in Fig. 3.1b. Such tracer matrix information is saved as output.

Because each tracer was initially injected at only one unique source height, the tracer matrix is equal to the transilient matrix, $\mathbf{C}(t, \Delta t)$, where $\Delta t = t - t_o$. This is a simple and direct, albeit computer intensive, method to measure transilient matrices for any case that can be simulated by LES.

3.2. Convective Mixed Layer

Ebert *et al.* (1989) simulated a convective mixed layer over a homogeneous surface, with surface kinematic heat flux of 0.06 K.m.s^{-1}, convective velocity scale $w_* = 1.46$ m.s^{-1}, convective temperature scale $T_* = 0.041$ K, initial mixed-layer depth $z_{io} = 1600$ m, convective time scale $t_* = z_{io}/w_* = 18.3$ min, and Brunt-Väisälä frequency in the capping stable layer of 0.0099 s^{-1}. The vertical and horizontal domain sizes for this model are $\Delta Z = 1.5\, z_{io}$ and $\Delta X = \Delta Y = 5 z_{io}$, respectively. Lateral boundary conditions are cyclic. For a fine-mesh version of this model (48 X 160 X 160 cells), Fig. 3.2 shows measurements of tracer dispersion for a few of the tracers.

Fig. 3.3 shows measured transilient matrices for a variety of time steps, using a medium-mesh version of the LES model (24 X 80 X 80 cells). Over short time intervals, tracers to disperse little from their starting locations, resulting in a transilient matrix with the largest elements near the main diagonal (Figs. 3.3a&b).

As the time interval becomes larger (Figs. 3.3c&d) the matrix becomes noticeable asymmetric. As discussed in section 2.4, the majority of air from mid- and high-level sources within the mixed layer is moving slowly downward, while a smaller amount of air from near-surface sources is moving a bit more rapidly upward. This is consistent with conventional measurements of vertical velocity distribution in free convection, with a few strong updrafts and many weak downdrafts.

Convective overturning is evident in Fig. 3.3e, and eventually the tracers become well mixed (Figs. 3.3f).

3.3. Neutral Boundary Layer

Schumann *et al.* (1989) simulated an idealized neutral boundary layer. The depth scale z_i is defined as the height where the mean flow speed is 99% of the asymptotic value. This depth scale and the friction velocity u_* are found from the simulation, and are used to scale the output. The source and destination heights are scaled by z_i, and $z_o/z_i = 10^{-5}$.

The neutral transilient matrices (Fig. 3.4) exhibit much more symmetry than for the convective mixed layer, which is not surprising because small and medium eddies are more important during mechanical mixing.

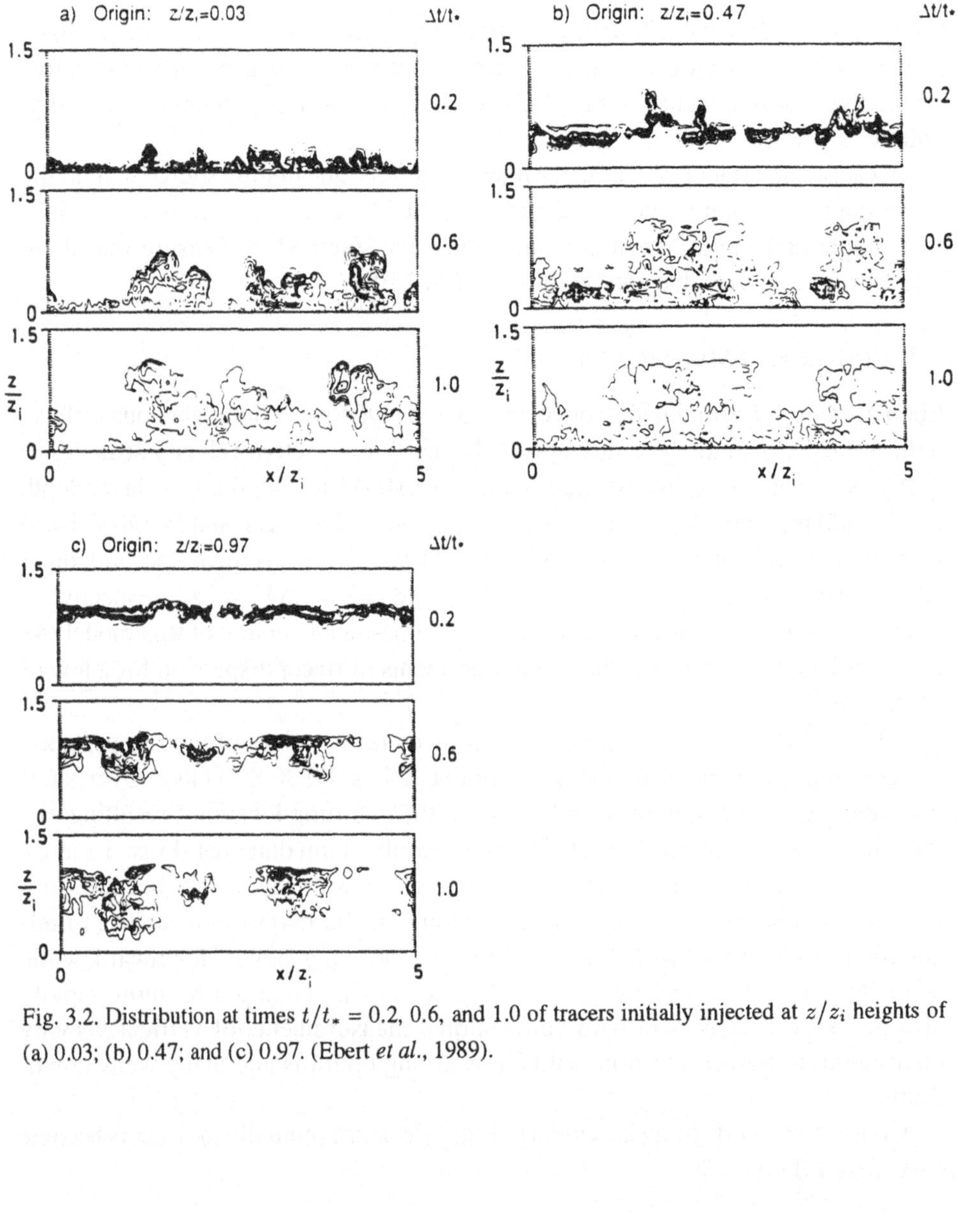

Fig. 3.2. Distribution at times t/t_{*} = 0.2, 0.6, and 1.0 of tracers initially injected at z/z_i heights of (a) 0.03; (b) 0.47; and (c) 0.97. (Ebert *et al.*, 1989).

3.4. Convection Over a Heterogeneous Surface

Another simulation was run almost identical to that described in section 3.2, except that the bottom surface was divided into two equal halves, where one half had a surface kinematic heat flux of 0.08 K.m/s and the other 0.04 K.m/s. A sea-breeze-like circulation developed in this simulation, as is shown in Fig. 3.5. However, when averaged over horizontal areas, the net nonlocal mixing was not much different from that for the homogeneous case (Fig. 3.6).

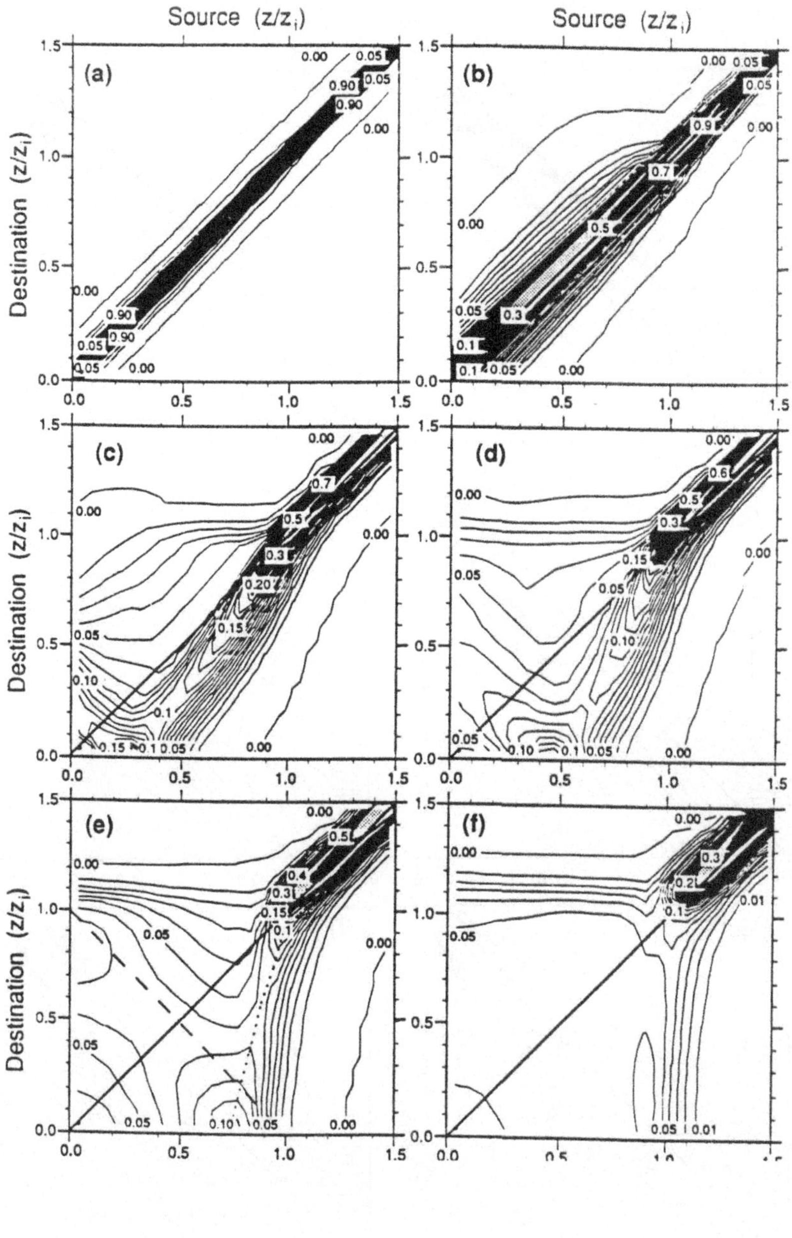

Fig. 3.3. Measured transilient matrices in a convective mixed layer for time intervals $\Delta t/t_*$ of (a) 0.02; (b) 0.2; (c) 0.6; (d) 1.0; (e) 2.0; and (f) 4.0. (Ebert *et al.*, 1989).

4. Nonlocal Statistics of Turbulence

4.1. Local Eulerian, Lagrangian, & Nonlocal Eulerian

Statistics such as vertical velocity variance, $\overline{w'^2}(z)$, or turbulence kinetic energy, TKE(z), can be calculated in a vertically-local Eulerian sense. For example, $\overline{w'^2}(z)$

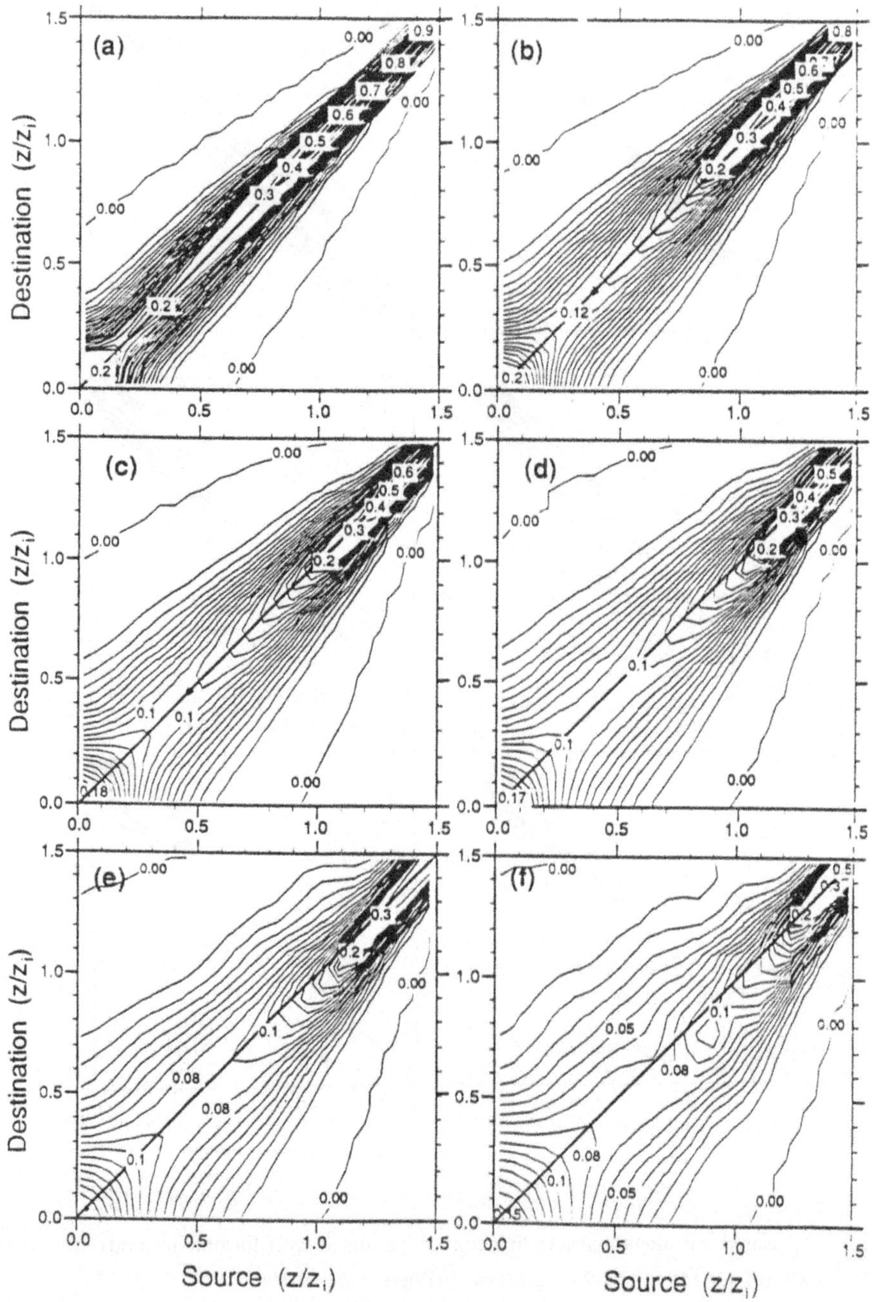

Fig. 3.4. Measured transilient matrices in a neutral boundary layer for time intervals $\Delta t \cdot u_*/z_i$ of (a) 0.005; (b) 0.05; (c) 0.125; (d) 0.25; (e) 0.5; and (f) 1.0. (Schumann *et al.*, 1989).

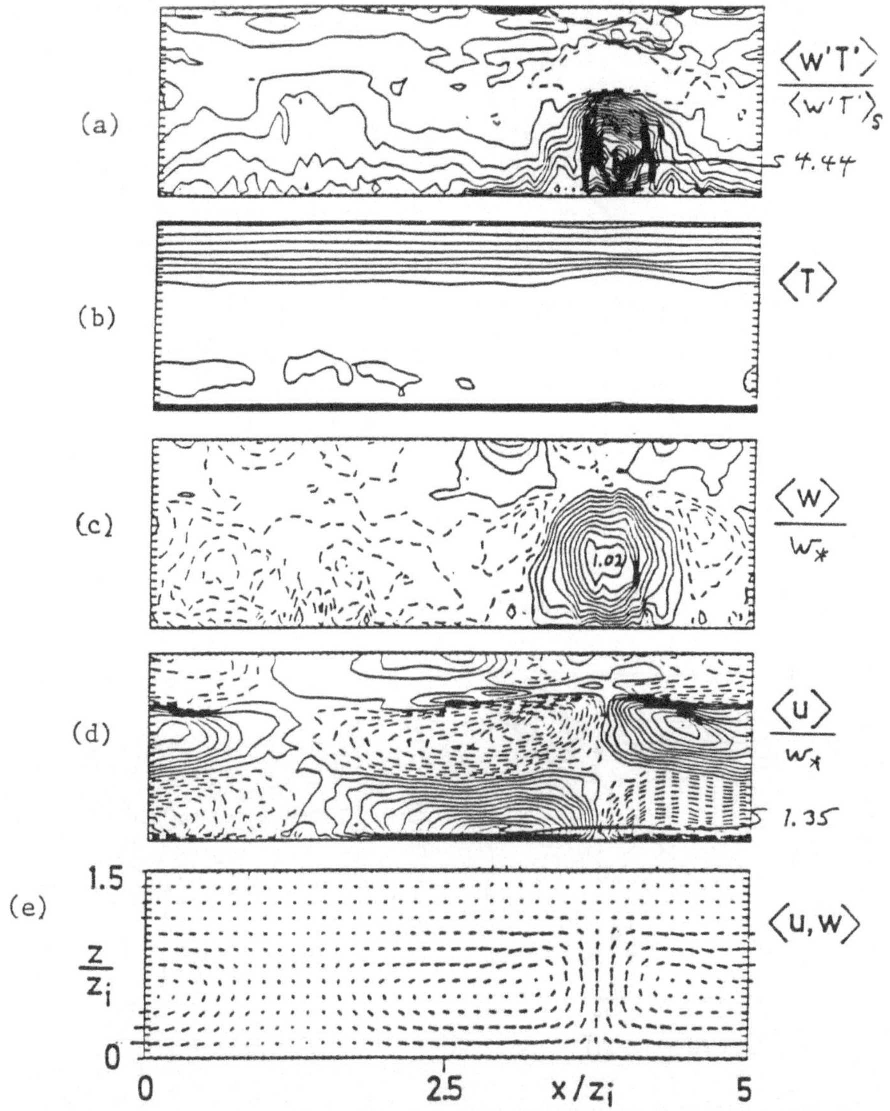

Fig. 3.5. LES of a convective boundary layer over a heterogeneous surface. Cross-sections of (a) resolved kinematic heat flux; (b) potential temperature; (c) vertical velocity; (d) horizontal velocity; and (e) streamlines. (Schumann, personal communic.).

can be calculated as a time average of w'^2 at a point at height z, an area average at height z at an instant in time, or a line average at height z over a finite time interval. By the ergodic hypothesis, these should all equal the hypothetical local-Eulerian ensemble average at z, under stationary homogeneous turbulent conditions. These statistics are local in height. They are based on the velocities and turbulent state at only that height.

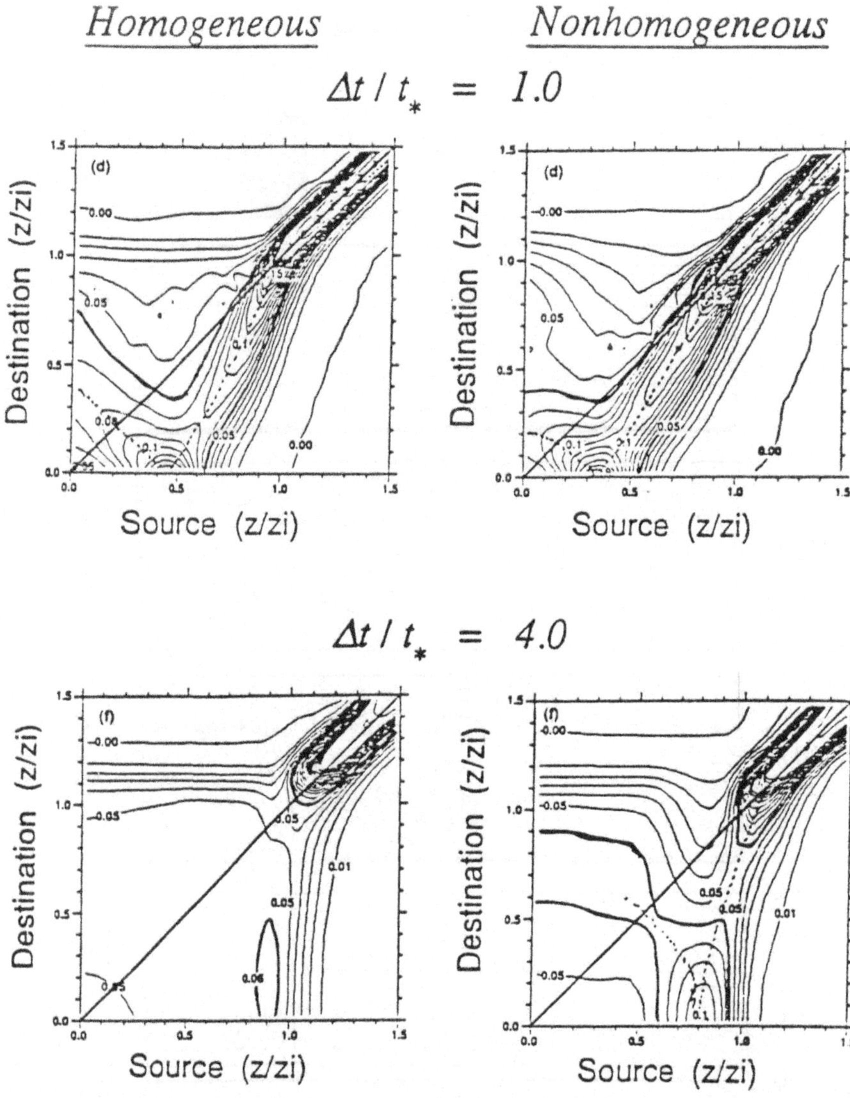

Fig. 3.6. Comparison of measured transilient matrices in convective mixed layers over homogeneous and inhomogeneous surfaces for two time intervals. (Schumann *et al.*, 1989).

Lagrangian statistics such as $\overline{w'^2}[Z(t)]$ follow tagged particles or parcels of air. They can be vertically-nonlocal, because the tagged particle can travel finite distances vertically $[Z(t)]$ and horizontally during a time interval. These statistics are closer to the basic conservation laws, such as Newton's second law which applies to, and follows, a moving mass. The Lagrangian framework is difficult to use for many practical situations where the turbulent state is needed at Eulerian positions relative to the Earth's surface (e.g., at smoke stacks, at bridges, at wind turbines, over crops, etc.).

Transilient matrices such as those measured with LES models are examples of vertically-nonlocal Eulerian statistics. First, the matrices are indeed statistics because they represent the net effect of an ensemble of turbulent structures. They are Eulerian because they represent the turbulence over a fixed spot on the ground. They are nonlocal because they are based on the vertical movement of air parcels from source to destination locations at different heights.

Local Eulerian, Lagrangian, and nonlocal Eulerian statistics are obviously related to each other, because they describe the same physical phenomenon. The relationship between local Eulerian and Lagrangian statistics have been discussed extensively elsewhere (Monin and Yaglom, 1971; Pasquill, 1974; Hinze, 1975; Tennekes and Lumley, 1972; Lumley and Panofsky, 1964), and will not be reviewed here. In the following subsections, additional nonlocal Eulerian statistics will be developed from the transilient matrix, and will be compared with their local Eulerian counterparts.

4.2. Nonlocal Flux

The vertical kinematic flux, $F_k(t, \Delta t)$ of S across any level $Z = k \cdot \Delta z$ that occurs during the time interval between t and $t + \Delta t$ is (Stull and Driedonks, 1987):

$$F_k(t, \Delta t) = \frac{\Delta z}{\Delta t} \sum_{i=1}^{k} \sum_{j=1}^{N} c_{ij}(t, \Delta t) \cdot [\bar{S}_i(t) - \bar{S}_j(t)] \tag{4.1}$$

which is equivalent to the recursion relation:

$$F_k(t, \Delta t) = F_{k-1}(t, \Delta t) + \frac{\Delta z}{\Delta t} \cdot \sum_{j=1}^{N} c_{kj}(t, \Delta t) \cdot [\overline{S_k}(t) - \overline{S_j}(t)] \tag{4.2}$$

with $F_k(t, \Delta t) = 0$ at $k = 0$ and $k = N$. It is also equivalent to (Ebert *et al.*, 1989):

$$F_k(t, \Delta t) = \frac{\Delta z}{\Delta t} \sum_{i=1}^{k} \sum_{j=k+1}^{N} c_{ij}(t, \Delta t) \cdot \bar{S}_i(t) - c_{ij}(t, \Delta t) \cdot \bar{S}_j(t)] \tag{4.3}$$

where the first subscript on c always represents the destination index, and the second is the source index. Also, while mean variables such as $\bar{S}_i$ represent layer averages and can be conceptually positioned within the center of a layer [i.e., at $z = (i - 0.5) \cdot \Delta z$ for equally-spaced grid layers], the flux occurs across the boundary between two layers; that is, across level $Z = k \cdot \Delta z$.

Remembering that transilient turbulence represents the advective-like transport associated with the turbulent movement of air parcels, it is obvious that there cannot be turbulent fluxes through the ground, unless the soil dances. As described in Stull (1988a), the sum of turbulent and molecular fluxes is the effective flux, which is what most investigators mean when they say surface flux. Be aware that the fluxes

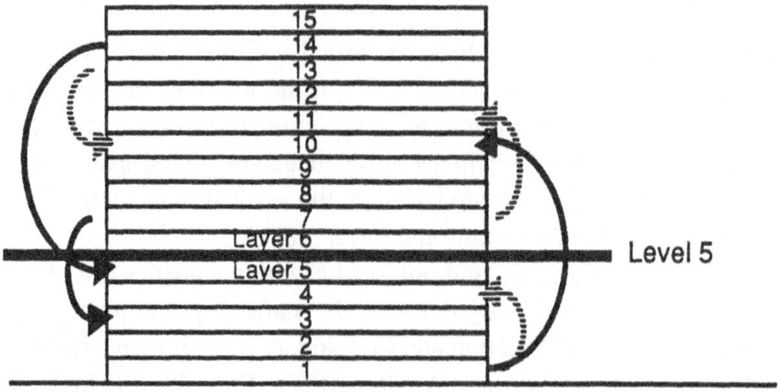

Fig. 4.1. Eddies that move across level 5 contribute to the vertical flux

of (4.1) to (4.3) are NOT effective fluxes, but only the true turbulent portion of the flux; namely, $\overline{w'S'}$.

Recursion relation (4.2) is efficient in computer programs, allowing the flux algorithm to step from level to level. If the initial step is from the ground where the flux is zero, then the final step into a region of zero turbulence should also give zero flux. This is a convenient check of the computer program.

Equation (4.3) states that fluxes across any level $Z = k \cdot \Delta z$ depend only on those eddies that start on either side of the level and end on the other side. This is physically reasonable, as sketched in Fig. 4.1.

Although (4.1) to (4.3) are very nonlocal with summations over a wide range of source and destination heights, the resulting fluxes are qualitatively and quantitatively identical to local fluxes, as in the example of Fig. 4.2. If an air parcel crosses some height enroute between source and destination locations, and carries with it scalar characteristics such as heat and moisture, then when that parcel crosses the height of interest, it will have a vertical velocity w' and a scalar perturbation s' that contributes to the local eddy-correlation flux. The nonlocal flux is thus more than equivalent to a local flux; it is identical. Nonlocal flux convergence of heat across a layer contributes to warming of that layer in the same manner as a local flux, and the same conservation constraints are obeyed.

The analytical form of the nonlocal flux equation is (Stull, 1984a):

$$F(Z,t) = -\int_{z=a}^{Z}\int_{\xi=a-z}^{b-z} \gamma(t,z,\zeta)\cdot[\bar{S}(t,z+\zeta)-\bar{S}(t,z)]d\zeta dz \qquad (4.4)$$

using the same notation as (2.12).

4.3. Transport Spectrum

The transport spectrum gives the contributions of different mixing-process distances (eddy sizes) to the total flux. Let $\gamma = m \cdot \Delta z$ be the separation distance between source and destination. The contribution to vertical kinematic S flux of all

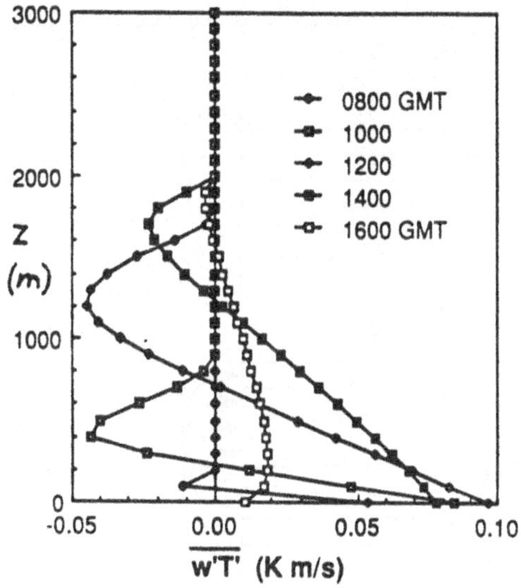

Fig. 4.2. Vertical profiles of kinematic flux for different times during the growth of a convective mixed layer, computed from transilient matrix coefficients using eq (4.1). (Stull and Driedonks, 1987).

source-destination pairs having equal separation distance γl, and that cross level $Z = k \cdot z$ is

$$F_{k|m}(t,\Delta t) = \frac{\Delta z}{\Delta t} \sum_{i=1}^{k} \sum_{j=k+1}^{N} \delta_{m,|i-j|} \cdot [c_{ji}(t,\Delta t) \cdot \overline{S_i}(t) - c_{ij}(\Delta t) \cdot \overline{S_j}(t)] \quad (4.5)$$

where $F_{k|m}$ is defined as a transport spectral component, and $\delta_{m,|i-j|}$ is the Kronecker delta. Fig. 4.3 shows an example of one such component.

The set of $F_{k|m}$ for all values of m gives the transport spectrum, and the sum over all these separation components yields the net transport (i.e., the total flux) across level Z:

$$F_k(t,\Delta t) = \sum_{m=1}^{N-1} F_{k|m}(t,\Delta t). \quad (4.6)$$

Transport spectral components have the same dimensions as the kinematic flux, and exist only when there is vertical transport between a source and destination.

Fig. 4.4 shows a transport spectrum of vertical heat flux in a convective mixed layer, where the individual spectral components have been grouped into bands of small, medium, and large separation distances. The sum of all bands gives the total kinematic heat flux. Small "eddies" (i.e., small separation distances) are important in the surface layer, but make little contribution in the interior of the mixed layer. Medium-size "eddies" transport most of the heat for this example. The largest eddies bring warm entrained air downward, and make a negative flux contribution throughout the mixed layer.

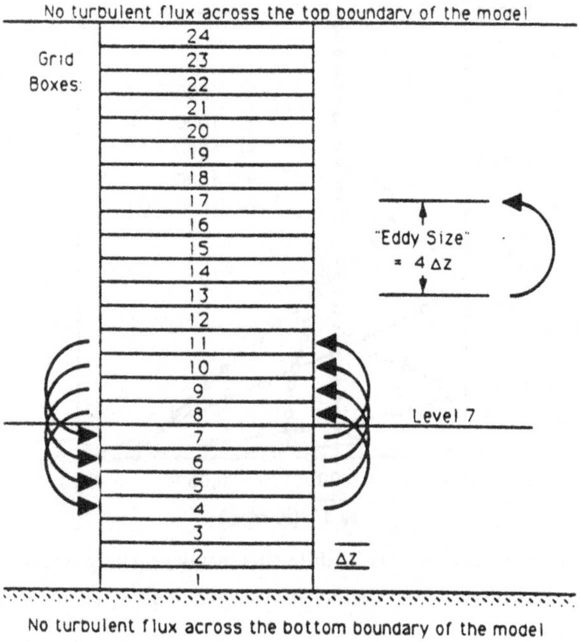

Fig. 4.3. Sketch of those source-destination pairs of separation distance $m = 4$ that contribute to the vertical kinematic flux across level $k = 7$. (Ebert *et al.*, 1989). These "eddies" contribute to just one spectral component of the transport spectrum.

Classical local cospectra and classical flux Fourier spectra give the contributions from various horizontal wavelengths to the vertical flux. The transport spectrum gives the contributions of vertical "wavelengths" to vertical flux.

Fig. 4.4 compares closely to classical horizontal Fourier spectra of flux, as shown in Fig. 4.5. Such similarity is expected because eddies tend to be isotropic. Namely, eddy circulations that are large in the vertical are also large in the horizontal, etc.

One can examine air movement independent of the scalars that are transported. The transport spectrum equation for air mass (without the scalar tracer) is

$$F^{air}_{k|m}(t, \Delta t) = \frac{\Delta z}{\Delta t} \sum_{i=1}^{k} \sum_{j=k+1}^{N} \delta_{m,|i-j|} \cdot [c_{ji}(t, \Delta t) - c_{ij}(t, \Delta t)]. \tag{4.7}$$

The sum of all air transport spectral components must equal zero, as required by continuity.

Fig. 4.6 shows the air mass transport spectrum for a convective mixed layer. Looking at Fig. 4.6b, near the top of the mixed layer ($z/z_i = 0.94$) small eddies (i.e., small source-destination separation distances) move air downward while medium eddies move air upward. Near the middle of the mixed layer ($z/z_i = 0.5$) medium eddies move air downward while larger eddies transport air upward.

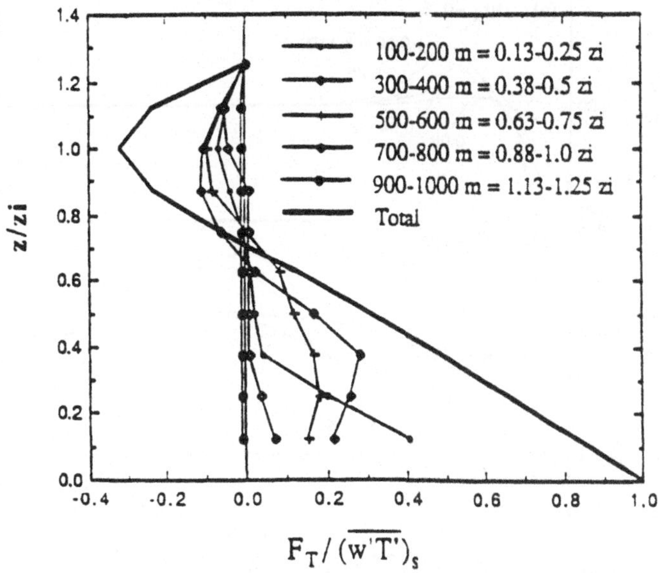

Fig. 4.4. Transport spectrum of heat from a parameterized convective mixed layer, where spectral components are grouped into small, medium, and large size ranges. The individual size bands sum to the total vertical kinematic heat flux (heavy solid line) at each height (Zhang and Stull, 1992).

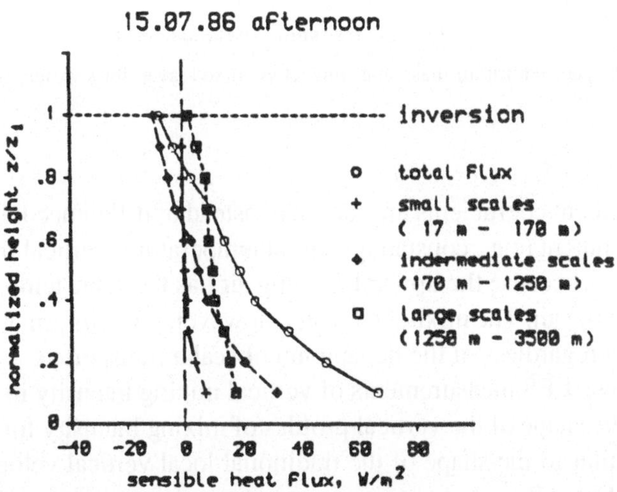

Fig. 4.5. Fourier decomposition of vertical kinematic heat flux as a function of height in a convective mixed layer, as measured from instrumented aircraft. (Jochum, 1988)

4.4. Mixing Intensity

Define a vertical nonlocal mixing intensity by:

$$P_k(t, \Delta t) = \sum_{i}^{k} = 1 \sum_{j=k+1}^{N} [c_{ji}(t, \Delta t) + c_{ij}(t, \Delta t)] \tag{4.8}$$

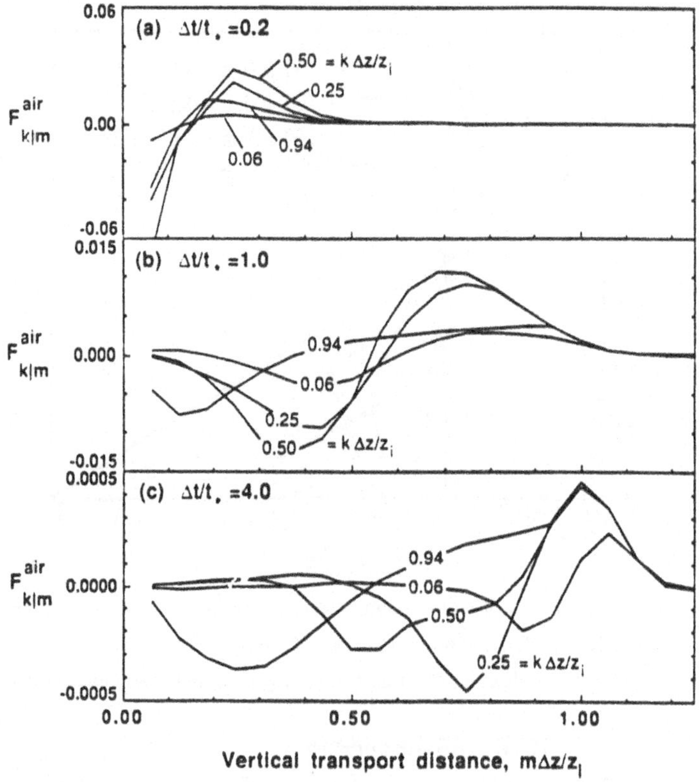

Fig. 4.6. Transport spectrum for air mass in a convective mixed layer, for a variety of time intervals. (Ebert *et al.*, 1989)

where the transilient coefficients are summed instead of differenced. In a vertically uniform environment (e.g., constant potential temperature) vertical mixing causes zero net heat flux because the upward-moving air has the same temperature as the downward-moving air. The mixing intensity, however, gives information about the vigor of mixing regardless of the net amount of scalar transported.

Fig. 4.7 shows LES measurements of vertical mixing intensity in a convective mixed layer. The shape of the vertical profiles of mixing intensity for various time intervals is similar to the shape of the traditional local vertical velocity variance, also plotted in Fig. 4.7.

In the limit of large times within a mixed layer, when the transilient coefficients equal each other within the ML domain, the mixing intensity approaches the following theoretical limit:

$$P_k(t, \Delta t \rightarrow \infty) = \frac{2k}{n_i}(n_i - k) \tag{4.9}$$

where n_i is the grid index at z_i, at the top of the ML.

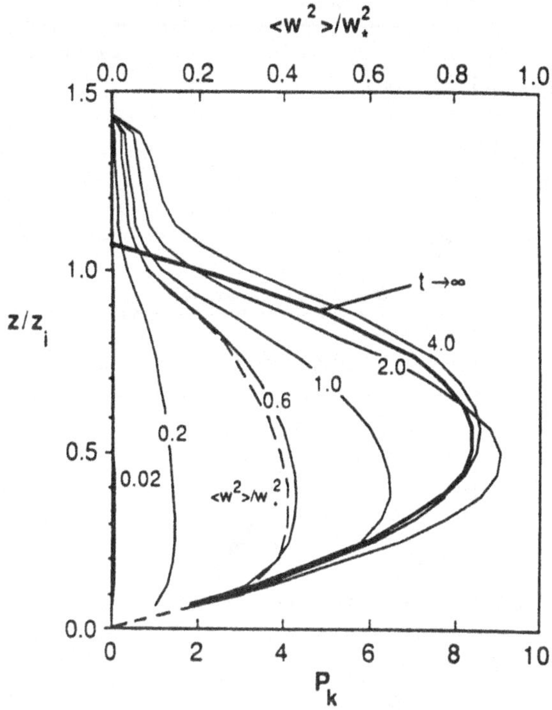

Fig. 4.7. Vertical profiles of mixing intensity in a convective mixed layer, for a variety of time intervals. (Ebert *et al.*, 1989)

4.5. Process Spectrum

Just as the flux can be segregated into transport spectral components, so can the mixing intensity be segregated into process spectral components:

$$P_{k|m}(t, \Delta t) = \sum_{i=1}^{k} \sum_{j=k+1}^{N} \delta_{m,|i-j|} \cdot [c_{ji}(t, \Delta t) + c_{ij}(t, \Delta t)] \tag{4.10}$$

and

$$P_k(t, \Delta t) = \sum_{m=1}^{N-1} P_{k|m}(t, \Delta t). \tag{4.11}$$

The first equation separates the total mixing vigor into contributions from various source-destination separation distances (i.e., eddy sizes).

Fig. 4.8 shows LES measurements of the process spectrum for a convective mixed layer. Middle-size eddies have the most vigor, but all eddy sizes contribute.

4.6. Mixing Lengths

Mixing length can be defined as the average distance that turbulence moves air in the process of mixing (Prandtl, 1925). Transilient coefficient $c_i k$ represents

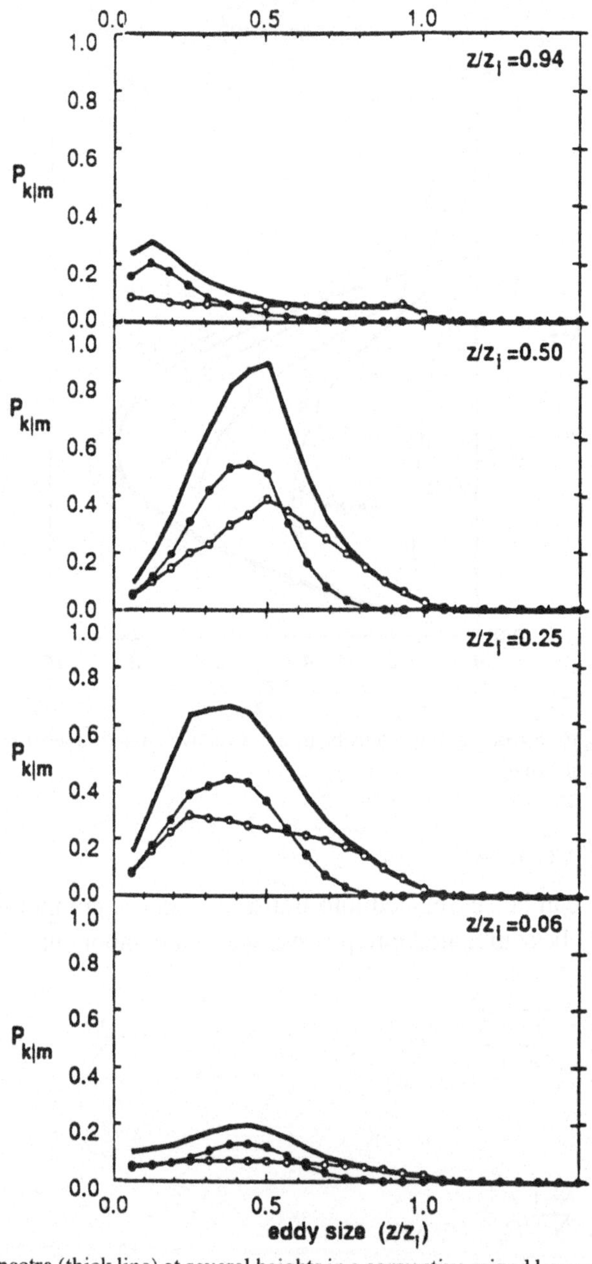

Fig. 4.8. Process spectra (thick line) at several heights in a convective mixed layer, valid for $\Delta t/t_* = 1.0$. The open and closed circles indicate the fraction of the process spectrum contributed by upward and downward mixing, respectively. (Ebert *et al.*, 1989)

the relative amount of air involved in mixing across distance $\Delta z \cdot |i - k|$. The weighted sum of both upward and downward mixing to and from level k from all possible sources and destinations gives the average mixing-length $l_k(t, \Delta t)$ at

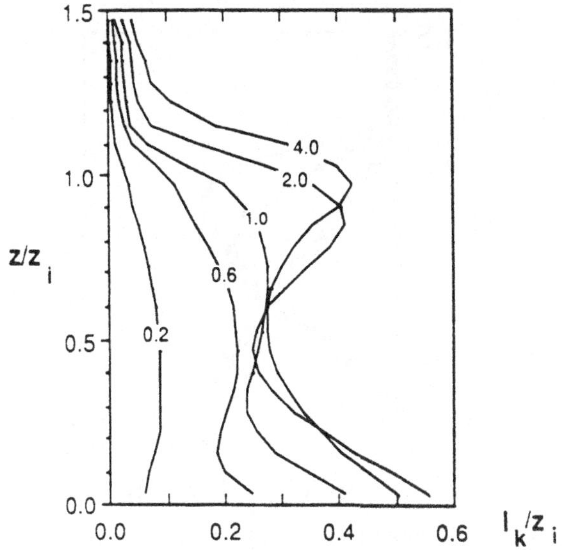

Fig. 4.9. Mixing-length profiles scaled by mixed layer depth z_i. Each curve corresponds to a different dimensionless time t/t_*, where t_* is the convective time scale. (Ebert *et al.*, 1989)

height $Z = k \cdot \Delta z$:

$$l_k = \Delta z \cdot \sum_{i=1}^{N} \frac{1}{2}[c_{ik}(t, \Delta t) + c_{ki}(t, \Delta t)] \cdot |i - k|. \tag{4.12}$$

Fig. 4.9 shows mixing-length profiles for a convective mixed layer as measured via LES. Mixing occurs across greater distances over greater time intervals, up to a limit given by a fraction of the ML depth, z_i. The largest mixing lengths occur near the surface and the top of the ML at large time intervals, because air parcels moving to or from those locations can come from or go to the greatest distance away.

4.7. Anisotropy in One-Dimension

If the statistics associated with mixing between some reference height and locations above it differ from mixing statistics between that height and locations below, then the turbulence is said to be "anisotropic in one-dimension" at that height. In the convective mixed layer, turbulence is usually anisotropic in the vertical.

As indicated in Fig. 2.2, the upper triangle of the transilient matrix holds only upward-mixing information, while the lower triangle holds downward-mixing information. By segregating many of the previously-defined nonlocal statistics into upward and downward moving components, one can examine the effects of 1-D anisotropy. For process spectra, such a segregation has already been shown in Fig. 4.8.

Mixing length can be segregated by upward and downward transport from

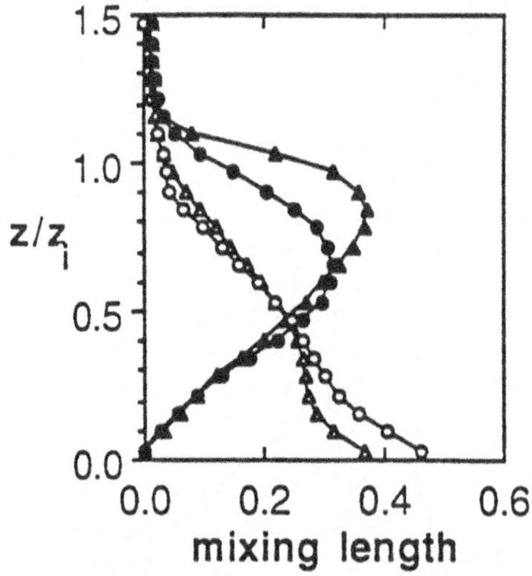

Fig. 4.10. LES-measured mixing length profiles scaled by mixed layer depth z_i for a convective mixed layer, segregated into upward and downward contributions, valid at $t/t_* = 1.0$. Plotted is mixing upward ○ and downward • **from** each **source** height, and upward ▲ and downward △ **to** each **destination.** (Ebert *et al.*, 1989)

source level j:

$$l_j^{\uparrow}(t, \Delta t) = \frac{\Delta z \cdot \sum_{i=j}^{N} c_{ij}(t, \Delta t) \cdot |i - j|}{\sum_{i=j}^{N} c_{ij}(t, \Delta t)} \tag{4.13a}$$

$$l_j^{\downarrow}(t, \Delta t) = \frac{\Delta z \cdot \sum_{i=1}^{j} c_{ij}(t, \Delta t) \cdot |i - j|}{\sum_{i=1}^{j} c_{ij}(t, \Delta t)}, \tag{4.13b}$$

and upward and downward transport to destination level i:

$$l_i^{\uparrow}(t, \Delta t) = \frac{\Delta z \cdot \sum_{j=1}^{i} c_{ij}(t, \Delta t) \cdot |i - j|}{\sum_{j=1}^{i} c_{ij}(t, \Delta t)} \tag{4.13c}$$

$$l_i^{\downarrow}(t, \Delta t) = \frac{\Delta z \cdot \sum_{j=1}^{N} c_{ij}(t, \Delta t) \cdot |i - j|}{\sum_{j=1}^{N} c_{ij}(t, \Delta t)}. \tag{4.13d}$$

Fig. 4.10 shows upward and downward mixing lengths for a convective mixed layer, valid at $t/t_* = 1.0$. From the surface layer, thermals can rise great distances until they hit the top of the mixed layer, while eddies can descend only a small distance until they hit the ground. Theories that assume mixing length increases linearly from the surface apparently give more weight to the curves in Fig. 4.10 with the solid symbols.

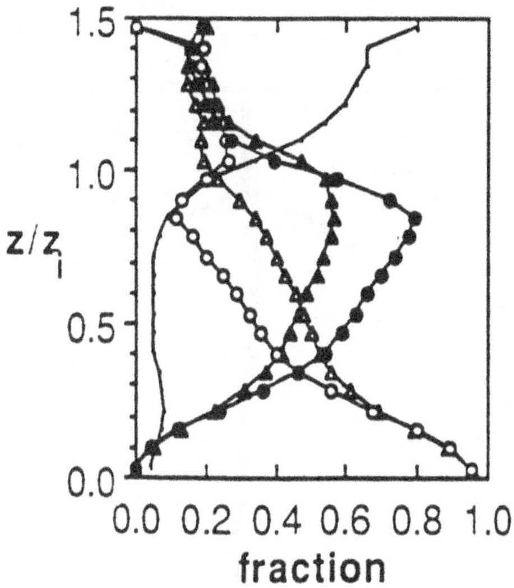

Fig. 4.11. Fraction of air moving up and down within a convective mixed layer as measured by a LES, valid at $t/t_* = 1.0$. Plotted is mixing upward o and downward • fractions **from** each **source** height, and upward ▲ and downward △ fractions **to** each **destination**. The solid curve without data points indicates the fraction of air not involved in resolvable transport. (Ebert *et al.*, 1989)

The fraction of air moving upward and downward to/from destinations/sources is also easily defined:

$$f_j^{\uparrow}(t, \Delta t) = \sum_{i=j+1}^{N} c_{ij}(t, \Delta t), \qquad f_j^{\downarrow}(t, \Delta t) = \sum_{i=1+1}^{j-1} c_{ij}(t, \Delta t) \tag{4.14a, b}$$

$$f_i^{\uparrow}(t, \Delta t) = \sum_{j=1}^{i-1} c_{ij}(t, \Delta t), \qquad f_i^{\downarrow}(t, \Delta t) = \sum_{j=i+1}^{N} c_{ij}(t, \Delta t). \tag{4.14c, d}$$

Fig. 4.11 shows these fractions for the same case as Fig. 4.10. The presence of such streams of upward and downward moving air that advectively transport tracers in a nondiffusive manner have been the motivating factor behind various top-down/bottom-up (Wyngaard and Brost, 1984; Wyngaard, 1987; Wiel, 1990) and two-stream turbulence models (Chatfield and Brost, 1987).

Fluxes can also be partitioned into upward and downward components, as shown in Fig. 4.12. Upward and downward transports of heat in the convective mixed layer for this case are very large, and nearly cancel. The total heat flux profile, also shown in Fig. 4.12, is a small difference between large contributions.

This is consistent with classical measurements of flux, such as the aircraft measurements shown in Fig. 4.13. Frequent occurrences of large magnitudes of $w'\theta'$ are observed, but many have opposite signs. Thus, the flight-leg average flux is a small number.

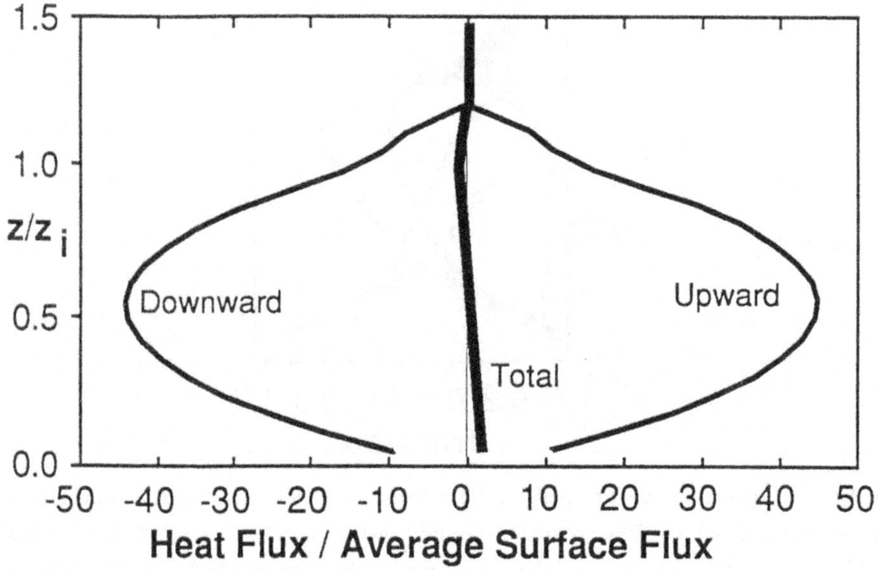

Fig. 4.12. Upward and downward nonlocal contributions to heat flux for a convective mixed layer, as measured by LES. Above the top of the ML the LES produced spurious results, which were deleted from this plot. (Ebert *et al.*, 1989)

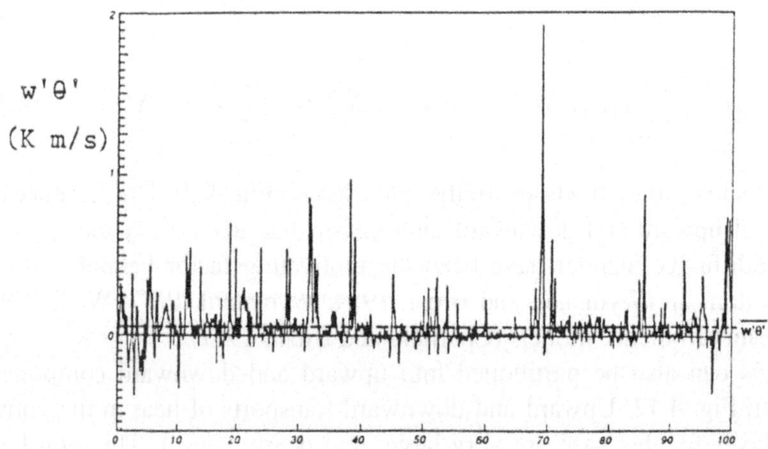

Fig. 4.13. Example of instantaneous values of $w'\theta'$ measured by an instrumented aircraft flying a straight, level flight leg in the surface layer of a convective mixed layer. (Stull, 1988a)

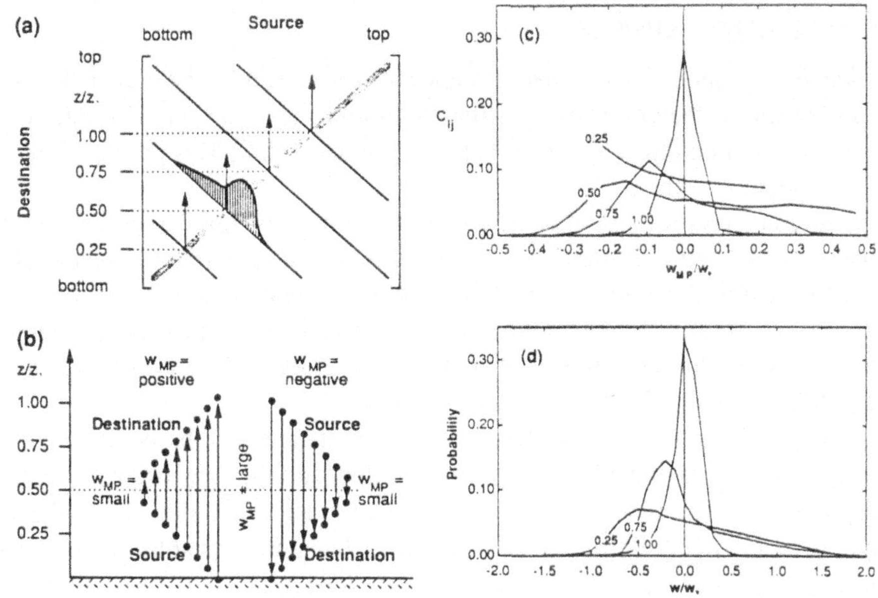

Fig. 4.14. (a) Schematic showing a cross-diagonal diagonal distribution of matrix-element values at reference height $z/z_i = 0.5$. (b) The physical meaning of elements along that cross diagonal, based on the definition of a mixing process velocity. (c) Plots of cross-diagonal values of matrix elements for a variety of reference heights, based on LES measurements. (d) Classical frequency distribution of layer-averaged w as measured by the LES model. (Ebert *et al.*, 1989)

Anisotropy in one dimension is apparent in the asymmetry of the transilient matrix relative to its main diagonal (Figs. 2.3 and 3.3). This effect is seen by extracting cross-diagonal elements from the matrix (Fig. 4.14a) for a variety of reference heights (Fig. 4.14b, for reference height $z/z_i = 0.5$), where the reference height is the height where the cross-diagonal crosses the main diagonal. Elements further from the main diagonal correspond to greater values of mixing-process velocity: $w_{MP} = (i - j) \cdot \Delta z/\Delta t$. These mixing process velocities are the same as implied in Fig. 2.3d.

The cross-diagonal elements from the transilient matrix, plotted in Fig. 4.14c, look similar to the classical local vertical-velocity frequency distribution observed for the convective mixed layer simulated by the LES (Fig. 4.14d). In particular, there is skewness of the vertical velocity distribution near the middle of the mixed layer, and a narrower symmetric distribution at the top.

5. Parameterization of Turbulence Closure

5.1. LeChatelier's Principle

LeChatelier's principle states that "if some stress is brought to bear upon a system in equilibrium, a change occurs, such that the equilibrium is displaced in a direction which tends to undo the effect of the stress" (Weast, 1968). This principle applies to the boundary layer.

Picture an equilibrium initial condition of a statically and dynamically stable boundary layer with no turbulence. Suppose that a short period of solar heating occurs to disturb the equilibrium by heating the ground, which in turn heats the surface layer. A turbulent circulation develops in response to this static instability that moves the warm air upward. Once the warm air has risen, the system is again in equilibrium, and turbulent circulations cease.

A similar evolution occurs when wind shears increase to the point where the Richardson number drops below its critical value, and the flow becomes dynamically unstable. Turbulence develops to mix the faster and slower air streams, resulting in a new equilibrium wind field that is dynamically stable with no turbulence.

In many real boundary layers, buoyant or dynamic stress is brought to bear on the boundary layer over long periods of time, such as solar heating during the daytime over land. In this situation, the boundary layer is continuously responding to the imposed stress, and is continuously turbulent as long as the imposed forcing continues.

When this principle is approximated in finite difference form, one can describe the destabilization and stabilization processes as happening in steps (Fig. 5.1). During the first part of a step, the externally imposed forcings and boundary conditions (called dynamics here) alter the mean lapse rate or vertical shear. If conditions are right, enough instability builds up to trigger turbulence. In the second part (called mixing here), turbulence partially undoes the instability by causing vertical mixing that reduces shears and buoyant instabilities.

5.2. Types of Closure

All turbulence closures consist of two components. One is the basic framework that represents the underlying physical assumptions made in the parameterization. By definition, such a parameterization contains unknown parameters. The second component involves choice of parameter values.

Examples of these two components are demonstrated below for (local) K-theory and for (nonlocal) transilient turbulence theory. Both are first-order closures.

K-theory assumes small eddies cause turbulent fluxes that flow down the local mean gradient, analogous to molecular diffusion. The parameter in this theory is K, the eddy diffusivity. When K is parameterized as a function of local mean flow variables, the result is first order-closure. The equations below describe this approach. Let height Z correspond to grid index i.

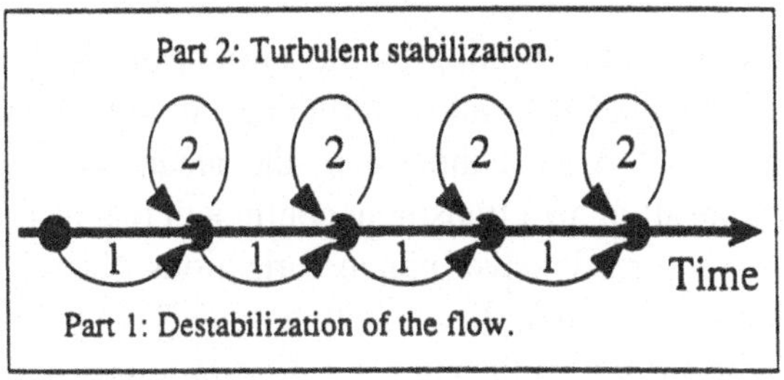

Fig. 5.1. Time line showing how the equations of motion with appropriate body forcings and boundary conditions are applied during the first part of each time step. During the second part, turbulence reduces any instabilities that have built up. If the flow was not destabilized during the first part of the step, then there is no turbulence and part two of the step does nothing. (Stull, 1987)

Local Physical Framework:

$$\bar{S}_i(t+\Delta t) = \bar{S}_i(t) + \Delta t \cdot (\, net\ source\ at\ Z) - \Delta t \cdot [\frac{\partial \overline{w's'}}{\partial z}(Z)] \tag{5.1}$$

$$\overline{w's'}(Z) = -K_i \cdot [\frac{\partial \bar{S}}{\partial z}(Z)]. \tag{5.2}$$

Local Closure Parameterization:

$$K_i = function[\bar{U}_i, \bar{V}_i, \bar{\Theta}_i, \partial \bar{U}/\partial z(Z), \partial \bar{\Theta}/\partial z(Z)...]. \tag{5.3}$$

Although (5.1) is shown as a forward difference for didactic purposes, in actuality a higher-order finite difference scheme would probably be used for numerical stability.

Transilient turbulence theory assumes that mixing between heights with finite separation can occur directly by air parcel movement, associated with coherent structures of a variety of sizes. The parameters in this theory are the elements of the transilient matrix. Each element of the matrix represents mixing between grid points i and j , and is parameterized by mean flow variables at both i and j.

Nonlocal Physical Framework:

$$\bar{S}_i^*(t+\Delta t) = \bar{S}_i(t) + \Delta t \cdot (\, net\ source\ at\ Z) \tag{5.4a}$$

$$\bar{S}_i(t+\Delta t) = \sum_{j=1}^{N} c_{ij}(t, \Delta t) \bar{S}_j^*(t+\Delta t). \tag{5.4b}$$

Nonlocal Closure Parameterization:

$$c_{ij} = function[\bar{U}_i, \bar{U}_j, \bar{V}_i, \bar{V}_j, \bar{\Theta}_i,\ T\bar{h}eta_j, ...]. \tag{5.5}$$

First, (5.4a) is evaluated to give $\bar{S}_i^*$, an intermediate value of the variable that includes the effects of dynamics and boundary conditions. Then (5.5) is solved, which determines the amount of mixing by the amount of instability in the mean flow. Finally (5.4b) is evaluated to allow turbulent mixing to partially undo the instability, as in Fig. 5.1. At each time step, the mean wind and temperature variables can change, resulting in a transilient coefficient that is a function of time.

Both (5.3) and (5.5) are "responsive" parameterizations. Namely, the value of the parameter is designed to respond to the mean flow state. This is necessary for most geophysical flow situations where turbulence is nonstationary and heterogeneous. "Unresponsive" parameterizations for K and C are described in the literature (Stull, 1984a, 1988a; Berkowicz and Prahm, 1980), but will not be discussed further here.

5.3. Mixing Potential

Some of the transilient parameterizations utilize a mixing-potential approach. Let A_{ij} be the potential for mixing from grid point j to grid point i. This potential is related to the amount of instability in the flow, which is assumed to govern the amount of turbulent mixing that results.

To utilize this in a parameterization for c_{ij}, define a row norm, RN_i, by (Stull, 1990):

$$RN_i = \sum_{j=1}^{N} A_{ij} \tag{5.6}$$

and an L_∞ matrix norm $\| A \|_\infty$ by:

$$\| A \|_\infty = max_i(RN_i). \tag{5.7}$$

By dividing a mixing potential by the matrix norm, one can build a transilient coefficient that satisfies the conservation constraints described in section 2.3. The off-diagonal elements are:

$$c_{ij} = \frac{A_{ij}}{\| A \|_\infty} \qquad for \quad i \neq j. \tag{5.8}$$

Each row of the matrix must sum to one because of air-mass conservation, which allows us to solve for the diagonal elements:

$$c_{ii} = 1 - \sum_{j=1; j\neq i}^{N} c_{ij}. \tag{5.9}$$

The mixing potential and the four equations above are a function of time and time step interval, and thus must be recomputed at each time step so as to respond to the mean flow. If the time steps are small, or if turbulence is quasi-stationary, the equations above need not be recomputed at each time step, although their result should be applied in the forecast model at each step.

For the atmosphere, this potential is parameterized as a function of the dynamic instability of the flow. Two parameterizations have been tested fairly extensively: one based on a nonlocal approximation to the turbulence kinetic energy (TKE) equation, and the other based on a nonlocal approximation to the Richardson number. These are described in the next subsections.

Mixing potential could be defined in terms of other instability measures for other flow situations. For example, it could be a function of the Reynolds number for flow over wings or in pipes, a function of Rayleigh number for organized convection between two plates, or a function of Rossby number for synoptic-scale eddies in planetary atmospheres.

5.4. Nonlocal Turbulence Kinetic Energy (TKE)

Stull and Driedonks (1987) used a nonlocal analogy to the turbulence kinetic energy (TKE) equation to parameterize the mixing potential for $i \neq j$:

$$Y_{ij} = \frac{\Delta t \cdot T_o}{(\Delta_{ij} z)^2} \cdot [(\Delta_{ij}\bar{U})^2 + (\Delta_{ij}\bar{V})^2 - \frac{g \cdot (\Delta_{ij} z) \cdot (\Delta_{ij}\bar{\Theta}_\nu)}{\bar{\Theta}_\nu \cdot R_c}] - \frac{D_Y \cdot \Delta t}{T_o} \quad (5.10)$$

where Y_{ij} is the flow-instability contribution to the mixing potential. The symbol Δ_{ij} represents a nonlocal difference; for example, $\Delta_{ij}\bar{\Theta} = \bar{\Theta}_i - \bar{\Theta}_j$. It is assumed for simplicity that $Y_{ji} = Y_{ij}$.

The initial guess values of Y_{ij} from (5.10) are then modified such that Y_{ij} increases monotonically both row-wise and column-wise from the corners of the transilient matrix toward the main diagonal. The amount of internal subgrid mixing within grid box i is then estimated by using the largest off-diagonal elements in the same row, and adding an additional estimate of unresolved subgrid mixing potential

$$Y_{ii} = max(Y_{i,i-1}, Y_{i,i+1}) + Y_{ref}. \quad (5.11)$$

There are four parameters in (5.10). The recommended values of these are: time-scale $T_o = 1000s$, critical Richardson number $R_c = 0.21$, dissipation factor $D_Y = 1.0$, and potential for unresolved subgrid mixing $Y_{ref} = 1000$. With these four parameters one can find N^2 transilient coefficients, where N is the number of grid layers in the model.

For equally-space grid points and no interference from plant canopies, the mixing potentials are:

$$A_{ij} = A_{ji} = Y_{ij}. \quad (5.12)$$

Equation (5.10) uses wind and temperature differences between heights i and j to determine the mechanical and buoyant production/consumption of turbulence. The last term parameterizes the effects of dissipation of TKE. The amount by which production exceeds consumption and dissipation determines the potential for mixing. If there is zero or negative potential for mixing, then A_{ij} is set to zero and there is no modeled turbulence. For simplicity, (5.12) makes the mixing potential matrix symmetric. These parameterized transilient matrices do not contain the asymmetries that are known to occur in the measured matrices (see subsection 3.2).

5.5. Nonlocal Richardson Number (R_i)

One of the first Richardson number (Ri) parameterizations for the transilient matrix required the solution of a complex matrix equation, which was cumbersome (Stull, 1984). More recently, Zhang and Stull (1991) have reframed a Richardson number parameterization into a mixing potential approach, which allows the direct solution of elements of the transilient matrix.

A nonlocal analogy to the bulk Richardson number is defined as

$$r_{ij} = \frac{g \cdot (\Delta_{ij}\bar{\Theta}_\nu) \cdot (\Delta_{ij}z)}{(\Delta_{ij}\bar{U})^2 + (\Delta_{ij}\bar{V})^2} \tag{5.13}$$

with $r_{ij} = r_{ji}$. If $r_{ij} > R_c$, then no turbulence is initiated, and Y_{ij} is set to zero. Otherwise, turbulence is assumed to cause mixing until the flow reaches a termination value of the Richardson number, R_T:

$$A_{ij} = w_{ij}[1 - (r_{ij}/R_T)], \qquad for \quad i \neq j \tag{5.14}$$

where w_{ij} is a distance weighting function given by:

$$w_{ij} = U_o \cdot \Delta t / [\Delta z \cdot |i - j|], \qquad for \quad i \neq j. \tag{5.15}$$

Both w_{ij} and A_{ij} are truncated to be between zero to one, the latter to prevent convective overturning. Also, both are assumed to be symmetric matrices.

The off-diagonal transilient elements are found from

$$c_{ij} = A_{ij}/N_d \tag{5.16}$$

where N_d is the number of grid points within the subdomain that is turbulent. The diagonal elements are found from (5.9).

Recommended values for the three parameters are: critical onset Richardson number $R_c = 1.5$, critical turbulence-termination Richardson number $R_T = 2.0$, and velocity scale $U_o = 0.5m/s$. With these three parameters, N_2 transilient coefficients can be found, where N is the number of grid layers in the model.

5.6. Unequal Grid Spacing

Many models utilize an expanding grid system with the smallest grid layers near the surface (see section 2.5). Such a scheme has been tested in a mesoscale model with the TKE transilient parameterization (Raymond and Stull, 1990). The following

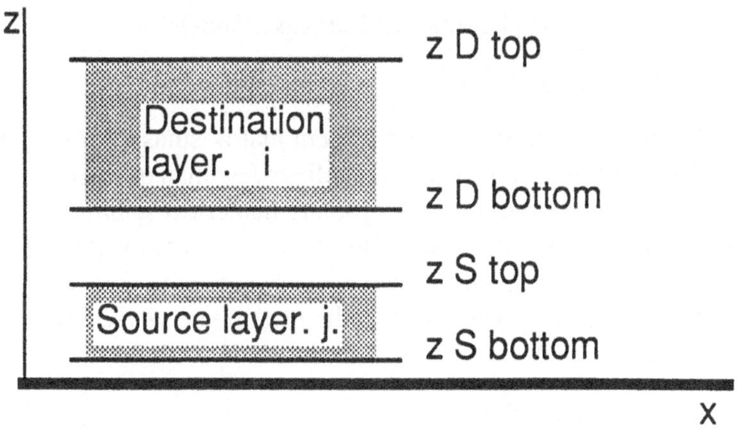

Fig. 5.2. Definition of top and bottom limits of source (j) and destination (i) layers for unequal grid spacing. (Stull, 1990).

approximate expression can be used in place of (5.12) for situations where the source and destination locations are relatively far apart:

$$A_{ij} = A_{ji} = m_j \cdot Y_{ij} \tag{5.17}$$

where m_j is the relative mass of air within source grid layer j.

In general, however, a more accurate equation (Stull, 1990) should be used in place of (5.17) whenever the source and destinations are close, and when they contain different masses of air:

$$A_{ij} = \frac{\int_{zDbottom}^{zDtop} \rho(z_D) \int_{zSbottom}^{zStop} \rho(z_S) \cdot Y_{ij}(z_D, z_S) dz_S dz_D}{\int_{zDbottom}^{zDtop} \rho(z_D) dz_D} \tag{5.18}$$

where ρ is air density, z_D is height above ground within destination layer i, and z_S is height above ground within source layer j (see Fig. 5.2). Also,the height difference in (5.10) should be written as $\Delta z_{ij} = z_D - z_S$.

Although the parameterizations of (5.17) and (5.18) and Fig. 2.4 make sense physically, Raymond and Stull (1990) found that it did not always work as well as desired.

Part of the problem is caused by the parameterization of (5.10), which causes too much mixing near the surface. The mass weighting scheme of (5.17) exacerbates the problem further. To correct for this, Raymond and Stull did not use the mass weighting of (5.17) even tough the grid spacing was unequally spaced. This correction worked only because the grid spacing in their mesoscale model increased monotonically with height.

6. Applications in Forecast Models

6.1. Absolute Numerical Stability

Because each row and column of the transilient matrix sums to one for equally-spaced grid points, all of the eigenvalue moduli are less than or equal to one. Stull (1986) shows that this condition yields absolute numerical stability for any size forward time step and any grid spacing. This numerical stability of the turbulence parameterization has indeed been realized in every numerical forecast made with transilient turbulence theory. Nevertheless, the most physically-realistic forecasts require a good choice for Δz and Δt, as is discussed in section 8.1.

6.2. One-Dimensional Models

Fig. 6.1 shows a shell of a main FORTRAN 77 program for making forecasts using transilient turbulence theory. A FORTRAN listing for the turbulence subroutine, "Turb" is given by Stull (1986), with the exception that the subgrid mixing code listed there should be corrected to utilize the Y_{ref} parameterization described by (5.11).

As implied in Fig. 6.1 and in the discussion of the previous subsection, surface boundary conditions are applied to only the bottom grid point. Molecular conduction, evaporation, and friction are assumed to alter only the bottom grid layer. These fluxes between the surface and the lowest layer of air can be parameterized by bulk aerodynamic formulae (Stull, 1988a), surface heat budgets with partitioning schemes as appropriate, or can be imposed directly as boundary conditions.

Any mixing of heat, moisture, or momentum higher into the model from the lowest grid point in the fluid is assumed to be driven primarily by turbulence, and is thus already handled by the transilient parameterization. Note that the transilient parameterization cannot cause turbulent mixing between the air and the ground, because it is assumed that moving air parcels stop just above the ground, resulting in zero turbulent flux at the bottom of the lowest grid layer. Thus, the effective flux profile that is usually given in the literature must use the molecular flux from the boundary-condition portion of the parameterization at the bottom of the lowest grid layer, and use transilient fluxes for all other levels.

Boundary-layer depth is not explicitly forecast by the transilient model. It must be inferred from the forecast output data. One way is to examine the subdomain within which turbulent mixing is occurring. This domain corresponds to the elements along the main diagonal of the transilient matrix that are not equal to one. Another is to find the height where the buoyancy flux is most negative (for convection), or where the buoyancy flux is some small fraction of its surface value (for stable boundary layers). For unstable boundary layers, one can also conceptually lift an air parcel from the lowest grid point, and determine at what altitude it crosses the sounding.

One of the strengths of the transilient parameterization is that it responds to destabilizations at all heights and by all dynamic stability causes. For example,

```
program model
C Declare arrays & other variables, & put global variables
C in common blocks.
      include Global        !Global variables in common blocks

C Initialize
      call SetMod(...)      !Set up the model (grid spacing,etc)
      call SetPhy(...)      !Set physical parameters, like Rc
      call GetICs(...)      !Get initial conditions from disk
      call GetBCs(...)      !Get boundary conditions from disk
      call SaveIn(...)      !Save initial state on disk

C Forecast
10    continue              !For each time step
        call Step(...)      !Increment counters, clock, etc.
        if (done) goto 20   !Determine if done
        call Dynam(...)     !Apply dynamics and boundary conds
        call Turb(...)      !Apply transilient mixing turbulence
        call SaveRe(...)    !Save results on disk or printout
      goto 10               !Loop back for another time step
20    continue              !Forecast is finished

C Post Processing
      call PostPr(...)      !Draw graphs, etc.
      stop
      end
```

Fig. 6.1. Shell of main program in FORTRAN 77 for a 1-D forecast model of oceans or atmospheres. The Dynam subroutine implements the destabilization portion of the split time step by applying mean flow dynamics, thermodynamics, advection, and other body forcings to all grid points and by applying boundary conditions to adjust the bottom and to grid points as appropriate. The Turb subroutine implements the restabilization portion of the time step by calculating the amount of instability, parameterizing the transilient matrix coefficients, and utilizing that matrix to calculate the effects of turbulent mixing (Stull, 1986).

a day during which there is both strong wind shear and strong heating is easily handled by the model, while a classical convective mixed-layer bulk model would give erroneous forecasts because of its lack of consideration of the wind-shear effects.

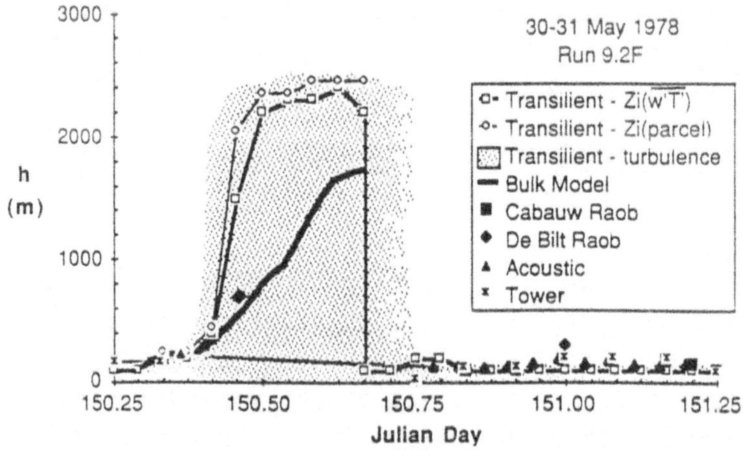

Fig. 6.2. Boundary-layer depth evolution simulated by the TKE transilient model, using initial conditions and boundary conditions observed near the Cabauw tower in the Netherlands, 30-31 May 1978. Three estimates of boundary-layer depth are estimated from the transilient model, and are compared to nearby radiosonde (raob) observations, acoustic sounder records, and tower measurements of temperature and turbulence (Stull and Driedonks, 1987).

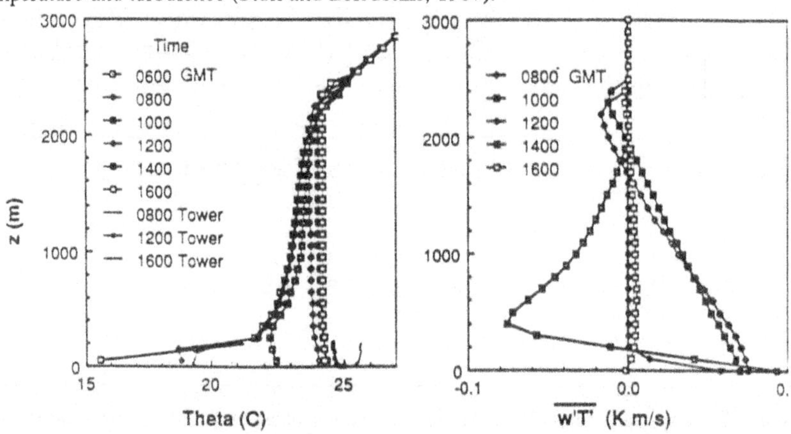

Fig. 6.3. Potential temperature (theta) and kinematic heat flux (w'T') profiles simulated for the same Cabauw case as Fig. 6.2. (Stull and Driedonks, 1987)

6.3. Convective Mixed Layer

Three days of observations from near the Cabauw tower (50 km south of the North Sea, at 51° 58'N, 4° 56'E) in the Netherlands were used by Stull and Driedonks (1987) to test the transilient model. Fig. 6.2 shows the boundary-layer depth on one of those days, 30-31 May 1978, and compares it to observations from radiosondes launched near Cabauw and DeBilt, from an acoustic sounder, and from tower data. The model utilizes the TKE parameterization approach. Fig. 6.3 shows the corresponding temperature profiles and heat flux evolution, Fig. 6.4 shows the specific humidity and moisture flux, and Fig. 6.5 shows wind component forecasts.

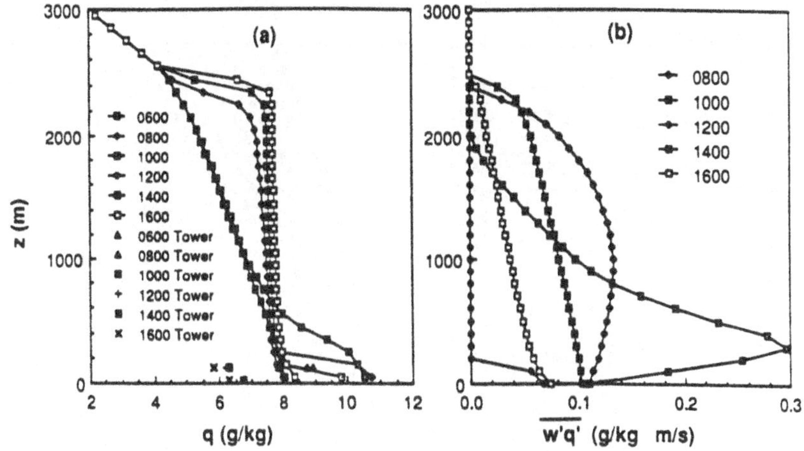

Fig. 6.4. Specific humidity (q) and kinematic moisture flux (w'q') profiles simulated for the case in Fig. 6.2. (Stull & Driedonks, 1987)

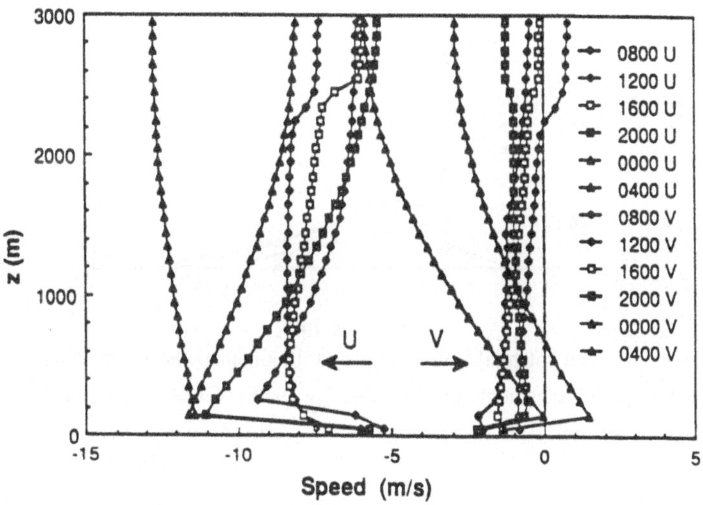

Fig. 6.5. Easterly (U) and northerly (V) wind profiles simulated for the same Cabauw case as Fig. 6.2. (Stull and Driedonks, 1987)

Another of the three Cabauw case study days was used by Wang and Albrecht (1990) to compare with their own mixed-layer models. All models they compared gave similar predictions for the mixed-layer growth.

Data from one day of the BLX83 field experiment in Oklahoma were used by Zhang and Stull (1992) to test both the TKE and the Ri transilient parameterizations. Fig. 6.6 shows temperature profiles and heat flux, using the Ri approach, while Fig. 6.7 shows the corresponding temperature and humidity results from the TKE approach. Figs. 6.8 & 6.10 show the verifications. Wind profiles are given in Fig. 6.9.

Mixed-layers simulated with the TKE transilient parameterization are compared in Fig. 6.11 with "measurements" from the LES model described in section 3. The

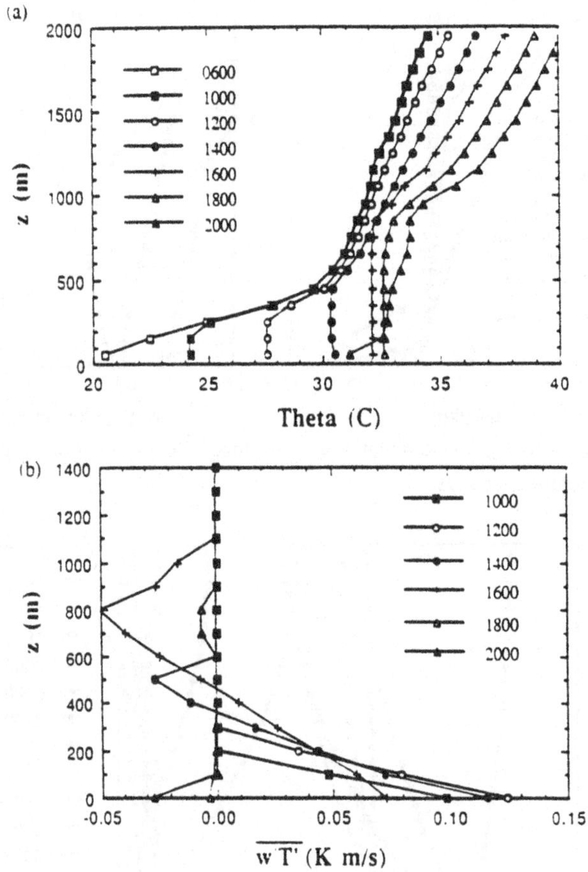

Fig. 6.6. Profiles of (a) mean potential temperature, and (b) kinematic heat flux, as a function of time (CDT) simulated by the Ri transilient model for 28 May 1983 in Oklahoma during the BLX83 field experiment. (Zhang and Stull, 1992)

TKE simulated temperature profiles exhibit a surface layer that is too deep. For the TKE transilient parameterization, Fig. 6.12 shows the fractions of air moving up and down, the mixing lengths, vertical mixing intensity, and transport spectrum, valid at $t/t_* = 1.0$. The TKE parameterized transilient matrix (Fig. 6.13) shows symmetry and relatively intense mixing near the surface that was not evident in the corresponding LES matrices shown in Fig. 3.3.

6.4. Neutral Boundary Layer

Tests of the TKE transilient parameterization for an idealized neutral boundary layer were presented by Stull and Driedonks (1987). Fig. 6.14 shows various aspects of the wind and momentum flux. A geostrophic wind of 5 m/s in the U direction, and drag at the ground, were the only imposed boundary conditions. The potential temperature profile is adiabatic over the whole depth of the model, and there is no surface heating or cooling. Stull and Driedonks (1987) found nearly logarithmic variation of the winds with height, with a smooth decrease of the momentum flux

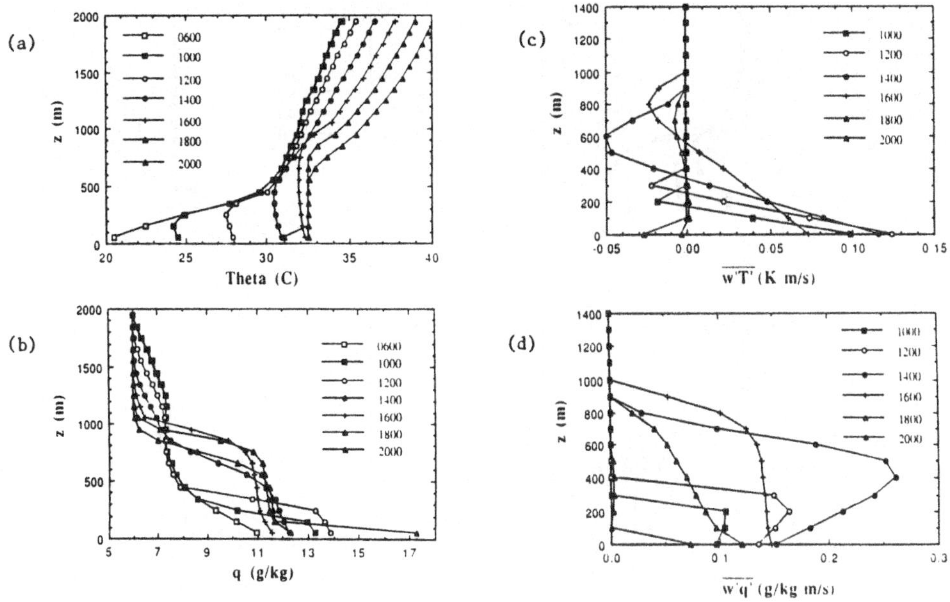

Fig. 6.7. Profiles of (a) mean potential temperature, (b) mean specific humidity, (c) kinematic heat flux, and (d) kinematic moisture flux, as a function of time (CDT) simulated by the TKE transilient model for 28 May 1983 in Oklahoma during the BLX83 field experiment. (Zhang and Stull, 1992)

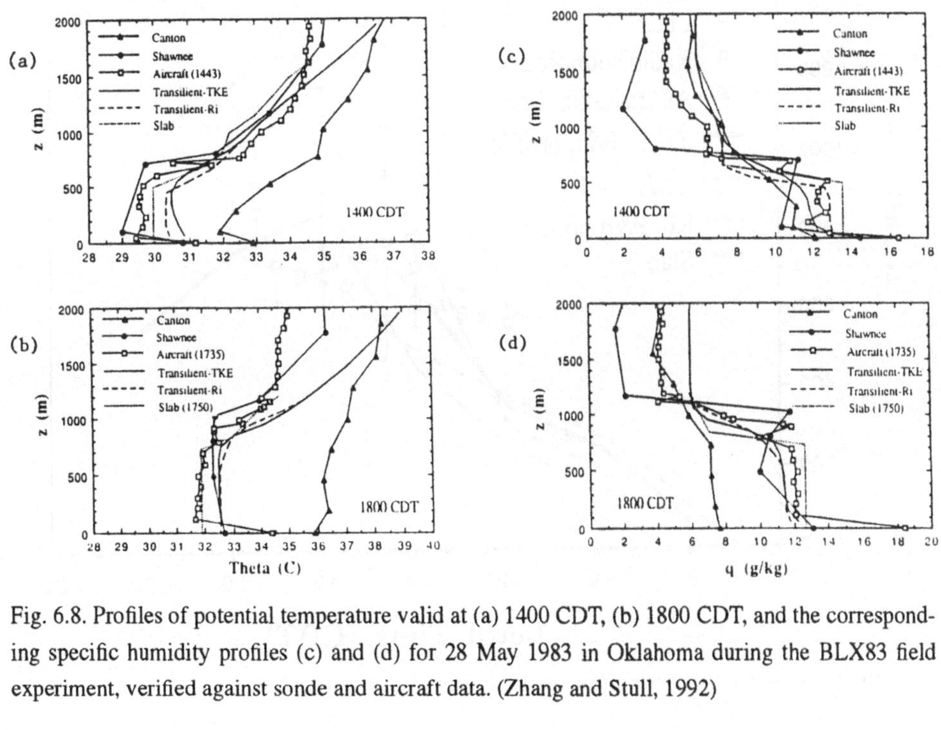

Fig. 6.8. Profiles of potential temperature valid at (a) 1400 CDT, (b) 1800 CDT, and the corresponding specific humidity profiles (c) and (d) for 28 May 1983 in Oklahoma during the BLX83 field experiment, verified against sonde and aircraft data. (Zhang and Stull, 1992)

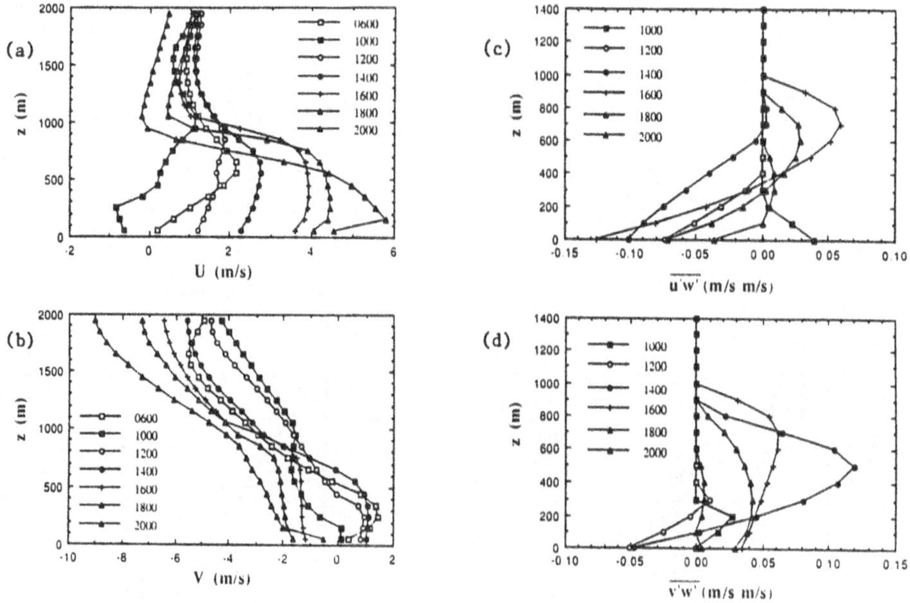

Fig. 6.9. Profiles of (a) U, eastward mean wind component, (b) V, northward mean wind component, (c) kinematic vertical flux of U, and (d) kinematic vertical flux of V, simulated by the TKE transilient model for 28 May 1983 in Oklahoma during the BLX83 field experiment. (Zhang and Stull, 1992)

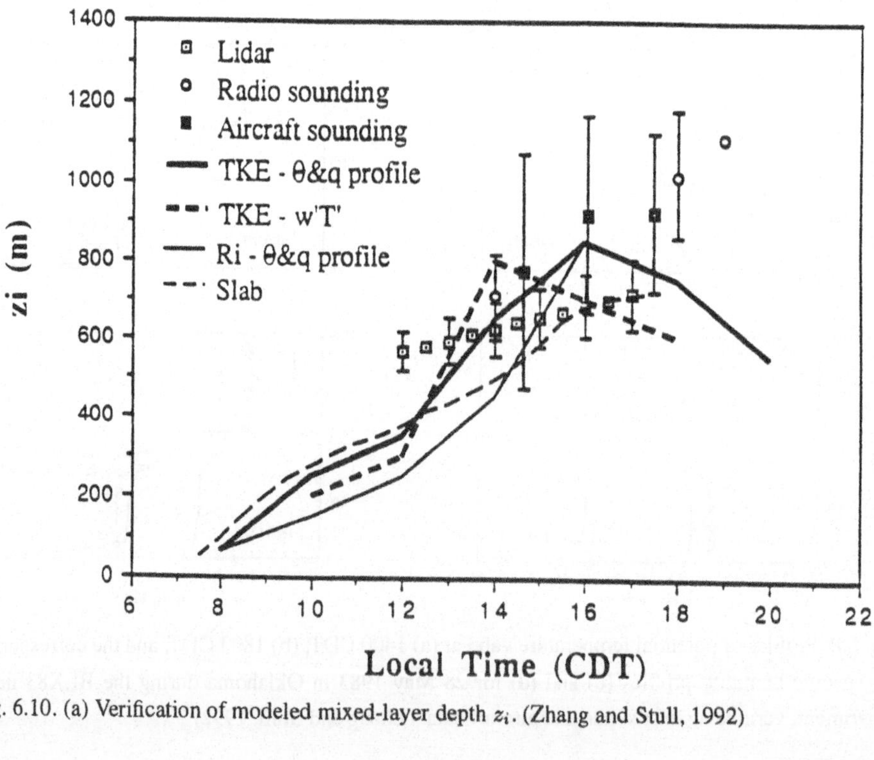

Fig. 6.10. (a) Verification of modeled mixed-layer depth z_i. (Zhang and Stull, 1992)

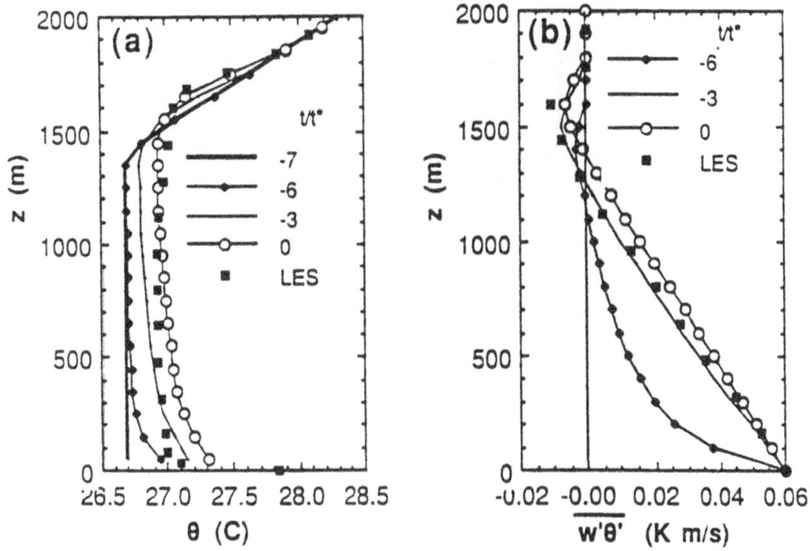

Fig. 6.11. Transilient (TKE) simulations for an idealized convective mixed layer, valid at various values of t/t_*, but all using $\Delta t/t_* = 1.0$. The $t/t_* = 0$ curve of the TKE transilient model is valid at the same time as the LES data. (a) Potential temperature; (b) kinematic heat flux. (Stull, 1991a)

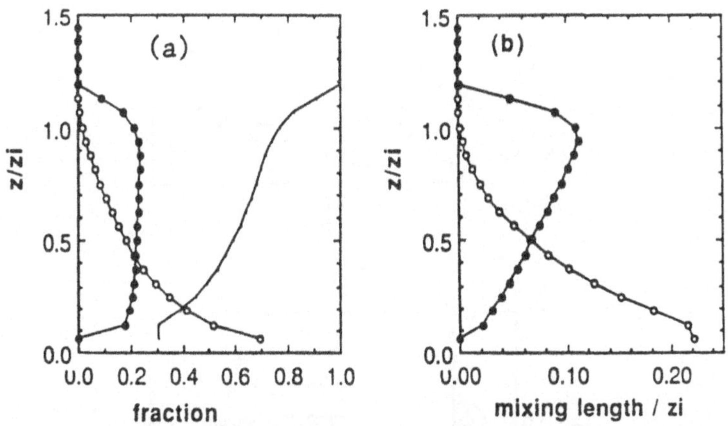

Fig. 6.12a, b. Transilient (TKE) simulations for an idealized convective mixed layer, valid at $t/t_* = 1$. (a) fraction of air moving upward (o), downward (•), and not moving (.); and (b) mixing length segregated for upward and downward moving air. (Stull, 1991a)

away from the surface. The depth of the modeled turbulent region was strongly affected by T_0, one of the parameters. For real boundary layers, there is usually a stable layer that limits the height of the mechanically mixed boundary layer.

6.5. "DRY" CLOUD-TOPPED BOUNDARY LAYER

Stratocumulus clouds are cooled by longwave radiation at cloud top, and can be warmed by solar radiation during the day in the upper part of the cloud. These create buoyant forcings in the interior of the boundary layer, rather than at the

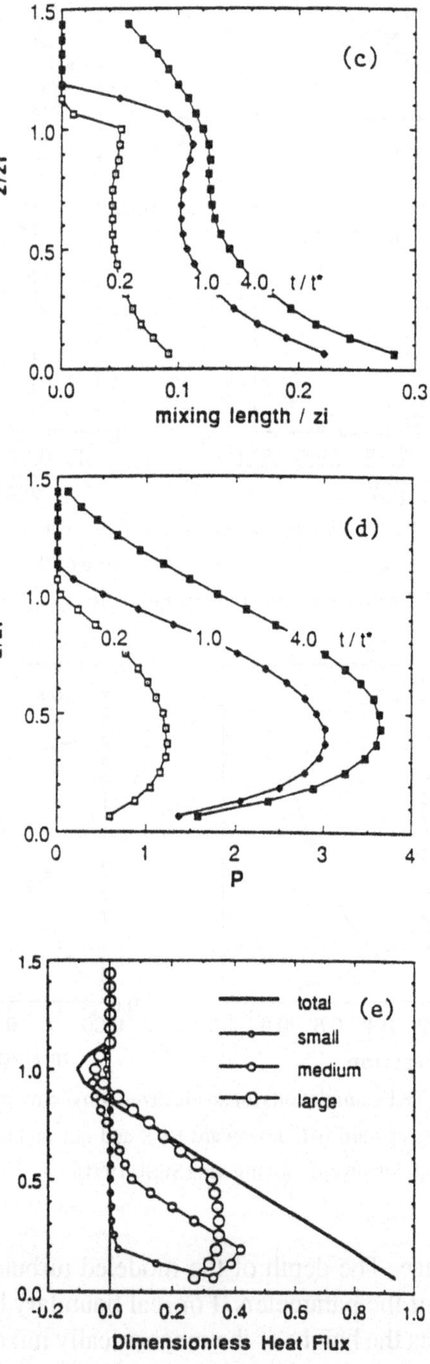

Fig. 6.12c-e. Transilient (TKE) simulations for an idealized convective mixed layer. (c) total mixing length; (d) mixing intensity P; and (e) transport spectra, valid at $t/t_* = 1$. These can be compared with similar figures in section 3. (Stull, 1991a)

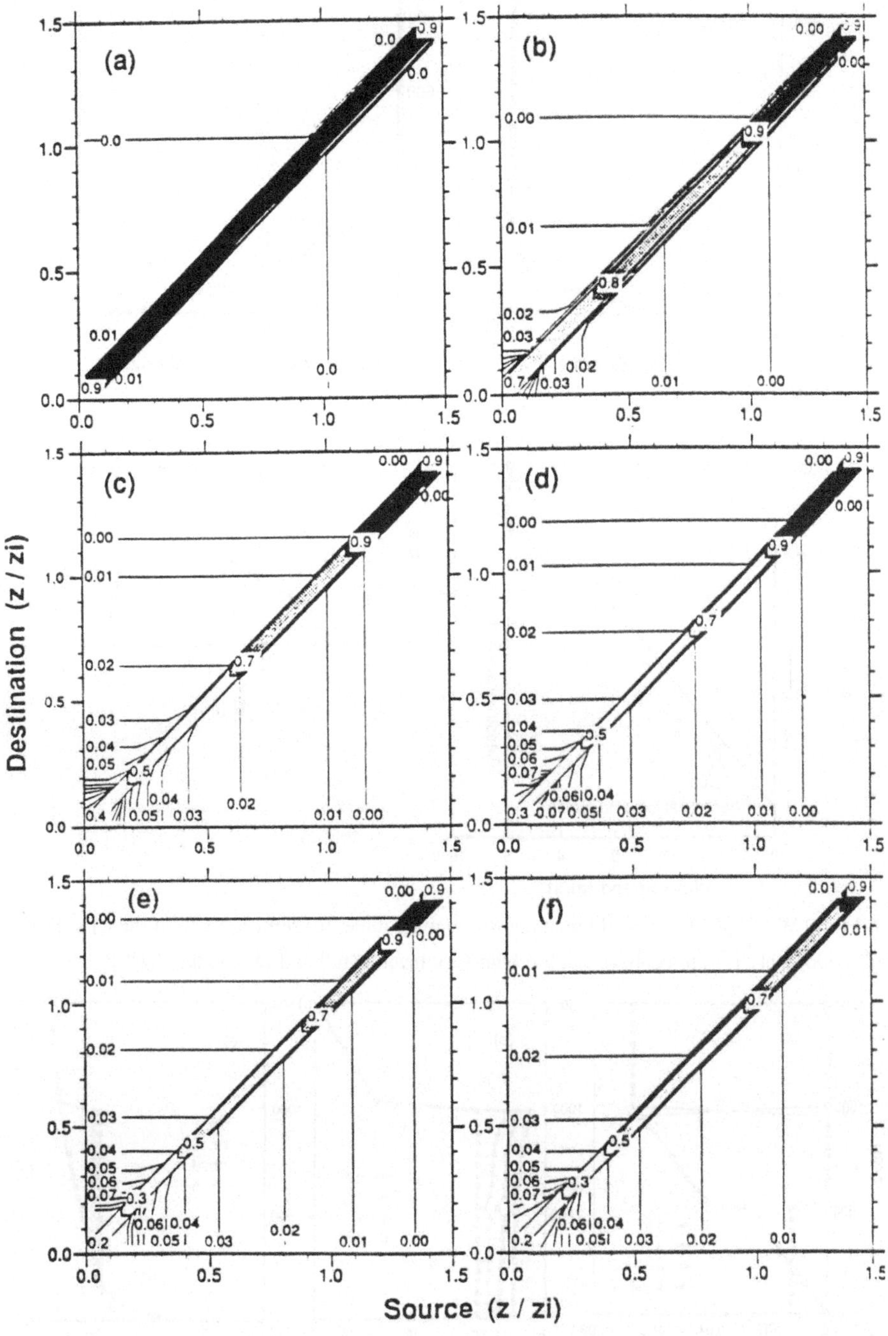

Fig. 6.13. Contoured transilient matrices for an idealized convective mixed layer parameterized by the TKE transilient approach, valid at $\Delta t/t_*$ of: (a) 0.02; (b) 0.2; (c) 0.6; (d) 1.0; (e) 2.0; and (f) 4.0. These can be compared to LES "measured" values in Fig. 3.3. (Stull, 1991a)

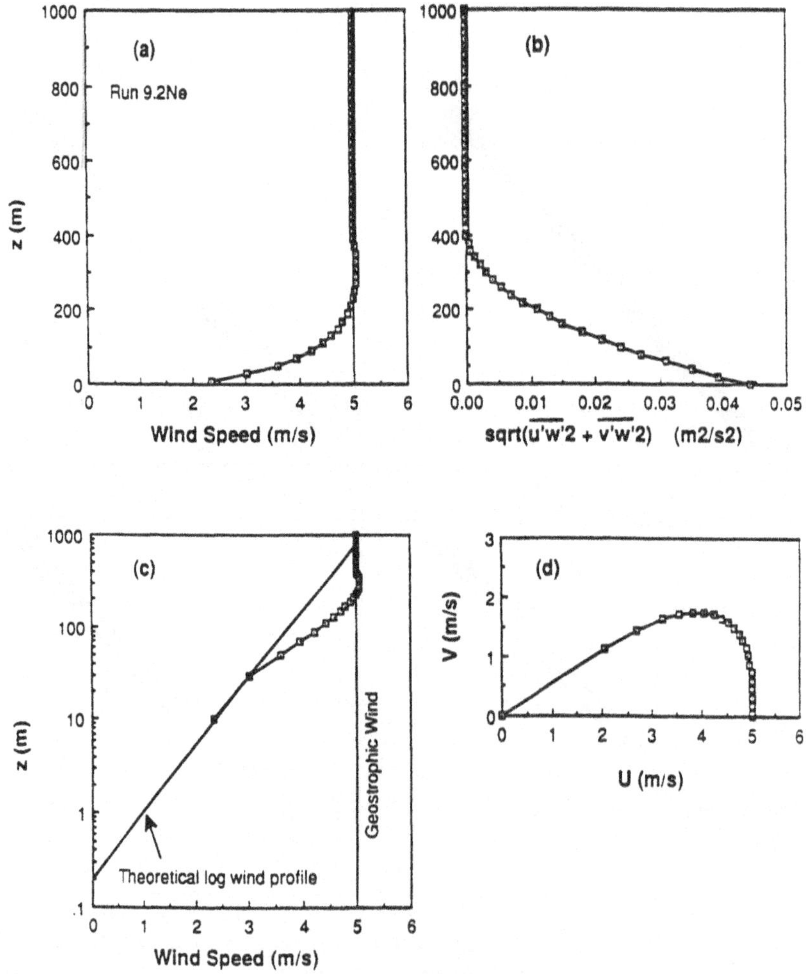

Fig. 6.14. (a) Wind speed; (b) local kinematic stress magnitude; (c) semi-log plot of the wind profile; and (d) hodograph, for an idealized neutral boundary layer. (Stull and Driedonks, 1987)

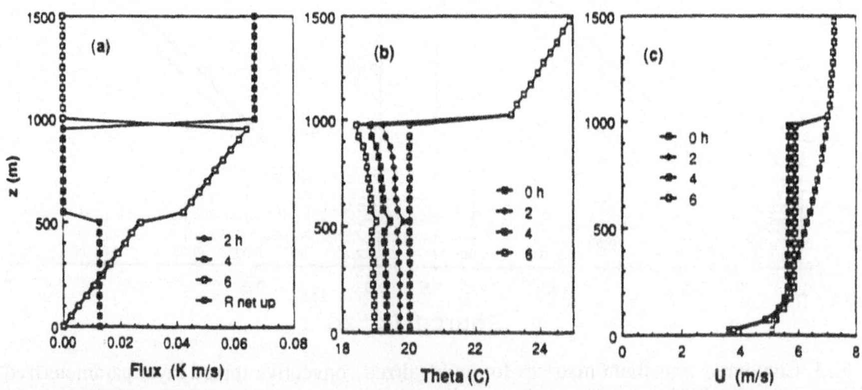

Fig. 6.15. Modeled (a) kinematic radiative and turbulent sensible heat fluxes; (b) potential temperature evolution; and (c) wind evolution in a "dry" stratocumulus cloud at night with longwave cooling at cloud top and heating at cloud base. (Stull and Driedonks, 1987)

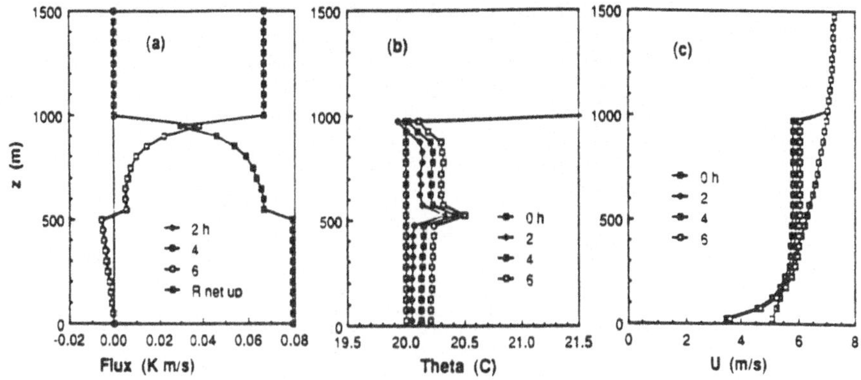

Fig. 6.16. Same as Fig. 6.15, except for daytime, which includes the addition of solar heating within the top half of the cloud. (Stull and Driedonks, 1987)

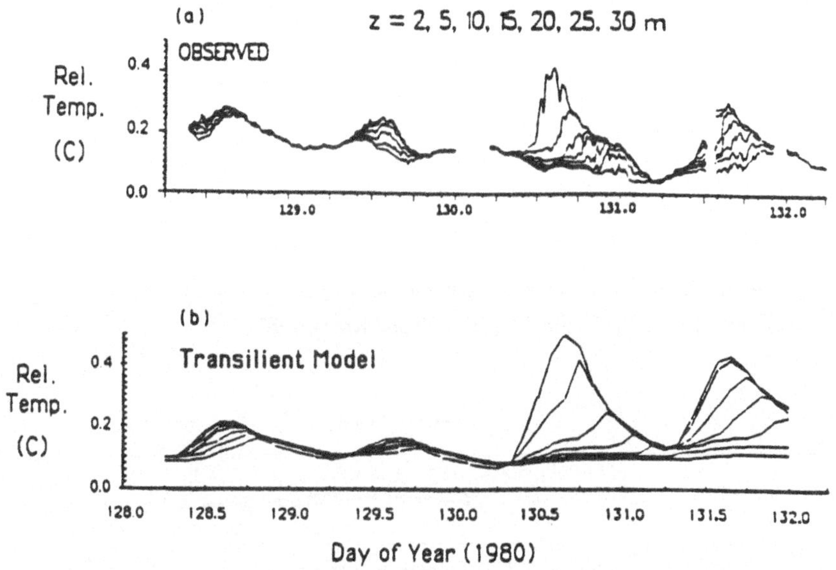

Fig. 6.17. Four day time series of relative temperature at a variety of depths in the Pacific Ocean: (a) measured; and (b) modeled using transilient turbulence theory (Stull and Kraus, 1987).

earth's surface as in typical cloud-free situations. Hence, this case is a good test for a turbulence parameterization scheme.

Stull and Driedonks (1987) simulated an idealized stratocumulus-topped boundary layer, using a so-called "dry" cloud or "dust" cloud. In this scenario, radiative forcings similar to those found in a real cloud are imposed on a cloud-free boundary layer. This allows close examination of the turbulence response without the overhead and complications of detailed water budgets and cloud microphysics.

Fig. 6.15 shows the initial temperature and wind conditions, and imposed radiative forcing profile for a nocturnal boundary layer, with no solar heating in the cloud. That figure also shows the atmospheric response in the vertical turbulent fluxes, and the evolution of wind and temperature. The TKE transilient parame-

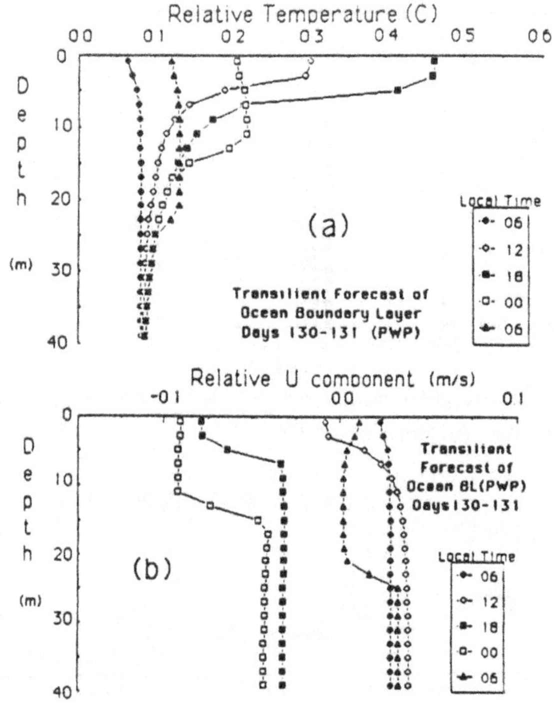

Fig. 6.18. Temperature and U-current component profile evolution simulated by the transilient model during one diurnal cycle in the Pacific Ocean (Stull and Kraus, 1987).

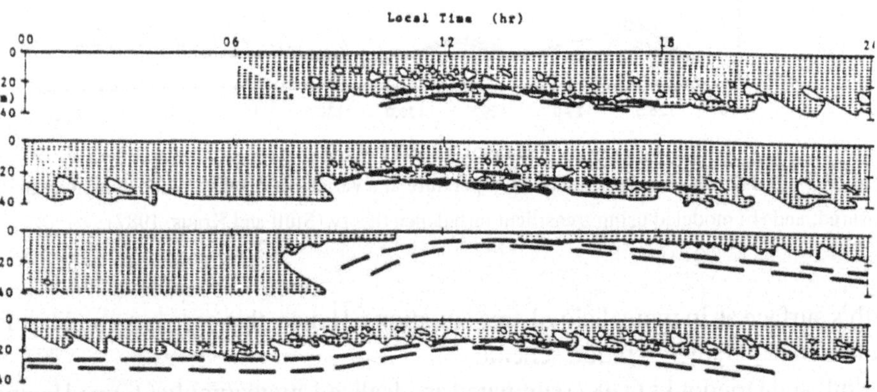

Fig. 6.19. Transilient modeled turbulent domain (shaded) and observed thermocline depth (between the heavy dashed lines) during four consecutive days in the Pacific Ocean. (Stull and Kraus, 1987).

terization responds well to the height-varying forcing, and creates a corresponding height variation in the turbulent flux. Similar results are shown in Fig. 6.16 for a daytime case with both longwave cooling and solar heating in the cloud.

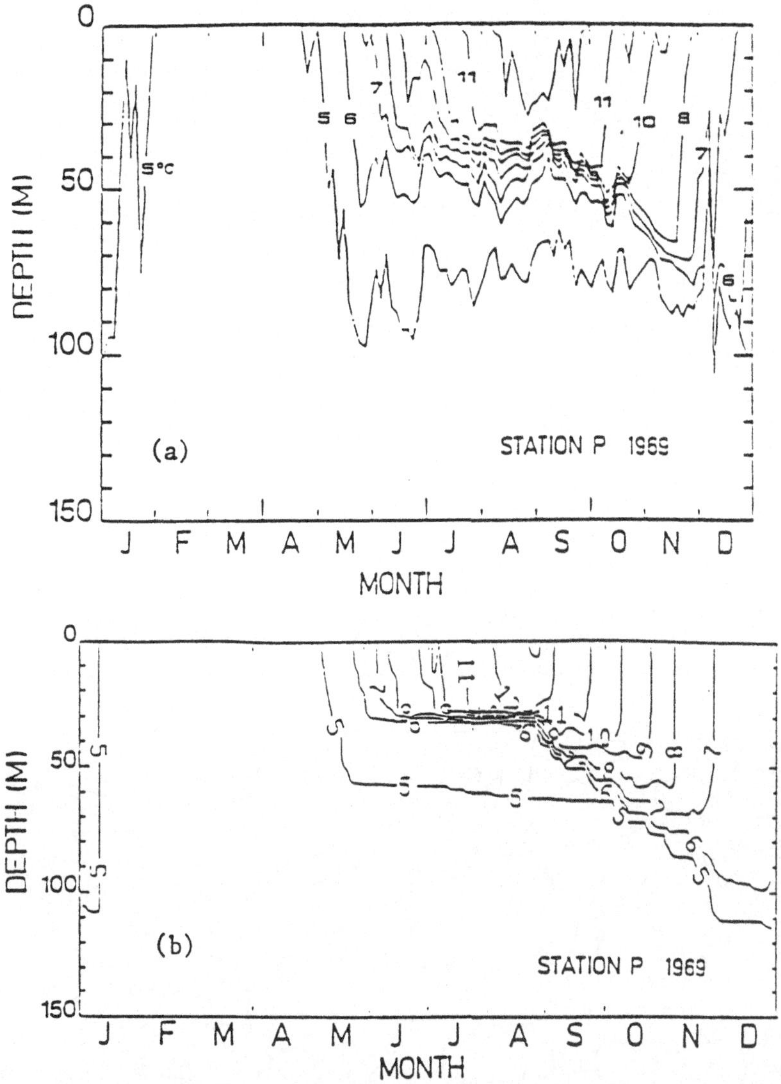

Fig. 6.20. Isotherm depths at weather ship Papa in the Pacific Ocean during 1969: (a) observed; (b) simulated using the TKE transilient parameterization. (Gaspar et al., 1988).

6.6. Ocean Models

With minor modification for the density and thermal expansion coefficient, the transilient parameterization can be applied to oceanic boundary layers. Data from the upper Pacific Ocean at (30.9°N, 123.5°W), about 400 km west of San Diego, were used by Stull and Kraus (1987) to test an early Richardson-number version of transilient turbulence theory. Four consecutive days during May 1980 were simulated, which included periods of daytime stabilization with light surface winds, periods of stronger winds that generated ocean waves and turbulence, and nighttime cooling of the sea-surface causing convection.

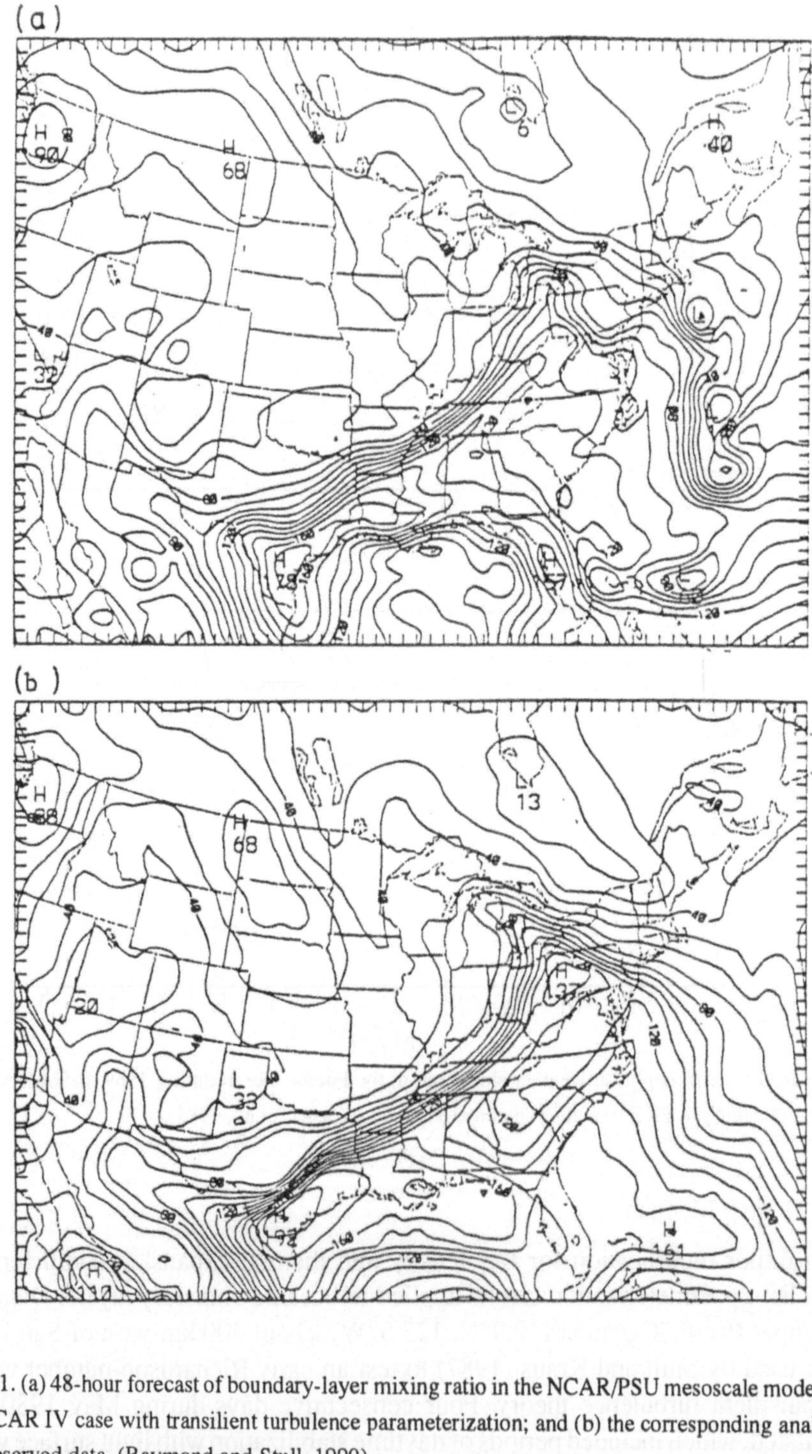

Fig. 6.21. (a) 48-hour forecast of boundary-layer mixing ratio in the NCAR/PSU mesoscale model for the OSCAR IV case with transilient turbulence parameterization; and (b) the corresponding analysis from observed data. (Raymond and Stull, 1990)

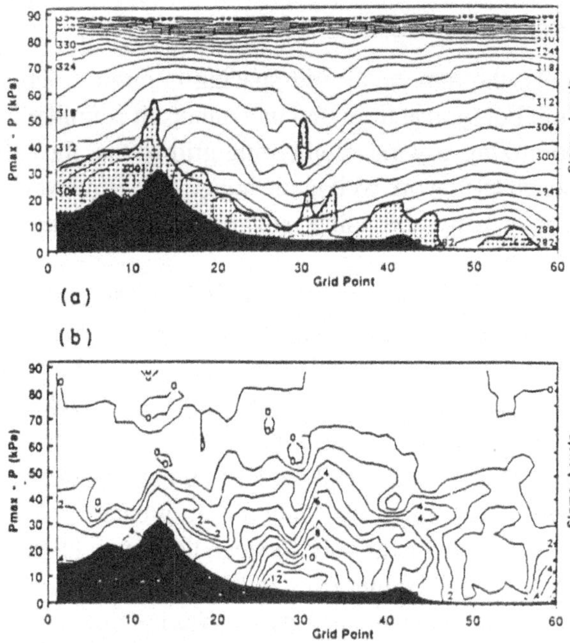

Fig. 6.22. East-west vertical cross-section across the middle of the mesoscale model, showing 21-hour forecast fields of (a) potential temperature (K); and (b) mixing ratio (g/kg). Turbulence is shaded in (a). The ground is black, showing the Rocky mountains in the west and the Appalachian mountains to the east. (Raymond and Stull, 1990).

Fig. 6.17 compares modeled and measured temperatures at seven depths, during the four days. Evolution of the temperature profile is shown in Fig. 6.18 for one of those days. The observed thermocline depth is compared with the modeled depth in Fig. 6.19. The model simulation compares well with observed ocean mixed-layer characteristics for this case study.

Gaspar et al. (1988) used a full year of data from weather ship Papa at (50°N, 145°W) in the Pacific Ocean (west of Vancover Island and south of the Gulf of Alaska) to test a variety of models of vertical mixing in the ocean. The observed bathothermograph evolution in Fig. 6.20a during 1969 is compared to the corresponding transilient model forecasts in Fig. 6.20b. Although the general seasonal variations in temperature and mixing are simulated, the transilient model tends to develop a thermocline that is too sharp and too close to the surface. Gaspar et al. attribute this to a poor parameterization for the transilient coefficients, and to inadequate knowledge of the mean current forcings, which the transilient model requires as boundary conditions. They also found that the transilient model takes considerable time to run on the computer.

6.7. THREE-DIMENSIONAL MESOSCALE MODELS

In mesoscale models, the horizontal grid spacing is often much larger than the vertical grid spacing. For typical time step increments, it is unlikely for turbulence to cause horizontal mixing beyond neighboring grid boxes, thus allowing simple K-theory mixing schemes to work well in the horizontal. However, vertical mixing can easily occur across many grid points, suggesting an application for transilient turbulence theory (T3).

Raymond and Stull (1990) incorporated the TKE transilient turbulence parameterization into the National Center for Atmospheric Research (NCAR) - Pennsylvania State University mesoscale model. They first removed most of the existing turbulence parameterizations from the model, including Blackadar's convective boundary layer, a bulk stable and neutral boundary layer, K-theory mixing in the vertical, dry convective adjustment, and the Kuo cumulus parameterization.

They then applied T3, column by column during each time step. Horizontal mixing was parameterized to occur only between those neighboring columns that were both vertically turbulent. Numerical stability of the host model was insured by a weak implicit filter applied in the horizontal; it was not maintained by increasing the "physical" turbulence. Thus, physics and numerics were separated as much as possible.

Horizontal model domain was 4880 km by 3680 km, and covered much of North America. Horizontal grid spacing was 80 km, time step was 2 minutes, and a variable grid spacing was used in the vertical with 15 levels between the surface and 100 mb. Simulations were verified against two case-study data sets, each 72 hours long: OSCAR IV in April 1981, and CAPTEX in September 1983.

Fig. 6.21 compares a 48 hour forecast of near-surface humidity with the corresponding analysis field from the OSCAR IV case. A cold front between Ohio and Texas is evident by the moisture contrast. It is obvious that a reasonable forecast is made with the T3 scheme, and that the model does not "blow up".

A vertical cross-section made east-west across the middle of the model (Fig. 6.22) shows a variable depth boundary layer, and non-boundary-layer turbulence aloft in the frontal region. The relative numbers of grid points participating in vertical turbulent mixing are shown in Fig. 6.23 to vary with a diurnal cycle, and to decrease with altitude.

With the transilient turbulence scheme, the mesoscale model took three times longer to run than the original model. Similar results were found for the CAPTEX case.

7. Dispersion Modeling

7.1. DETERMINISTIC VS. RANDOM INFLUENCES

Willis and Deardorff (1976) discovered a deterministic vertical oscillation of the center of mass of passive nonbuoyant tracers in a convective mixed layer. When

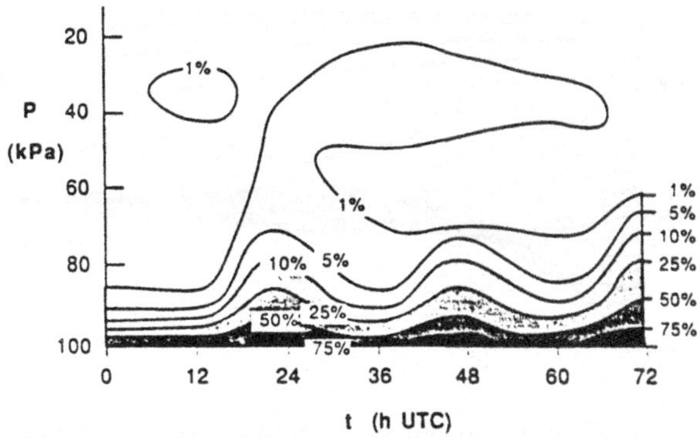

Fig. 6.23. Time-height diagram showing the fraction of grid points within any horizontal level that participate in turbulent vertical mixing. (Raymond and Stull, 1990)

scaled with respect to free convection velocity and height scales, the resulting similarity theory is universal. Superimposed on this deterministic oscillation are random tracer movements which are best described by statistics on plume spread relative to the center of mass.

Measurements of transilient matrices can be used in a diagnostic sense to calculate center of mass movement, dispersion, and other statistics as shown in the subsequent subsections. However, parameterizations of the transilient matrix should not be used to make forecasts of center of mass movement, as discussed here.

In a reply to questions from Deardorff (1985), Stull (1985) demonstrated how a transilient matrix can duplicate the deterministic movement of the center of mass when the forecast time step is carefully chosen to be a multiple of the convective time scale. However, additional research since then has revealed that it is not appropriate to use a transilient matrix to model the deterministic vertical movement of the plume center of mass, particularly when arbitrary time steps could be taken.

Vertical oscillations of the tracers are associated with negative eigenvalues of the transilient matrix. If shorter time steps are desired, then one must take the root of the matrix (see eq 2.5), which means taking the root of the eigenvalue. The resulting imaginary eigenvalues are difficult to apply to the physical situation of tracer dispersion. Thus, a transilient forecast of oscillatory tracer movement depends on the numerics of a strategic choice of time step, not on the physics. For this reason, it is not recommended to use transilient parameterizations to forecast the deterministic tracer oscillations.

7.2. Center of Mass

Center of mass $\overline{Z_j}$ for tracers originating at height $z = (j - 1/2) \cdot \Delta t$ may be diagnosed from measured transilient matrices via:

$$\overline{z_j}(\Delta t) = -0.5 \cdot \Delta z + \Delta z \cdot \sum_{i=1}^{N} c_{ij}(\Delta t) \cdot i \tag{7.1}$$

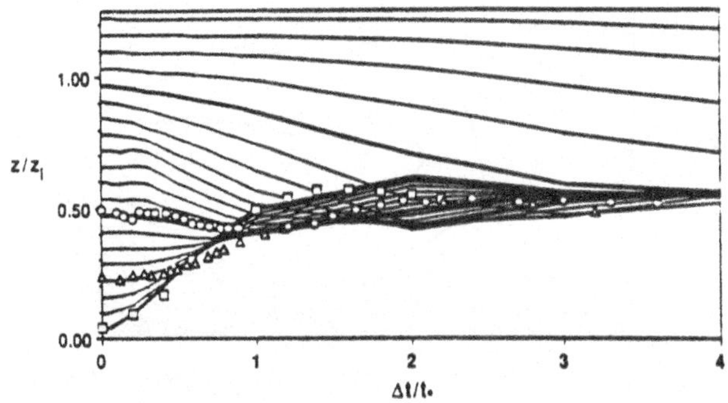

Fig. 7.1. Center-of-mass height of tracers originating at each level, j, as a function of time (solid lines). Experimental results from Willis and Deardorff are also plotted for sources at heights z/z_i = 0.04, 0.24, and 0.49 (Ebert et al., 1989).

where Δt must be the time since tracer emission. Fig. 7.1 compares the center of mass diagnosed from the LES-measured transilient matrices with observations by Willis and Deardorff (1976, 1978, 1981).

Although different than the center of mass, the location of the concentration maximum (Fig. 7.2) also has a deterministic oscillation. Fig. 7.2 also shows isolines of c_{ij} values as measured by LES for a convective mixed layer. Each diagram in this figure was constructed from a single column (j) of the transilient matrix, sampled over different time intervals.

7.3. Vertical Dispersion

Absolute dispersion, as quantified by the variance of tracer heights σ_z^2, can be diagnosed from (Stull, 1986; Ebert et al., 1989):

$$\sigma_z^2(t) - \sigma_z^2(0) = N_2 \cdot t \tag{7.2}$$

where N_2, the second moment of the transilient coefficients, is:

$$N_2(t) = \frac{(\Delta z)^2}{\Delta t} \cdot \sum_{i=1, j=const}^{N} (i-j)^2 \cdot c_{ij}(t, \Delta t) \tag{7.3}$$

for the discrete version, and is:

$$N_2(t) = \int_{\zeta=a-z}^{b-z} \gamma(t, z, \zeta) \cdot \zeta^2 d\zeta \tag{7.4}$$

for the analytical version. Equation (7.2) is valid over time intervals during which N_2 is relatively constant. For longer periods, it must be evaluated in steps, with N_2 recomputed at each step based on the current value of the transilient matrix.

Fig. 7.3 shows a diagnosis of vertical absolute dispersion diagnosed from the LES-measured transilient matrices, for the special case of $\Delta t = t$.

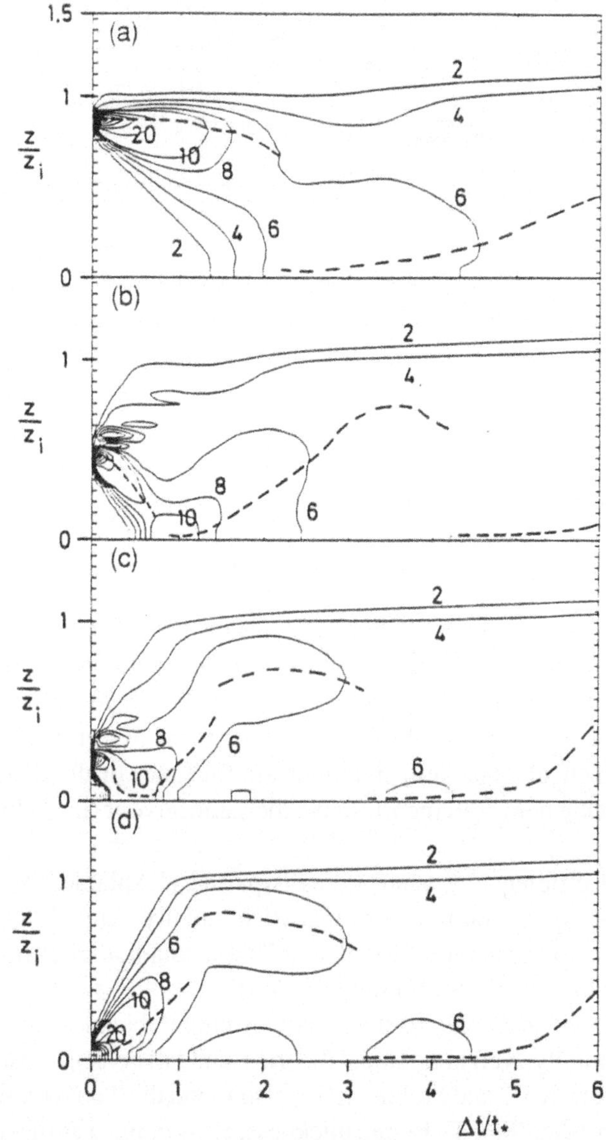

Δt/t*

Fig. 7.2. Transport diagram showing isolines of $c_{ij}(\Delta t)$ as a function of $\Delta t/t_*$ for various source heights z/z_i = (a) 0.91, (b) 0.47, (c) 0.22, and (d) 0.03. Contours are in percent of the original concentration. Concentration maximum is shown with the heavy dashed line. (Ebert et al., 1989).

7.4. Dispersive Power Relationships

Equation (7.2) for nonlocal turbulence gives a linear increase of σ_z^2 with time for all times when N_2 in constant (i.e., stationary turbulence and constant c_{ij}). This differs from the statistical theory of Taylor (1921), which suggests that σ_z^2 should increase with the square of time initially and then change to a linear increase for

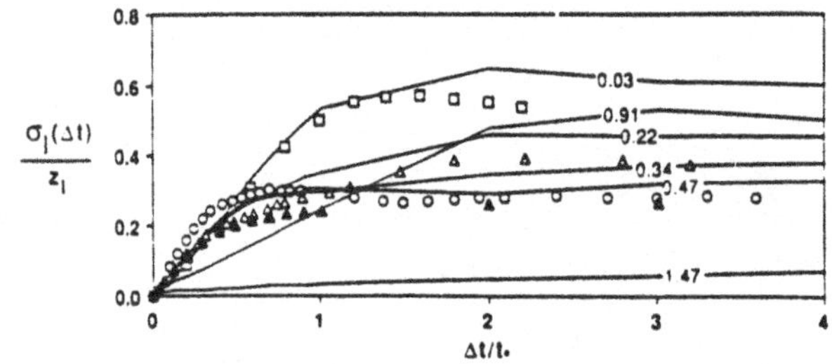

Fig. 7.3. Absolute dispersion from sources at selected normalized source heights, z/z_i. Open data points correspond to the same Willis and Deardorff data as in Fig. 7.1, while the solid triangles are from atmospheric measurements by Eberhard et al. (1988). (Ebert et al., 1989)

longer times:

$$\sigma_z^2 = \sigma_w^2 \cdot t^2 \qquad \text{Statistical Theory, for } t << t_L \tag{7.5a}$$

$$\sigma_z^2 = (2 \cdot t_L) \cdot \sigma_w^2 \cdot t \qquad \text{Statistical Theory, for } t >> t_L, \tag{7.5b}$$

where t_L is the Lagrangian time scale, and σ_w^2 is the vertical velocity variance.

There has been ongoing debate about the initial-dispersion-rate difference between nonlocal and statistical theories (Troen et al., 1980a,b; Prahm and Berkowicz, 1980; Sawford, 1988; Stull, 1988b). This debate is mostly of academic interest, because most practical dispersion situations are for times or distances downwind of the source stack where both theories give the same answer (i.e., a linear increase of σ_z^2 with time).

The apparent difference at small times is partially explained by the fact that plume concentration distribution is not Gaussian initially. Stull (1984) shows concentration distributions that initially look more like a double-exponential, but which later change to a Gaussian distribution (Fig. 7.4).

Such a change in shape allows the square of plume thickness to vary with the square of time initially even in nonlocal theories, and later change to a linear variation (Fig. 7.5). Berkowicz and Prahm (1979) also found this behavior using spectral diffusivity theory (their Fig. 4). Plume thickness, z_e, is defined as the distance from plume center where the concentration drops to a specified fraction (such as 10%) of the centerline concentration $[S(z_e)/S(0) = 0.1]$. Thus, statistical and nonlocal dispersion theories agree at all times, when plume thickness is considered:

$$z_e = t^2, \qquad \text{for } t << t_L \tag{7.6a}$$

$$z_e = t, \qquad \text{for } t >> t_L. \tag{7.6b}$$

Field data can be misinterpreted and biased toward statistical theory if it is assumed that the plume concentration distribution is Gaussian. It is better to fit a general distribution shape to the data, such as

$$S(z)/S(0) = e^{a \cdot |z|^r} \tag{7.7}$$

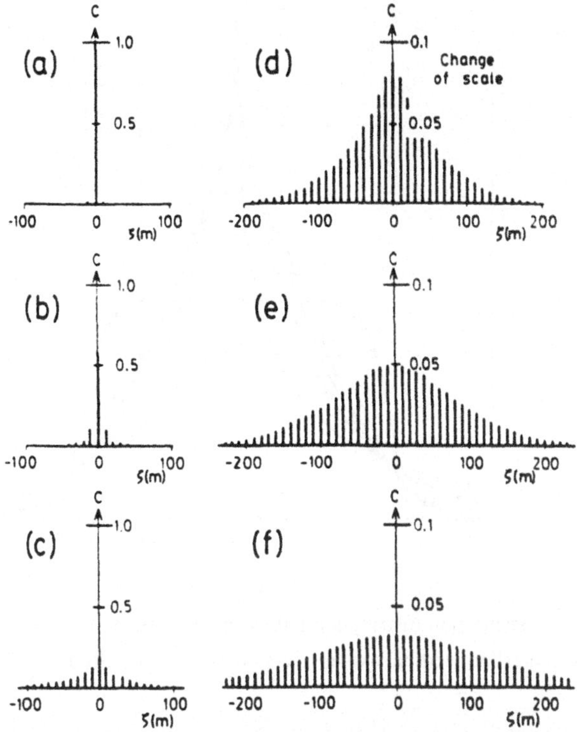

Fig. 7.4. Time variation of 1-D plume concentration distribution from a point source, as evaluated from the transilient matrix based on stationary turbulence with an inertial subrange. Vertical axis gives value that is proportional to tracer concentration; horizontal axis gives vertical distance from plume centerline. (a) 1.17 s; (b) 1.25 min; (c) 5 min; (d) 10 min; (e) 20 min; (f) 40 min. (Stull, 1984)

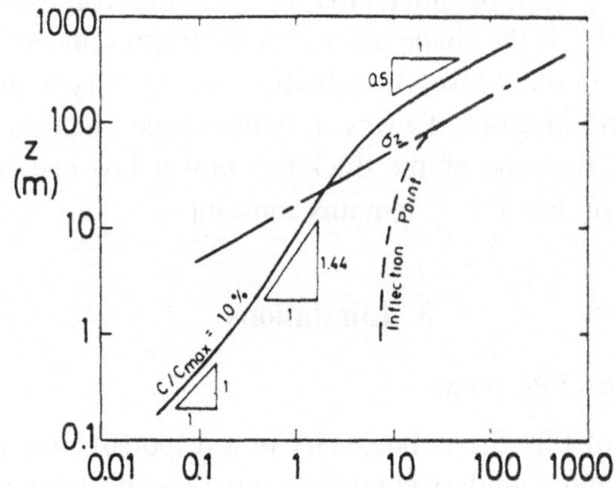

Fig. 7.5. Solid curve shows evolution of plume thickness based on a nonlocal turbulence model with an inertial subrange. Dash-dot curve shows the variation of standard deviation of tracer position in the plume. Dashed curve shows the location of the inflection point in the tracer distribution. (Stull, 1984)

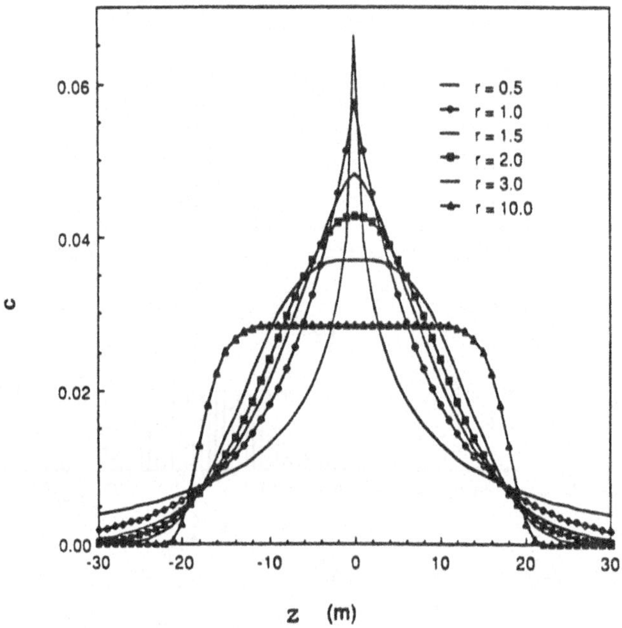

Fig. 7.6. Variation of concentration distribution with shape parameter, r, in (7.7). Horizontal axis is distance from plume centerline; vertical axis is tracer concentration. (Stull, 1988b)

for one-dimensional diffusion, where $S(0)$ is the centerline concentration of the plume, a is a spread parameter, and r is a shape parameter. Fig. 7.6 shows the variety of shapes that modeled using (7.7), including the Gaussian distribution when $r = 2$. For real plumes, r is often less than 2, as shown by Stull (1988b).

For a one-dimensional plume, the ratio of plume width to standard deviation is shown as a function of shape parameter in Fig. 7.7. Relationships for 3-D plumes are given by Stull (1988b). If the shape parameter for tracer concentration is constant with time (Gaussian or otherwise), then both z_e and σ_z^2 vary with the same power of time or downwind distance. However, if the shape changes with time, as is apparently the case, then the plume thickness power law can change with time even though the power law for σ_z^2 remains constant.

8. Limitations

8.1. Destabilization Problem

The split time step of Fig. 5.1 is necessary in a responsive parameterization because the dynamics and boundary conditions must first destabilize the flow before the transilient coefficients can be parameterized. Unfortunately, the amount of destabilization during the first part of each split time step depends partly on the discretization.

For example, heating of the lowest layer in the model by conduction from the surface during one time step (part 1) destabilizes the boundary layer by making

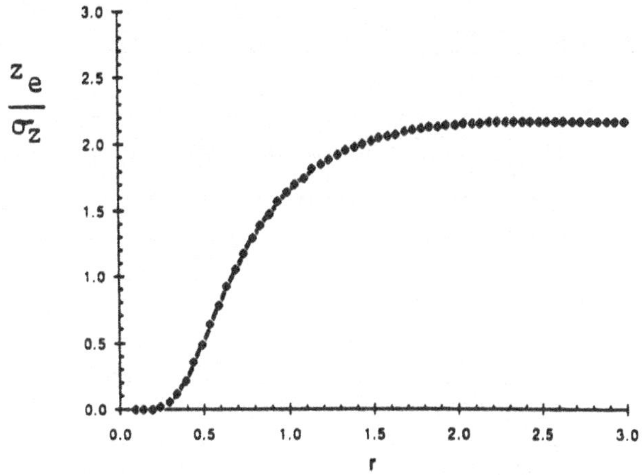

Fig. 7.7. Vertical axis gives ratio of plume edge thickness to standard deviation of tracer position. Horizontal axis is the shape parameter in (7.7), which equals 2 for a Gaussian distribution. (Stull, 1988b)

the lowest grid point warmer than many others above it in the mixed layer. The nonlocal parameterizations described in the previous sections generate mixing (part 2) between this lowest grid point and the points above it in an attempt to restabilize the boundary layer.

If the grid spacing were halved and all else remain unchanged, then the temperature change of the lowest grid point during any one time step (part 1) would double, and the warm surface-layer air would be parameterized (part 2) to mix higher in the boundary layer. The result is a numerical solution which depends in part on the discretization, which is an undesirable artifact of the model. To minimize the affects of this artifact, Stull and Driedonks (1987) recommend that the discretization be chosen such that $\Delta z/\Delta t > 0.1 m/s$.

In some cases, it is difficult to satisfy these descritization recommendations. For this situation, dynamics and/or boundary conditions can be applied for a fraction of the time step. Then the transilient coefficient matrix can be determined. Next, the dynamics and boundary conditions are applied for the remainder of the time step. Finally, the previously-determined transilient matrix is applied to perform the mixing.

8.2. Convective Structure Memory

Although one might guess from (2.5) that knowledge of the transilient matrix for any time interval Δt_1 allows one to determine the matrix for any other time interval Δt_2, unfortunately such is not always the case. The problem is most noticeable for convective conditions, where tracers that are caught in updrafts or downdrafts remain in those structures for long periods of time and affect the dispersion statistics. Hence, the name "convective structure memory" is given for this problem.

As an example, Fig. 8.1 shows an idealization of updrafts and downdrafts. If

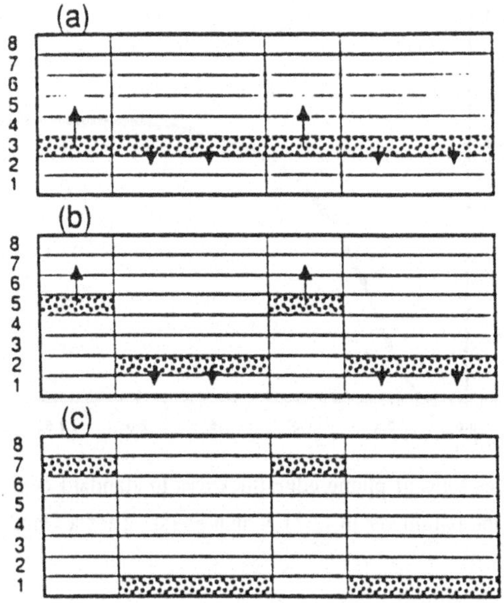

Fig. 8.1. Idealized convective mixed layer with updrafts and downdrafts. (a) Initial tracer introduction into layer 3. (b) Tracer locations after about $\Delta t = 0.2t_*$. (c) Tracer locations an equal time step later, at about $\Delta t = 0.4t_*$. (Stull, 1990)

a layer of tracer is introduced at height index 3 in this figure, then some of the tracer is in updrafts, and some in downdrafts. These move with time as sketched in Figs. 8.1b & c. The corresponding transilient matrices for these cases are shown in Fig. 8.2. One can think of these matrices as representing measurements of the true dispersion for this idealized case.

One might guess that the matrix in Fig. 8.2b, which describes the mixing during the time interval between Fig. 8.1a and 8.1b, can be applied twice to get directly from Fig. 8.1a to the final state in Fig. 8.1c. Such is not the case, because the transilient matrix in Fig. 8.2b represents layer averages by definition, as sketched in Fig. 8.3c. Thus, information about which portions of the tracers are in updrafts and downdrafts is lost at the end of each time step. Application of the transilient matrix twice yields the modeled results in Figs. 8.3e and 8.4b, not the "true" results as sketched in Figs. 8.1c and 8.2c.

The effect of convective structure memory was tested by Ebert et al. (1989) using measured transilient matrices from a LES of the convective mixed layer. Fig. 8.5 shows measured transilient matrices for various time intervals that were taken to the appropriate powers to bring them to the same effective time interval. If (2.5) were valid, then all of the matrices would have looked identical. Their obvious dissimilarity is caused by the convective structure memory in the real atmosphere that is not captured by the transilient matrix.

Convective structure memory and the destabilization problem are related. In order to retain the maximum amount of information about matrix asymmetry due to convection, it is best to choose a time step of the same order as the convective

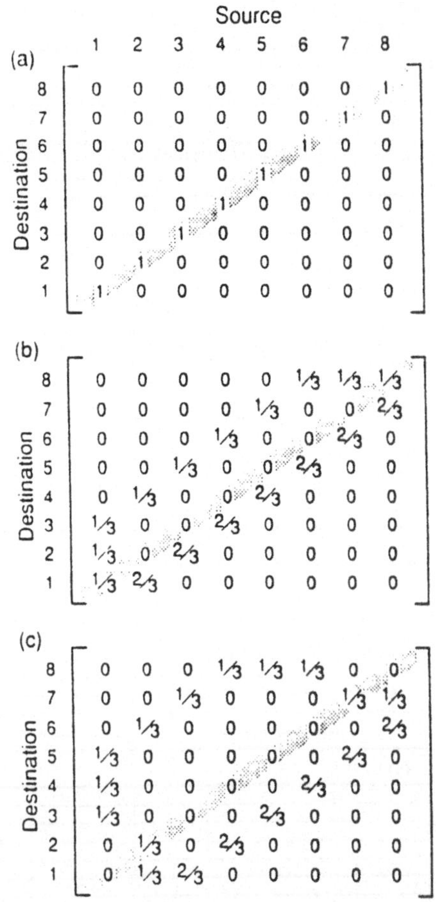

Fig. 8.2. Transilient matrices corresponding to the convective dispersion sketched in Fig 8.1. The matrix of (c) represents one big step from the initial to the final condition, not two short steps.(Stull, 1990)

time scale (t_*); namely, $\Delta t = t_* = 10$ min. According to the previous section, the best grid interval is thus $\Delta z > 60m$ for a convective case.

8.3. Other Considerations

Different parameterizations for transilient matrices introduce different peculiarities. Chrobok (1988) and Zhang and Stull (1992) found that the TKE parameterization leads to surface layers that are unrealistically thick and well mixed, and mixed layers that are not sufficiently well mixed. This is not an intrinsic problem with nonlocal methods in general, but is an artifact of the particular parameterization. For example, Zhang and Stull (1992) found that the Richardson number parameterization does not have this problem.

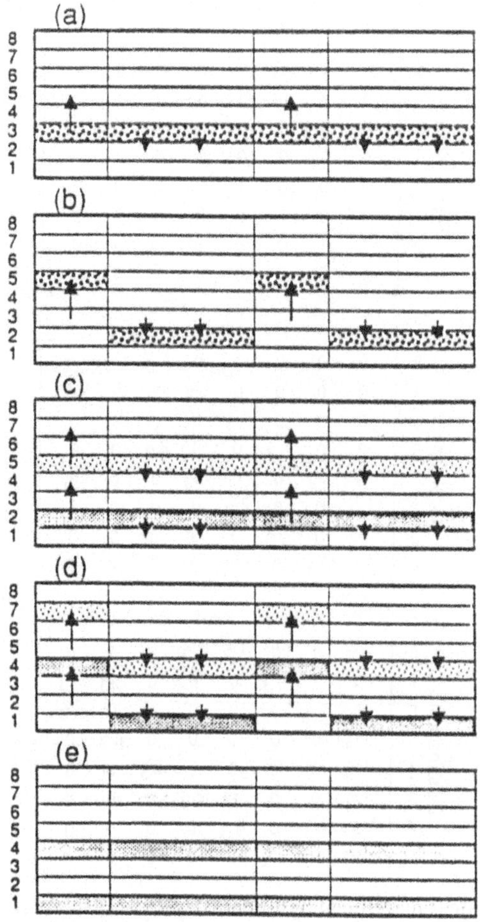

Fig. 8.3. Same scenario as Fig. 8.1, except that the idealization shows what the transilient matrices represent when taking two small time steps instead of one big step. (a) and (b) Same as in Fig. 8.1. (c) The transilient matrix does not retain information about which portions of the tracers are in updrafts or downdrafts at the end of the first time step, but holds only a statistic about how much tracer is in the whole layer. When the next step is taken starting from (c), the result is shown in (d), and the resulting transilient matrix holds this result as layer averages (e). Thus, the result of two short steps in (e) differs from the "true" dispersion shown in Fig. 8.1c. (Stull, 1990)

9. Additional Developments

This section documents work in progress, ideas, theories, and potential applications.

9.1. Potential Application to Plant Canopies

Looking first at temperature and heat flux, forest canopy tops are often heated by the sun during the day, causing the air near the canopy top to be warmer than the air near the ground (Fig. 1.4). This causes a subadiabatic lapse rate below the canopy

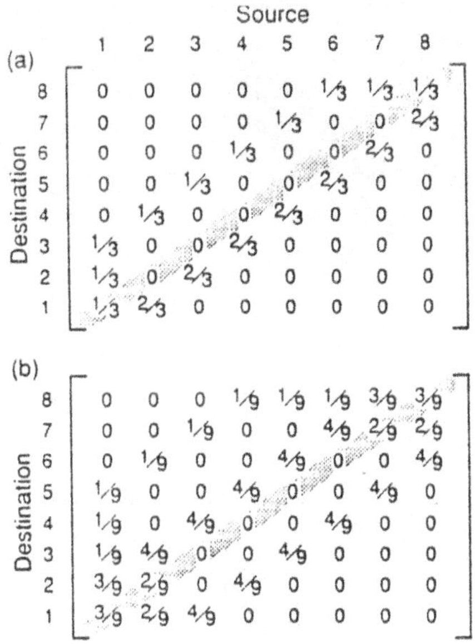

Fig. 8.4. Transilient matrices in (a) and (b) corresponding to (c) and (e) of Fig. 8.3, respectively. (Stull, 1990)

top. However, heat fluxes are measured to be positive (upward) throughout the canopy; implying a countergradient transport. Also, there is occasional bursting of turbulence below the canopy top, and ramp temperature variations near the surface. As plotted in Fig. 1.4, these effects could be explained by parcels of cooler and faster-horizontal-velocity air from higher in the mixed layer than convective sink (hence positive heat flux) through the canopy top, causing the bursts of wind speed and turbulence and temperature ramps.

Such nonlocal effects are easily handled by transilient turbulence theory. An example of a preliminary experiment with a transilient turbulence parameterization is shown in Fig. 9.1, based on work performed by de Bruin and colleagues at Wageningen Agricultural University. Heating applied as an interior forcing at a grid point corresponding to canopy top causes the subadiabatic lapse rate below, yet the heat flux is positive at all heights.

Looking next at momentum and turbulence, plant structures can generate turbulence when the wind blows, and can damp turbulence and restrict convective transport. These opposite effects should be modeled separately (Stull, 1990).

Turbulence is generated by wakes behind trunk, branch, and stem structures. The rate of TKE production might be modeled as the cube of the wind speed divided by a length scale such as the average horizontal distance between wake-generating plant structures. This term should be added to the turbulence parameterization (5.10) as a body source term in the interior of the model at each grid level where there are plant structures. Thus, this effect can enhance vertical nonlocal mixing.

Plants also induce a net form drag on the mean flow, which can be modeled

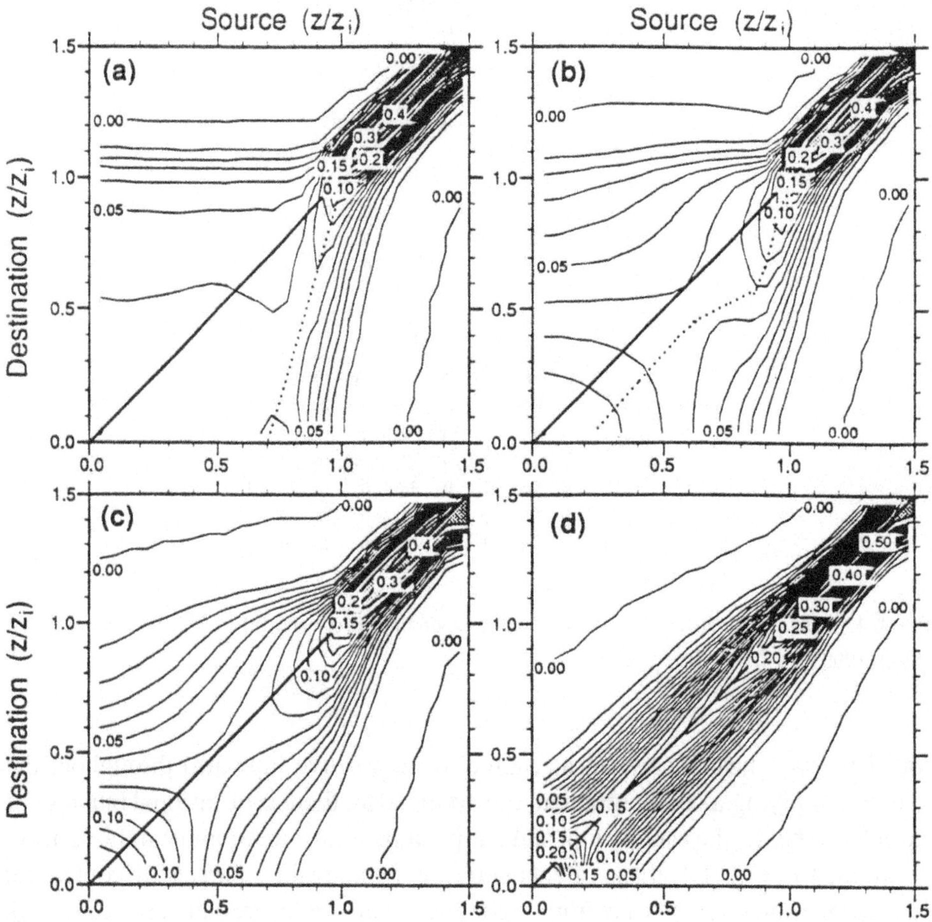

Fig. 8.5. Contours of transilient matrices after the "measured" matrices were taken to the appropriate powers necessary to bring them to an effective time step of $\Delta t_2 = 2t_* = 36.53$ min. The time steps of the original matrices are: (a) 1.0; (b) 0.5; (c) 0.2; and (d) 0.02. The corresponding powers are: (a) 2; (b) 4; (c)10; and (d) 100.(Ebert et al., 1989)

by a drag coefficient multiplied by the square of the wind speed. This term should be included in the mean wind equations at each grid point where plant structures exist (i.e., not only at the surface). The drag coefficient is not the usual skin drag coefficient used in surface-layer parameterizations, but is a form drag related to flow past a cylinder (around plant limbs) or past a plate (around leaves). In addition to this term applied at interior grid points, the usual skin-drag term should continue to be applied at the bottom grid point to represent the effects of surface friction. Although these terms affect the mean momentum equations, they do not alter the nonlocal mixing parameterization.

Plant canopies and other obstructions reduce vertical eddy transport. Fig. 9.2 sketches interference to vertical nonlocal mixing. Define an interference coefficient,

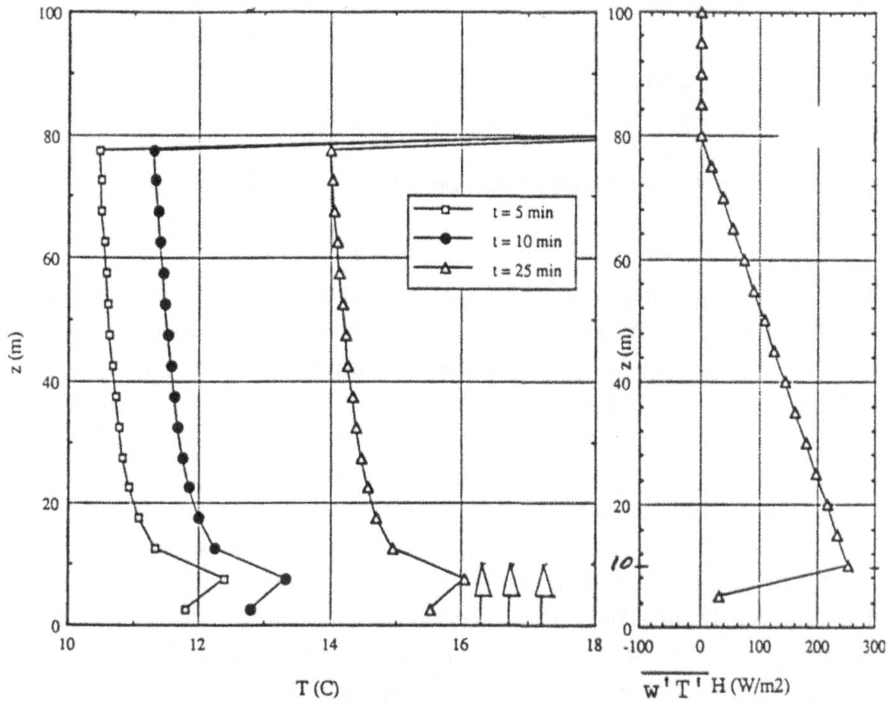

Fig. 9.1. Temperature and heat flux simulated by a TKE transilient parameterization, where canopy-top heating is approximated by imposed warming at the second grid point above the ground.

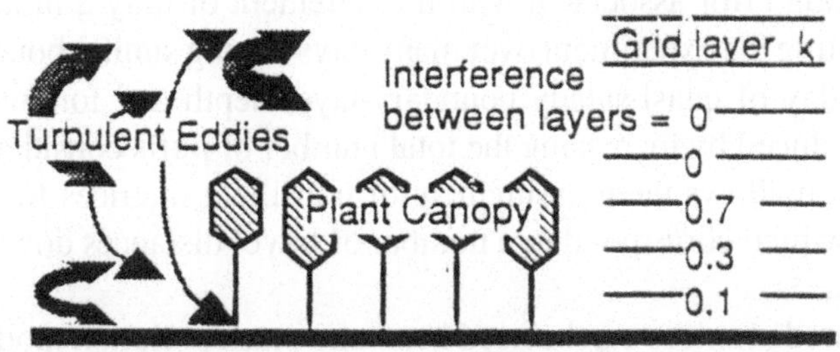

Fig. 9.2. Interference caused by plant structures reduces vertical motions. The interference coefficient indicates the relative amount of blockage. In this sketch, the reduction to vertical mixing is indicated by the thinness of the lines representing eddies. (Stull, 1990)

γ_k, that varies between 1.0 for total interference by plants, to 0.0 for no interference of vertical eddy motions. The interference is not quite like a resistance, which varies between zero and infinity. The interference coefficient is a measure of obstructions to mixing between grid layers k and $k+1$. For finite-sized eddies between non-neighboring layers, the total interference is modeled as the product of interferences across the mixing distance. When plant canopies are present, (5.12) and (5.17) might be modified to be

$$A_{ij} = m_j \cdot Y_{ij} \cdot \prod_{k=i}^{j-1} (1 - \gamma_k) \tag{9.1}$$

for $j > i$. If $i > j$, then the product will be from $k = j$ to $k = i - 1$.

9.2. Potential for Measurements in the Atmosphere

Although it is virtually impossible to inject layers of tracers into the real atmosphere as was done in the LES "measurements", an alternative practical method is possible for the special case of quasi-stationary, quasi-homogeneous boundary layers.

Inject a sequence of individual small smoke puffs into the boundary layer from one height on a tower. Track the movement of the center of each puff by lidar (Eloranta, 1991 personal communication). Vertical locations of each puff as a function of time can be grouped by time interval (i.e. time since puff generation), and the results displayed as a frequency distribution of puff destinations, from the one source height, for the common time interval. When the vertical distribution is grouped into height bins, and normalized to give a total area under the distribution equal to 1.0 (i.e. 100% of the smoke puffs are at some height in the vertical), the result is a sample of a single column of the transilient matrix.

When the same experiment is done for puffs from different heights on the tower, or for other days when the boundary layer is of different depth, samples for other columns of the transilient matrix are obtained. Each of these columns of data contains sampling error associated with measurement of only a finite number of puffs. By repeating the experiment over many days having similar boundary layers, or during one day of quasi-steady boundary-layer depth and forcings, sampling errors can be reduced by increasing the total number of puffs considered.

This approach allows the measurement of transilient matrices for a variety of time intervals, which correspond to a number of travel distances downwind of the source location.

Such measured matrices can be used in smoke-stack pollution models, because the column of the matrix corresponding to the stack height gives the expected distribution of tracers downwind of the stack corresponding to the time interval of the matrix. By using matrices for different time intervals, the dispersion of emissions as a function of downwind range is easily determined. Such matrices intrinsically include both the deterministic effects of plume centerline movement identified by Willis and Deardorff, and the dispersion relative to that centerline.

George Young (personal communication) experimented with a different way to "measure" transilient matrices: namely, "backing out" the transilient coefficients from measurements of the change of the mean boundary-layer state. This data inversion technique proved impossible, because the problem is underspecified. There are more unknowns (transilient coefficients) than knowns (independent measurements of mean state in the atmosphere).

Acknowledgements

Hadassah Kaplan and Nathan Dinar at the Israel Institute for Biological Research are gratefully acknowledged for encouraging me to write this review. The unequal-grid spacing work was performed at the German Aerospace Research Establishment (DLR) at the invitation of Anne Jochum, and was tested in the USA in a mesoscale model by William Raymond at the University of Wisconsin (UW). Ralph Dlugi and Renate Forkel at the University of Munich, Dieter Etling and his colleagues at Hannover University in Germany, and Henk de Bruin and his colleagues at the Wageningen Agricultural University in the Netherlands provided insight on the plant canopy problem. Ulrich Schumann at DLR and the University of Munich is thanked for his important measurements of transilient matrices using large eddy simulation (LES). Beth Ebert at UW performed much of the analysis of the resulting LES-measured matrices. Jim Purser (formerly at the British Met. Office) and Steve Jascourt, both at UW, contributed ideas about the basis matrices, and about measures of dispersion, while George Young from Penn State University and Ed Eloranta from UW contributed ideas about measurement of transilient matrices in the atmosphere. Qing Zhang at UW made tests of various parameterizations against atmospheric case studies. Eric Kraus of CIRES in Colorado, and Philip Gaspar at CNRM in Toulouse, France, encouraged me to apply transilient theory to the ocean. Ad Dreidonks at the Royal Netherlands Meteorological Institute provided one of the first opportunities for me to devote a major effort to the parameterization of transilient matrices, and many of the other scientists there encouraged broad testing of the result. Ruwim Berkowicz and Lars Prahm at RIS in Denmark did much of the pioneering work on the characteristics on nonlocal methods, and absorbed much of the initial debate. Comments and suggestions by Jim Deardorff of Oregon State University and Brian Sawford of CSIRO in Australia stimulated more work. Thanks also go to the United States National Science Foundation for grants ATM-8508759 and ATM-8822214; and for United States Air Force Geophysical Laboratory grant F-1962-87-K-0035. The American Meteorological Society, American Geophysical Union, and Kluwer Academic Press are gratefully acknowledged for permision to reproduce many of the figures appearing in their journals. Finally, Takehiko Hasagawa, from Osaka Japan, is thanked for initiating my interest in nonlocal mixing by asking for information about the best K-theory parameterization for heavy-gas dispersion.

References

Berkowicz, R., 1980: *On the spectral turbulent diffusivity theory for homogeneous turbulence*. J. Fluid Mech., **100**, 433–448.

Berkowicz, R., 1984: Spectral methods for atmospheric diffusion modeling. Bound.-Layer Meteor., **30**, 201–220.

Berkowicz, R. and L.P. Prahm, 1979: *Generalization of K-theory for turbulent diffusion. Part I: Spectral turbulent diffusivity concept*. J. Appl. Meteor., **18**, 266–272.

Berkowicz, R. and L.P. Prahm, 1980: *On the spectral turbulent diffusivity theory for homogeneous turbulence*. J. Fluid Mech., **100**, 433–448.

Berkowicz, R. and L.P. Prahm, 1984: *Spectral representation of the vertical structure of turbulence in the convective boundary layer.* Quart. J. Roy. Meteor. Soc., **110**, 13–34.

Blackadar, A.K., 1979: *Modeling pollutant transfer during daytime convection.* Preprints, 4th Symp. on Turbulence, Diffusion and Air Pollution, Reno, Amer. Meteor. Soc., 443–447.

Boudreau, B.P. and D.M. Imboden, 1987: *Mathematics of tracer mixing in sediments: III. The theory of nonlocal mixing within sediments.* Amer. J. of Science, **287**, 693–719.

Chatfield, R.B. and R.A. Brost, 1987: *A two-stream model of the vertical transport of trace species in the convective boundary layer.* J. Geophys. Res., **92**, 13,263–13,276.

Chollet, J.-P. and M. Lesieur, 1981: *Parameterization of small scales of three-dimensional isotropic turbulence utilizing spectral closures.* J. Atmos. Sci., **38**, 2747–2757.

Chrobok, G., 1988: *Zur numerischen Simulation konvektiver Grenzschichten mit Integralen SchlieBungsansätzen.* Diplomarbeit im Fach Meteorology, November 1988, Inst. für Meteorologie und Klimatologie der Universität Hannover. 92pp.

Deardorff, J.W., 1985: *Comments on 'Transilient turbulence theory, Part I.'* J. Atmos. Sci., **42**, 2069.

Eberhard, W.L., W.R. Moninger and G.A. Briggs, 1988: *Plume dispersion in the convective boundary layer. Part. I. CONDORS field experiment and example measurements.* J. Appl. Meteor., **27**, 599–616.

Ebert, E.E., U. Schumann and R.B. Stull, 1989: *Nonlocal turbulent mixing in the convective boundary layer evaluated from large-eddy simulation.* J. Atmos. Sci., **46**, 2178–2207.

Eringen, A.C., 1972: *On nonlocal fluid mechanics.* Int. J. Engng. Sci., **10**, 561–575.

Estoque, M.A., 1968: *Vertical mixing due to penetrative convection.* J. Atmos. Sci., **25**, 1046–1051.

Fiedler,B.H., 1984: *An integral closure model for the vertical turbulent flux of a scalar in a mixed layer.* J. Atmos. Sci., **41**, 674–680.

Fiedler, B.H. and C.-H. Moeng, 1985: *A practical integral closure model for mean transport of a scalar in a convective boundary layer.* J. Atmos. Sci., **42**, 359–363.

Fortak, H.G.: *Non-Markovian turbulent dispersion in the atmosphere.* Preprints of the 16th International Tech Meeting on Air Pollution Modelling and its Applications, 6–10 April, Lindau, F.R. Germany.

Gaspar, P., Y. Gregoris, R.B. Stull and C. Boissier, 1988: *Long-term simulations of upper ocean vertical mixing using models of different types.* Small-scale Turbulence and Mixing in the Ocean, Proceedings of the 18th Leige Colloquium on Ocean Hydrodynamics. J.C.J.Nihoul and B.M. Jamart (Ed.), Elsevier Oceanography Series, **46**, 542pp.

Hinze, J.O., 1975: *Turbulence*, 2nd Ed., McGraw-Hill. pp790.

Hostetler, C.J. and B.E. Opitz, 1982: *Simulation of solute transport: A Markov model.* Report from Pacific Northwest Lab, P.O. Box 999, Richland, WA 99352. pp14.

Imboden, D.M., 1981: *Tracers and Mixing in the Aquatic Environment: A critical discussion of diffusion models and an introduction into concepts of non-Fickian transport.* Habilitation Thesis. Swiss Federal Institute of Technology (ETH), Zurich. pp137.

Jenkins, A.D., 1985: *Simulation of turbulent dispersion using a simple random model of the flow field.* Appl. Math Modelling, **9**, 239–245.

Jochum, A.M., 1988: *Turbulent transport in the convective boundary layer over complex terrain.* Preprints from the Eighth Symposium on Turbulence and Diffusion, April 25–29, 1988, San Diego. Amer. Meteor. Soc. 417–420.

Klemp, J.B. and D.K. Lilly, 1978: *Numerical simulation of hydrostatic mountain waves.* J. Atmos. Sci., **35**, 78–107.

Kraichnan, R.H., 1959: *The structure of isotropic turbulence at very high Reynolds numbers.* J. Fluid Mech., **5**, 497–543.

Kraichnan, R.H., 1966: *Isotropic turbulence and inertial-range structure.* Phys. Fluids, **9**, 1728–1752.

Kraichnan, R.H., 1971a: *An almost-Markovian Galilean-invariant turbulence model.* J. Fluid Mech., **47**, 513–524.

Kraichnan, R.H., 1971b: *Inertial range transfer in two and three-dimensional turbulence.* J. Fluid Mech., **47**, 525–535.

Kraichnan, R.H., 1976: *Eddy-viscosity in two and three dimensions.* J. Atmos., Sci., **33**, 1521–1536.

Leonard, A., 1974: *Energy cascade in large-eddy simulations of turbulent flows*. Adv. Geophys., **18A**, 237–248.

Leslie, D.C., 1973: *Developments in the Theory of Turbulence*. Clarendon Press. pp 368.

Lumley, J.L. and H.A. Panofsky, 1964: *The Structure of Atmospheric Turbulence*. Interscience. Wiley. pp239.

Monin, A.S. and A.M. Yaglom, 1971: *Statistical Fluid Mechanics: Mechanics of Turbulence*. The MIT Press. (Originally published in Russian in 1965). pp769.

Morse, P.M. and H. Feshbach, 1953: *Chapt. 7. Greens functions. Methods of Theoretical Physics*, McGraw-Hill. 997pp.

Narasimhan, M.N.L. and E.A. Saibel, 1989: *Turbulence in journal bearings from a nonlocal point of view*. Int. J. Engng. Sci., **27**, 219–236.

Pasquill, F., 1974: *Atmospheric Diffusion: The Dispersion of Windborne Material from Industrial and other Sources*. 2nd Ed. Wiley. pp429.

Prahm, L.P. and R. Berkowicz, 1980: *Reply (to Troen et al. 1980a)*. J. Appl. Meteor., **19**, 118.

Prahm, L.P., R. Berkowicz and O. Christensen, 1979: *Generalization of K-theory for turbulent diffusion. Part II: Spectral diffusivity model for plume dispersion*. J. Appl. Meteor., **18**, 273–282.

Prandtl, L., 1925: *Bericht über Untersuchungen zur ausgebildeten Turbulenz*. Ztschr. f. angew. Math. und Mech., **5**, 136–139.

Raymond, W.H. and R.B. Stull, 1990: *Application of transilient turbulence theory to mesoscale numerical weather forecasting*. Mon. Wea. Rev., **118**, 2471–2499.

Roberts, P.H., 1961: *Analytical theory for turbulent diffusion*. J. Fluid Mech., **11**, 257–283.

Romanof, N., 1982: *Application of Wiener-Hermite expansions to the turbulent diffusion*. Meteor. and Hydrology, **2**, 25–31.

Romanof, N., 1987: *A nonlocal model for the diffusion of pollutants released by instantaneous sources*. Preprints of the 16th International Tech Meeting on Air Pollution Modelling and its Applications, 6-10 April, Lindau, F.R. Germany. 10 pp.

Romanof, N., 1989: *Nonlocal models in turbulent diffusion*. Z. Meteorol., **39**, 89–93.

Saffman, P.G., 1969: *Application of the Wiener-Hermite expansions to the diffusion of a passive scalar in homogeneous turbulent flow*. Phys. Fluids, **12**, 1786–1798.

Sawford, B.L., 1988: *Comments on 'Transilient turbulence theory. Parts I and III'*, J. Atmos. Sci., **45**, 2092–2093.

Schumann, U., R.B. Stull and E.E. Ebert, 1989: *Nonlocal turbulent mixing in boundary layers evaluated from large-eddy simulations*. Preprints from the Seventh Symposium on Turbulent Shear Flows, Stanford Univ., August 21-23, 1989. 29.1.1–29.1.6.

Speziale, C.G. and A.C. Eringen, 1981: *Nonlocal fluid mechanics description of wall turbulence*. Computers & Math. with Applic., **7**, 27–41.

Stanišić, M.M., 1985: *The Mathematical Theory of Turbulence*. Universitext. Springer-Verlag. pp429.

Stull, R.B., 1984: *Transilient turbulence theory.*
Part I: The concept of eddy-mixing across finite distances. J. Atmos. Sci., **41**, 3351–3367.

Stull, R.B., 1985: *Reply to Deardorff's "Comments on "Transilient turbulence theory, Part I"*. J. Atmos. Sci., **42**, 2070–2072.

Stull, R.B., 1986: *Transilient turbulence theory, Part III: Bulk dispersion rate and numerical stability*. J. Atmos. Sci., **43**, 50–57.

Stull, R.B., 1987: *Transilient turbulence algorithms to model mixing across finite distances*. Environ. Software, **2**, 4–12.

Stull, R.B., 1988a: *An Introduction to Boundary Layer Meteorology*. Kluwer Academic Publishers, Dordrecht. 666pp.

Stull, R.B., 1988b: *A reevaluation of two dispersion theories*. J. Atmos. Sci., **45**, 2082–2091.

Stull, R.B., 1990: *Nonlocal turbulent mixing: measurement and parameterization of transilient matrices*. Preprints of the Ninth Symposium of Turbulence and Diffusion. April 30 – May 3, 1990. Roskilde, Denmark. Amer. Meteor. Soc., 348–351.

Stull, R.B., 1991a: *A comparison of parameterized vs. measured transilient mixing coefficients for a convective mixed layer*. Bound.-Layer Meteor., **55**, 67–90.

Stull, R.B., 1991b: *Static stability – an update*. Bull. Amer. Meteor. Soc., **72**, 1521–1529.

Stull, R.B. and A.G.M. Driedonks, 1987: *Applications of the transilient turbulence parameterization to atmospheric boundary layer simulations*. Bound.-Layer Meteor., **40**, 209–239.

Stull, R.B., E.E. Ebert, S.Jascourt, and J. Purser, 1992: *Convective structure memory.* (manuscript in preparation, to be submitted to J. Atmos. Sci.)

Stull, R.B. and T. Hasagawa, 1984: *Transilient turbulence theory. Part II: Turbulent adjustment*. J. Atmos. Sci., **41**, 3368–3379.

Stull, R.B., and E.B. Kraus, 1987: *The transilient model of the upper ocean.* J. Geophys. Res. – Oceans, **92**, 10745–10755.

Tennekes, H. and J.L. Lumley, 1972: *A First Course in Turbulence.* The MET Press. pp300.

Troen, I., T. Mikkelsen and S.E. Larson, 1980a: *Comments on 'Generalization of K-theory for turbulent diffusion. Part II'*. J. Appl. Meteor., **19**, 117–118.

Troen, I., T. Mikkelsen and S.E. Larson, 1980b: *Note on spectral diffusivity theory.* J. Appl. Meteor., **19**, 609–615.

Wang, S. and B.A. Albrecht, 1990: *A mean-gradient model of the dry convective boundary layer.* J. Atmos. Sci., **47**, 126–138.

Weast, R.C., 1968: *CRC Handbook of Chemistry and Physics,* 49th edition. The Chemical Rubber Co., 2106 pp.

Weil, J.C., 1990: *A diagnosis of the asymmetry in top-down and bottom-up diffusion using a Lagrangian stochastic model.* J. Atmos. Sci., **47**, 501–515.

Willis, G.E. and J.W. Deardorff, 1976: *A laboratory model of diffusion into the convective planetary boundary layer.* Quart. J. Roy. Meteor. Soc., **102**, 427–445.

Willis, G.E. and J.W. Deardorff, 1978: *A laboratory study of dispersion from an elevated source within a modeled convective planetary boundary layer.* Atmos. Environ., **12**, 1305–1311.

Willis, G.E. and J.W. Deardorff, 1981: *A laboratory study of dispersion from a source in the middle of the convective boundary layer.* Atmos. Environ., **15**, 109–117.

Wyngaard, J.C., 1987: *A physical mechanism for the asymmetry in top-down and bottom-up diffusion.* J. Atmos. Sci., **44**, 1083–1087.

Wyngaard, J.C. and R.A. Brost, 1984: *Top-down and bottom-up diffusion of a scalar in the convective boundary layer.* J. Atmos. Sci., **41**, 102–112.

Zhang, D. and R.A. Anthes, 1982: *A high-resolution model of the planetary boundary layer – sensitivity tests and comparisons with SESAME-79 data.* J. Appl. Meteor., **21**, 1594–1609.

Zhang, Q., 1990: *Test of Transilient Turbulence Theory Against a Field Experiment.* Masters Thesis. Dept. of Meteorology, University of Wisconsin – Madison. pp88.

Zhang, Q. and R.B. Stull, 1992: *Alternative nonlocal descriptions of boundary layer evolution.* (Submitted to J. Atmos. Sci.)

BUOYANT PLUME RISE IN A LAGRANGIAN FRAMEWORK

HAN VAN DOP
IMAU, University Utrecht, P.O.Box 80005, Utrecht, The Netherlands.

(Received October 1991)

Abstract. For the dispersion of buoyant material, the interaction with the environment by entrainment forms a serious obstacle for a formulation in a Lagrangian framework. Nevertheless an outline is given here on how buoyant plume rise in a Lagrangian sense could be described. Though the method contains a number of heuristic elements, it has all the advantages of a Lagrangian formulation. It is shown that it is possible to formulate a Lagrangian model which both is able to recover the classical formulations for plume rise in a calm environment and to accomodate more recent Eulerian formulations in a turbulent environment. Moreover, the method offers excellent possibilities to include the turbulent characteristics of the plume's environment and arbitrary stratifications of the boundary layer. These facts make it attractive for various practical applications. Some examples are given which illustrate this.

1. Introduction

The concept of the random walk in order to describe turbulent transport of a tracer in a turbulent flow has some attractive features: the equations are relatively simple and the statistical properties of the turbulent environment can be incorporated as indicated by Thomson (1984), Van Dop *et al.* (1985) and Thomson (1987). Though the approach has already been known for a long time in homogeneous turbulence (Taylor, 1921), its application to diffusion in e.g. inhomogeneous turbulence in the convective atmospheric boundary layer, led to some striking results (see De Baas *et al.* (1986), which could not be obtained with the widely used 1st-order closure Eulerian diffusion theory. They were found to be in fair agreement with experimental results (Willis and Deardorff, 1974, 1976 and 1978).

Though a number of Lagrangian diffusion models appeared to yield erroneous asymptotic solutions when applied in inhomogeneous turbulence, Thomson (1984, 1987) and Van Dop *et al.* (1985) reformulated the equations and successfully solved this problem in principle.

Also, studying the dispersion of particle pairs and concentration fluctuations, Lagrangian methods have been applied with success (Durbin, 1980; Kaplan and Dinar, 1988; Thomson, 1990). The last author succeeded in formulating a 3-D model which is comparable, if not superior to other approximate methods to describe concentration statistics (see Thomson, 1990, and references given there).

So far the majority of the applications of Lagrangian models have been limited to describe turbulent motion of passive tracers only. Zanetti (1984) and Cogan (1985) formulated Lagrangian models for buoyant dispersion, however, on a somewhat ad hoc basis. In this paper we explore the possibility to describe turbulent motion of buoyant tracers in a more fundamental way.

Boundary-Layer Meteorology **62:** 97–105, 1993.

There are two major aspects that distinguish buoyant and passive dispersion: (i) buoyant fluid particles "create" their own turbulent field in an environment which may be laminar or turbulent and (ii) the exchange processes between the plume particles and the (turbulent) environment should be included in the dynamics.

For a review and a detailed analysis of entrainment and turbulent transfer between the plume and its environment we refer to Netterville (1990). For a summary of the formulation of Lagrangian dynamic equations, we refer to Van Dop (1992).

The equation of motion for a buoyant plume particle can be formulated by including the buoyant acceleration as follows:

$$dW = -\frac{W}{T_w}dt + Bdt + \varepsilon_w^{1/2} d\omega_{w'} \tag{1}$$

where B is defined by

$$B \equiv \frac{g}{T}(\Theta - \Theta_a),$$

where Θ and Θ_a are the particle's and the ambient temperatures, respectively. We may formulate the dynamic behaviour of B as

$$dB = -\frac{B}{T_B}dt - N^2 W dt + \varepsilon_B^{1/2} d\omega_B, \tag{2}$$

where N is the Brunt-Vaïssala frequency, $\varepsilon_{W,B}$ the viscous dissipation for momentum and temperature and $T_{W,B}$ the corresponding Lagrangian timescales. The random increments $d\omega_{W,B}$ have the property $\overline{d\omega_{W,B}} = 0$ and $\overline{d\omega_{W,B}(t) d\omega_{W,B(t')}} = \delta_{t',t}dt$. Together with the trajectory equation, $dZ = W dt$, Eqs. (1,2) form the basic set of equations. We now investigate some properties of the set of eqs. (1,2).

2. Mean Characteristics

First we consider a simple case with $T_B = T_W = n^{-1}(= constant)$. Eqs. (1,2) yield with the initial conditions W(0)=W_o, B(0)=B_o and Z(0)=0 the equation for the mean velocity

$$\frac{d^2\bar{W}}{dt^2} + 2n\frac{d\bar{W}}{dt} + (n^2 + N^2)\bar{W} = 0, \tag{3}$$

which can be solved to yield

$$\bar{W}(t) = \frac{B_o}{N} e^{-nt} \sin Nt + W_o e^{-nt} \cos Nt. \tag{4}$$

For the mean plume height $\bar{Z}$, assuming W_o=0, we obtain

$$\bar{Z}(t) = \frac{B_o}{n^2 + N^2}\{1 - e^{-nt}[\cos Nt + \frac{n}{N} \sin Nt]\}. \tag{5}$$

The exponential behaviour follows from the Lagrangian equations using constant timescales $T_{B,W}$, for the plume turbulence.

In order to simulate a power law behaviour, which, in the early stage of plume rise in calm environments is confirmed by experimental evidence (see for example Turner, 1973), the Lagrangian equations should be modified. An obvious change is to make $T_{B,W}$ proportional to t, resulting directly in a power law behaviour for the mean plume height or the temperature. This would also be consistent with the idea that the turbulent timescales increase in expanding plumes. Assuming $T_{B,W} = A(t + t_o)$ we get for the mean buoyancy the solution

$$\bar{B} = B_o\{1 + \frac{t}{t_0}\}^{-1/A}, \tag{6}$$

which converges ($t \rightarrow \infty$) to the similarity solution (Csanady, 1973) provided that

$$A = 3/4 \tag{7}$$

and

$$t_o = (2R_o/(3\beta B_o))^{1/2},$$

where R_o is the initial plume radius and β an entrainment constant. Through the definitions of the plume particle buoyancy and the heat output of the source, Q, R_o and t_o can be related to the buoyancy flux parameter, F, by

$$\frac{9}{4}B_0^3 t_o^4 = \frac{1}{\beta^2}\frac{F}{U'}$$

yielding the familiar asymptotic solution for B:

$$\bar{B} = \{\frac{4F}{9\beta^2 U}\}^{1/3} t^{-4/3}.$$

Thus the Langevin formulation can be made to correspond asymptotically ($t \rightarrow \infty$) to the classical plume rise formula in a still environment by selecting the Lagrangian timescale proportional to t and choosing appropriate values for the constants.

Plume rise in a turbulent environment was recently addressed in detail by Netterville (1990), who introduced an additional turbulent exchange mechanism, "extrainment", generated by the ambient turbulence. The physical idea is that if the plume turbulence dominates (which is always true in a still environment), the turbulent timescales of the plume should be applied in Eqs. (1,2) whereas if the environmental turbulence dominates (which is the case in the initial and final stage of a line thermal with zero initial vertical velocity), the ambient time scale, T_E, should be used. These views are fairly reflected in a simple though approximate expression for n which is

$$n \simeq \frac{1}{T_E} + \frac{1}{T_{B,W}}, \tag{8}$$

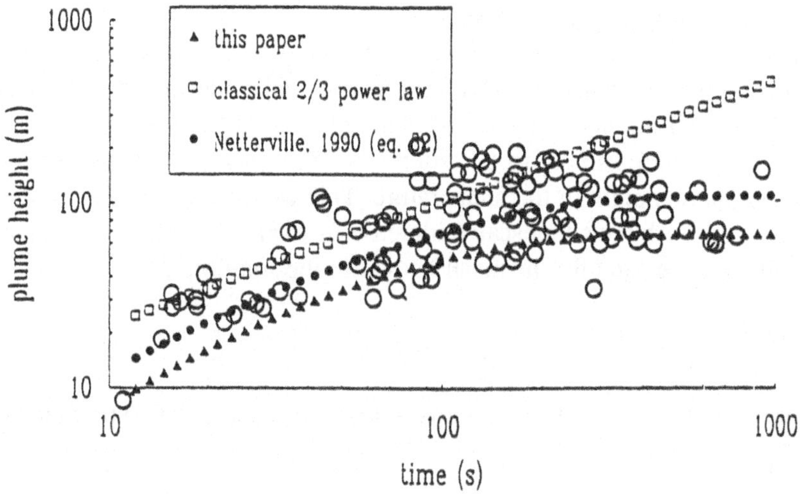

Fig. 1. Plume height as a function of time using the Nanticoke parameters (see Netterville, 1990).

with $T_{B,W} = A(t + t_o)$. Using this, the Lagrangian formulation is compared with Netterville's expression and the classical 2/3 law in Fig. 1, using values of 0.7641 ms^{-2} and 2.781 s for B_o and t_o respectively. These values correspond to the Nanticoke data as presented by Netterville (1990) (see Fig. 1).

Though the Lagrangian formulation produces lower values than Netterville's, it has the same leveling off behaviour in the final stage. (For the entrainment constant β the value of 0.65 was used for comparison with Netterville's formulation).

The approximate expression (Eq.8) offers the advantage of various straightforward applications, such as making T_E dependent on the actual height z, or introducing different forms for $T_{B,W}$ as indicated by Van Dop (1992).

3. The Evaluation of Plume Width

3.1. Still Environment

So far, we have only considered mean plume rise. The Lagrangian equations provide also for an independent evaluation of the plume variance or plume width. This requires, however, explicit expressions for ε_W and ε_B in Eqs.(1,2). Here, we assume

$$\varepsilon_W = c_1 W^2 / T_W \tag{9a}$$

Similarly, we propose

$$\varepsilon_B = c_2 B^2 / T_B \tag{9b}$$

where c_1 and c_2 are constants O(1), and T_W and T_B are assumed to be identical and equal to $T_{B,W}$ as defined in the previous section. There is of course an obvious

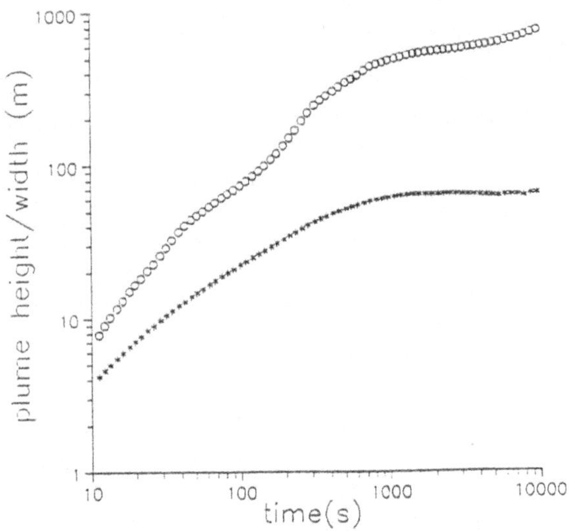

Fig. 2. Height (*) and width(o) of a plume adopting turbulent characteristics of the environment. $B_o = 1ms^{-2}$ and $t_o = 5s$.

choice for $\varepsilon_{W,B}$ by imposing the similarity requirement. In that case we have $\varepsilon_{W,B} \sim (F/U)^{2/3} t^{-5/3}$. Here, however we proceed with Eqs. (9a,b).

3.2. Turbulent Environment

The ambient turbulent characteristics also play a role in the rate of increase of the plume width, which should be reflected both in the plume particle's timescale and dissipation. For the dissipation we simply propose a weighted interpolation between the plume and ambient dissipation.

$$\varepsilon_W = \frac{W^2 T_W + (T_W/T_E)^2 \varepsilon_E}{1 + (T_W/T_E)^2} \tag{10}$$

assuming that $\varepsilon_E = 2u_*^2/T_E$. The form of Eq.(10) assures that $\varepsilon_W \rightarrow \varepsilon_E$ if $T_W >> T_E$ and $\varepsilon_W \cong W^2/T_{B,W}$ for $T_W/T_E \rightarrow 0$. The square of the weighting factor T_W/T_E was introduced for a rapid transition to environmental characteristics.

Eqs.(8,10) have been used for a simulation of plume rise in a homogeneous, neutral, turbulent environment with T_E = 100 s and $\varepsilon_E = 1.8.10^{-3} m^2 s^{-3}$. The source was located at z=0. For numerical convenience we assumed that when the buoyancy dissipation ε_B dropped below $10^{-8} m^2 s^{-3}$, it was set to zero. In Fig. 2 the plume width and height are presented showing that when $t \geq T_E$, the mean plume rise tends to a constant value. Also the growth of the plume variance is strongly inhibited after that time.

Using the Nanticoke parameters we made a simulation of the plume width. In Fig. 3 we compare the plume variance with the prediction of plume radius according to Netterville's formulation. For comparison also the mean plume height is

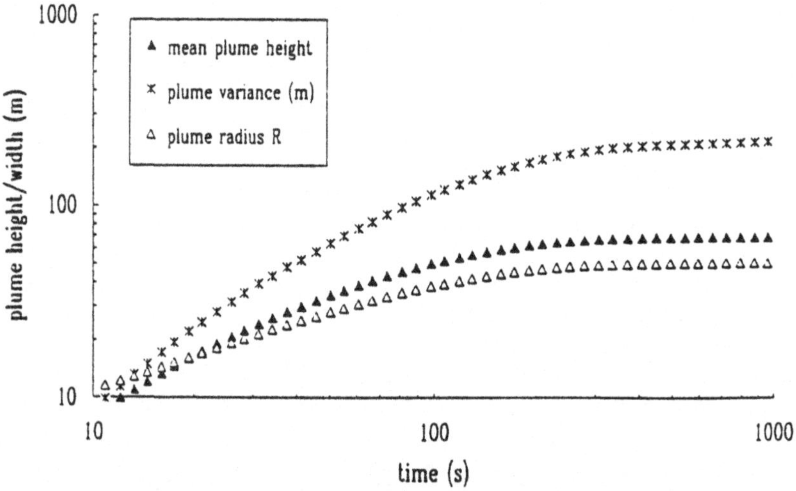

Fig. 3. Plume width (square root variance) and radius R as a function of time, for the Nanticoke parameters. For comparison also the plume height is given.

presented in Fig. 3. The plume variance exceeds the plume radius by approximately a factor of two. They could be made in better agreement by adopting smaller values for the dissipation constants c_1 and c_2, which for this simulation were taken equal to 1.

An interesting feature and a significant difference with the Eulerian formulation is the following observation: the leveling of of the mean plume height implies also a leveling off of the plume radius R, since it is directly proportional to it. This is an unsatisfactory consequence of the Eulerian theory, since one might expect that the plume, though not anymore significantly buoyant, still increases in width due to entrained ambient turbulence. According to the Lagrangian theory plume height levels off, but plume width continues to increase (though slightly only (cf Fig.3)). For weakly buoyant plumes, however, one might expect an underprediction of the plume radius in strong ambient turbulence.

4. Some Applications

We shall conclude with some examples of numerical evaluations which illustrate the versatility of the method.

Dispersion in a non-adiabatic environment will be considered. First plume rise in a stable environment with uniform stratification (constant N) will be evaluated. In addition results will be presented of dispersion in a neutral boundary layer, capped by an inversion.

4.1. DISPERSION IN A STABLE ENVIRONMENT WITH CONSTANT N

Three values of N were selected, respectively 0.0129, 0.0408 and 0.2 s^{-1}, corresponding with lapse rates of $5.10^{-3}, 5.10^{-2}$ and 1.2 K m^{-1} in order to test the

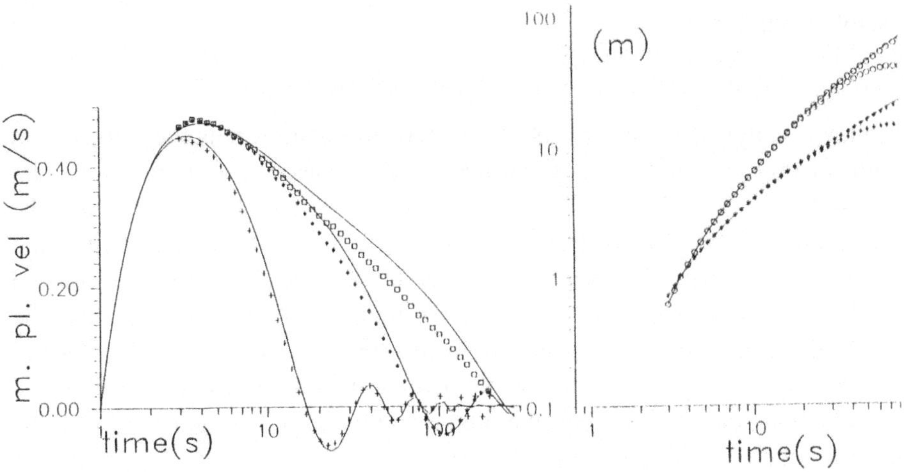

Fig. 4. Characteristics of plume dispersion in a stable environment. (a) mean plume velocity for stratification N=0.0129 (p), 0.0408 (*) and 0.2 (+) per s. (b) Mean plume height and width for N=0.0129 (o) and 0.0408 (*) per s. Of both pairs in the figure, the upper curves represent height and the lower ones, width respectively. The solid lines depict the exact solutions for the mean plume height.

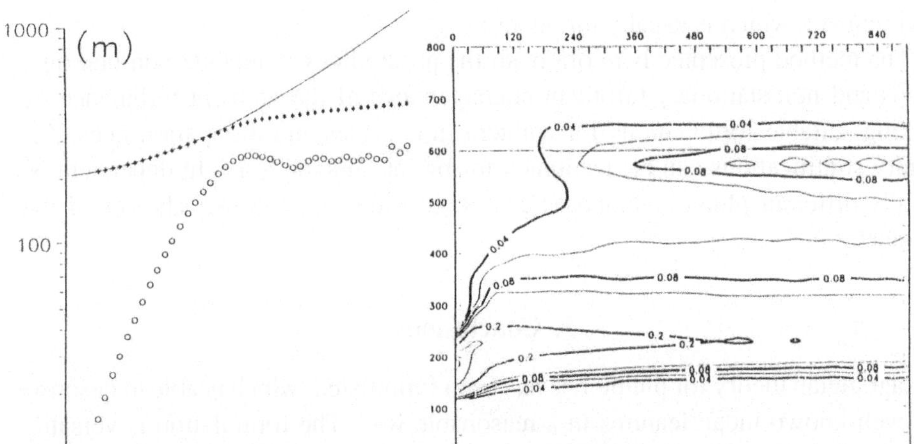

Fig. 5. Plume characteristics in a boundary layer capped by an inversion at an altitude of 600 m. (a) plume height (*) and plume width (o); (b) normalized concentration contours of a source emitting at 175 m. The vertical axis is the height (m) and the horizontal axis the time (s) since release.

numerical accuracy in a wide range of stabilities. In Fig. 4 plume height and rise are shown for N=0.0129 s^{-1} (upper pair) and N=0.0408 s^{-1} (lower pair). As a reference the exact solutions are also given (solid lines). The divergence of the numerical solution from the exact solution for the mean height for $t > 100$ s is due to numerical innacuracies. Plume rise as well as plume width are strongly limited in a stable environment.

4.2. PLUME RISE IN A NEUTRAL LAYER CAPPED BY AN INVERSION

Buoyancy dominated plume rise is considered in a neutral boundary layer with a height of 600 m. For $t_o = 5$ s, particles were released at a height of 175 m, with a buoyancy $B_o = 1ms^{-2}$. The inversion was simulated by assuming that the stratification at 600 m changed from 0 to $5.10^{-2}km^{-1}$. In Fig. 5a plume height and plume width are given. We observe that in the initial stage plume rise is according to the 2/3 law (solid line). After $\sim$ 100 s the plume starts to level off. Also the plume width, which first rapidly increased, converges to a constant value. Due to the overshoot of the plume in the stable inversion, some (artificial) oscillations occur, which are clearly visible in the plume width and in the figure showing the concentration plot as a function of time (Fig. 5b). The concentration profile has its main maximum at around 230 m and a secondary maximum just under the inversion height. It is also noted that the plume does not extend very much below the source height, since in this example only dispersion in a quiescent environment is considered, and that the intensity of the plume's turbulence is, due to entrainment, soon insufficient to cause further downward dispersion. Therefore, a more realistic description of plume rise should also include the dynamic characteristics of the environment, which is usually turbulent.

The method presented here offers all the possibilities to include non-homogeneous and non-stationary turbulent characteristics of the ambient turbulence by making suitable formulations for ambient time scales and dissipation rates. Arbitrary stratifications can be accounted for by introducing a height dependent N. Finally different plume geometries can be considered by proper choices of the parameter A.

5. Conclusions

A Lagrangian theory for plume rise has been formulated, which is able to describe the well-known mean features in a reasonable way. The formulation is versatile and allows easy inclusion of environmental characteristics such as thermal stratification (inversion layers) and environmental turbulence. Also different entrainment assumptions (e.g., those of Nieuwstadt (1991)) can be accomodated.

Based on ad hoc assumptions for the dissipation of temperature and velocity, independent estimates for plume width were obtained. The rate of change of the plume width in this formulation is faster ($\sim t$) than the mean plume rise ($\sim t^{2/3}$), which would lead to higher predictions of maximum surface concentrations from buoyant point sources. Unfortunately sufficiently accurate data on plume rise or width are not available to make further adjustments on the empirical formulation presented here. Large eddy simulation models might provide more detailed data of buoyant plume dynamics which could be used to develop the Lagrangian formulation further.

A drawback of the Lagrangian method is that in order to remove the statistical noise, a large number of flow realisations should be evaluated, for which in some cases a main frame might be required instead of a convenient personal computer which was used for all calculations presented in this paper.

References

Cogan, J.L. (1985). *Monte Carlo simulation of buoyant dispersion* Atmos. Envir., **19**, 867–878.

Csanady, G.T. (1973). *Turbulent diffusion in the environment.* Kluwer (Dordrecht) pp. 248.

De Baas, A.F., H. Van Dop and F.T.M. Nieuwstadt (1986). *An application of the Langevin equation for inhomogeneous conditions to dispersion in a convective boundary layer.* Quart. J. Met. Soc., **112**, 165–180.

Durbin, P.A. (1980). *A stochastic model of two particle dispersion and concentration fluctuations in homogeneous turbulence.* J. Fluid Mech. **100**, 279–302.

Kaplan, H. and N. Dinar (1988). *A stochastic model for dispersion and concentration distributions in homogeneous turbulence*, J. Fluid Mech., **190**, 121–140.

Netterville, D.D.J. (1990). *Plume rise, entrainment and dispersion in turbulent winds*, Atmos. Envir., **24a**, 1061–1081.

Nieuwstadt, F.T.M. (1991). *A large eddy simulation of a line source in a convective atmospheric boundary layer, part II: Dynamics of a buoyant line source*, submitted to Atmos. Envir.

Taylor, G.I. (1921). *Diffusion by continuous movements*, Proc. London Math. Soc., **20**, 196

Thomson, D.J. (1984). *Random walk modelling of diffusion in inhomogeneous turbulence*, Quart. J. R. Met. Soc., **110**, 1107–1120

Thomson, D.J. (1987). *Criteria for the selection of stochastic models of particle trajectories in turbulent flows*, J. Fluid Mech., **180**, 529–556.

Thomson, D.J. (1990). *A stochastic model for the motion of particle pairs in isotropic high-Reynolds-number turbulence, and its application to the problem of concentration variance*, J. Fluid. Mech., **210**, 113–153.

Turner, J.S. (1973). *Buoyancy effects in fluid.* Cambridge University Press, London.

Van Dop, H., F.T.M. Nieuwstadt and J.C.R. Hunt (1985). *Random walk models for particle displacements in inhomogeneous unsteady turbulent flows*, Phys. Fluids, **28**, 1639–1653.

Van Dop, H. (1992). *A note on plume rise in a Lagrangian framework.* To be published in Atmos. Envir.

Willis, G.E. and J.W. Deardorff (1974). *A laboratory model of the unstable planetary boundary layer*, J. Atmos. Sci, **31**, 1297–1307.

Willis, G.E. and J.W. Deardorff (1976). *A laboratory model of diffusion into the convective boundary layer*, Quart. J. Met. Soc., **102**, 447–445.

Willis, G.E. and J.W. Deardoff (1978). *A laboratory study of dispersion from an elevated source within a modeled convective planetary boundary layer*, Atmos. Envir., **12**, 1305–1311.

Zannetti, P. and Nazik Al-Madani (1984) *Simulation of transformation, buoyancy and removal processes by Lagrangian particle methods*, Procs of the 14th International Technical Meeting on Air pollution Modeling and its Application (ed. Ch. de Wispelaere) Plenum Press, New York., 733–744.

MIXING EFFICIENCY OF A STABLY STRATIFIED FLUID

K. ODUYEMI
Dundee Inst. of Techn., Dundee DD1 1HG, Scotland, UK.

(Received October 1991)

Abstract. Universally valid expressions for the momentum and mass diffusion coefficients in stratified shear flows do not exist. Similarly, the effect of molecular diffusion on the mixing efficiency of a stably stratified has not been investigated previously. In this paper the mixing efficiency in stably stratified flow is considered. Numerical investigations of unsheared stratified flows provide qualitative insight into the turbulence characteristics of stably stratified turbulent flows. Different types of stratified shear flows give qualitatively similar results for the mixing efficiency in the gradient Richardson number, Ri, range of 0.01–1.0. Mixing effcciency is proportional to Ri and its magnitude appears to depend on stratification in the fluid column. Suspended solids appear to affect the turbulence exchange processes in a different manner to those of salt anf heat. The ratio of the diffusivity of solute (or mass) to the diffusivity of momentum is less than 1 and can be taken as 0.25 for sediment-laden turbulent shear flows. This ratio is about a factor of 2–3 larger when heat- and salt-induced stratification exist, the difference possibly due to the settling nature of suspended solids.

1. Introduction

The understanding and qualification of mixing processes in some geophysical (e.g., estuarine and groundwater flows) water quality problems are sometimes affected by stratification and molecular diffusion. For example, in a reach of a partially-mixed estuary there is scope for continuous solute stratification in the vertical direction, and in the period leading to slack water the moderately low Reynolds number turbulence decays and molecular diffusion is expected to influence the mixing efficiency of the resulting flow.

The influence of stratification on the vertical turbulent diffusion processes in stably stratified flows has been previously considered (Rossby and Montgomery, 1935; Odd and Rodger, 1978; Riley *et al.*, 1981). A theoretical expression exists for the vertical turbulent diffusion coefficient in an unstratified turbulence shear flow. However, in continuously, stably stratified flow the vertical diffusion of both momentum and mass is inhibited by the stratification and hence, the existing theoretical expression is not valid. In such a situation, the turbulence momentum diffusion coefficient is larger than the turbulence diffusion coefficient of heat and mass. At present, universally valid expressions for the coefficients do not exist.

To the author's knowledge the effect of molecular diffusion on the mixing efficiency of a stably stratified fluid has not been investigated previously, though the turbulence statistics that could aid the progress of such investigation have been provided by some researchers (Gerz *et al.*, 1988; Komori *et al.*, 1983; Webster, 1964; Oduyemi and Britter, 1989). This paper briefly considers the effect of stable stratification and molecular diffusion on the mixing efficiency of:

i) unsheared stratified flows;

Boundary-Layer Meteorology **62:** 107–115, 1993.

ii) stratified shear flows.

The former should permit an understanding of stratified flows at a more simplistic level than the latter. A stratified shear flow is a more relevant flow, since shear is mostly present in real turbulent flows.

Under item (ii) above, this paper will attempt to provide reasons for the non-existence of a universally valid expression for the dependency of the ratio of the turbulent momentum diffusion coefficient to the turbulent diffusion coefficient of heat or mass, β, on the gradient Richardson number (a parameter used for quantifying stratification in density stratified flows).

2. Turbulence Statistics In An Unsheared Stratified Flow

2.1. METHODOLOGY

The numerical simulations were performed by using a modified version of the code developed by Riley *et al.* (1981). The calculation uses 32^3 uniformly spaced points and was carried out on a DEC Vax machine. The code numerically solves the continuity, Navier-Stokes and scalar equations in three spatial directions. A pseudo-spectral numerical method was used in the code. The method involves expanding the dependent variables in truncated Fourier series and evaluating spatial derivatives in Fourier space. The use of Fourier series implies periodic boundary conditions. The pseudo-spectral method has been successfully used in the past for problems with periodic boundary conditions (Oduyemi and Britter, 1989; Orszag and Patterson, 1972; Shirani *et al.*, 1981).

2.2. INITIAL CONDITIONS

The velocity field was first initialised and as the velocity derivative skewness built up to values typical of laboratory experiments, the density field was initialised (at a non-dimensional time $(u_0/L_0)t$=0.8, where L_0/u_0 is the initial time scale). The initial value of the turbulence Reynolds number, based on the lateral Taylor microscale, was 25. The initialisation of the velocity field is similar to that used in previous work (Riley *et al.*, 1981). For brevity, the initialisation of the velocity field will not be discussed here.

2.3. DISCUSSION OF RESULTS

After the initial adjusting period for the velocity derivative skewness ($0 < (u_0/L_0)t < 0.8$), the non-dimensional longitudinal and vertical mean square velocity fluctuations and kinetic energy dissipation rate for the passive scalar numerical experiments decay in a manner similar to that found in experimental grid turbulence (Oduyemi and Britter, 1989).

The investigation of mixing efficiency of stably stratified flows was performed by varying the initial value of the turbulence Froude number ($Fr = R_i^{0.5} u_0/(NL_0)$, where L_0/u_0 and N are the initial time scale and Brunt Vaisala frequency respectively and R_i the Richardson number). Initial values of Fr used are 0.24, 0.4, 0.8,

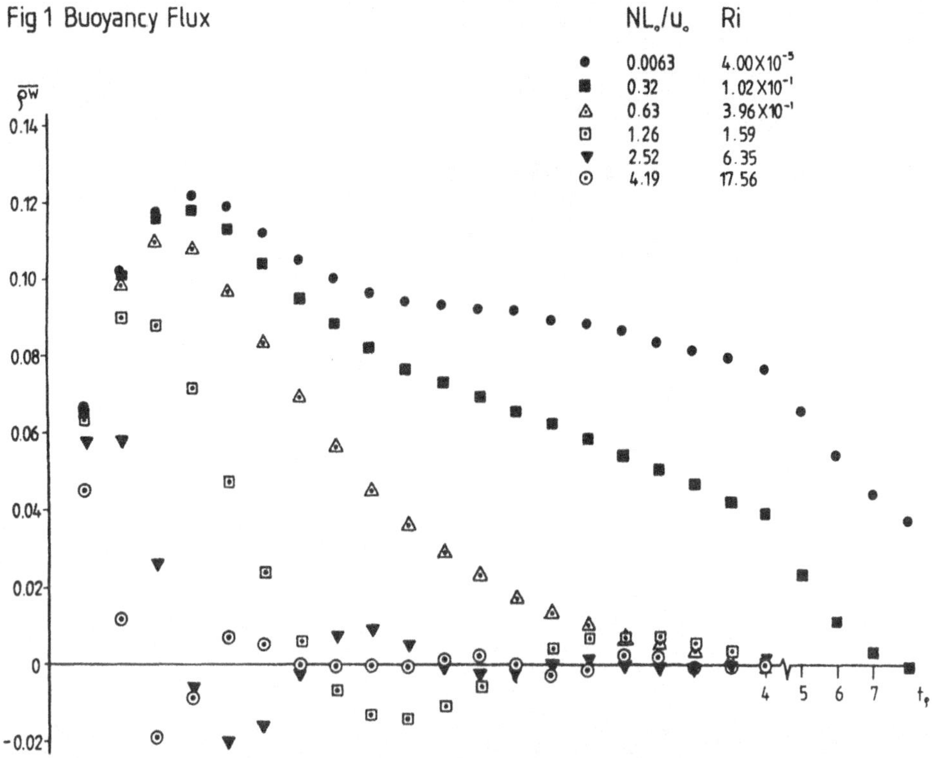

Fig. 1. Buoyancy Flux

1.6, 3.1 and 159, whilst the values of molecular Prandtl number used are 0.2, 1.0 and 3.0.

Fig. 1 shows the buoyancy flux increasing as it would in the absence of stratification and then decreasng as the effect of stable stratification becomes significant.

Fig. 2 illustrates the degree of mixing in a stably stratified fluid. The mixing efficiency is represented by a flux Richardson number, which is defined as the ratio of the net change in potential energy (the time integral of buoyancy flux) to the sink of the turbulent kinetic energy (arising from viscous dissipation and density stratification; Orszag and Patterson, 1972). The mixing efficiency is found to be initially proportional to Ri^1 (Ri raised to the power of one) and it levels off at $1 \leq Ri \leq 18$ as would be expected if there is partition of energy between viscous and density dissipation.

The trend in numerical results in Fig. 2 is similar to that of the laboratory data, except that the magnitudes of the laboratory data at $Ri > 0.1$ are about a factor of 3 less than those for the simulations (Rottman and Britter, 1986; Britter, 1988). The reasons for the differences in the magnitudes may be attributed to the different conditions that exist in both cases:

a) the Prandtl number in the case of the simulations varies between 0.2 and 3.0, whilst in the laboratory the Prandtl (or Schmidt) number could be as high as

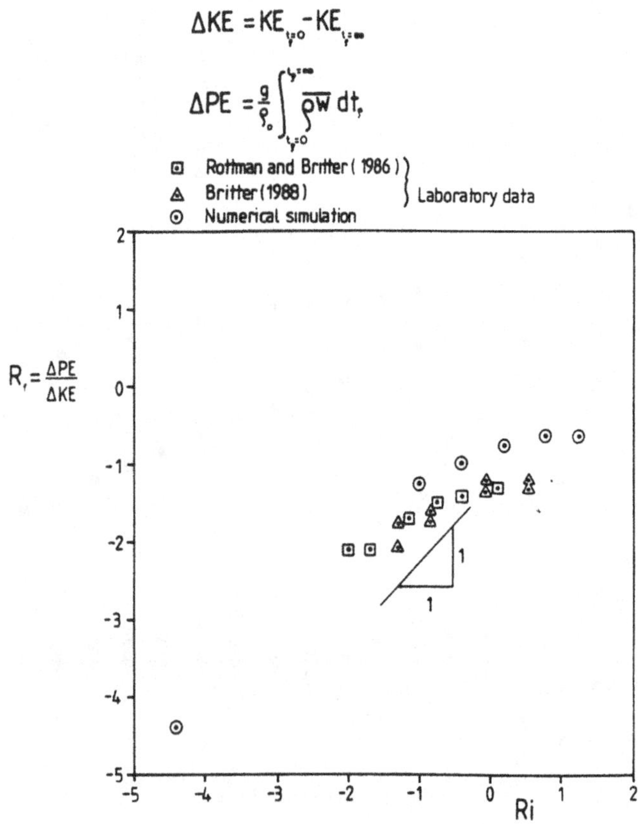

Fig. 2. Mixing efficiency in stably stratified flows

1000 for salt stratified experiments and could be higher than 3.0 for heat stratified experiments;
b) the Reynolds number defined by $u_0 L_0/\nu$ in the simulations is 59 whilst in the laboratory the Reynolds number $(u'M/\nu)$ is 1000 or more, where M is the mesh size for laboratory grid turbulence;
c) the definition of the denominator of the mixing efficiency in the laboratory is in terms of drag coefficient (CdU^2), which is different from the definition adopted for the simulation cases $(1.5u'^2)$. (U- mean velocity, u'-rms velocity, Cd - drag coefficient).

The numerical data show that the mixing efficiency reaches a maximum of $\approx$0.2 between a Ri value of 1 and 18. This suggests that the relative influence of buoyancy flux in mixing a fluid is small.

In order to consider the effects of Prandtl number on the mixing efficiency it is necessary to first consider the effect of Prandtl number on the distribution of buoyancy term. At present, well established work exists on the influence of molecular diffusion on turbulence parameters in the near non-stratified conditions (Oduyemi and Britter, 1989), but the effect of molecular diffusion on turbulence

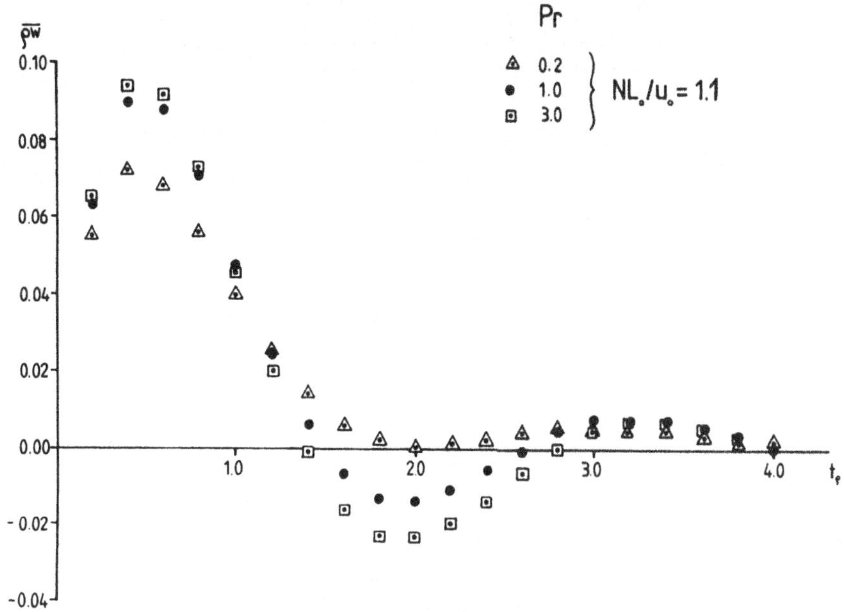

Fig. 3. Buoyancy flux as a function of time after initialisation of density field

parameters in the time range [1] $N_t > 1.0$, where the stratification effect is significant, is not well understood.

Fig. 3 shows that the buoyancy flux for $N_t > 1.0$, in general, oscillates with a period $N_t \simeq 2.5$ about the abscissa. Observations also show that the minimum value of buoyancy flux decreases with molecular diffusion and that a counter-gradient buoyancy flux exists for some period after $N_t \simeq 1.0$ (see Fig. 3). These observations can be attributed to the less positive peakedness in the buoyancy flux and lower amplitude of cyclic distribution of buoyancy flux as the diffusivity increases. It is worth mentioning that the fine scales in the density field may not be fully resolved for $Pr = 3.0$ and may lead to an underestimated value for ρw, which is a buoyancy flux term..

The effect of Prandtl number on the mixing efficiency was only investigated for $NL_0/u_0 = 1.26$ and the difference in mixing efficiency over a Pr range of 0.2 to 3.0 is about 10% (0.19 - 0.21). The distribution of the buoyancy term can be used in explaining this observation, since the integral of this flux term is the numerator of the mixing efficiency expression. The net negative integral flux between two cases of Pr and the net positive integral flux almost balances out (see Fig. 3); that is why the Pr effect is very minimal.

[1] N_t= Brunt-Vaisala frequency multiplied by time (non-dimensional time for stratified flow experiment)

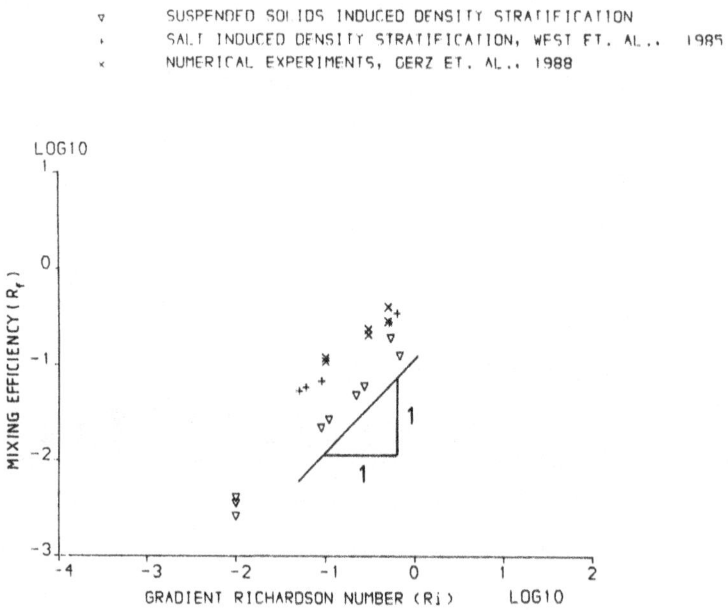

Fig. 4. Mixing efficiency in stably stratified shear flows

3. Turbulence Statistics in a Stratified Shear Flow

The field results presented in this section were undertaken in 1984/85 in estuaries around the United Kingdom. The field techniques and analyses have been previously described and can be found in West and Oduyemi (1989). The mixing efficiency of a stratified fluid is one aspect of the estuarine turbulence research that has not been previously dealt with adequately, and is supposed to be one of the most interesting, in that it thus influences the water quality aspect of the estuarine/river flow.

The field results are presented in Fig. 4, with the flux Richardson number plotted against the gradient Richardson number. The mixing efficiency (flux Richardson number) and gradient Richardson number for a stratified shear flow are as defined in most literature (see for example Dyer, 1986).

The interest in a flux Richardson - gradient Richardson curve stems from the suggestion that such a curve may give an insight into the behaviour of a linear vertical density profile flow structure, viz: the relative importance of the diffusivity of a mass (or heat) to that of momentum.

The mixing efficiency of the suspended solids-induced density-stratified fluid (see Fig. 4) increases monotonically between values of Ri of 0.01 and 1 and is found to be proportional to Ri^1. The maximum value of the mixing efficiency for this set of field data is approximately 0.25 and it occurred when $Ri \simeq 1$. There is therefore a suggestion that the ratio of the eddy diffusivity of suspended solids to the eddy viscosity, β, is less than 1 and equal to 0.25 between Ri values of 0.01

and 1. This average value of $\simeq 0.25$ compares with the value of 0.2 for a sandy estuary as observed by Soulsby *et al.* (1985).

The field observations thus quantify the value of β that could be used by modellers for predicting the concentration of suspended solids in a suspended solids-induced density-stratified flow. A β value of 1 has often been used in the past for sediment concentration prediction in a sediment-laden flow, because of lack of general agreement on the relationship between β and Ri.

West *et al.* (1985) have undertaken a field study of a salt-induced stratified shear flow. Ri values in their field experiments vary between 0.05 and 0.7. The data obtained from West *et al.*, Fig. 4, show that the mixing efficiency, in a similar fashion to the results of the suspended solids-induced density-stratified shear flow, is proportional to Ri^1. The mixing efficiency values for the salt-induced stratified flow are however a factor of 2-3 larger than those for suspended solids-induced stratified flow.

A reason for the difference in magnitude of the mixing efficiency values may be the different characteristics of the mass inducing the stratification in the vertical column of water. Suspended solids are known to settle in a water column and hence, would be expected to affect the vertical turbulence exchange processes in a different manner to the dissolved salt. A relationship for β has already been proposed for a salt-induced stratified shear flow (West *et al.*, 1985) and interested readers should refer to that paper.

Gerz *et al.*(1988) have similarly undertaken a numerical study of a stratified turbulent shear flow. Ri values in their 'numerical' experiments vary between 0.01 and 1.0 and the Prandtl numbers of the simulated flows are 0.7 and 5.0. The numerical results shown in Fig. 4 are plotted in the form of mixing efficiency against Ri. The results correspond to a Prandtl number of 0.7. These results compare reasonably well with the field results for salt-induced density-stratified shear flows, but are about a factor of 2-3 larger than those for suspended solids-induced density-stratified shear flows.

The results of Gerz *et al.* (1988) have been previously shown to compare well with the experimental data of Komori *et al.*(1983) in heat-induced density-stratified water flow and of Webster (1964) in heat-induced density-stratified air flow. These results show that the manner in which salt and heat inhibit the turbulence exchange processes are similar, whilst suspended solids inhibit the processes in a different manner, possibly because of their settling nature.

4. Conclusions

A study of mixing efficiency in stably stratified flows has been undertaken. Two types of stably stratified flows have been considered. Investigations of unsheared stratified flows have provided a qualitative insight into the turbulence characteristics of stably stratified turbulent flows, though it is recognised that most real stably stratified flows are shear flows.

Stratified shear flows gave qualitatively similar results for the mixing efficiency in the Ri range of 0.01-1.0. Mixing efficiency is proportional to Ri^1 and its magnitude appears to depend on the nature of masses inducing the stratification in the water column. Suspended solids appear to affect the turbulence exchange processes in a different manner to salt and heat.

The ratio of the diffusivity of pollutants to the diffusivity of momentum, β, is less than 1 in the Ri range of 0.01 to 1.0 and can be taken as 0.25 for sediment-laden shear flow. β values for salt- and heat-induced density-stratified flows are a factor of 2-3 larger than the above value in the same Ri range.

'Numerical' experiments are a useful tool for validating laboratory and field studies and should be used wherever possible.

Acknowledgements

The author gratefully acknowledge the support of the European Research Community of Flow Turbulence and Combustion and the useful discussion with Dr. R. Britter of Cambridge University, UK.

References

Britter, R.E., (1988). *Laboratory experiments on turbulence in density-stratified fluids*, 8th Symposium of Turbulence and Diffusion, USA.

Dyer, K. R. (1986) *Coastal and estuarine sediment dynamics*, John Wiley and Sons, 71–73.

Gerz, T., Schumann, U., Elghobashi, S.E., (1988). *Direct numerical simulation of stratified homogeneous turbulent shear flows*, J. Fluid Mech.

Komori, S., Ueda, J., Ogino, F., and Mizushina, T. (1983) *Turbulence structure in stably stratified open-channel flow*, J. Fluid Mech. **130**, 13–26.

Odd, N.V.M., and Rodger, J.G., (1978). *Vertical mixing in stratified tidal flows*, Proceedings of the American Society of Civil Engineers. Journal of the Hydraulics Division, vol. **104**, No. HY3, 469–486

Oduyemi, K. O., and Britter, R.E. (1989). *Molecular and stratification effects on the evolution of concentration fluctuations in decaying homogeneous turbulence with transverse mean-concentration gradient*, Advances in Turbulence 2 (ed. H. H. Fernholz and H. E. Fiedler), Spring-Verlag, Berlin, 186–191.

Orszag, S. A., and Patterson, G. S. (1972). *Statistical models and turbulence*, Lecture Notes in Physics, Springer, New York, Vol. 12.

Riley, J. J. Metcalfe, R. W., and Weissman, M. A. (1981). *Direct numerical simulations of homogeneous turbulence in density-stratified fluids*, Nonlinear Properties of Internal Waves (ed. B. J. West), AIP Conference, **76**, 79–112.

Rossby, C. G. and Montgomery, R. B.,(1935). *The layer of friction influence in wind and ocean currents*, Papers in Physical Oceanography and Meteorology, vol. 3, No. 3.

Rottman, J.W., and Britter, R.E. (1986). *The mixing efficiency and decay of grid generated turbulence in stably stratified fluids*, 9th Aust. Fluid Mechanics Conf., New Zealand.

Shirani, E., Ferziger, J. H., and Reynolds, W. C. (1981). *Mixing of a passive scalar in isotropic and sheared homogeneous turbulence*. Thermosciences Div., Dept. Mech. Eng., Stanford University, California, 1981, Tech. Report TF-15.

Soulsby, R. L., Salkield, A. P., Haine, R. A., and Wainwright, B. (1985). *Observations of turbulent fluxes of suspended sand near the sea bed*, Euromech 192: Transport of Suspended Solids in Open Channels, Neubiberg, 183–186.

Webster, C.A.G. (1964) *An experimental study of turbulence in a density stratified shear flow*, J. Fluid Mech. **19**, 221–245.

West, J.R., Knight, D. W., and Shiono, K. (1985) *A note on the determination of vertical turbulent transport coefficients in a partially mixed estuary*, Proceedings of the Institution of Civil Engineers, Part 2. **79**, 235–246.

West, J. R., and Oduyemi, K. O. (1989). *Turbulence measurements of suspended solids concentration in partially mixed estuaries*, J. Hydraulic Eng,., American Soc. of Civil Engineers. 457–474.

POSSIBILITY OF SIMULATING GEOPHYSICAL FLOW PHENOMENA BY LABORATORY EXPERIMENTS

H. BRANOVER, A. BERSHADSKII, A. EIDELMAN and M. NAGORNY
Center for MHD studies, Ben Gurion University of the Negev,
P.O.B. 653, Beer-Sheva 84105, Israel

(Received October 1991)

Abstract. Possibilities of laboratory simulation of two different atmospheric layers – stratospheric and boundary layer – are considered. The laboratory simulation is performed by fully developed turbulent flows of mercury ($Re \sim 7\text{x}10^4$) in strong magnetic fields. The processes of direct and inverse transfer of energy and passive scalar are investigated.

1. Strongly Anisotrophic Turbulence in the Stratosphere and in Mercury Flows Exposed to a Magnetic Field

1.1 Studies of liquid metal flows exposed to external magnetic fields, performed at the Center for MHD Studies of Ben Gurion University, revealed a number of most unusual flow phenomena, specifically in relation to strongly anisotropic turbulence (Branover and Sukoriansky, 1988; Henoch et al., 1991). In some cases the observed turbulence was almost two-dimensional in a plane perpendicular to the external magnetic field. Unique features of this type of turbulence resemble very closely the anisotropic turbulence which has been detected and measured by a number of researchers in geophysical flows – in the atmosphere and in oceans. Of course the nature of the forces causing such phenomena are different in geophysical cases (Coriolis forces, buoyancy forces due to density stratification) and in magneto-hydrodynamic flows (electromagnetic forces). However, the structure of turbulence disturbances and the transfer properties are identical.

In recent spectral data for kinetic energy and temperature obtained in the stratosphere by means of 6900 flights in the framework of "Global Atmospheric Sampling Program" (GASP) (Gage and Nastrom, 1986) we discovered a large region in which the spectral energy density has the power "–7/3" (Fig. 1). These data were obtained for large-scale pulsations in the region of wavelengths from 480 to 4800 km.

The data are not in agreement with direct prediction of two-dimensional turbulence theory ('–3'-spectrum for kinetic energy, see Kraichnan and Montgomery. 1979), so it became necessary:

a) to create a theoretical model which would better explain the experimental results than the ideal two-dimensional turbulence model. Such a model has to take into account the movement along a coordinate normal to the two-dimensional turbulence plane;

b) to carry out an experimental study of such turbulence created under laboratory conditions.

Stratospheric spectra are in agreement with the theoretical model developed by us for anisotropic turbulence when its hydrodynamic characteristics change slowly along one of the coordinates. This model with a "slow change" coordinate (designated in the following as 'SC') substantially differs from hypothetical two-dimensional turbulence and is also in agreement with experimental data obtained in our laboratory. In these experiments turbulence in mercury flow was created in the presence of a uniform magnetic field. In this way there was a possibility to simulate energy and passive scalar turbulent transfer processes in the stratosphere.

It is noteworthy that quasi-two-dimensional turbulence was already simulated in the laboratory by means of magnetohydrodynamic experiments (Sommeria and Moreau, 1982). But for simulation of SC type anisotropic turbulence it is necessary to introduce special modifications of the experimental conditions. The main such modification in our facility is the narrowing of the channel in the direction of the magnetic field.

1.2 In real stratospheric conditions the hydrodynamic gradient along certain coordinates is substantially less than the gradient in the plane perpendicular to the above coordinate. To realize two-dimensional turbulence the following condition is necessary

$$\frac{\bar{\varepsilon}}{L} >> |\overline{\frac{\partial \varepsilon}{\partial z}}| \tag{1}$$

where z is the SC coordinate;

$$\bar{\varepsilon} = \frac{\int_s (\varepsilon \mathbf{r}) d\mathbf{r}}{S}$$

is the average rate of turbulent energy dissipation in the plane perpendicular to the axis z; L – integral turbulent scale in this plane. If this condition is not satisfied, the energy flow between planes perpendicular to the axis z can be more substantial for the formation of a spectrum than $\bar{\varepsilon}$ in the plane.

In this case, instead of the "two-dimensional" inertial law for an energy transfer interval (Kraichnan and Montgomery, 1979), namely

$$E = C_1(\bar{\varepsilon})^{2/3} k^{-5/3} \tag{2}$$

the spectral law

$$E = C_2 |\overline{\frac{\partial \varepsilon}{\partial z}}|^{2/3} k^{-7/3} \tag{3}$$

can be obtained from analogous considerations of dimensionality. Accordingly, for an interval of enstrophy energy transfer instead of a "two-dimensional" law

$$E = C_3(\overline{\varepsilon_\omega})^{2/3} k^{-3} \tag{4}$$

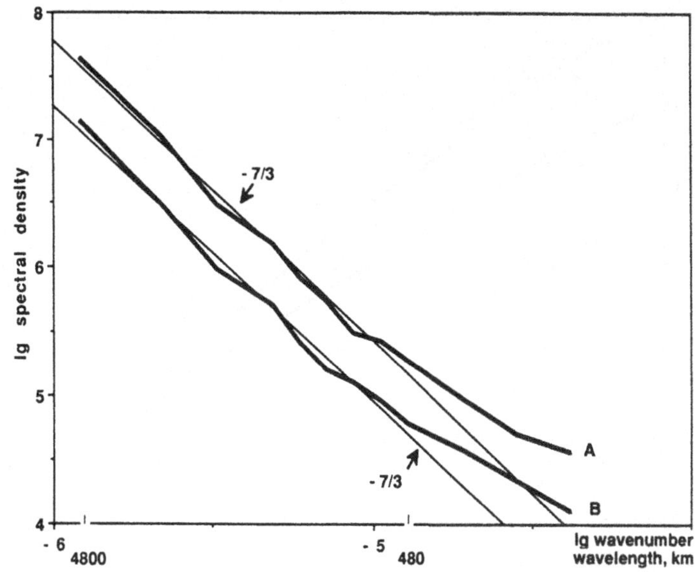

Fig. 1. Spectra from the GASP stratospheric flights at least 4800 km long: A – kinetic energy, B – temperature

we can obtain a spectral law

$$E = C_4 |\overline{\frac{\partial \varepsilon_\omega}{\partial z}}|^{2/3} k^{-11/3}, \tag{5}$$

where

$$\bar{\varepsilon}_\omega = \overline{\partial (rot \quad \mathbf{u})^2 / \partial t}.$$

Thus in a more physical turbulent model with an SC coordinate, both energy transfer processes and spectral laws substantially differ from those for a two-dimensional model.

Analogous calculations can be made for a passive scalar. Recent GASP spectral measurements (Gage and Nastrom, 1986) for velocity and passive scalar (temperature) in the stratosphere are presented in Fig. 1. Straight lines are shown to compare the data with expression (3) in log-log scales.

Spectral data obtained with our experimental MHD facility (Branover and Sukoriansky, 1988), are shown in Fig. 2. These data are obtained at different values of the magnetic field and at different distances from the generating grid. Straight lines are shown to compare with expressions (3) and (5) for turbulence with the "slow change" coordinate. The characteristic property of this facility is that the channel is narrow in the direction of the magnetic field. As a result, large-scale fluctuations of $\bar{\varepsilon}$ along the z axis (which is parallel to the magnetic field) are substantial and turbulence with a "slow change" coordinate is created rather than the two-dimensional turbulence.

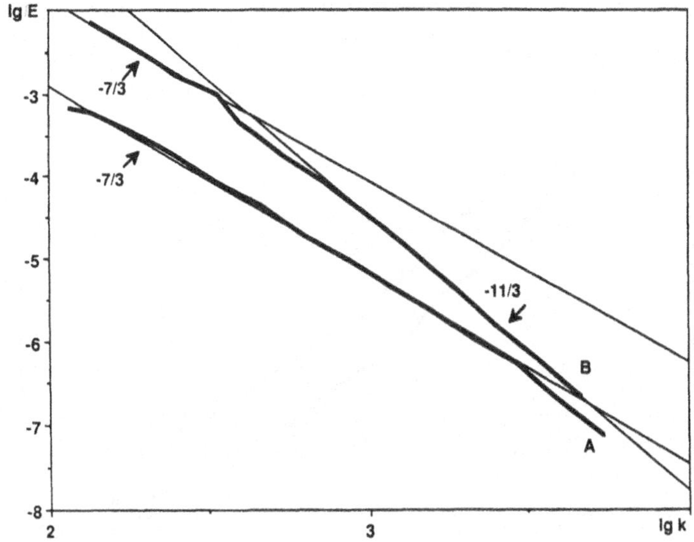

Fig. 2. Turbulent spectra in the channel with bars parallel to the magnetic field for different values of Ha/Re∗1000: A – 4.8; B – 10.

Note that in an analogous MHD experiment (Platnieks and Seluto, 1989) with a wide channel the energy spectra in weak magnetic fields (B=$0 \cdot 2 \div 0 \cdot 4$T) also have a tendency to the (3) type spectra. In a strong magnetic field B=$0 \cdot 6$T, spectra of two-dimensional turbulence like (4) are formed (Fig. 3).

1.3 It is of great importance to be able to establish the effective value of the coefficient of turbulent diffusion K∗, used in almost all calculations of contaminant dispersion in the atmosphere. The classical Richardson's law

$$K^* \sim (\bar{\varepsilon})^{1/3} R^{4/3}, \tag{6}$$

where R, the effective scale of a passive scalar cloud, is not always observed in the stratosphere.

For interval of enstrophy transfer in ideal two-dimensional turbulence, the theoretical law for passive scalar transfer is known:

$$K^* \sim (\overline{\varepsilon_\omega})^{1/3} R^2 \tag{7}$$

which follows from using dimensionality considerations similar to those for Eq. 6.

Expressions (6) and (7) correspond to spectral laws (2) and (4), respectively. But obviously the stratospheric spectral law (3) is more adequate. This law corresponds to SC turbulence. For this case, considerations of dimensionality, similar to those from which classical laws (6) and (7) were obtained, yield

$$K^* \sim |\frac{\partial \overline{\varepsilon}}{\partial z}|^{1/3} R^{5/3}. \tag{8}$$

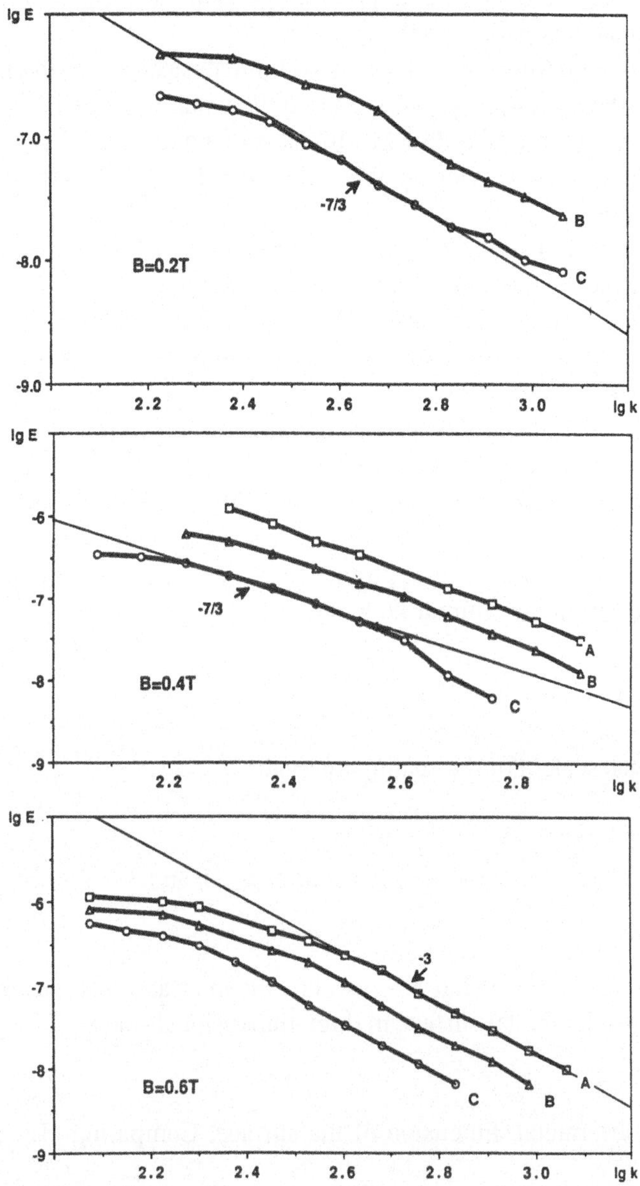

Fig. 3. Turbulent spectra measured at different distances from the grid with bars parallel to the magnetic field with different values of B, x/M:A – 9; B – 13.2; C – 21.5

Empirical observations of an effective diffusion coefficient of a passive scalar in the stratosphere are complex since accompanying factors which are difficult to take into account are always present. As a rule the exponent in the empirical function

$$K^* \sim R^\beta \tag{9}$$

is in the range $4/3 \leq \beta \leq 2$. It is important to obtain the value β from laboratory simulations of stratospheric turbulence.

In laboratory conditions it is possible to eliminate accompanying factors which are present in the stratosphere and to obtain the value for β. This value can then be used for calculation of turbulent diffusion of impurities taking into account accompanying factors. Obviously this value will be β=5/3 (see Ruo-Shan and Maxworthy, 1990).

1.4 The fractal structure of passive scalar clouds in a turbulent atmosphere attracts great interest (see for example Sreenivasan, 1991). This is mainly due to the influence of the fractal structure on the spread of different waves in the atmosphere and on the performance of some physico-chemical processes sensitive to fractal properties of the medium. In this sense the influence of peculiarities of the surface of passive scalar clouds is very important. Its fractal structure determines the properties of turbulence.

Let us label $p(x)$ the density of the probability distribution for a vortex with characteristic size x ($x = r/L$ -dimensionless scale, r -dimensional scale of a vortex). Let us consider the average total square of the surface which separates the turbulent liquid from the laminar flow part:

$$S = \int_{\eta}^{L} x^2 p(x) dx \tag{10}$$

here η – viscous scale. For the scaling interval:

$$p(x) \sim x^{-a}. \tag{11}$$

Substituting (11) into (10) we obtain that at $L >> \eta$ and $3 - a < 0$

$$S \sim (r/L)^{3-a} \tag{12}$$

and as $r/L \to 0$, it follows that $S \to \infty$, i.e., the interface becomes fractal. As is known for the fractal surface in three-dimensional space

$$S \sim (r/L)^{2-D_\sigma} \tag{13}$$

where D_σ is the fractal dimension of the surface. Comparing (12) and (13) we obtain

$$a = D_\sigma + 1 \tag{14}$$

i.e.,

$$p(x) \sim x^{-D_\sigma - 1}. \tag{15}$$

From the definition of $p(x)$ it follows that the volume of the active turbulent liquid can be represented as $p(x)x^3 dx$. On the other hand, if turbulent energy is additive for the volume of turbulent liquid, then

$$p(x)x^3 dx \sim \overline{du_x^2} \tag{16}$$

where

$$\overline{u_x^2} = \int_{x^{-1}}^{\infty} E(k)dk. \tag{17}$$

In Kolmogorov's case (2)

$$\overline{u_x^2} \sim x^{2/3} \tag{18}$$

and

$$D_\sigma = 7/3. \tag{19}$$

Such fractal dimensions were observed in numerous laboratory experiments and in atmospheric research when the characteristic scale $\leq 10^3$km (see Sreenivasan, 1991).

In the two-dimensional case, similar reasoning leads to the fractal dimension of the cloud perimeter

$$D_p = 4/3 \tag{20}$$

in the case of spectrum (2). This fractal dimension (20) was observed in laboratory experiments.

Let us consider now SC turbulence. Formula (15) does not depend on space dimensions. But in this case, D_p in the plane normal to the "slow change" coordinate is used instead of D_σ. However, the law of additivity of energy has to be considered in 3-dimensional space, i.e., using formula (16) for spectrum (3)

$$D_p = 5/3. \tag{21}$$

As one can see in Fig. 1, in the stratosphere the transition from spectral law (2) to spectral law (3) occurs at $x \sim 10^3$ km. That is why the fractal dimension (21) could be observed only at scales $x \geq 10^3$ km. Observations of a fractal dimension of passive scalar clouds with such scales are unknown to the authors.

Obviously such large-scale observations are very difficult. That is why the role of laboratory simulation is especially important. The data obtained in laboratory simulations with stratified rotating liquids are shown on Fig. 4 (Ruo-Shan and Maxworthy, 1990). Here D_p=4/3 at small scales is marked by a straight line. We note that in the experiment described in Ruo-Shan and Maxworthy (1990), the situation is dynamic (propagation of the front) and that is why measurements of spectral characteristics are complex. It is still possible to obtain the fractal dimension D_p from spectral characteristics of the quantity $\partial u^2/\partial x$ (or using Taylor hypotheses, from the quantity $\partial u^2/\partial t$).

2. Inverse Energy Transfer in Atmospheric and Laboratory Turbulence

2.1 More than twenty years ago it was observed that in ocean turbulence flows with inverse energy transfer exist (Starr, 1968; Ozmidov et al., 1970). Namely,

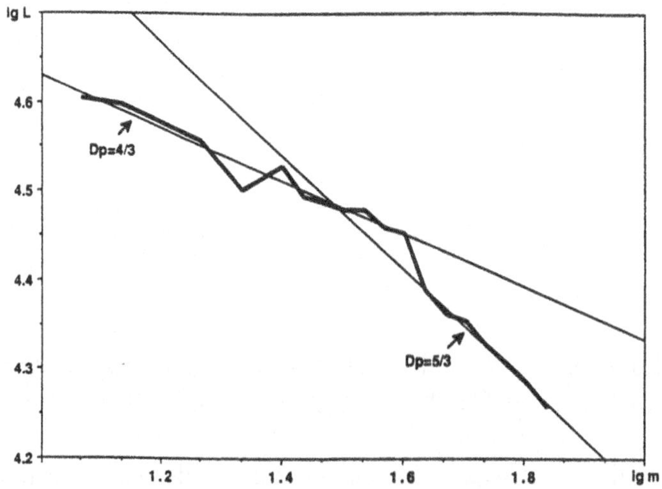

Fig. 4. Interface length L measured with a certain gauge length m versus the value of m itself

turbulent energy is transferred from smaller to larger eddies and ultimately to the mean flow. This was termed the negative eddy viscosity. In recent experiments with mercury flow in transverse magnetic fields, not only was the inverse energy transfer clearly detected, but also the increase of the energy of the mean flow (instead of dissipation) due to the influx of energy from two-dimensional disturbances (Henoch et al., 1991). Besides velocities of the mean flow and of turbulent disturbances, heat transfer has also been measured. The measurements indicated very substantial differences in heat transfer properties between the "normal" three-dimensional and the quasi two-dimensional turbulence caused by electromagnetic forces.

If the system cannot follow the fine details of the processes of energy input, it starts to average these details over scales. And if it is a self-similar process, then such averaging manifests itself as a physical process of energy transfer from small scales to large ones.

For example, let a trial Brownian particle move in a turbulent fractal field. It is known, that if its characteristic size is small enough, this particle cannot leave the turbulent area (Townsend, 1966). If in this case the fractal dimensionality of particles inside trajectories D_w is more than the fractal dimensionality of the turbulent fractal $D(D_w > D)$, then we have two conjugated effects:

1. Fractal turbulent properties will be manifested for more and more scales (inverse transfer of geometrical properties).
2. Movement of the trial particle will asymptotically (at an increase of scale) follow the liquid Lagrangian particle.

2.2 The mean distance l of particle displacement during the time $(t - t_o)$ can be estimated as follows (Sokolov, 1986):

$$l \sim (t - t_o)^{1/D_w}; \qquad (D_w > 2). \tag{22}$$

Let us divide the flow region into cells with characteristic scale equal to the Kolmogorov length η. The number of cells which will be visited in its random walks on the turbulent fractal is

$$N \sim l^D. \tag{23}$$

Then the probability to find the particle in a particular cell at the moment t is

$$p(t) \sim N^{-1} \sim (t - t_o)^{D/D_w} \tag{24}$$

assuming that the particle can visit each cell with equal probability and $l >> \eta$.

The probability of finding the particle in a particular cell at some moment within the time period T is

$$p(T) \sim \int_{t_o}^{t_o+T} p(t)dt \sim T^{1-D/D_w}\,. \tag{25}$$

The integral in (25) exists if

$$D < D_w\,. \tag{26}$$

In this case the relation (25) can be written in terms of frequency $\omega \sim T^{-1}$

$$p(\omega) \sim \omega^{1-D/D_w}\,, \tag{27}$$

where $p(\omega)$ can be interpreted as the probability density of motion characterized by frequency ω. It follows from (26), (27) that the motions with smaller frequencies are more probable than those with larger frequencies. Since in general energy is transferred to the most probable states, relation (26) can be considered as the condition for the existence of an inverse energy transfer.

2.3 It follows from (27) that the spectrum of a passive scalar in the low frequency range has the following form

$$E_\theta(\omega) \sim \omega^{D_f-1} \tag{28}$$

where D_f=2D/D_w is the so called fracton dimension (Sokolov, 1986). In the universal class of Alexander-Orbach (Sokolov, 1986), D_f=4/3 and from (28)

$$E_\theta(\omega) \sim \omega^{1/3}. \tag{29}$$

It is noteworthy that atmospheric spectral measurements of temperature provide examples of low frequency power ranges not very far from (29). A recent example of such measurements is shown on Fig. 5 (Tjernstrom, 1990). Formulae (28), (29) can be satisfied in the space of any dimension only if condition $D < D_w$ (i.e., $D_f < 2$) is satisfied.

2.4 Another example of application of the above principle is connected with "Burgerization" of the Navier-Stokes equation in the case of inverse energy transfer in the large-scale region.

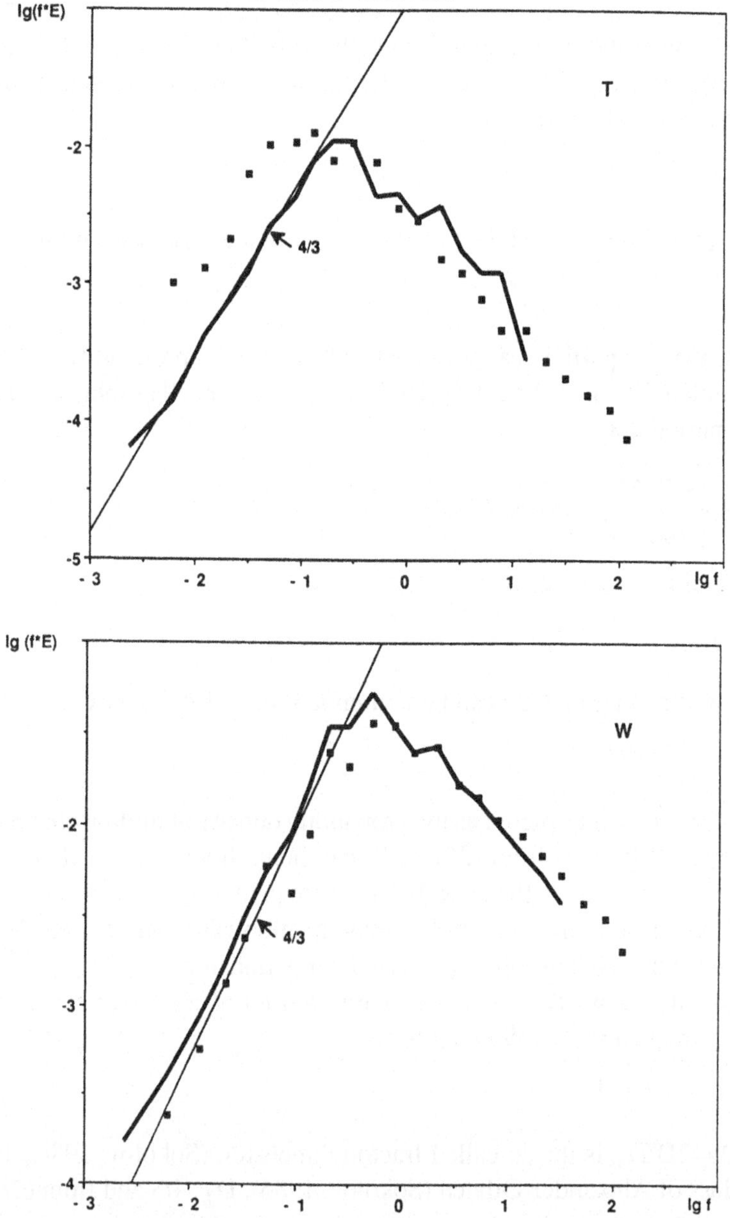

Fig. 5. Spectra for temperature and wind from tower (squares) and simultaneous aircraft (solid line) measurements, arbitrary units.

If D_p is the fractal dimension of isobar surfaces and D_∇ is the fractal dimension of isoenergetic surfaces, then at $D_p > D_\nabla$ inverse energy transfer takes place for those scales where the above noted inequality is satisfied. The pressure forces are introducing energy into the volume through its surface and inertia forces $(\mathbf{u}\nabla)\mathbf{u}$ are distributing this energy inside the volume. Therefore at $D_p > D\nabla$ it follows from the above formulated general principle that inverse energy transfer takes place.

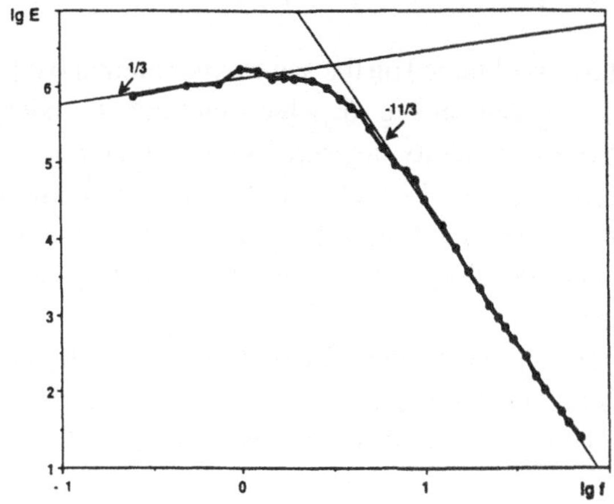

Fig. 6. Longitudinal velocity spectrum behind a honeycomb in a transversal magnetic field (B=1T)

Let us estimate the integrals

$$\int_{v_r} \Delta p d\mathbf{r} \sim \oint_{\Sigma_r} \nabla p d\sigma \sim r^{(x+D_p)} \tag{30}$$

$$\int_{v_r} div(\mathbf{u}\nabla)\mathbf{u}d\mathbf{r} \sim \oint_{\Sigma_r} (\mathbf{u}\nabla)\mathbf{u}d\sigma \sim r^{(y+D_\nabla)} \tag{31}$$

here v_r – volume of liquid with characteristic size r; Σ_r-surface of this volume; and y – parameters to be determined (scaling hypothesis). Then for an incompressible liquid

$$x + D_p = y + D_\nabla \tag{32}$$

hence

$$x - y = D_\nabla - D_p. \tag{33}$$

Let us estimate

$$\frac{\int_{v_r} \nabla p d\mathbf{r}}{\int_{v_r} \mathbf{u} \nabla \mathbf{u} d\mathbf{r}} \sim r^{x-y} \sim r^{D_\nabla - D_p}. \tag{34}$$

From (34) it follows that if $D_p > D_\nabla$ at $r \to \infty$ (large scales) the term ∇p can be neglected in comparison with $(\mathbf{u}\nabla)\mathbf{u}$ in the Navier-Stokes equation. In other words if inverse energy transfer takes place, the Navier-Stokes equation is transferred into the Burgers equation without ∇p for large scales also. It follows that for large scales the equation for a passive scalar and for the velocity the field have similar forms and their scale spectra can coincide.

Indeed in atmosphere observations (Fig.5, Sokolov, 1986) spectrum (29) is observed also for components of the velocity field for the same scales as for a passive scalar. An analogous situation takes place also for atmospheric observations shown on Fig. 1.

Our experimental data obtained on the same experimental facility as data shown above (on Fig. 2) are given on Fig. 6, which presents the energy spectrum for longitudinal pulsations of velocity measured in a flow behind a honeycomb in a transvers magnetic field ($B \cong 1T$). Data in Fig. 6 include the lower frequency range which was not reflected in Fig. 2. Spectral data for the low frequency range are well approximated by (29) which testifies to the existence of inverse energy transfer in this range.

All the above indicates clearly that a laboratory magnetohydrodynamic experiment can serve as a reliable simulation of geophysical flow phenomena, measurements of which are generally very difficult, costly and time-consuming.

References

Branover, H., Sukoriansky, S., 1988. *Turbulence Peculiarities Caused by Interference at Magnetic Fields with the Energy Transfer Phenomena.* AIAA, Progress in Astronautics and Aeronautics, Vol. 112, pp. 87–99.

Gage, K.S., Nastrom, G.D. 1986. *Theoretical Interpretation of Atmospheric Wavenumber Spectra of Wind and Temperature Observed by Commercial Aircraft During GASP.* J. Atmosph. Science, Vol. 43, pp. 729–740.

Henoch, C., Hoffert, M., Branover, H., and Sukoriansky, S. 1991. *Anisotropic Turbulence: Analogies Between Geophysical and Hydromagnetic Flows.* AIAA, Progress in Astronautics and Aeronautics (to be published).

Kraichnan, R.H., Montgomery, D. 1979. *Two-dimensional Turbulence* Phys. Fluids, Vol. 43, pp. 547–619.

Ozmidov, R.V. et al. 1970. *Some Features of the Turbulent Energy Transfer and the Transformation of Turbulent Energy in the Ocean.* Atm. Ocean. Phys., Vol. 6.

Platnieks, I., Seluto, S.F. 1989. *The Effect of Initial Boundary Conditions Upon the Formation and Development of MHD Turbulence Structure.* In Liquid Metal Magnetohydrodynamics, pp. 433–439.

Ruo-Shan, T., Maxworthy, T. 1990. *The Dynamics and Geometry of a Two-dimensional turbulent front.* Phys. Fluids A (27), pp. 1224–1230.

Sokolov, I.M. 1986. *Dimensionalities and other geometric critical exponents in percolation theory.* Sov. Phys. Uspekhi, Vol. 150, pp. 924–945.

Sommeria, J., Moreau, R. 1982. *Why, How and When MHD Turbulence becomes Two-Dimensional* J. Fluid Mech., Vol. 118, p. 507–518.

Sreenivasan, K.R. 1991. *Fractals and Multifractals in Fluid Turbulence* Annu. Rev. Fluid Mech., Vol. 23, pp. 539–660.

Starr, V.P. 1968. *Physics of Negative Viscosity Phenomena.* Mc. Graw Hill, New-York.

Tjernstrom, M. 1990. *On the Use of Pressure fluctuation on the random of a subreliner aircraft for boundary layer turbulence measurements.* Ninth Symposium on Turbulence and Diffusion, pp. 123–126.

Townsend, A.A. 1966. *The mechanism of entrainment in free turbulent flows.* J. Fluid Mech., Vol. 26, pp. 689–715.

INVESTIGATION OF THE PLANETARY BOUNDARY LAYER HEIGHT VARIATIONS OVER COMPLEX TERRAIN

R. LIEMAN and P. ALPERT
Tel-Aviv University, Tel-Aviv, Israel.

(Received October 1991)

Abstract. The effects of sea-breeze interactions with synoptic forcing on the PBL height over complex terrain are investigated through the use of a 3-D mesoscale numerical model. Two of the results are as follows. First, steep PBL height gradients – order of 1500 m over a grid interval of 10 km – are associated with the sea-breeze front and are enhanced by the topography. Second, a significant horizontal shift in the maximum PBL height relative to the mountains, is induced by a corresponding displacement of the thermal ridge due to the mountains, in the presence of large scale flow.

1. Introduction

The realistic distributions of planetary boundary layer (PBL) heights over complex topography are of particular importance for a good prediction of the dispersion of air pollutants. Observational studies by radiosondes/minisondes (Dayan *et al.*, 1988) and recently also by lidar (laser-radar), e.g. Hasmonay *et al.* (1990) are of limited value over complex terrain because of the relatively large expected spatial variance of PBL height. In contrast, the analytical studies focus mostly on flat terrain, e.g. Stull (1988), and the main tool over complex terrain remains the three dimensional (3-D) numerical model, e.g. Anthes and Warner (1978), Segal *et al.* (1982,1985), Alpert and Getenio (1988) and Alpert *et al.* (1988). The 3-D models, however, became quite expensive to operate particularly when high horizontal and vertical resolutions are required, e.g. Seaman *et al.* (1989), and therefore relatively few such experiments have been performed over the meso-$\tilde{A}$ scale ($\breve{S}_x \sim 5 - 10\ \ km$). Also, these studies focused mostly on the flow fields, and less on PBL height variations (Alpert, 1988).

Hence, the purpose of the present study is to focus on the temporal and spatial variations of the convective PBL height in association with the relevant dynamical processes over complex terrain. For this, a 3-D high resolution mesometeorological model was applied over Israel for two realistic cases with opposing geostrophic flow and different synoptic situations. One is a warm summer case with a generally onshore westerly current, while the other is a cold winter case with an easterly offshore current (not shown in detail in this work).

Boundary-Layer Meteorology **62:** 129–142, 1993.

2. Model and Method

2.1. Model and Mesh

The recent MM4 version of the PSU/NCAR mesoscale model as described by Anthes and Warner (1978) and Anthes *et al.* (1987) was applied over a region of 300x360 km in the southeast Mediterranean, Fig. 1, with horizontal resolution of $\check{S}x = \check{S}y = 10\ km$. The vertical coordinate is a normalized pressure $\bar{I}$ and the model has 16 $\bar{I}$-levels whose altitudes above the ground are approximately given by: 0, 20, 40, 80, 160, 320, 640, 1280, 2000, 13000, 4000, 6000, 8000, 10000, 12000 and 16000 m. The model equations and the numerical schemes are reviewed by Anthes *et al.* (1987). The present simulations were carried out without moisture and could be partly justified by the clear and relatively dry events that have been chosen.

2.2. Initalization

The static initialization – e.g., Haltiner and Williams (1980) –was chosen since it was found that the model adjustment to strong local forcings, thermal as well as topographical, were quite fast. The Cressman (1959) interpolation over isobaric surfaces was performed adopting the Goodin *et al.* (1979) radius of influence R,

$$R = 1.66\sqrt{A/N} \tag{1}$$

where A is the area of the large-domain, see Fig. 1, and N the number of available radiosondes. The variables were then vertically interpolated to the $\bar{I}$-levels and the vertically averaged divergence of the horizontal wind was required to vanish. This was done in order to prevent initial gravity wave disturbances, e.g. Pielke (1984). Six neighbouring radiosondes were used for the initialization in successive applications of the Cressman interpolation (more details are found in Lieman,1990).

The model was quite sensitive to the lateral boundary conditions and following a few experiments, the sponge boundary condition was applied with a varying boundary weight for the synoptic large-scale flow increasing gradually from 1 at the boundary to 0 at the fifth inner domain point.

2.3. Surface Properties

Table I summarizes some of the surface parameters for our simulations. The surface temperature is calculated through a surface thermal energy balance following Deardorff (1978).

2.4. Model Topography

Envelope orography was used, where the standard deviation of the topographical height as calculated over 1 km horizontal resolution, was added to the average altitude at each point. Since the simulation grid interval was 10 km, 100 points were involved in the calculation for each point. The resulting topography along with surface stations used for model verifications are shown in Fig. 2, which

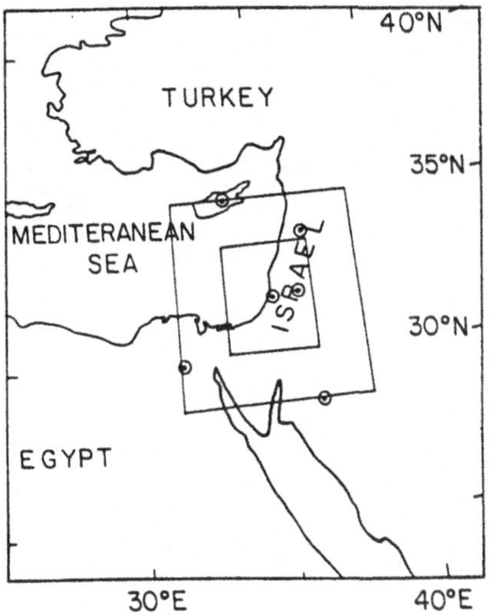

Fig. 1. General view of the area. The large-domain for the initialization and the smaller domain illustrating the simulation area. The available radiosondes are indicated by ⊙.

TABLE I
Surface properties in the model simulations

	Albedo	Soil Moisture (%)	Roughness Length (cm)	Soil heat capacity (cal cm^2 K^{-1} sec$^{-0.5}$)	Emissivity at 9 μ in (%)
Land	0.25	2-5	10	0.02	85
Sea	0.08	100	10^{-4}	0.06	98

shows two major north to south mountain ranges at distances of about 40 and 100 km from the Mediterranean coast respectively. Fig. 3 shows the west to east topographical cross-section along 32°N (ordinate 21 in Fig. 2) with and without the envelope orography along with the original topography based on the 1 km resolution. In order to find the optimal horizontal grid interval which captures most of the topographical variance, the cross-section height function (Fig. 3) was decomposed into its Fourier components, following the suggestion by Young and Pielke (1983). From the distribution of topographical variance as a function of wavelength, one finds that a 10 km interval represents the major part (over 95%) of the variance based on the 1 km resolution. A considerable reduction in the topographical variance occurs when the grid interval doubles to only 20 km. This result led to our decision to use a 10 km grid interval.

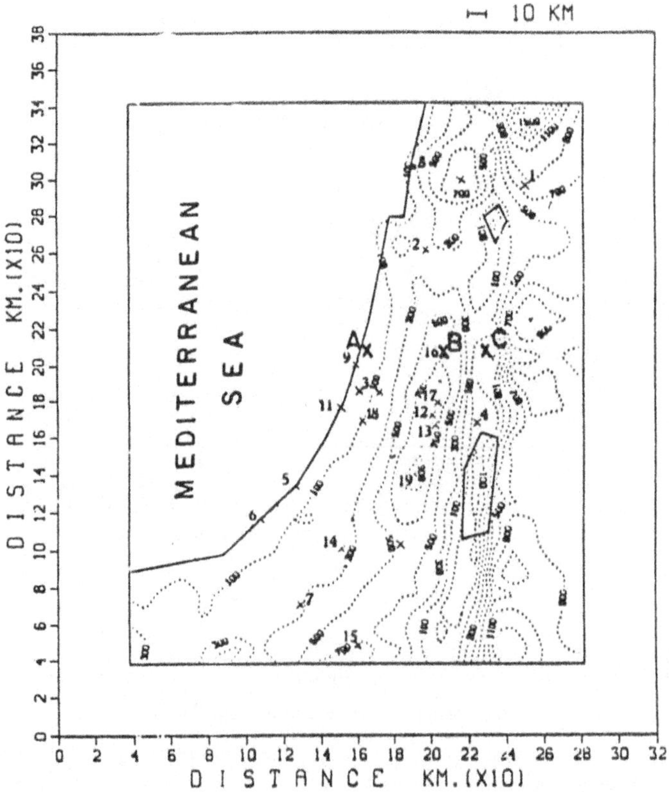

Fig. 2. Model topography with contour interval of 200 m. The surface stations used for model verification are indicated. Points A, B and C are referred to later.

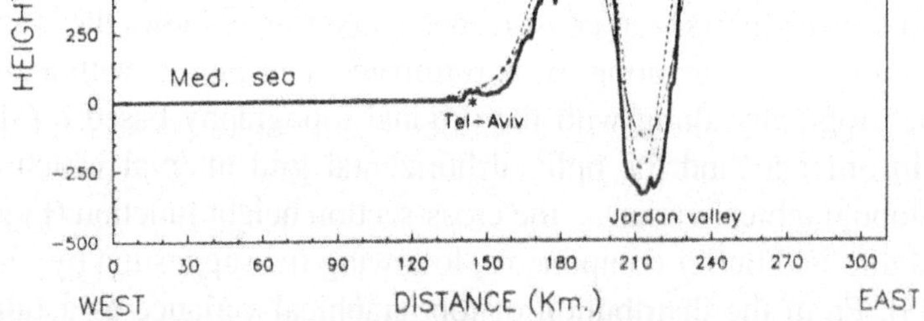

Fig. 3. Topographical cross-section along latitude 32°N (full) by 1 km resolution. The other lines are with 10 km resolution representing: (i) simple average by 10 points (dotted); (ii) as in (i) with smoothing (dashed-dotted); and (iii) as in (i) with the addition of the standard deviation to obtain envelope orography (dashed).

2.5. PBL HEIGHT

The turbulent fluxes in the PBL are calculated through the high resolution parameterization of Zhang and Anthes (1982), who follow the Blackadar (1977) formulation. The study focuses on events with clear days and strong solar radiation so that the two convection criteria are met. These are:

i) $h/L > 1.5$

ii) $Ri_b < 0$

where h is the PBL height, L the Monin-Obukhov length and Ri_b is the bulk Richardson number. Fig. 4 illustrates the vertical structure of the 1-D PBL model, where Z_1 is top of surface layer, Z_2, Z_3, Z_4 etc., are the higher PBL levels at the aforementioned model altitudes, Θ_{vg}, Θ_{va}, are the virtual potential temperatures of the ground surface and the lowest model level respectively. The PBL height h is then calculated through the thermodynamic approach by comparing the positive (P) to the negative (N) buoyancy regions, see Fig. 4. The height h defined in this manner is slightly higher than the zero buoyancy height or the level of most negative heat flux (other definitions that are commonly used for defining the mixed-layer height). The current approach does not neglect the turbulent entrainment at the top of the PBL, but assumes a constant entrainment coefficient (E=0.2) as discussed in Anthes *et al.* (1987). Stull (1988) pointed out that this turbulent entrainment accounts for about 10-20% of the observed variation of the mixed-layer depth.

3. Model Results –Summer Case

3.1. SYNOPTIC BACKGROUND

During summer, the eastern Mediterranean is dominated by a Subtropical ridge extending from the north-African shores to the east, and by the Persian trough extending from the monsoonal low through the Persian Gulf to the northeast Mediterranean and Turkey, e.g. Alpert *et al.* (1991). Fig. 5a, b show the 1000 and 500 HPa ECMWF analyzed charts for such a typical summer case on the 17 June 1987, 12 UTC. The Persian trough induced a northerly-northwesterly synoptic near-surface wind flow (Fig. 5a,5b) slightly weaker than the summer average along with a 500 HPa (not shown) upper-level weak trough to the east. Fig. 6 shows the vertical cross-section along latitude 32° N based on the ECMWF initialized dataset showing significant downward motion due to upper-level subsidence. The downward motion which persists through the summer is believed to be responsible for the quite steady inversion or stable layer found at an average altitude of 764 m (Dayan *et al.*, 1988). Although the Bet-Dagan radiosonde for 17 June 1987 12 UTC (14 LST), Fig. 7, does not show the common summer inversion, a relatively stable layer (1000-2000 m) can be noticed which is associated typically with drier air aloft the mixed layer. It was a clear sunny day with near-surface maximum temperatures of 29-30°C.

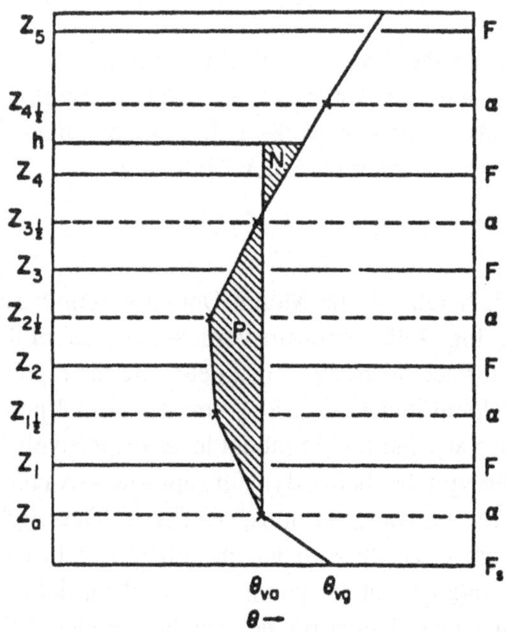

Fig. 4. The vertical structure of the 1-D PBL model where Z_1 is top of the surface layer, Z_2, Z_3, Z_4 etc. are the higher PBL model levels. The virtual potential temperatures of the ground and the lowest model level are Θv_g, Θv_a respectively. The shaded regions P and N represent positive and negative buoyancy layers respectively, that are associated with a parcel of air originating at z rising to h, the top of the PBL. Prognostic variables denoted by $\tilde{A}$ are calculated at half-levels (surface, 1.5, 2.5...) while fluxes (F) are computed at the levels 1,2 etc. For further details see Anthes *et al.* (1987).

3.2. Model Simulation

The simulation started on 06 UTC, and the large thermal forcing at this time led to a relatively short adjustment period (Segal *et al.*, 1985) of only about a couple of hours. Fig. 8a shows the surface winds after 6h, at 12 UTC, indicating that the eastern Mediterranean (EM in brief) region is dominated by a westerly-northwesterly 5-7 m/s flow as observed, (the observed wind vectors are also shown), see Segal *et al.* (1985), Alpert (1988). Above the mountain ridges winds accelerate to about 10 m/s and are associated with the major growth of the PBL height up to about 3000 m as indicated by the h isolines. The upper level PBL flow at 925 HPa –about 750 m above MSL or 100-200 m above the western mountain ridge –shows stronger winds above the mountain ridge; these winds are enhanced by the sea-breeze (SB) front. The SB front is typified by a relatively sharp horizontal temperature gradient (Fig. 8b), and even more interesting is the very sharp drop of the PBL height from 3000 m to less than half (1500 m) through a distance of only one grid point –10 km. To the south, the horizontal PBL height gradient is smaller, and is probably associated with the weaker SB due to the gentler topographic slopes in the south as well as the increased distance of the mountains from the coastline. Some boundary effects cannot be ruled out as well. Another factor enhancing the a

b

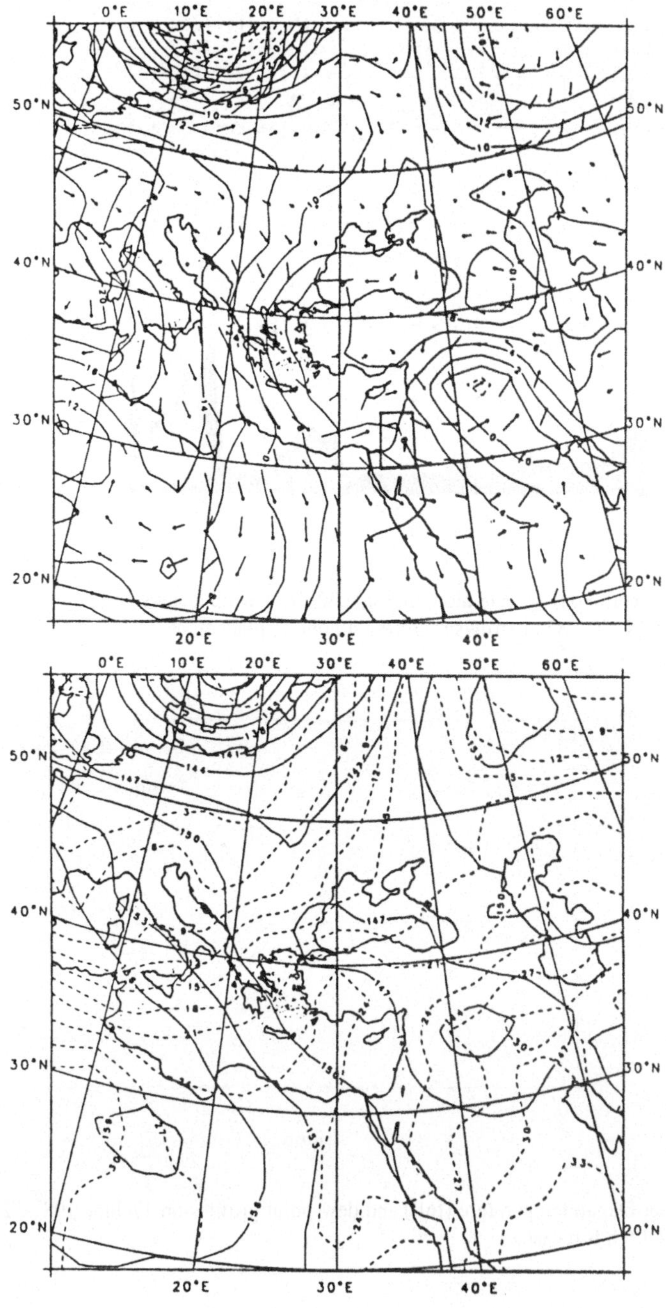

Fig. 5. The 1000 (5a) and 850 (5b) HPa ECMWF analyzed charts for the 17 June 1987 12 UTC. Wind arrows represent 12 hour displacements. Contour interval is 2 and 3 dm for (a) and (b) respectively. In (b), dashed lines are isotherms at $3°C$ intervals. In (a) model domain is drawn.

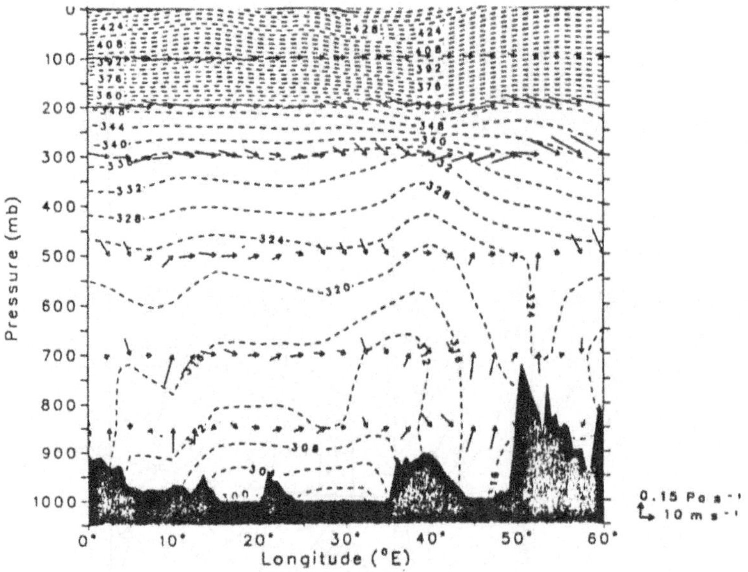

Fig. 6. The vertical cross-section along latitude 32°N of wind arrows and potential temperature on the 17 June 1987 UTC. Wind arrows represent 4 hour displacement.

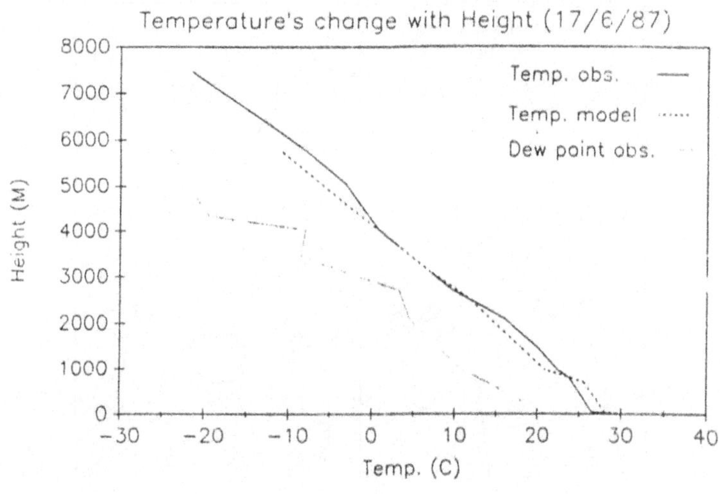

Fig. 7. The Bet-Dagan temperature (full) and dew-point profiles on 17 June 1987 12 UTC. Dashed line is the model 6 h forecast.

SB front is the anabatic flow from the Jordan Valley and Dead Sea which opposes the westerly cooler SB flow as illustrated by Alpert *et al.* (1982).

Special emphasis will be given here to the two warm thermal ridges strongly associated with the mountains oriented south to north. Reaching temperatures as high as 29°C they lead to a 925 HPa isobaric temperature gradient of about 5-6° C over a short distance of about 40-50 km and enhance the SB front (see fig. 8b). Of particular interest and important in the contex of the PBL height, to be discussed later, notice the slight easterly downwind deviation of the thermal ridges from their probable source - the mountains. This easterly shift is more pronounced above the western ridge, where the SB strengthens the synoptic induced winds.

3.3. PBL Height Variations

The spatial cross-sectional variations of the PBL height (along y-axis point 21 in Fig. 8), are shown in Fig. 9 for four times: 08, 10, 12 (corresponding to Fig. 8) and 14 UTC. At 08 UTC, heating has just started, and the PBL height nearly follows the topography, except in the Jordan Valley between the mountain ridges where the PBL is thinner (400 m). As the SB penetrates land at 10 UTC (dashed line) and the inland convection increases, a drastic change in PBL height occurs. At the coast (point A), the PBL height dropped by 50% to 600 m, while over the mountain (point B) and the Valley (point C), the opposite happens, i.e., the PBL height increases sharply to 3000 and1500 m respectively at B and C as compared to those 2 h earlier. At 12 UTC this process continues but much more slowly as the SB front moves inland (from point X=19 to X=20), and the PBL height grows farther to the east. When the surface starts cooling, the PBL height drops through the full cross-sectional line at 14 UTC. The Mediterranean SB penetration into the valley at that time causes the PBL height drop to be particularly significant over the valley - from 3200 to 1500 m within a couple of hours. The abrupt wind increase (5 to 15 m/s) associated with this phenomenon, was discussed by Alpert *et al.* (1982).

The temporal variation at the three points A, B and C is summarized in Fig. 10. The main conclusions follow:

(i) In the coastal region (point A), the PBL height drops early in the morning as the SB penetrates while in the other regions it continues to grow with the enhanced convection.

(ii) At 1200 UTC the SB reaches the mountain top (point B), causing the PBL height to drop from 2000 to 1400 m in within one hour.

(iii) Over the Jordan Valley (point C), the PBL drop starts only after 1300 and is the sharpest as it occurs in coincidence with the reduction in daily heating.

4. Discussion and Conclusions

The depth of the mixed layer over complex topography is strongly influenced by the mesoscale circulations and their interaction with the synoptic forcing as well as the quite well-known factors over flat infinite homogeneous terrain i.e. the surface/upper-level heat fluxes, and the lapse rate. In the present paper, special emphasis is given to the interaction between the mesoscale SB circulations over topography and the large-scale forcings in two synoptic situations with reversed large-scale flow. That is the reason that we also did a 'winter case' simulation with easterly synoptic flow (not shown here). The following general conclusions from both simulations can be drawn. First, in the coastal region, the PBL is not

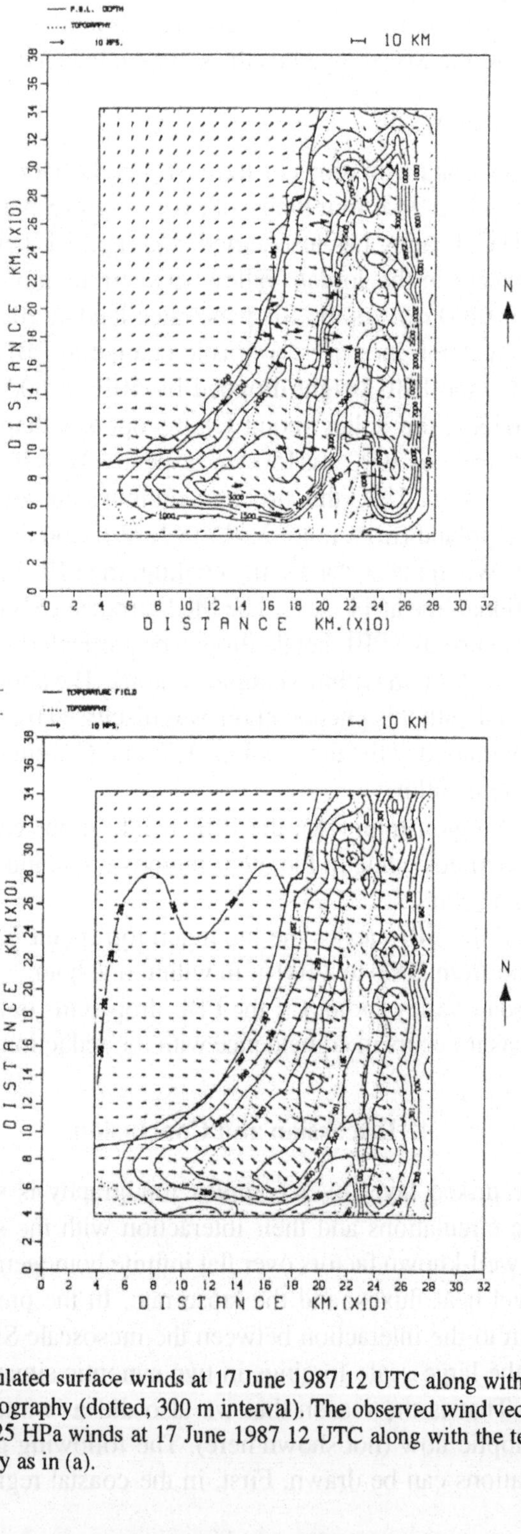

Fig. 8. (a) Model-simulated surface winds at 17 June 1987 12 UTC along with PBL height, h (full 500 m interval) and topography (dotted, 300 m interval). The observed wind vectors are also shown. (b) Model-simulated 925 HPa winds at 17 June 1987 12 UTC along with the temperature (full, 1K interval) and topography as in (a).

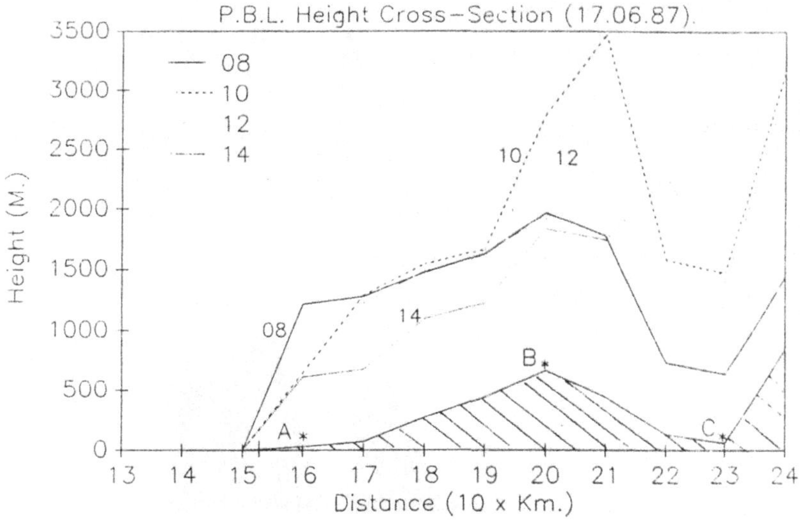

Fig. 9. The spatial cross-sectional variations of the model PBL height (m) along latitude 21 of Fig. 8 for 08 (full line), 10 (dashed), 12 (light dotted) and 14 (heavy dotted) UTC.

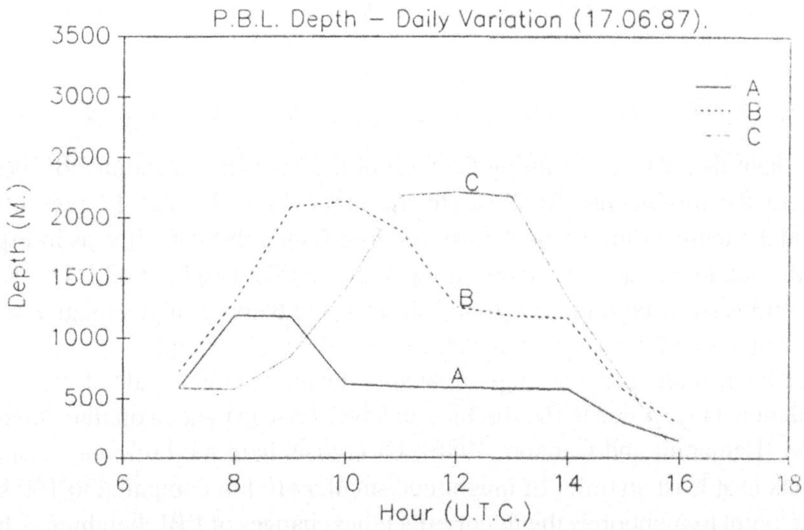

Fig. 10. The time variation of the model PBL height (m) at the three points A (full), B (dashed) and C (dotted). Location of the points is shown in Figs. 2 and 9.

as deep as it is inland and it tends to become even shallower at about 11 LST. In the inner regions the maximum PBL occurs about 3 h later. Second, the thermal ridges over the mountains become convergent lines for the horizontal flow. At the same time, being elevated mixed layers, they contribute to the sharp increase of the PBL height. The locations of the upper level thermal ridges are strongly affected by the interaction of the large-scale flow with the SB, and this becomes

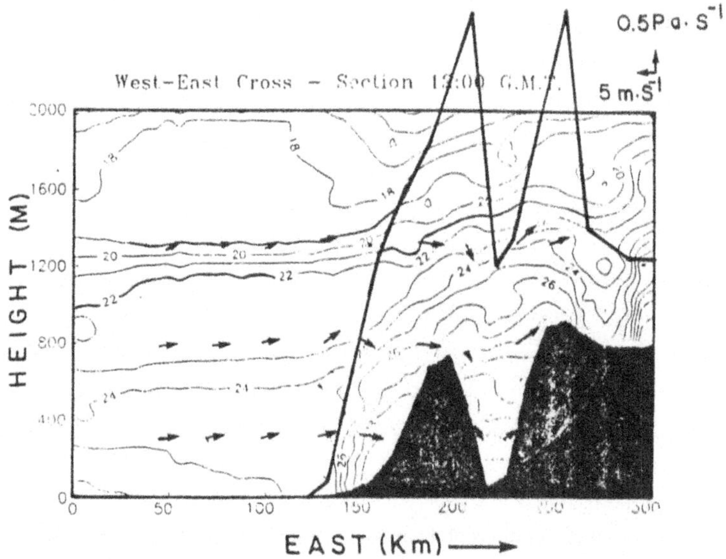

Fig. 11. the vertical cross-sections of the model simulated temperature (thin full line, 1°C interval) and wind arrows on 17 June 1987 12 UTC. Heavy full line shows the PBL height variation.

the dominant factor in determining the shift of the lines of maximum PBL height relative to the mountains. To illustrate the latter, Figs. 11 and 12 present the wind and the temperature vertical cross-sections (along the same line as in Fig. 6) during the summer and winter cases along with the PBL height at 12 UTC. These cross-sections show both the horizontal shifts of the thermal ridges relative to the mountain peaks and the associated dislocations of the PBL height.

These latter findings are strongly related to the much larger-scale effects of the elevated mixed layers due to the Rockies and the Mexican plateau on thunderstorm initiation (Benjamin and Carlson, 1986), Of course, here the horizontal scale of the effects is at least an order of magnitude smaller (10 km compared to 100 km). Another point to mention is the major effect that changes of PBL height may have on the dispersion of air-pollution. In this context we should mention recent results obtained during the recent SCCCAMP (South-Central Coast Cooperation Aerometric Monitoring Program) of lidar observations of mixing-layer over complex terrain by McElroy and Smith (1991). Their study area over southern California is very similar to that of our present work, both in the terrain and in the synoptic forcing, and their findings are also very reminiscent of ours. In particular, Fig. 13 which is Fig. 8 from McElroy and Smith (1991), shows how the mixing layer aloft generated by the mountain, strongly influences the PBL height and consequently the dispersion of air pollutants, which may be presented in such a coastal area. Fig. 13 should be compared with our simulation result in Fig. 12.

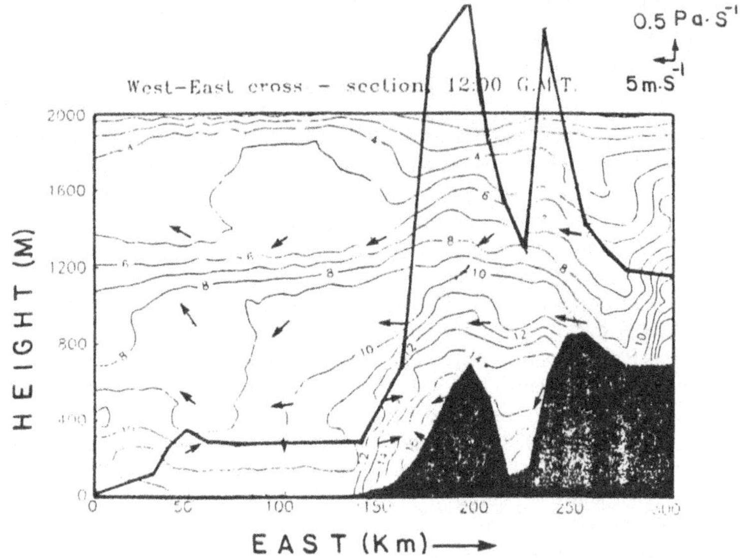

Fig. 12. As in Fig. 11 but for the winter case.

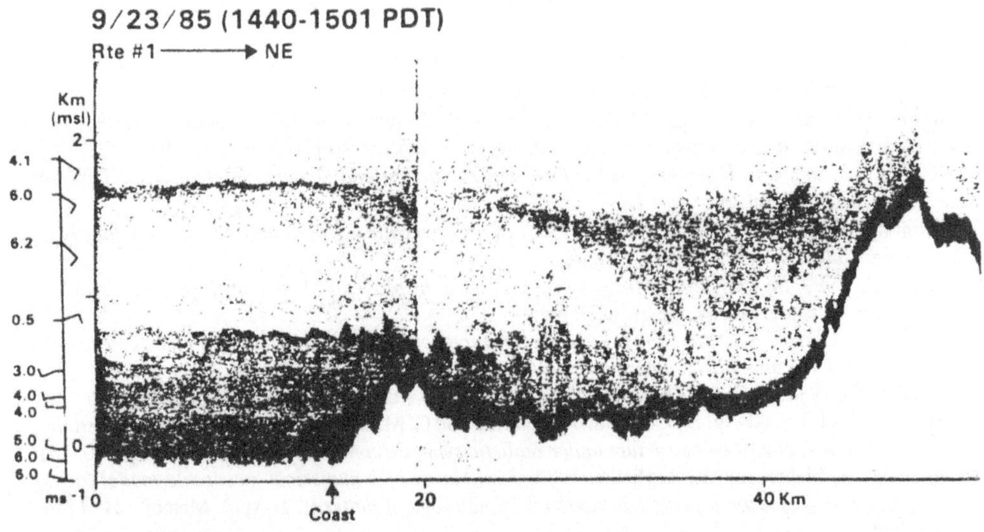

Fig. 13. Grey-scale representation of lidar backscatter signal return (at 0.532 $\bar{I}$m) for 23 September 1985, 1440-1501 PDT, with the wind sounding for station NDT (Pt. Mugo, South California) at 1459 PDT. From McElroy and Smith (1991, Fig. 8).

Acknowledgement

Support to Mr. Lieman was provided by the Israel-USA Binational Science Foundation Grant Nos. 8600230 and 8900186. Data for the model simulations is from the ECMWF and IMS (Israel Meteorological Service). Thanks to NCAR and B. Kuo for support in the adaptation of the PSU/NCAR model at TAU. Thanks to

Rachel Duani for typing and to A. Dvir for helping in drafting the figures.

References

Alpert, P., A. Cohen, J. Neuman and E. Doron, 1982. *A model simulation of the summer circulation for the Eastern Mediterranean past Lake Kinneret in the Jordan valley*. Mon. Wea. Rev. **110**, 994–1006.

Alpert, P., 1988. *The combined use of three different approaches to obtain the best estimate of meso-A surface winds over complex terrain*. Bound. Layer Meteor., **45**, 291–305.

Alpert, P. and G. Getenio, 1988. *One level modelling for diagnosing surface winds over complex terrain*. Part I: *Comparison with a 3-D modelling in Israel*. Mon. Wea. Rev. **116**, 2025–2046.

Alpert, P., B. Getenio and R. Rosental, 1988. *One level modelling for diagnosing surface winds over complex terrain*. Part II: *Applicability to short range forcasting*. Mon. Wea. Rev. **116**, 2047–2061.

Alpert, P., Abramski R. and Neeman B.U., 1991. *Persian trough or subtropical high –the prevailing summer synoptic system in Israel*. Israel J. of Earth Sci., (in press).

Anthes, R. A. and T.T. Warner, 1978: *Development of hydrodynamic models suitable for air pollution and other mesometeorological studies*. Mon. Wea. Rev., **106**, 1045–1078.

Anthes, R.A., E.-Y. Hsie and Y.-H. Kuo, 1987. Description of the Penn State/NCAR mesoscale model version (MM4)., 66 pp.

Benjamin, S.G., and T.N. Carlson, 1986. *Some effects of surface heating and topography on the regional severe storm environment*. Mon. Wea. Rev., **114**, 307–329.

Cressman, G. P., 1959. *An operative objective analysis scheme*. Mon. Wea. Rev. **87**, 367–374.

Dayan, U., R. Shenhav and M. Graber, 1988. *The spatial and temporal behavior of the mixed layer in Israel*. J. Appl. Meteor. **27**, 1382–1394.

Deardorff, J. W., 1978. *Efficient prediction of ground surface temperature and moisture with inclusion of a layer of vegetation*. J. Geophys. Res. **83**, 1889–1903.

Goodin, W.R., G.J. McRae and J.H. Seidenfeld, 1979. *A comparison of interpolation methods for sparse data: Application to wind and concentration fields*. J. Appl. Meteor., **18**, 761–771.

Haltiner G.J. and R.T. Williams, 1980. *Numerical Prediction and Dynamic Meteorology*. J. Wiley, New York, 477 pp.

Hasmonay, R., Dayan U. and Cohen A., 1991. *Lidar Observation of the atmospheric boundary layer in Jerusalem*. J. Appl. Meteor., (in press).

Lieman R., 1990. *Investigation of the mesoscale flow in the planetary boundary layer over Israel* – a numerical study. (M. Sc. thesis, in Hebrew).

McElroy J.L., and T.B. Smith, 1991. *Lidar descriptions of mixing-layer thickness characteristics in a complex terrain/coastal environment* J. Appl. Met., **30**, 585–597.

Pielke, R.A., 1984. *Mesoscale Meteorological Modelling*, Academic Press, New York, 612 pp.

Seaman N.L., F.L. Ludwig, E.G. Donall, T.T. Warner and C. M. Bhumralkar, 1989. *Numerical studies of urban planetary layer structure under realistic synoptic conditions*. J. Appl. Met. **28**, 760–781.

Segal, M., Y. Mahrer and R. A. Pielke, 1982. *Application of a numerical mesoscale model for the evaluation of seasonal persistent regional climatological patterns*. J. Appl. Meteor., **21**, 1754–1762.

Segal, M., Y. Mahrer, R.A. Pielke and R.C. Kessler, 1985. *Model evaluation of the summer daytime induced flows over southern Israel*. Isr. J. Earth. Sci., **34**, 39–46.

Stull, R.B., 1988. *An introduction to boundary layer meteorology*, Kluwer publ. 666 pp.

Young, G. S. and R. A. Pielke, 1983. *Application of terrain height variance to Mesoscale Modelling*. J. Atmos Sci., **40**, 2555–2560.

Zhang. D. and R. A. Anthes, 1982. *A high resolution model of the planetary boundary layer. Sensitivity tests and comparisons with the SESAME-79 data*. J. Appl. Meteor., **21**, 1594–1609.

THE IMPORTANCE OF MESOSCALE PHENOMENA ON THE DISPERSION OF POLLUTANTS IN THE VICINITY OF A POINT-SOURCE EMISSION

(RESEARCH NOTE)

L. RANTALAINEN
Imatran Voima OY, Vantaa, Finland

(Received October 1991)

Abstract. The effects of mesoscale meteorological phenomena on the transport of pollutants were analyzed by 3-dimensional dispersion calculations in case of a hypothetical nuclear accident. The results were in agreement with the earlier results obtained by Kesti *et al.* (1988) clearly indicating the importance of mesocale meteorological phenomena in short-range dispersion of pollutants.

1. Introduction

The majority of the bigger fossil fuel plants and both the nuclear power plants in Finland are located in the coastal area of the southern part of the country. The transport of the emissions of the fossil fuel plants and the releases of radioactive material of the nuclear power plants into the atmosphere are normally estimated by conventional straight-line dispersion models. The mesoscale wind and rain fields in Finland were analyzed by Kesti *et al.* (1987). On the basis of these data the mesoscale meteorological phenomena were analyzed by a 3-dimensional trajectory model in the vicinity of the release points by Kesti *et al.* (1988). The results indicated that mesoscale phenomena were significant for short range dispersion. The present study attempts to elaborate and justify these results by modelling a hypothetical nuclear accident using the same meteorological data used by Kesti *et al.* (1987).

2. Calculation Models

Two calculation models TUULET and 3DIM were used to estimate the dispersion.

2.1. TUULET CODE

The TUULET code (Saikkonen, 1991) is a straight-line Gaussian plume model where a finite 3-dimensional space gamma cloud integration is performed by a rapid line convolution method, originally generated by Vuori (1978). The TUULET program is tested against other codes presented in Vuori (1978) and Rantalainen (1988), and the results of the TUULET code are found to be the same within a tolerance of ±20%.

Boundary-Layer Meteorology **62:** 143–150, 1993.

2.2. 3DIM CODE

The 3DIM code was generated in Imatran Voima Oy during the years 1983-86. No publication is available from the 3DIM code. The model is a quite simple 3-dimensional Gaussian puff model, which calculates the concentration of each nuclide and makes the dose-rate and dose estimates. The 3-dimensional wind data was assessed by 3-component wind data at four heights. Rain data were two-dimensional, assuming that the rain is not height dependent.

The 3DIM code uses a Cartesian coordinate system. The horizontal grid point interval is fixed at 250 m. The square grid covers an area of 20 km x 20 km. The internal time step interval of calculation was one minute.

The effects of turbulence were estimated according to the dispersion coefficients for Pasquill's stability classification (Slade, 1969).

The washout factor Λ was supposed to be the same for all nuclides except for the noble gases, which were not washed out.

$$\Lambda[s^{-1}] = 10^{-4} \cdot \{I[mm/h]\}^{0.6} \tag{1}$$

where I is the precipitation rate.

The constants of the washout factor were selected according to Puhakka *et al.* (1988) and Jylhä (1991) after the Chernobyl accident.

Dry deposition was estimated by assuming that the deposition rate is dependent on the concentration of the nuclides in the air at ground level multiplied by the constant deposition rate of 10^{-2} m/s for the nuclides.

The plume dilution was estimated in standard ways, so that the radioactive decay of each nuclide was taken into account, as well as the purification of the plume caused by dry and wet deposition.

The gamma cloud shine is calculated by a simple semi-infinite cloud model. The external exposure is estimated without using shielding factors (the value of the shielding factor was one).

The doses were assessed from external exposure caused by the plume, by the activity deposited on the ground, and by the internal exposure caused via inhalation.

The temporal variation of the input wind and rain data was taken into account by updating the data every 15 minutes. Four layers were used; 0-100, 100-500, 500-1100 and greater than 1100 m. No interpolation in space or time was used.

The 3DIM code has been tested by computing the dose rates and doses in the case of a constant activity release rate and with homogeneous wind and rain fields. Those estimates have been compared with the estimates of the TUULET program. According to the test runs, the 3DIM code gives results of reasonable accuracy. The centre-line dose rates for a plume transported at a height of 60 m were conservative by factor of two.

3. Emission of Radioactive Material

The emission data were selected from a hypothetical nuclear accident type of FK2 according to DRK,1979, modified for the Loviisa LWR Nuclear Power Plant (NPP). The thermal power of the NPP is 1375 MW, which determines the activity of the reactor core and thus the potential release amount.

The selected accident is severe. It was assumed that the release started very soon after the shutdown of the reactor, and the duration of the release was three hours. The release fraction of the total inventory in per cent of reactor core was: noble gases 100, iodine 40, cesium 29 and strontium 3. The probability of a severe accident of this kind in the Loviisa NPP has not been analyzed via probabilistic safety analyses but is certainly extremely small because of the many redundant systems ensuring the safety of the NPP, including full containment with an ice condenser. This very severe accident was selected to clarify the effects of mesoscale dispersion, because in a more realistic accident, only noble gases would be released and thus the doses would be caused almost entirely by external radiation from the plume.

The release was supposed to escape at a constant rate through a break in the containment wall, so the release point is at a height of 30-60 m. The release plume is rapidly mixed by the wake effect caused by the buildings of the NPP. Because the containment includes an ice condenser, the thermal energy of the plume was supposed to be so small that effectively no significant plume rise took place and the actual release height was about 60 m.

4. Mesoscale Wind and Rain Fields

The mesoscale phenomena considered were as follows: boundary-layer jet stream, roll vortices, sea breeze front, rain showers and a gust front. According to Kaurola *et al.* (1989) the cumulative time probability of these phenomena is 40% in Southern Finland.

Examples of the meteorological phenomenon were analyzed for the southern coast of Finland by Kesti *et al.* (1987). As the Loviisa Nuclear Power Plant is also located on the southern coast, the wind and rain field data taken from Kesti *et al.* (1987) are well suited for the Loviisa area.

The data were introduced into the 3DIM model so that the starting time of the release occured at the time when the wind data of Kesti *et al.* (1987) began for each phenomenon. The temporal variation of the wind and rain field data was taken into account by updating the input data every 15 minutes. In all cases other than the gust front and the roll vortices, the original data of Kesti *et al.* (1987) were used in their original form. Thus the input data were kept constant as long as new data were available.

The boundary-layer jet stream was too large scale to be fully analyzed by the 3DIM code.

The shower situation did not include a complete wind field. Instead, the wind velocity was assumed to be horizontally constant and in a steady state.

The wind field data for the roll vortices were extended in time from one hour to three hours by assuming that the structure of the phenomenon does not change during the time in question. The wind field for the roll vortices was estimated by Kesti *et al.* (1987) to flow from the north at a velocity of 1 m/s. Assuming that the wind field does not change, the extension of the time span in fact led to a situation where the field repeated itself after an hour at neighboring grid points to the south.

A constant turbulence intensity was used at all points of the grid. In all cases the turbulent intensity estimate was made according to Pasquill stability class D. In the case of the gust front, the local turbulence behind the nose of the front was increased by averaging the winds.

Wet deposition was calculated assuming that the rain washed through the whole plume. According to Kesti *et al.* (1987), the lower edge of the rain clouds in the case selected (12.9.1978) was at a height of 900 m and thus this estimate was quite realistic.

5. The Calculation Results

The integration time of the doses was the time of interest. The internal dose is the whole body effective dose equivalent commitment during the following 50 years caused by the activity inhaled during the time duration of 0-3 hours.

Some examples of the results of the 3DIM code and the wind field according to Kesti *et al.* (1987) in the case of a sea breeze are presented in Figure 1, and in the case of roll vortices in Figure 2.

The sea breeze has an interesting effect in the case where the sea breeze front is just above the release point, see Figure 1 (c) at time 45 minutes. The wind speed is low, which should increase the dose rate in the vicinity of the release point compared with the earlier (30 minutes, see Figure 1 (b)) dose rates with a higher wind speed. The vertical wind lifts up the plume so that the dose rate is as a whole almost unaffected.

The dose rate after 1 hour is almost the same as after 3 hours in the case of the roll vortices, because the roll vortices keep their structure almost unchanged during the time. They flow to the south at a speed of only about 3 km/h.

The different maximum dose-rates during 0-3 hours and doses after 3 hours from the starting of the release are presented in Table I.

By a factor of two, the results of the 3 DIM code are larger than the results of the TUULET program in the same conditions. This result is caused by the differences between the codes. According to the results of 3DIM, the spatial dose distribution, estimated in a complete wind field, differs remarkably from that estimated using one-point wind data. The maximum dose in the case of a comprehensive wind field, was somewhat smaller than that in the case of one-point wind data. It is uncertain whether this is always the case. In some cases when a mesoscale phenomenon is involved, the doses can be larger than in the case of a homogenous wind field. This

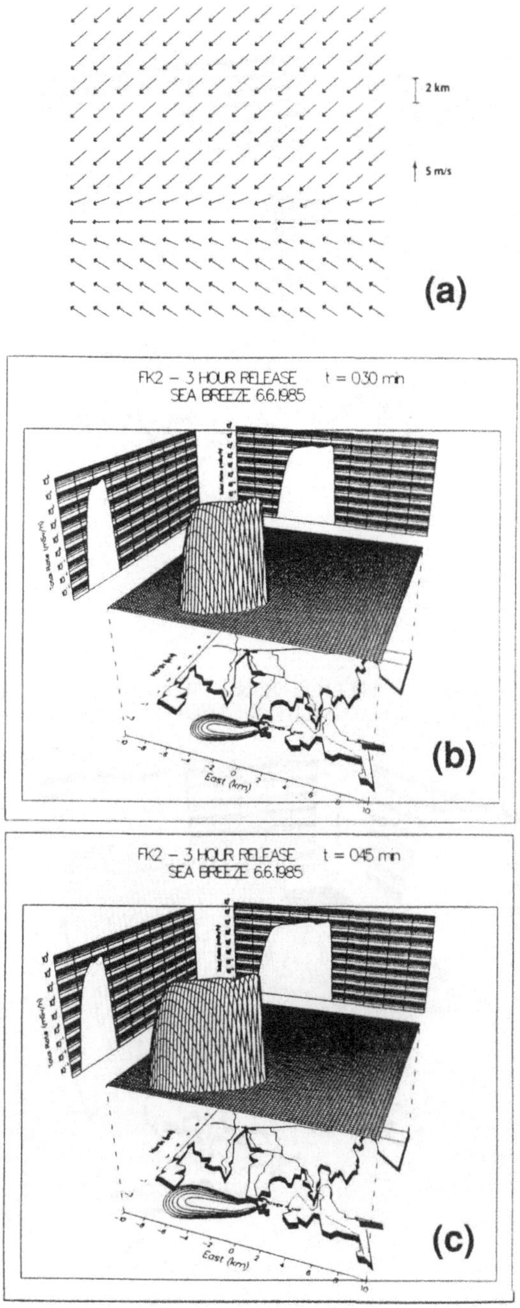

Fig. 1. (a) Sea-breeze wind field at a height of 10m,14.00 local time 6.6.1985. The data were taken from Kesti *et al.* (1987). (b) The dose rate 30 minutes after the start of the release. The release started at 14.00 o'clock at a height of 60m. The release includes 100% of noble gases, 40% of Iodine, 29% of Cesium, and 3% of Strontium of the core. The release rate is constant and the duration of the release is 3 hours. (c) The dose rate 45 minutes after the start of the release. Other data as in Figure 1 (b)

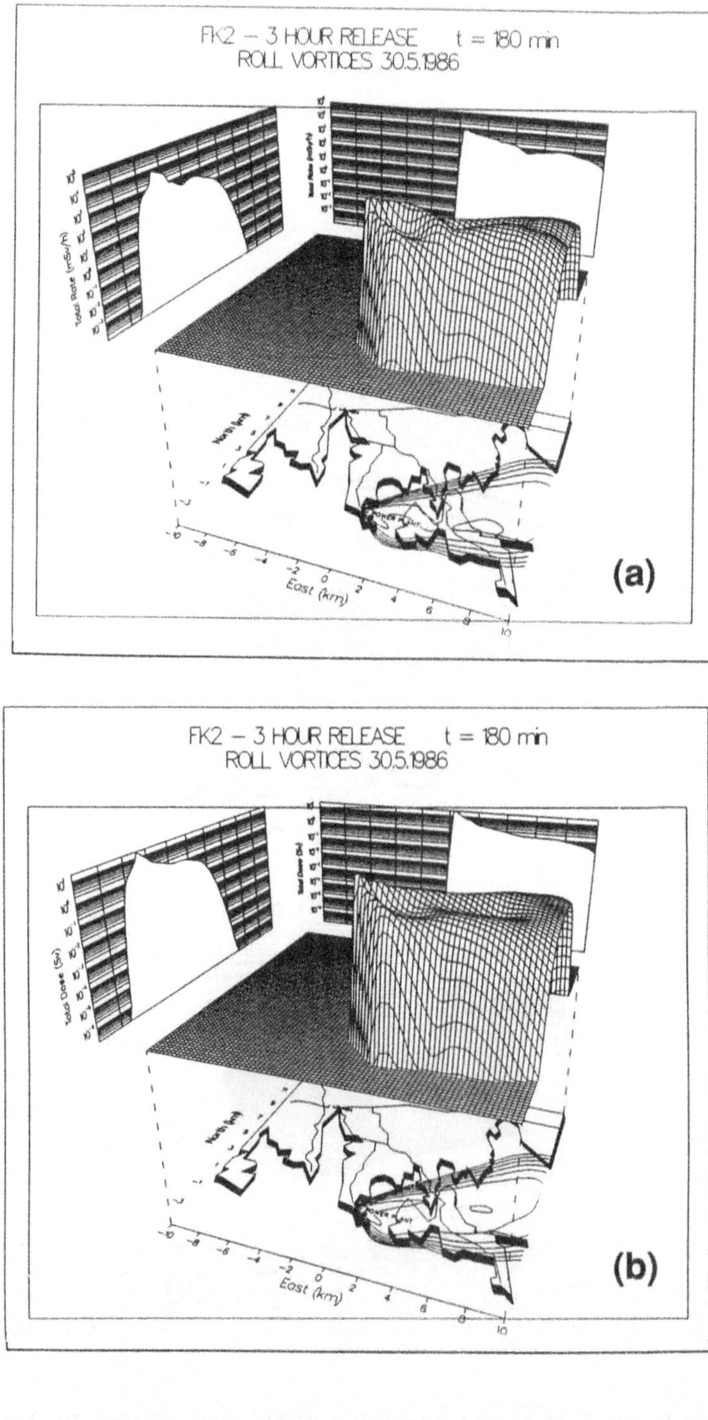

Fig. 2. (a)The dose rate 3 hours after the start of an accident in the case of the roll vortices 30.5.1986. The other data as in Figure 1 (b). (b)The total dose 3 hours after the start of the accident in the case of the roll vortices 30.5.1986. The other data as in Figure 1 (b).

TABLE I

The doses in sieverts (abbreviated as Sv) after 3 h, and dose-rate (Sv/h) during 0-3 h, according to the 3DIM code. Dose assesment according to the Pasquill D-stability class, and wind velocity 5m/s by the TUULET and 3DIM codes have been marked by 'D5, slight rain'. The release takes place during 0-3 h, at a height of 60m, and is affected by the wake of the buildings ($3600m^2$), and consists of 100% of noble gases, 40% of iodine, 29% of cesium, and 3% of strontium of the inventory of the Loviisa Nuclear Power Plant.

Phenomenon	Code	Maximum Dose Rate (Sv/h)	Maximum Dose (Sv)
Roll vortices	3DIM	10	20
Sea-breeze front	3DIM	10	10
Shower situation	3DIM	8	10
Gust front	3DIM	15	10
D5, slight rain	3DIM		20
D5, slight rain	TUULET		9

may take place for instance in a shower situation. The descending air masses in addition to the precipitation can increase the doses locally. On the other hand, the upcurrents in connection with shower clouds can diminish the doses.

It was not possible to analyze this feature because no comprehensive wind field data were available. However, the variation is probably greater in nature than the variations in this study or in the studies of Kesti *et al.* (1988).

The trajectories were already analyzed in Kesti *et al.* (1988) and the differences between the results of the complete wind field estimations and single-point wind estimation were found to be important. When a mesoscale phenomenon occures, the average direction estimate error after one hour of transport was 10 deg estimated on the basis of the data studied by Kesti *et al.* (1988) in the South of Finland.

6. Conclusions

Mesoscale phenomena have a strong effect on the transport of the pollutants. According to Kaurola *et al.* (1989) the probability of these phenomena occurring is relatively high, about 40% in Finland. Thus the effects of mesoscale phenomena should be taken into account in dispersion studies, but this cannot be done without complete wind and rain field data.

The maximum dose rates and doses caused by a nuclear accident were not found to be higher in the case of a mesoscale phenomenon than those estimated in the case of a homogeneous wind and rain field, but there were doubts whether this was perhaps not always the case in nature.

According to this study and those of Kesti *et al.* (1988) and Kaurola *et al.* (1989), it seems important to recognize situations where mesoscale phenomena may be involved, and to try to take them into account when estimating dispersion.

References

Deutche Risikostudie Kernkraftwerke- (1979): *Eine Untersuchung zu dem durch Stofälle in kerkraftwerken verursachten Risiko*. Bonnr, der Bundesminister fur Forschung und Technologie. 262 p.

Jylhä K., 1991: *Empirical scavenging coefficients of radioactive substances released from Chernobyl*, Atmospheric Environment, vol. 25A, No. 2, 263–270.

Kaurola J., Koistinen J., Leskinen M., Puhakka T. and Saarikivi P., 1992: *The Effect of Mesometeorological Factors on the Dispersion of Pollutants, Especially on the Coast of the Gulf of Finland, Part III, Abundance of Mesoscale Phenomena in the Southern Part of Finland and the Possibilities of Doppler-radar to Detect the Phenomena*, University of Helsinki, Department of Meteorology, Helsinki 1989, in Finnish, to be published in English.

Kesti P., Jylhä K., Koistinen J., Puhakka T. and Saarikivi P., 1988:*The Effect of Mesometeorological Factors on the Dipersion of Pollutants, Especially on the Coast of the Gulf of Finland, Part II, Estimation Errors Caused by Mesoscale Phenomena*, University of Helsinki, Department of Meteorology, 58 p., in Finnish, to be published in English.

Kesti P., Puhakka T., Jylhä K., Koistinen J. and Saarikivi P., 1987: *The Effect of Mesometeorological Factors on the Dispersion of Pollutants, Especially on the Coast of the Gulf of Finland, Part I, the Wind and Rain Fields Associated with Some Mesoscale phenomena*, University of Helsinki, Department of Meteorology, 54 p., in Finnish, to be published in English.

Puhakka T., Jylhä K., Saarikivi P., Koistinen J. and Koivukoski J., 1988: *Meteorological Factors Influencing the Radioactive Deposition in Finland after the Chernobyl Accident*, Report 29, Department of Meteorology, University of Helsinki, 49 p.

Rantalainen L., 1988: *The external dose caused by the release of radioactive noble gases from the nuclear power plant*, AMAPI, Thesis for Lic. Tech., Helsinki University of Technology, 118p.

Saikkonen T., 1991: *A calculation model TUULET to estimate the doses caused by the radioactive releases*, Imatran Voima Oy, DY1-G710-0035, In Finnish.

Slade D. H., Editor, 1969: *Meteorology and atomic energy*, U.S. Atomic Energy Commission, Division of Technical Information, Oak Ridge, 445 p.

Vuori S., 1978: *Fast Correction of Cloud Dose data Files Due to Chance in Dispersion Parameters*, Health Physics, vol. 34, June, Pergamon Press Ltd., 727–730.

A NUMERICAL MODEL FOR ESTIMATION OF THE DIURNAL FLUCTUATION OF THE INVERSION HEIGHT DUE TO A SEA BREEZE

YIZHAK FELIKS
Israel Institute For Biological Research P.O.B. 19 Ness-Ziona, Israel

(Received October 1991)

Abstract. Vertical profiles of temperature, measured over the sea in the summer near the Eastern coast of the Mediterranean, show significant diurnal fluctuation in the height of the marine inversion. During the day, the inversion moved down and during the night it moved up. The fluctuation was about 250 m. Numerical simulations of the daily fluctuation in the height of the inversion during the summer day shows the following: Over the sea, during daytime, the inversion base sinks by 250 m, and during the night, it rises back to its original height. The developing sea breeze during the day causes the air over the sea to move downward adiabatically. At night, the inversion rises mainly due to advection of cool stratified air (including an inversion at 480 m) from a long distance over the sea. Such diurnal fluctuations are observed 100 km off shore. This scale is determined by the scale of the sea breeze. Comparison of some of the model vertical profiles with the temperature profiles measured over the sea show a similar diurnal oscillation. The amplitude of the oscillation is the same.

1. Introduction

An inversion behaves like a lid on the atmosphere below it, and so prevents upward motion. This results in a large concentration of pollution and water vapor in the atmosphere below the inversion, and large gradients in the inversion. The concentration below the inversion base increases as the base of the inversion drops.

The Eastern Mediterranean is affected by a subtropical anticyclone during most of the year. In the summer the anticyclone moves northward. Thus the Eastern Mediterranea is closer to its center. As the air in the anticyclone descends, it is heated adiabatically. This causes a subsidence inversion in the middle and lower layers of the atmosphere. During the summer the subsidence is strong and persistent. This results in strong inversions (low base, great depth and large negative lapse rate). The daily inversions over the coast of Israel have been studied by Shaia and Jaffe (1976). Their analysis was based on 10 years of observations of temperature profiles measured by the afternoon radiosond (12 GMT) at Bet Dagan (7 km inland from the central coast of Israel). According to their statistics, inversions occurred on 81% of summer days (June-August). The base height of most of these inversions was between 500–1000 m and their mean thickness was about 400 m. In some of the inversions, the lapse rate was -5^0C to -7^0C per 100 m.

Due to lack of measurements nothing is known about the daily fluctuation in the marine inversion over the sea.

In the summer of 1987 two series of measurements over the Eastern Mediterranean near the Israeli coast were conducted. Vertical profiles of temperature, humidity and velocity were taken every four hours. A detailed description of the measurements is given in Barkan and Feliks (1991). In these profiles a prominent diurnal oscillation in the height of the marine inversion was observed. During the daytime the inversion moved adiabatically downward, reaching its lowest altitude in the afternoon. During the night the inversion moved upward, the total vertical movement being about 250 m.

In the summer, in coastal regions, sea and land breezes were frequently observed. Many studies have been devoted to simulating sea and land breezes by analytical and numerical models; for a comprehensive review see Atkinson (1981). In some of the models only a very weak inversion at a high altitude was introduced in the initial condition. The evolution of the inversion such as its diurnal oscillation, caused by the breeze, was not studied at all. In the following we utilize a 2-dimensional numerical model to study the diurnal oscillation of the marine inversion under the influence of diurnal sea and land breezes.

2. The Model

The numerical model used in this work is the model of Huss and Feliks (1981). Only a brief description is given here.

2.1. Model Equations

The horizontal components of the equations of motion for the levels are:

$$\frac{\partial u}{\partial t} = -V\nabla u + fv - \frac{1}{\rho}\frac{\partial p}{\partial x} + \frac{\partial}{\partial z}(K_m\frac{\partial u}{\partial z}) + K_h\frac{\partial^2 u}{\partial x^2} \tag{1}$$

$$\frac{\partial v}{\partial t} = -V\nabla v - fu - + \frac{\partial}{\partial z}(K_m\frac{\partial v}{\partial z}) + K_h\frac{\partial^2 v}{\partial x^2} \tag{2}$$

where $\nabla = (\frac{\partial}{\partial x}, \frac{\partial}{\partial z})$, V=(u,w), f is Coriolis parameter at $32^0 N$, K_m and K_h = $10^8 cm^2$/s are the vertical and horizontal eddy diffusivity coefficients.

The third component of the equation of motion is replaced by the hydrostatic approximation:

$$\frac{\partial p}{\partial x} = -\rho g \tag{3}$$

where g=981cm/s^2, ρ is the air density.

The incompressible continuity equation is assumed.

$$\frac{\partial u}{\partial x} + \frac{\partial w}{\partial z} = 0. \tag{4}$$

The thermodynamic energy equation for air is written in terms of potential temperature θ

$$\frac{\partial \theta}{\partial t} = -V\nabla\theta + \frac{\partial}{\partial z}(K_e\frac{\partial \theta}{\partial z}) + K_h\frac{\partial^2 \theta}{\partial x^2} \tag{5}$$

where Ke is the vertical eddy exchange coefficient for heat.

The equation of state in the atmosphere is:

$$\rho = \frac{p}{RT} = \frac{p_0^{\frac{R}{c_p}} p^{\frac{c_v}{c_p}}}{R\theta} \tag{6}$$

in the usual notation.

2.2. Exchange Coefficients

The closure of the model equations requires an explicit formulation of the exchange coefficients in terms of the dependent variables. The formulation adopted follows Pandolfo et al. (1973). We assume three stability regimes, which are identified by the local Richardson number, i.e.

$$R_i = \frac{\frac{g}{\theta}\frac{\partial\theta}{\partial z}}{(\frac{\partial V_h}{\partial z})^2}. \tag{7}$$

We adopt the concept of mixing length, without insisting on its physical content.

$$l = k(z + z_0)(1 + 4(\frac{z + z_0}{z_m})^{1.25})^{-1} \tag{8}$$

where z_m is the altitude at which l attains its maximum. According to an observational study by Lettau (1962) we choose $z_m = 250$ m. The roughness length z_0= 1cm over land and 0.1cm over sea.

The exchange coefficients in the 3 regimes are:

$$0 \leq R_i \leq \frac{1}{\mid\alpha\mid} : K_m = K_e = l^2(1 + \alpha R_i)^2 \mid \frac{\partial V_h}{\partial z} \mid \tag{9}$$

$$R_{i_c} \leq R_i \leq 0 : \quad \begin{aligned} K_m &= l^2(1 - \alpha R_i)^{-2} \mid \tfrac{\partial V_h}{\partial z} \mid \\ K_e &= K_m(1 - \alpha R_i)^{-1} \end{aligned} \tag{10}$$

$$R_i \leq R_{i_c} : \quad \begin{aligned} K_e &= \tfrac{l^2}{k^2} h \mid \tfrac{g}{\theta}\tfrac{\partial\theta}{\partial z} \mid^{\frac{1}{2}} \\ K_m &= A \mid R_i \mid^{-\frac{1}{6}} K_e \end{aligned} \tag{11}$$

where h=0.9 is Priestley's heat convection constant, k=0.4 von Karman's constant, α=-0.3, A=0.56, Ri_c=-0.16.

2.3. Boundary and Initial Conditions

The wind components u, v and w are assumed to vanish at the sea and land surface. At the upper boundary u and v are taken to be equal to the specified geostrophic wind. The upper boundary is assumed to be a material surface. Its height, h, is given by

$$\frac{\partial h}{\partial t} = w(z = h). \tag{12}$$

The potential temperature and pressure at the upper boundary are assumed to be constant. At the sea surface θ was constant during the integration. On the land surface θ was prescribed as a function of time:

$$\theta = 22 + 12sin(15t - 120), \tag{13}$$

a simple harmonic component with a period of 24 hours (t is in hours).

At the lateral boundaries for inflow, the normal derivative is assumed to be zero. For outflow, let A be one of the predicted variables u,v,h or θ, then it is assumed that

$$\frac{\partial A}{\partial t} = -c\frac{\partial A}{\partial x} \tag{14}$$

where c=150 m/s is the phase velocity of external gravity waves in the domain.

In the following numerical simulation the initial condition of horizontal homogenity was assumed, since only one vertical profile was available to us every 4 hours. The vertical profile was taken from the measurments at 8:00 (Fig. 6a). In this profile a significant inversion observed. Its base is at an altitude of 480 m, and its thickness is 500 m. The lapse in the inversion was -3.5^0C per 100 m. A 1-dimensional version of the numerical model was run until a steady state was reached. The resulting profile was taken as the initial condition for the model.

2.4. Numerical Aspects

The equations are applied to a grid contained within a vertical plane extending in the x-coordinate perpendicular to the assumed coast line. A staggered grid is used in the horizontal coordinate. u,v are computed on the grid points, θ, p, w and h are computed on the staggered grid points. The horizontal grid interval was taken as $\Delta x = 5\ km$ and 160 grid points including the boundaries are assumed at each level. 18 levels are taken at heights of 0,2.5, 5, 10, 20, 40, 80, 160, 320, 480, 640, 760, 960, 1280, 1600, 1920, 2240, 2560 m. A forward time scheme is used to approximate the time derivative. The "component by component splitting method" is used to solve the prognostic equations (Marchuk, 1975). Horizontal advection terms are approximated by the Lax-Wendroff scheme, and horizontal diffusion is solved explicitly. Vertical diffusion and advection are solved implicitly to allow large time steps. u,v,w, θ and p are computed at the 18 levels while K_e, K_m and R_i are computed at the midlevels. The time interval was $\Delta t = 60sec.$

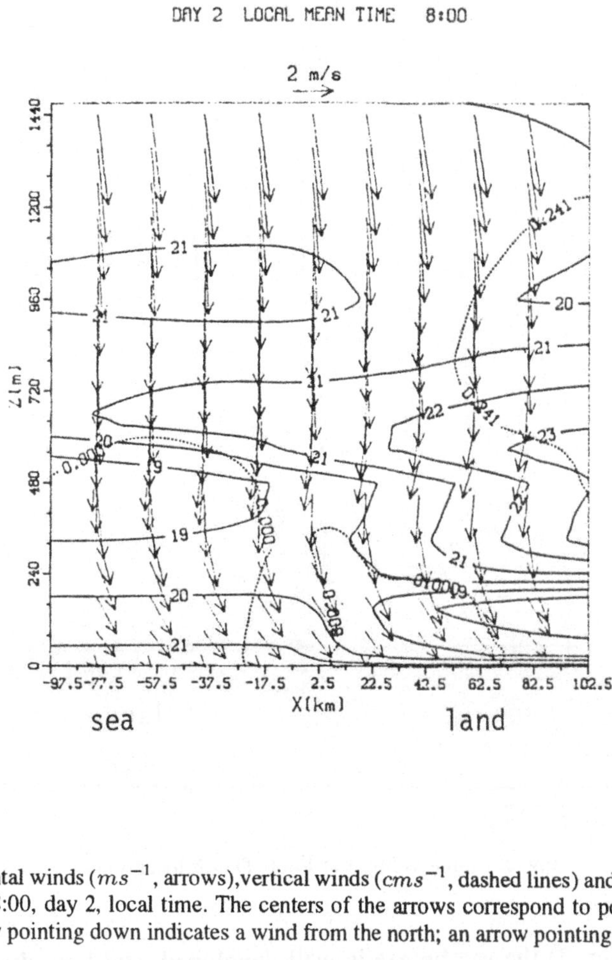

Fig. 1. Horizontal winds (ms^{-1}, arrows),vertical winds (cms^{-1}, dashed lines) and temperature (^0C, full lines), for 8:00, day 2, local time. The centers of the arrows correspond to position in the Z-X plane. An arrow pointing down indicates a wind from the north; an arrow pointing right, winds from the west.

3. The Numerical Experiment

3.1. The Wind Field

The integration started at 8:00 in the morning when the temperature contrast between sea and land is zero, and continued for 48 hours. The general features during the first and the second day of the integration are similar. However, some differences between the two days were observed. In the following we concentrate on the results of the second day since the first day is influenced by the assumed initial conditions of horizontal homogenity. Those initial coniditions are far from the situation we expect to find in a typical summer day in a coastal region.

At 8:00 (Fig. 1), the nocturnal inversion over land is observed up to 250 m. At higher altitudes strong winds can be seen.

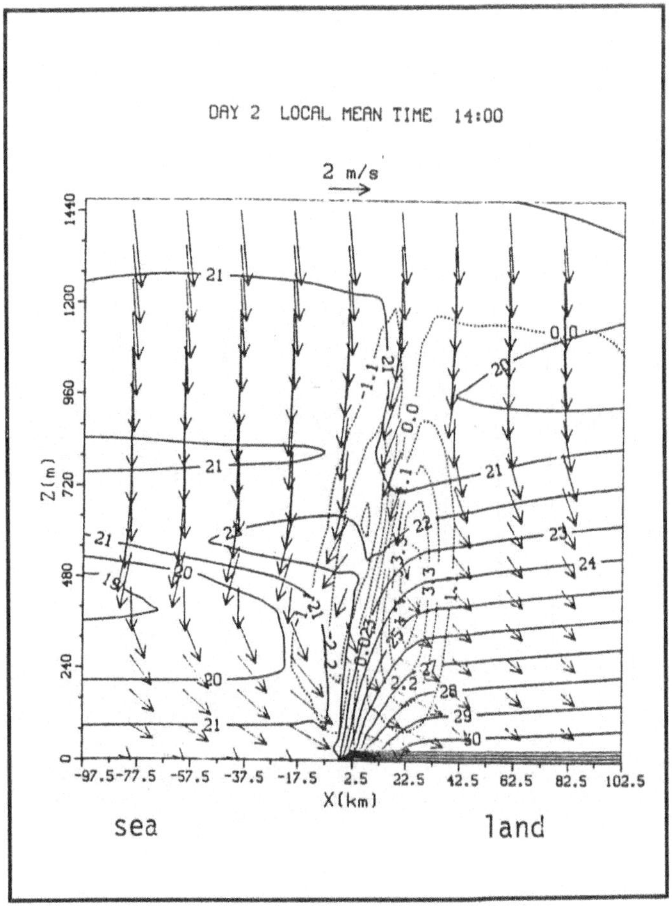

Fig. 2. Same as Fig. 1 at 14:00, Day 2, local time.

At 14:00 (Fig. 2) the sea breeze is well developed, and has advanced 20 km inland. Its front is indicated by the convergence of the horizontal velocity and by the sharp slope of the isotherms toward the sea. Above the front the largest vertical velocities are observed. Behind the front descending air is noted.

At 18:00 (Fig. 3), the front has advanced about 50 km inland. The sea breeze reaches an altitude of about 400 m. Ahead of the front at altitudes of 540–1050 m the horizontal wind strengthens, and vice versa behind the front. This is explained by adiabatic cooling of the ascending air over the front (the closed isotherm of 20^0C at 1000 m). This cool heavy air creates a horizontal pressure gradient, which intensifies the wind ahead of the front and develops the return flow behind it.

3.2. The Inversion

Over the land during day time the air in the lower layers is unstable, due to heating by convection. The marine inversion observed in the early morning hours whose

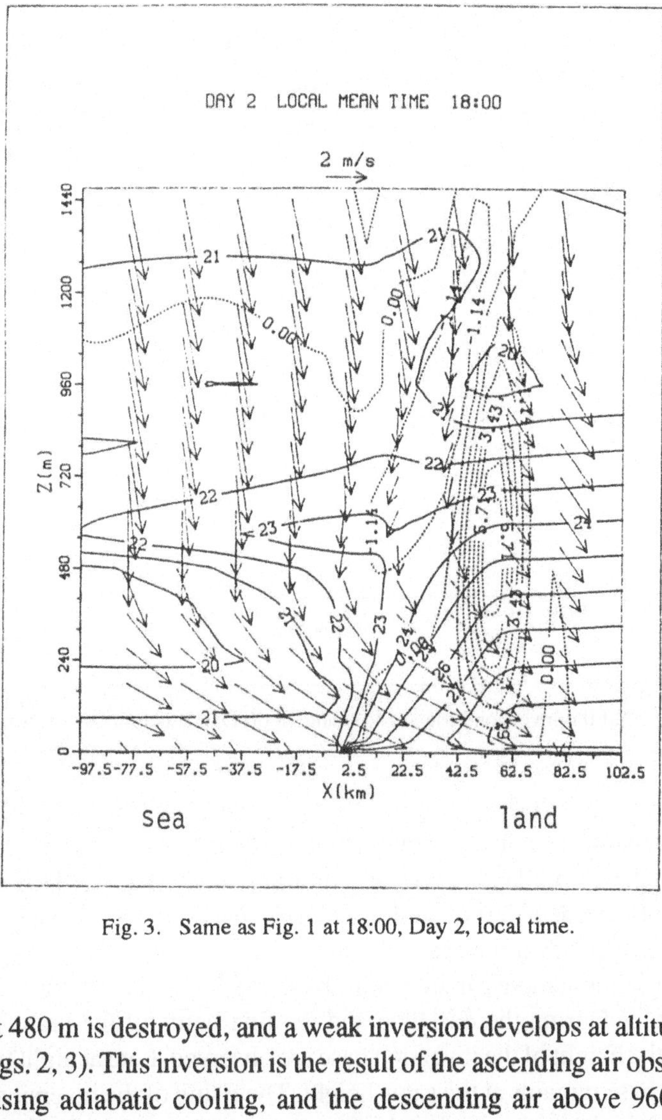

Fig. 3. Same as Fig. 1 at 18:00, Day 2, local time.

base was at 480 m is destroyed, and a weak inversion develops at altitudes of 960–1200 m (Figs. 2, 3). This inversion is the result of the ascending air observed below 960 m causing adiabatic cooling, and the descending air above 960 m causing adiabatic heating.

Over the sea descending air is observed during the day time. The vertical wind has its maximum strength near the coast and gradually decreases with increasing distance from the coast. This descending air causes adiabatic movement of the marine inversion over the sea. The maximum movement is observed at altitudes of 250–500 m where the descending current is stronger. In Fig. 4 the height of the inversion base at 14:00 and 18:00 is shown as a function of distance from the coast. Over the sea within 80 km of the coast, the height of the inversion base declines significantly towards the coast, dropping to 200 m in the afternoon hours.

Over the land during the day time, a very sharp change is observed in the height of the inversion base in the first 15 km (Fig. 4). During the night above the

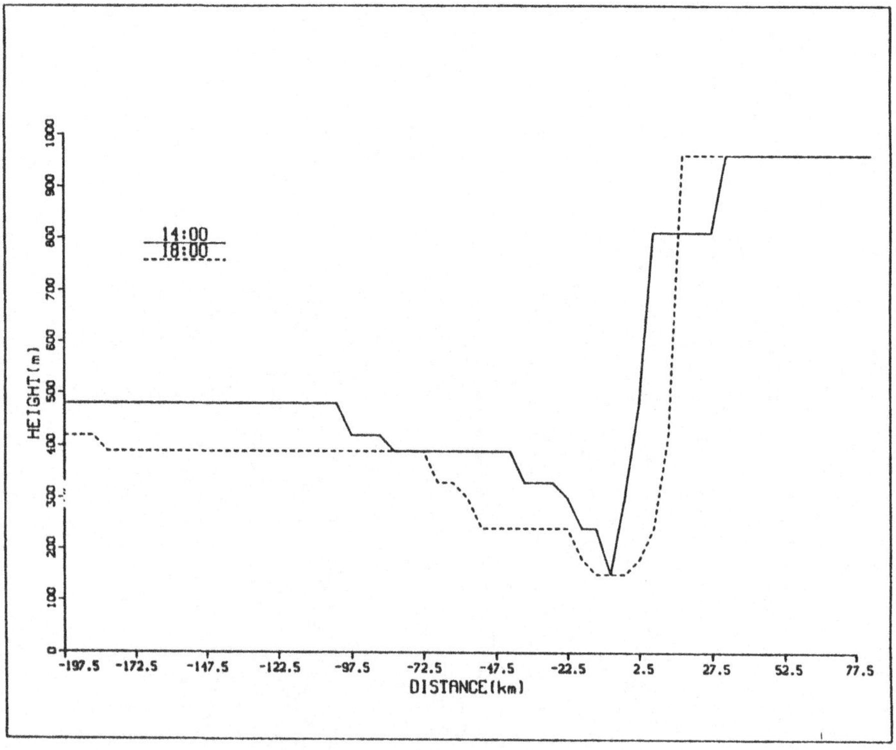

Fig. 4. The height of the inversion base at 14:00 and 18:00 Day 2 as function of distance from the coast.

nocturnal inversion we note the penetration of the marine inversion from the sea inland reaching 70 km inland in the early morning hours (Fig. 1). This is the result of advection of stratified air (containing a marine inversion whose base is at 480 m) from longe distances over the sea.

The diurnal fluctuations of inversion bases at 7.5 and 32.5 km offshore are shown in Fig. 5. During the day time and the first hours of the night, significant variations in the height of the inversion base are observed. Late at night till morning the inversion base remains at the same height. The rising of the inversion base over the sea to its initial position during the first hours of the night is mainly the result of advection of stratified air from long distances over the sea where the initial stratification hardly changes. The convergence of the horizontal wind near the coast during the night causes weak ascent of the inversion.

Some verifications of model results were made by comparing the computed profiles of the second day with the measured profiles over the sea. In Fig. 6 (a–d) the measured and the computed profiles are shown at different distances offshore at different hours. The general features of the measured and computed profiles are similar, and the height of the inversion base is almost the same. The diurnal fluctuation in the height of the inversion base as observed over the sea is well simulated by the model. Differences in the vertical gradient of the temperature

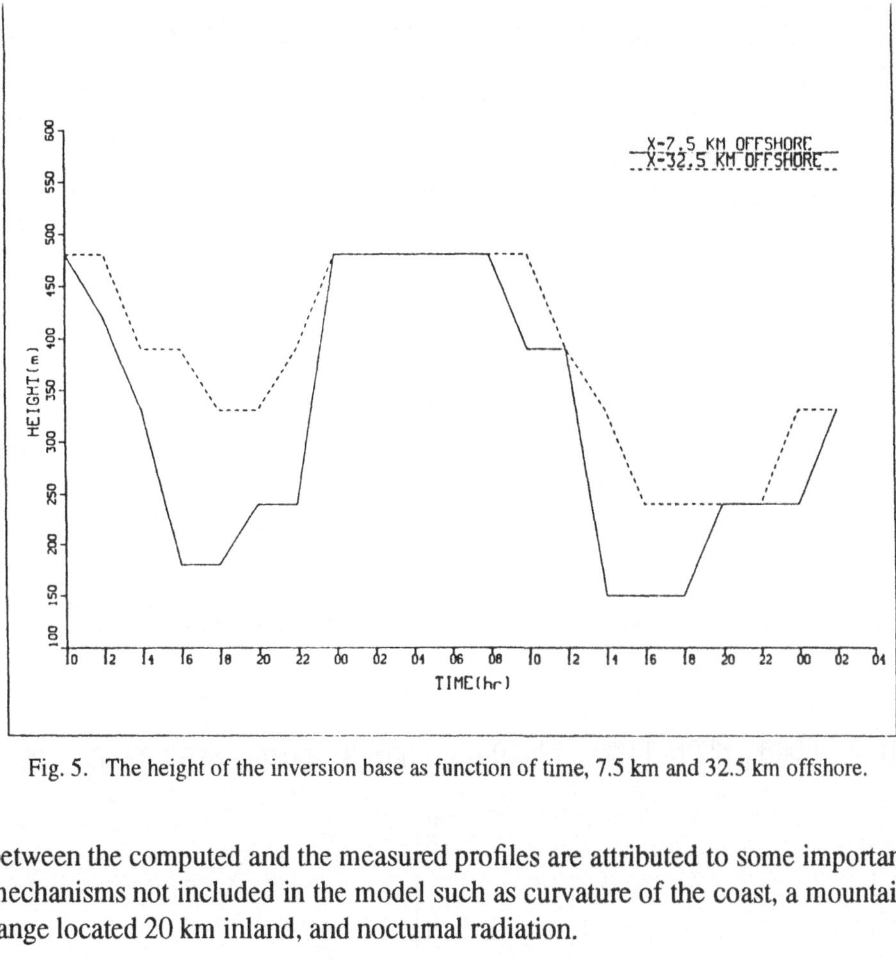

Fig. 5. The height of the inversion base as function of time, 7.5 km and 32.5 km offshore.

between the computed and the measured profiles are attributed to some important mechanisms not included in the model such as curvature of the coast, a mountain range located 20 km inland, and nocturnal radiation.

4. Summary

In the above numerical simulation we found a significant diurnal fluctuation in the height of the inversion base over the sea. This oscillation is the result of the coastal breezes.

During daytime the sea breeze circulation causes ascending air over the sea due to divergence of the breeze there. The vertical velocity causes adiabatic lowering of the inversion mainly at altitudes between 150–600 m. This results in the lowering of the inversion base by about 250 m near the coast. So the height of the inversion base in the afternoon is found at an altitude of about 200 m near the coast. The horizontal scale of the fluctuation in the height of the inversion is about 100 km. This scale is determined by the scale of the sea breeze. Over the land within 15 km of the coast a very sharp change in the height of the inversion base is observed. This change has a maximum in the afternoon hours of about 700 m. Farther inland the marine inversion whose base is at altitude of 480 m is destroyed while at an altitude of 960 m a weak inversion develops.

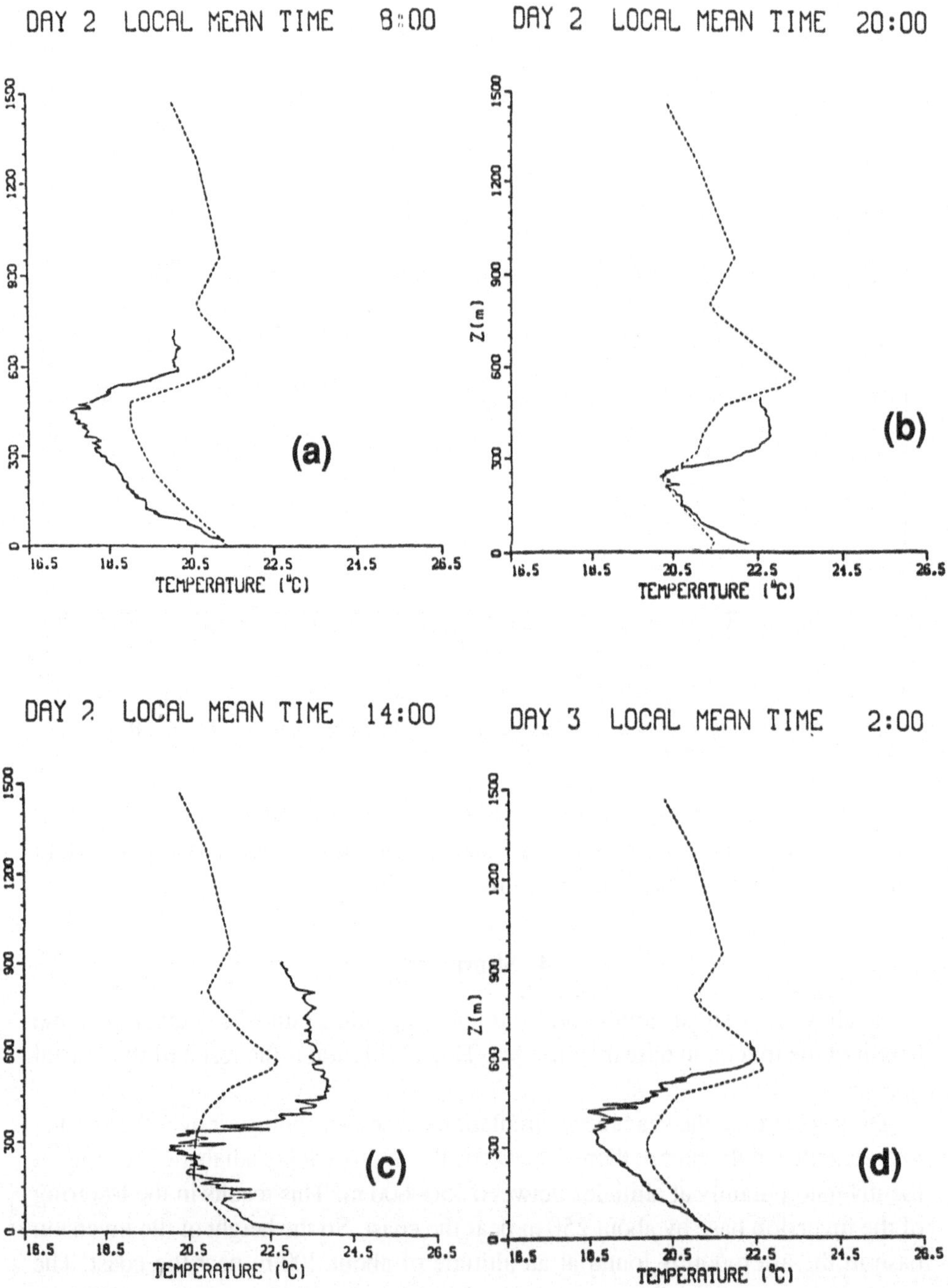

Fig. 6. Vertical profiles of the temperature, measurement full line, model simulation dashed line. (a) at 8:00, Day 2, 16.1 km offshore. (b) at 14:00, Day 2, 11.6 km offshore. (c) at 20:00, Day 2, 16.7 km offshore. (d) at 2:00, Day 3, 6.9 km offshore.

During the night the vertical velocities are very weak. The changes in the height of the marine inversion over the sea near the coast are mainly due to advection. At altitudes above 300 m the flow is towards the land. This flow advects a stratified air mass which contains the marine inversion whose base is at an altitude of 480 m.

A comparison of some of the computed temperature profiles with the measured profiles taken over the sea near the coast shows a similar diurnal oscillation, lowering of the inversion during the day and moving upward during the night. The amplitude of the oscillation in the observations and in the simulation is similar.

References

Atkinson, B. W. 1981: Meso-scale Atmospheric Circulations, London Academic Press, 405 pp.

Barkan Y. and Y. Feliks, 1992: Observations of the diurnal oscillation of the marine inversion near the Israeli coast. *Boundary-Layer Meteorology* **62**, 393–409, 1993 (this volume).

Huss , A., and Y. Feliks, 1981: A mesometeorological numerical model of the sea and land breezes involving sea-atmospheric interactions. Beitr. Phys. Atmos., **54**, 238–257.

Lettau, H. H., 1962: Theoretical wind spirals in the boundary layer of a barotropic atmosphere. Beitr. Phys. Atmos., bf 35, 195–212.

Marchuk, G. I., 1975: Methods of numerical mathematics., Springer, 316 pp.

Pandolfo, J. P., and C. A. Jacobs, 1973: Tests on urban meteorological pollutant model using co-validation data in Los-Angeles metropolitan area. Center for Environment of Man, 172pp CEM Rep. 490a.

Shaia, J. S., and R. S. Jaffe, 1976: Midday inversions over Beit Dagan. Isr. Meteor. Serv., Series A, no. 33, Beit Dagan.

SYNOPTIC AND MESOSCALE WEATHER CONDITIONS DURING AIR POLLUTION EPISODES IN ATHENS, GREECE

GEORGE KALLOS[1], PAVLOS KASSOMENOS[1] and ROGER A. PIELKE[2]
(1) University of Athens, Dept. of Applied Physics, Meteorology Lab., Ippocratus 33, Athens 10680, Greece.
(2) Colorado State University, Dept. of Atmospheric Sciences, Fort Collins, CO 80523, U.S.A.

(Received October 1991)

Abstract. Based on regular climatological and air quality data from the Greater Athens Area (GAA), the air pollution episodes observed in Athens during the period 1983–1990 were analysed and classified. The main characteristics of atmospheric conditions during days with high air pollution concentrations are summarized too. Model simulations show that the worst air pollution episodes in Athens occur during days with a critical balance between synoptic and mesoscale circulations and/or during days with warm advection in the lower troposphere.

1. Introduction

The cities of Athens and Pireus (and their suburbs) are located in a basin surrounded by mountains from three directions and open to the sea from the fourth (see Fig. 1). The main axis of the basin is SSW to NNE and is approximately 25 km in length. Its width is approximately 17 km. There are three main mountains, Hymettus to the E, Pendeli to the N-NNE and Parnitha to the N-NNW with elevations up to 1400 m. To the W of the basin is mountain Aegaleo with its peak elevation at 450 m. These mountains are physical barriers with small gaps between them. The opening of the basin to the sea is toward Saronic Gulf. Almost the entire basin of Athens could be considered as an urban area. The population of Athens, Pireus and their suburbs is approximately 3,600,000. To the E of mountain Hymettus is the plain of Mesogea while to the W of mountain Aegaleo is the Thriasion plain. The mountains around Athens are mostly covered with bushes. Only a small portion of their surface is covered by pine forests. There are three hills up to 200 m inside the basin (Pnyka, Lycabettus and Tourcovounia). In an area like Athens with such complicated physiographic characteristics, the flow fields and the Planetary Boundary Layer (PBL) depth show temporal and spatial variations. The flow fields and the PBL depth are crucial parameters for dispersion in an area (Pielke *et al.*, 1983; Glendening *et al.*, 1986; McKendry, 1989; Ulrickson and Mass, 1990a,b). There are seasonal and diurnal variations of the wind fields and PBL depths. Seasonal variations are mainly related to the persistent synoptic weather patterns and surface conditions. Synoptic conditions occuring during air pollution episodes will be described below. Surface conditions such as soil moisture, vegetation cover, etc. play an important role in the development of local circulations and of PBL depth (Segal *et al.*, 1989a,b; 1992; Avissar and Pielke, 1989; Pielke and Avissar, 1990). Because of the topographic, land-water distribution and land cover characteristics

Boundary-Layer Meteorology **62:** 163–184, 1993.

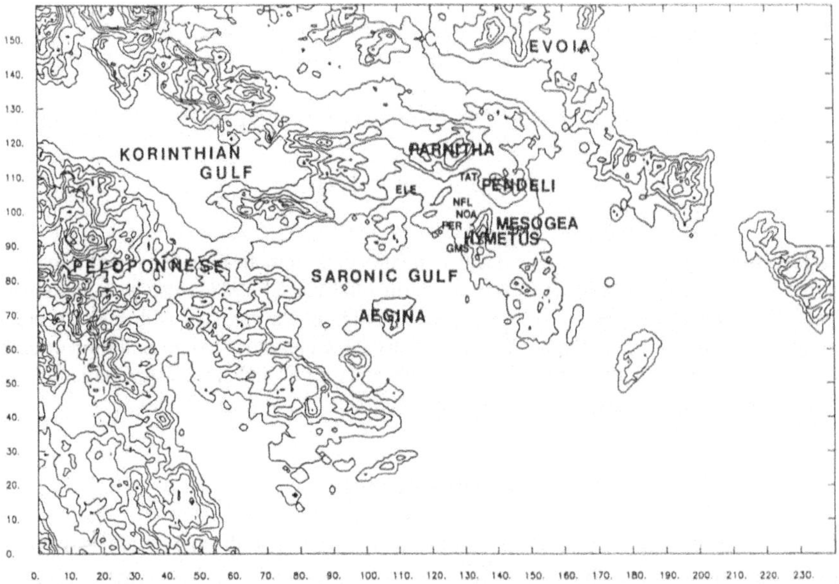

Fig. 1. The topography of southeastern Greece. Contours are every 250 m. The numbers at the axes show the distance in km from the southwestern corner of the domain.

of the area, local circulations such as sea (land)-breezes, drainage and upslope flows usually develop. Clear-sky conditions help in to the formation of temperature inversions during night-hours.

The main sources of air pollutants in the GAA are automobiles, industry and central heating during the cold months (Lalas *et al.*, 1982).

More than a million automobiles of all types are operating in the region. The lack of a high speed peripheral road network and the high number of automobiles operating are the main reasons for the slow traffic observed every day. The automobiles are considered as the main source of photochemical pollutants observed in the region. These were the reasons for the traffic restrictions imposed by the local authorities during the last eight years. According to these restrictions, only civilian cars with odd or even ending numbers on their tags are allowed to enter the center of Athens alternatively during the day-hours. During days with very high concentrations of air pollutants, some additional restrictions in the traffic of civilian cars and taxis are imposed for the Athens basin.

The industrial zones are located in the SSW part of the city of Athens and in the Thriasion plain. Some other sources are located in the harbour of Pireus while the contribution of air pollutants from ships and aircraft cannot be negligible.

The combination of emissions (mainly from traffic) with the appropriate atmospheric conditions is responsible for air pollution episodes in the GAA. Air pollution episodes occur during all seasons on a significant number of days. During these days, the concentrations of various air pollutants exceed the state limits. However, air quality in Athens cannot be considered worse than observed in other cities in Europe having the same problem.

2. Weather Conditions During Air Pollution Episodes

2.1. Synoptic conditions

The climate of the NE Mediterranean cannot easily be characterized as maritime or continental. Summer-months (June, July, August and September) are characterized as dry with almost no rain while winter (November, December, January and February) is the rainy season. Spring and autumn (March, April, May and October) are transient seasons where summer and winter-type weather patterns are interchanging. Transient seasons are not symmetrical but show some common characteristics.

Based on the above separation for each season, an attempt was made to classify the weather patterns which favor poor dispersion conditions in the Athens basin.

2.1.1. Summer season.

A typical weather pattern which appears during summer is the following: A high pressure system covers the Eastern Mediterranean and Balkan area up to the Black sea. Over the West and Central Mediterranean there is a ridge. A thermal low over the Anatolian Plateau is evident during day-hours because of heating of the dry land. The balance between these two systems defines the weather conditions over eastern Greece and the Aegean Sea. When the high pressure system is strengthened, it extends in an easterly direction and the pressure gradient across the Aegean is weakened. The synoptic circulation is weak from the N and therefore local circulations develop. During these days, thermal circulations dominate in the area of Athens. In contrast, when the high pressure system weakens, the thermal low over the Anatolian Plateau extends toward W and the pressure gradient over the Aegean becomes stronger. During these days the winds over the Aegean Sea are from the N, stronger during the day and weaker during the night. This kind of wind pattern is called Etesians or meltemi (Carapiperis, 1951). The synoptic flow dominates the sea breeze over Athens and the winds are from the N. This phenomenon lasts from two to five days and the dispersion conditions over Athens are good. Poor dispersion conditions occur when the synoptic flow is in near equilibrium with the sea breeze (almost stagnant conditions). When the sea breeze dominates the synoptic circulation, the dispersion conditions are better than the stagnant case but worse than on days with Etesians.

2.1.2. Winter season.

Winter is characterized by a low index circulation and the passage of low pressure systems over Greece is relatively frequent. When a low is over the Ionian Sea or western Greece, the winds over Athens are from the S to SE directions. The same occurs when Athens is within the warm sector of the low. After the passage of the cold front, the winds are from the NW to N. Similar veering of the winds occurs with the passage of a cold front moving from the NW toward Greece. This front is usually called Balkan front. During these circumstances, warm advection is observed in the lower troposphere, the atmosphere is stable because of the elevated temperature inversions observed during days with clear sky. During these days, dispersion conditions in Athens are poor. This phenomenon occurs for one to two days except for cases with a stationary low over South Italy and the Ionian Sea which may remain in the same position for four to five days. During these days, regularly, the southerly flow is relatively strong and dispersion conditions in Athens are good. Following the passage of the low or cold front, the sky is usually clear, and during the night, surface inversions form; therefore dispersion conditions are poor especially during the morning hours. In the case of the formation of a strong pressure gradient over the Aegean Sea (or the formation of a low in the area of Cyprus) and after the passage of the cold front, the dispersion conditions in Athens are good because of the relatively strong winds and the generation of mechanical turbulence. A similar pattern for dispersion was found by Pielke *et al.* (1987, 1991); Yu and Pielke (1986), for the United States.

The development of a high pressure system over the Central Mediterranean which extends toward the E and covers the area of Greece is the weather pattern associated with poor dispersion conditions in Athens. Sometimes this sytem has its center over the Balkan area and remains stationary for several days. This is a common pattern observed usually during mid-January for approximately two weeks. These days with clear sky conditions, a weak northerly synoptic flow and dry air are called Halkyon days (Dikaiakos, 1983). During these days, strong temperature inversions form during the night which, sometimes, do not breakup even around noon (Katsoulis, 1988a,b).

2.1.3. Transient seasons.

As was mentioned above, during these seasons the weather type regularly changes between summer and winter conditions, relatively quickly during spring and slower during autumn. Poor dispersion conditions over Athens occur during days with an anticyclone covering the Central and Eastern Mediterrranean and the Balkan area as well. Weak pressure gradients must occur over the Aegean Sea. The worst air pollution episodes observed during these months are associated with advection of warm air-masses from North Africa over the relatively cool Mediterranean Sea. During these days, the air near the ground is relatively cool and warmer aloft. These stable atmospheric conditions, in combination with weak flow, do not permit the development of a deep mixing layer and consequently air quality in Athens becomes poor. Such synoptic conditions usually occur when a trough is over the Iberian Peninsula and the Western Mediterranean and a ridge is over the Central Mediterranean. For some of these cases, warm advection is observed very low in the troposphere and is difficult to identify even on the 850 hPa synoptic map.

The characteristics of the synoptic conditions decribed above are summarized in Table I. The classification scheme presented in Table I has similarities with the scheme developed by Pielke *et al.* (1987) but its structure is according to the specific synoptic patterns usually observed in the region of Greece and Aegean Sea.

2.2. SYNOPTIC CLASSIFICATION

In order to classify these weather patterns, regular climatic (surface and upper-air), air quality data and synoptic maps were analyzed for the period 1974–1990. There are seven meteorological stations operating in the GAA, five inside the Athens basin, one in the Thriasion (ELE) and one in the Mesogea (SPA) plains. The upper-air station is located at the airport of Athens and is operated from the Greek Meteorological Service (GMS), while the other four surface stations are in Pireus (PER), the National Observatory of Athens (NOA), Nea Philadelphia (NFL) and Tatoi (TAT). The position of each station is shown in Fig. 1. In the Athens basin there is a network of eight air pollution monitoring stations operating since 1983. For these stations hourly concentrations of primary and secondary air pollutants are recorded.

The 3-hour observations from the meteorological stations were analysed for 1974–1990. Surface wind analysis showed that, for some of the stations, calms account for more than 50% during night-hours while during day-hours this percentage does not exceed 10%. Light winds (1–5 m/s) are quite frequent for all stations, day and night, while strong winds are observed mainly during day-hours. The prevailing wind directions are from the N to NE during night while during day-hours these are from SW to SSW and N to NE for all stations in the Athens basin. For the two stations outside of the basin (ELE and SPA), the winds show a preference for the N and S directions. The wind roses estimated from the radiosonde observations (GMS) at heights of 10, 500 and 2000 m are shown in Fig. 2. As it is seen, this station is influenced by local circulations because of its position near the coast. Moderate and strong winds from the NW or N are observed during all hours mainly at upper levels. W to SW winds are observed in the upper levels and near ground only during day-hours.

As was found from the analysis of the climatological data recorded in the above mentioned stations, Athens is characterized by its clear-sky conditions. Almost 45% of nights and 15% of days are with clear-skies and approximately 8% with overcast ones. During summer months there are usually no days with overcast skies while during the transient seasons 10% are typically overcast. These climatic conditions allow for the development of deep mixing layers during day-hours and quite shallow boundary layers during the night.

Using the detailed radiosonde data from station GMS, the mixing heights were estimated for day and night-hours. The afternoon mixing height calculations were based on the radiosondes at 14:00 LT while the night mixing heights were from

TABLE I
Synoptic classification for days with air pollution episodes in Athens.

	I	II	III	IV
GENERATION	High pressure system starts to develop over northeast Mediterranean and/or south Europe. Over Central Mediterranean there is a ridge which sometimes is associated with a through over Mediterranean and Iberian Peninsula. Warm air masses are observed initially over Central Mediterranean and south Italy.	A low pressure system is located over Central Mediterranean and moves eastward. The anticyclonic system over the Balkan peninninsula starts weakening. Usually this low is relatively shallow.	A cold front moves from the NW toward the Balkan Peninsula and Greece. Ahead of the cold front the synoptic circulation is relatively weak.	An anticyclonic system covers most of the Mediterranean and northeast Europe. A thermal low covers the Anatolian plateau during the day-hours. This combination establishes a pressure gradient over the Aegean Sea and consequently over Athens.
EVOLUTION	The anticyclonic system moves eastward toward Greece. Warm air masses are advected toward Greece and the lower tropospheric layers are very stable.	Greece is inside the warm sector of the low. The advection of warm air masses stabilizes the lower troposphere.	After the passage of the cold front, relatively cool air masses are observed over Greece. The synoptic flow is from the NW and is relativelty weak. Clear sky conditions support the formation of surface temperature inversions.	When the anticyclonic system extends eastward, or becomes stronger, the pressure gradient over the Aegean Sea weakens. Local circulations starts to develop.
END	High pressure systems over Central Europe extend eastward and a cold front passes over the Balkan peninsula or Greece establishing a NW or N flow.	After the passage of the cold front, a northerly flow (dry and cool air) is observed over Greece and the Aegean Sea near the surface and aloft. These lows move toward the ESE, E or NE suring the different seasons.	For some cases, the pressure gradients strenghten and a strong northerly flow occurs over Greece and the Aegean Sea. For others, a low pressure system moves toward Greece.	When the anticyclonic system weakens, the thermal low extends westward and the pressure gradient over the Aegean Sea becomes stronger. This is a typical weather pattern during the summer

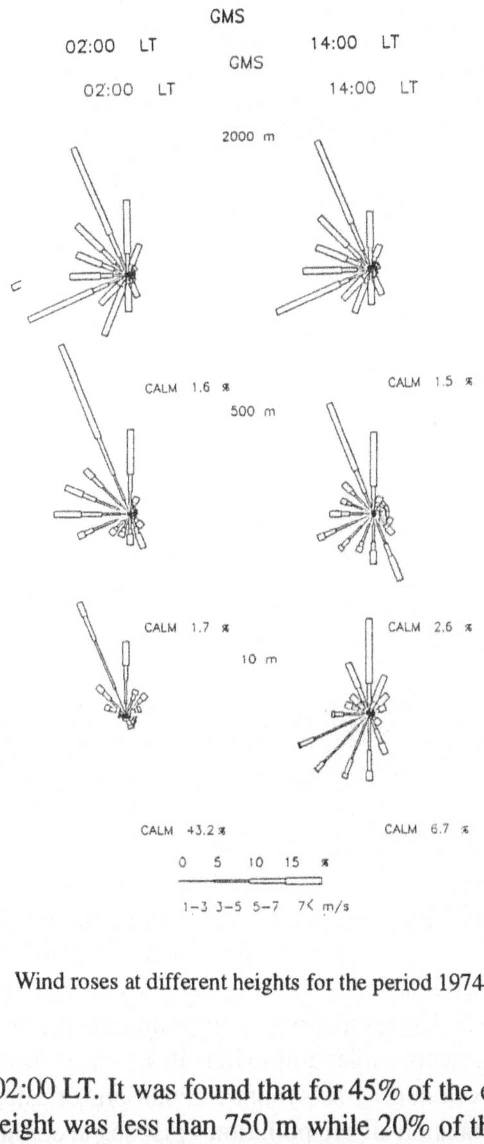

Fig. 2. Wind roses at different heights for the period 1974–1990.

the radiosondes at 02:00 LT. It was found that for 45% of the examined cases, the afternoon mixing height was less than 750 m while 20% of the time it was more than 1500 m. In general, afternoon mixing heights are higher during the warm season and lower during the cold one. For night-hours, mixing heights of less than 250 m account for the 31% of the cases while cases with heights between 250 and 750 m account for 36% of the total. For a small percentage of the total cases the night-hour mixing height is higher than 1500 m. Estimated mixing heights of less than 50 m are considered as not realistic and were rejected.

Based on air quality data from the monitoring network, the days with high concentrations characterized as air pollution episodes were selected. This selection was made according to the following criteria: The imposed state limits must be violated in at least two monitoring stations, for at least two hours, for at least two

TABLE II
Analysis of occurence of air pollution episodes in Athens for different seasons and synoptic categories for the period of 1983–1990

	WINTER	SUMMER	TRANSIENT	TOTAL
I	14	6	16	36
II	10	4	14	28
III	3	1	4	8
IV	–	8	–	8
TOTAL	27	19	34	80

TABLE III
Classification of air pollution episodes in Athens according to the appearence of warm advection in the lower troposphere. Strong warm advection is considered when there is a rise in temperature greater than 4° during the previous three days from the first day of the episode. Time period 1983–1990.

SEASON	STRONG WARM ADVECTION	WEAK OR NOT AT ALL
WINTER	17	10
SUMMER	13	6
TRANSIENT	19	15
TOTAL	49	31

consecutive days, and for at least two recorded air pollutants. For the years 1983–1990, 80 episodes were selected for a total number of 210 days. The distribution of these episodes for each season and each synoptic category described in Table I, is shown in Table II. The largest number of air pollution episodes was found during the transient seasons with winter following. It was also found that most of the episodes were observed during the synoptic conditions of category I and II. Table III shows the classification of air pollution episodes according to the existence (or not) of warm advection in the lower troposphere. The synoptic charts and the detailed radiosonde observations at GMS were used for this classification. As is seen in this table, for most of the cases, an increase of temperature of more than 4° C ocurred during the previous three days from the first day of the episode. Warm advection occurs mainly during the transient and winter seasons. Table IV shows the classification of the air pollution episode days according to each season and the strenght of the surface pressure gradient. For most of the days with violation of the imposed state limits for air quality, a very weak or relatively weak pressure gradient was found in the area of Greece. This is especially true during transient and winter seasons. A few days with relatively strong pressure gradients and poor

TABLE IV
Classification of the pollution episodes in Athens according to the observed pressure gradients in the area of Greece: (A) Very weak pressure gradients (weaker than 5 hPa per 1100 km), (B) relatively weak (5 hPa per 550–1100 km), (C) relatively strong (5 hPa per 100–550 km), and (D) strong (greater than 5 hPa per 100 km). Time period 1983–1990.

	PRESSURE GRADIENT			
SEASON	A	B	C	D
WINTER	23	23	15	4
SUMMER	14	17	20	1
TRANSIENT	34	26	31	2
TOTAL DAYS	71	66	66	7

air quality were found. The high concentrations of air pollutants recorded during these days are related to the existence of air pollutants from the previous days.

2.3. Local Conditions

Due to the orientation of the Athens basin and the mountains surrounding it, the flow field shows a preference in two directions: one from the N or NE and the other from the SW. Northerly winds usually persist when the synoptic-scale flow is stronger than the sea-breeze. This usually occurs when there is a strong pressure gradient over the Aegean Sea and Greece. Northerly winds are usually strong and favour the ventilation of the Athens basin. Southwesterly winds usually occur during days with local circulations stronger than the synoptic one or with strong SW flow because of the existence of a low over the Central Mediterranean and Ionian Sea. Strong SW synoptic winds usually occur during the cold period of the year. Sometimes, the low over the Central Mediterranean advects warm air masses over Greece. When the low weakens, the advected warm air masses stabilize the lower trophosphere and poor dispersion conditions might occur in Athens.

During days with relatively weak synoptic flow, local circulations are usually observed. Such local circulations are sea (land) breezes, and upslope and drainage flows. There are three main cells of sea-breezes which develop in the area of Attiki. One is from the Saronic Gulf toward Athens. Near the surface it blows from the WSW to S during day-hours while during night-hours it is mainly from the N. The second cell forms over Mesogea plain (East of Hymettus) where winds near the surface vary from NE to E and SE during day-hours and from the W-NNW during the night. The third cell forms over the Thriasion Plain (West of Aegaleo) and blows mainly from the S during day-hours and from the N during the night. The sea-breeze circulation over Athens and the Saronic Gulf was known from ancient times. The Athenian General Themistocles took advantage of the sea-breeze in order to defeat the Persian fleet near the island of Salamis. References to this event are given in Steyn and Kallos (1992). The depth of the onshore flow is 500 to 1000 m above the ground, where the return flow usually merges with the synoptic one which is usually from the NW quadrant. The sea-breeze over most of the Athens basin shows an anti-clockwise rotation as was found by Steyn and Kallos (1992). These three sea-breeze cells interact through the gaps between the mountains. These interactions define the conditions under which polluted air masses from the area of Athens are exiting the basin or not. During most day-hours, air masses from Athens move out of the basin through the gaps between Pendeli and Parnitha, and Hymettus and Pendeli. In contrast, through the gap between Parnitha and Aegaleo, polluted air-masses from the industrialized area of the Thriasion Plain move toward Athens. Under certain circumstances, polluted air-masses from the Thriasion Plain still move over Aegaleo toward Athens during the day or over the Saronic Gulf during the night and later, with the aid of the sea breeze, over Athens (Asimacopoulos *et al.*, 1991). This kind of transport also occurs during days with S or SW synoptic flow. The western suburbs of Athens are affected by this kind of transport.

Typically, during summer-months, the sea-breeze over Athens is relatively strong (Prezerakos, 1986). As was found from analysis of wind data from several places within the Athens basin, the winds are from the SSW to SW 4–6 m/s during days with fully developed sea breeze cells. Because of the orientation of the coastline and the mountain slopes, the sea breeze occupies all the basin quickly in the morning and significant amounts of polluted air are transported out through the gaps between the mountains. There is a significant number of days where the synoptic flow is relatively strong and the sea-breeze progresses slowly or does not penetrate through the entire basin (Helmis *et al.*, 1987). During these days, the observed SWly winds are lighter than those recorded during days with full development of the sea breeze, and the air quality in Athens is poor. For days with full development of sea-breezes, air quality is poor during morning and evening hours.

Sea-breeze circulations develop not only during summer and transient seasons but also during the winter (Carapiperis and Katsoulis, 1977). This is due to soil type (rocky mountain slopes), lack of vegetation, orientation of the slopes and the insolation. Because of this, the land becomes warmer than the sea and a pressure gradient forms during day-hours with clear-sky conditions. Of course, this kind of sea-breeze is weak and usually does not penetrate deeply inside the Athens basin.

The analysis of surface winds, at various locations, for the days characterized as air pollution episodes, is shown in Fig. 3. As is seen, during the night, calms or very light winds are observed while during day the winds are mainly from the WSW to SSW with speeds rarely exceeding the 5 m/s in the Athens basin. At the stations NFL and TAT which are located at the northern part of the Athens basin, a significant number of days with NNE or NE winds was found. This must be due

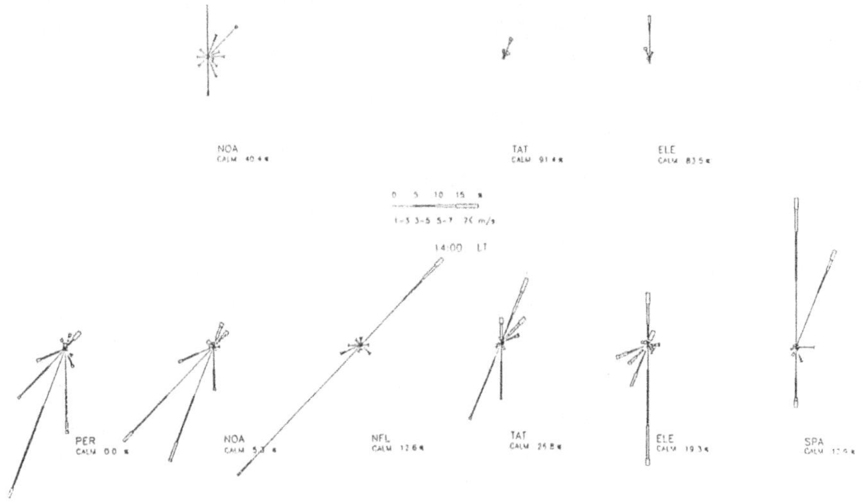

Fig. 3. Wind roses constructed for the air-pollution episode days at various stations in the GAA.

to the fact that for these days, the sea breeze is not strong enough to penetrate deeply inside the Athens basin. For the other two stations ELE and SPA, located in the Thriasion and Mesogea plains rescpectively, the winds show a preference for southerly and northerly directions. This is due to the orientation of these plains. Fig. 4 shows the wind roses for the air pollution episode-days, at different heights, estimated from the radiosondes released during night and day-hours from the station GMS. This analysis shows that, during days with high concentrations of air pollutants, the synoptic winds are from the W and N sectors.

3. Flow Field Simulations

Several model simulations for the Athens area have been performed during the last ten years. Most of these simulations are limited to a small domain around Athens and only for typical summer sea-breeze cases (e.g. Kallos and Kassomenos, 1991 and references therein). This was mainly due to the limited computer resources available and due to the belief that air pollution episodes in Athens are associated with sea-breezes. As was shown above, during days with full development of sea-breeze cells, the air quality of Athens is considered as poor during morning and evening-hours. Poor dispersion conditions are associated with stagnant conditions (synoptic-scale flow in near equilibrium with local circulations) or/and warm advection aloft.

In this study, the results of a model simulation will be shown. The case is May 25, 1990 with weak pressure gradients over Greece and warm advection aloft. During

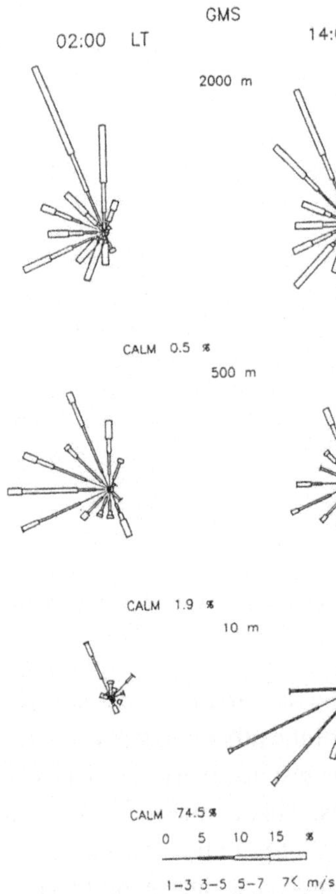

Fig. 4. Wind roses constructed for the air-pollution episode days at different heights for the upper air station GMS.

this day, the sea-breeze was developed but it was not as strong as it usually is during summer. The synoptic flow was from the W-WNW at the lower atmospheric layers and NW aloft with speeds ranging from 2 to 7 m/s. At noon, the flow was from the WSW in the lower layers. A strong temperature inversion was observed up to 920 mb during the night; the inversion did not break-up during day-hours (Kallos and Kassomenos, 1991). This inversion was due to warm advection at the layers of the trophosphere and did not allow the vertical development of local circulations in the Athens area. The sea-breeze started late in the morning as is shown in Fig. 5. It started later than usual and was not as strong. During night and morning hours the winds were very light at almost all stations while at noon they are from the SSW veering to S in the afternoon. Mixing height calculations during this day showed

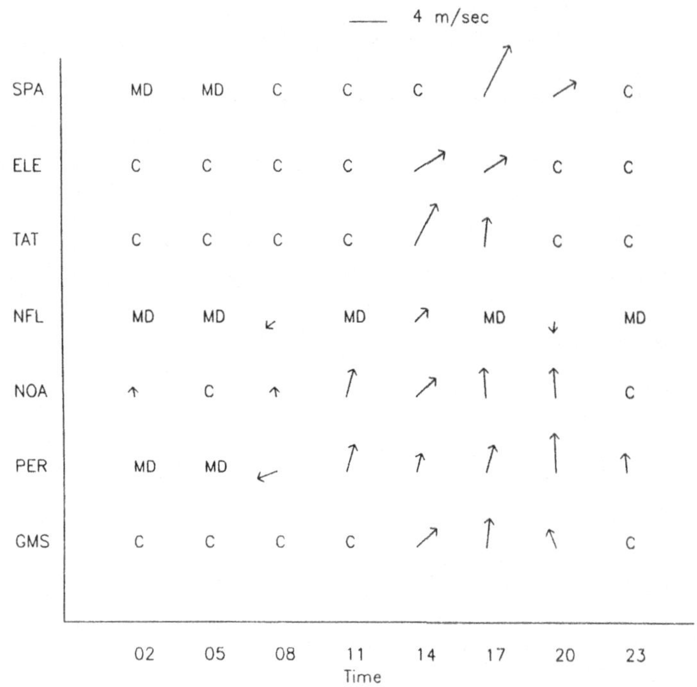

Fig. 5. Observed surface winds at different locations in Attiki during May 25, 1990. Missing data and calms are shown by "MD" and "C" respectively.

that the afternoon mixing height was about 150 m while the night mixing height was approximately 60 m. The same conditions also occured one day before and one day after. This air pollution episode was one of the worst in Athens. On May 27 the mixing heights were 600 m and 175 m, respectively, and the episode ended.

The model used for this simulation is the Colorado State University Regional Atmospheric Modelling System (RAMS) which was developed from the groups of R. Pielke and W. Cotton (e.g. see Pielke *et al.*, 1992; Xian an Pielke, 1991). This model has nesting capabilities and several options for lateral and top boundary conditions, turbulence, surface, radiation and cloud parameterizations. Three model domains were used (see Fig. 6). The coarsest domain covered all of Greece, the Aegean Sea and a portion of Turkey with grid increments of 16 km. The intermediate domain covered the NE part of Peloponnese, the Saronic Gulf and a large portion of mainland Greece and the island of Evoia with horizontal grid increments of 4 km. The finest grid covered most of the Saronic Gulf and the area of Attiki with grid separation of 2 km. The three domains had 45x43, 38x38 and 36x32 horizontal grid points and 27 vertical levels distributed up to 11 km. The simulation started at 2:00 LT and ended 24 hours later. The data used for initialization were upper-air data (radiosonde) from the airport of Helliniko in Athens (station GMS).

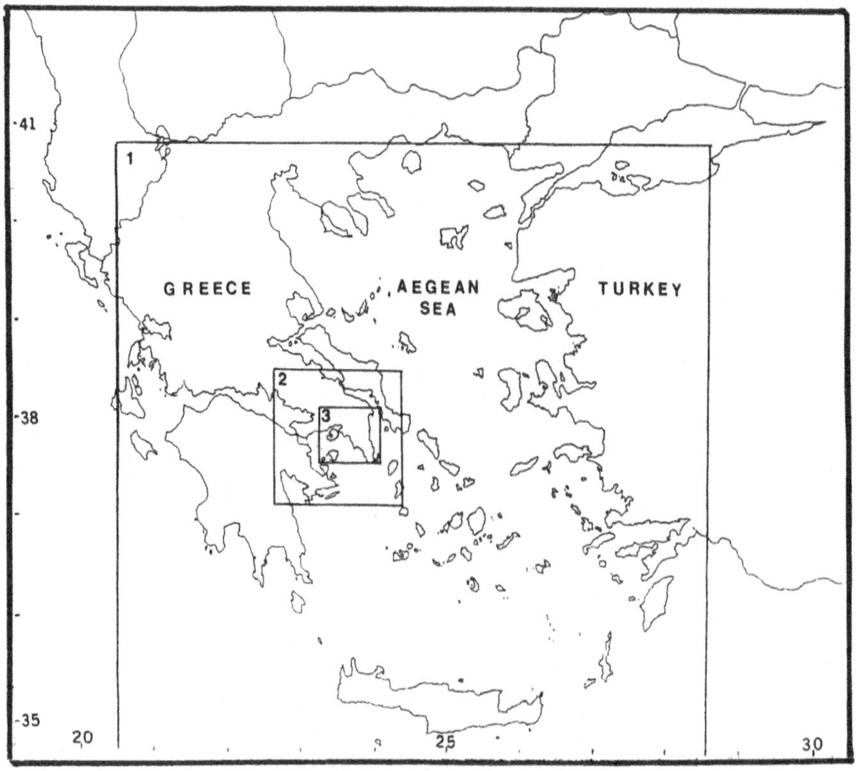

Fig. 6. Map of Greece and the surrounding area. The three model-domains used for the simulation are also shown.

Land-use data, such as vegetation index, urban areas, soil-type, etc. were derived from satellite images and cartographic maps while the roughness length over land was determined at each grid cell according to the land-use. Other capabilities of the model used for this simulation are the second-order turbulence closure, absorbing layer (Rayleigh friction) at the top five model levels and zero-gradient lateral boundary conditions for the coarsest grid.

Fig. 7 shows the wind fields 40 m above the ground at the three model domains, at 6:00 LT. The flow is NW to W over the mountainous areas of Greece; it becomes W in the area S of Peloponnese and SW to S over Saronic Gulf. Light winds are observed over the Aegean Sea (Fig. 7a). Only a portion of the coarsest grid is presented here in order to improve plotting resolution. The numbers at the X and Y axes show the distance in km from the center of the coarsest grid. Fig. 7b shows the wind fields at the 4 km grid domain. At this domain, the flow over the Saronic Gulf is from SSE veering to SE near the island of Aegina and E over the western part. The flow over Athens is mainly from the NE with speeds of 1–2 m/s. Over the sea between Attiki and Evoia, the flow is mainly from the N. Fig. 7c shows the flow at the finest The main features are the same as in the 4 km grid domain but with more detail. The flow is described in detail at the gaps between the mountains. Later in

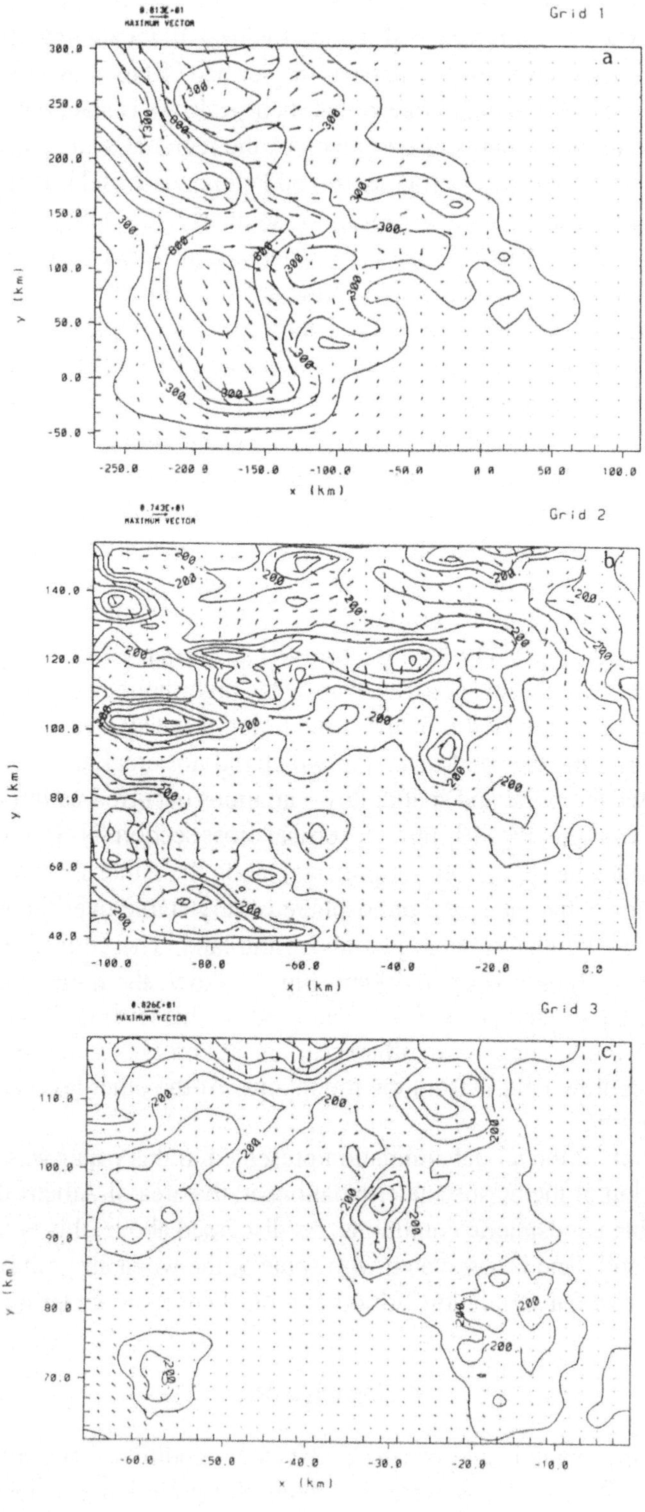

Fig. 7. Wind fields at the three model-domains, 40 m above ground, at 06:00 LT. (a) 16 km grid increments, (b) 4 km grid increments, and (c) 2 km grid increments.

the morning, when the sun rises and the land is heated, local circulations start to develop. Fig. 8 shows the flow fields near the surface (40 m above the ground) at 14:00 LT when the local circulations are at their peak. As is shown in Fig. 8a, the topographic features of mainland Greece block the regional-scale circulation. The flow is stronger near the coasts, it goes around Peloponnese and is directed toward the Saronic Gulf. Inside the Saronic Gulf, the southerly flow splits and one portion is directed through the gap between the Hymettus and Keratea mountains toward the Mesogea plain, a second toward the Athens basin while the third is directed to the western Saronic and passes over the Isthmus of Korinth (Fig. 8b). Air masses from the Athens basin exit through the two gaps at the northern edge (Fig. 8c). Wind speeds over the Saronic Gulf and Athens basin are within the range of 2–5 m/s. The sea breezes diminish late in the afternoon and in the evening-hours, light drainage flows start to appear. Fig. 9 shows the wind fields 40 m above the ground at the three model domains, at 22:00 LT. The regional-scale flow is still from the SW to S over the Saronic Gulf and the sea E of Peloponnese (Fig. 9a). Over the Athens basin, the flow is very weak with speeds of 1–2 m/s from variable directions(Fig. 9b,c). Over the northern part of Saronic Gulf the flow is from the SE. A relatively strong SWly flow is observed at the gap between Hymettus and Keratea mountains and over Mesogea. During the night-hours wind speeds are even lower. A general characteristic of the flow during this day was its small vertical extent because of the elevated inversion which did not break-up at all during the day. Most of the temporal and spatial flow variations occurred within the lowest 1 km of the troposphere. Fig. 10 shows a vertical cross section (N-S) of the potential temperature and wind (v and w components) along the Athens basin at 14:00 LT. This figure shows that the lower atmosphere is very stable over the sea and land except for a few hundred meters just above the land. The vertical extent of the inflow part of sea breeze is up to 400 m. Fig. 11 shows the horizontal wind field 1100 m above the ground at 14:00 LT. The winds at this level are from the W with speeds around 4 m/s. There is no influence of the sea breezes at this level. The only influence in the flow field is from the mountains. At the other levels above it, the winds veer to NW.

The intention of the model simulation presented in this paper was to show the diurnal variation of the mesoscale circulations in the area of Athens during a day with very stable atmospheric conditions. As discussed above, this is a case which appears frequently and is associated with some of the worst air pollution episodes in Athens. Additional simulations for other typical cases are under way.

4. Conclusions

In this work, the synoptic and mesoscale weather conditions occuring during air-pollution episodes in Athens were described. Air-pollution episodes in Athens area appear for a significant number of days during all seasons. Usually, their duration is two to four days and rarely last longer. For a small number of days,

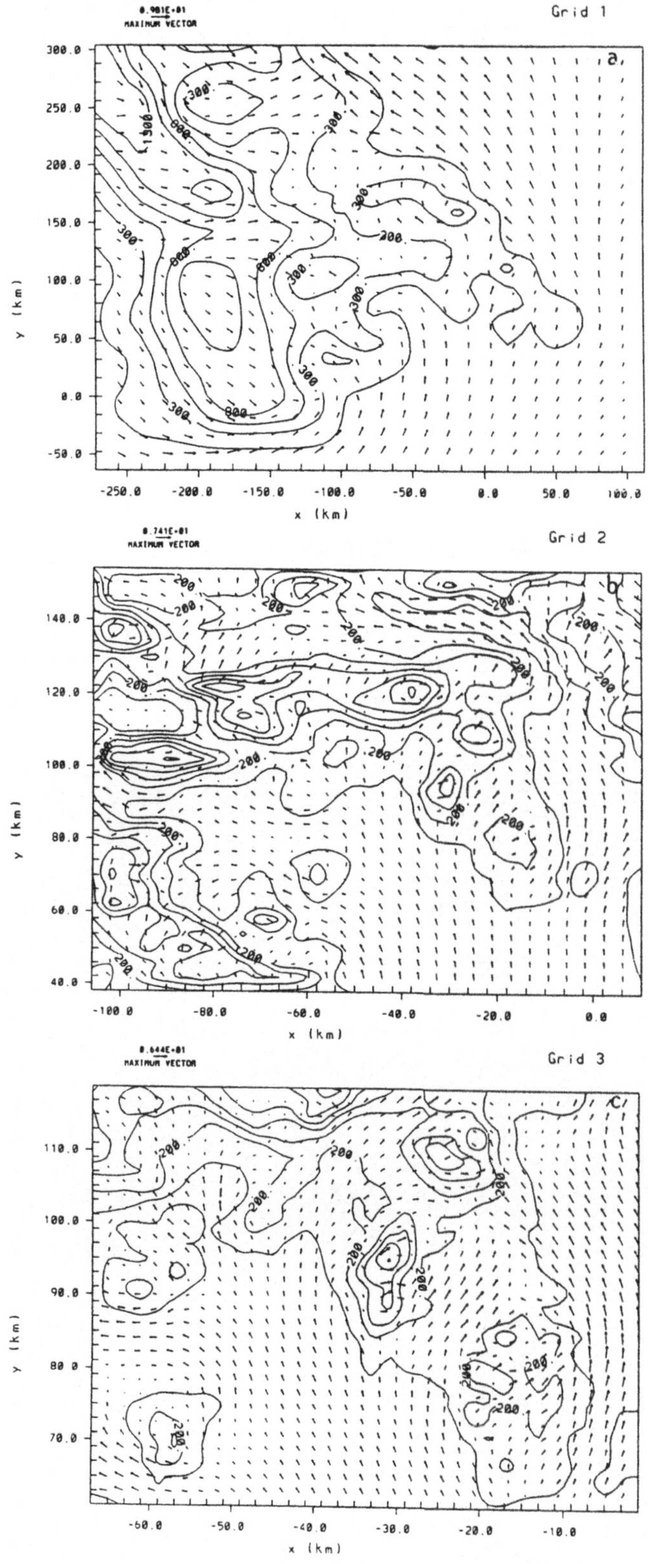

Fig. 8. Wind fields at the three model-domains, 40 m above ground, at 14:00 LT. (a) 16 km grid increments, (b) 4 km grid increments, and (c) 2 km grid increments.

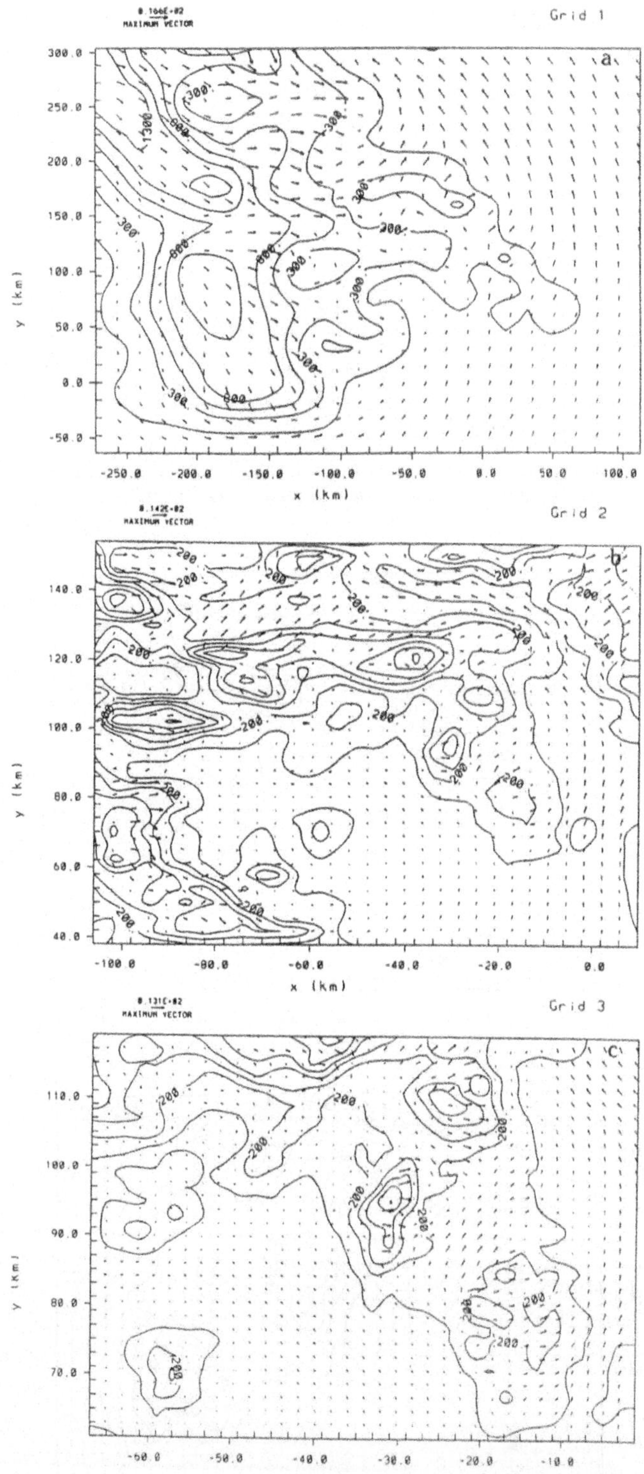

Fig. 9. Wind fields at the three model-domains, 40 m above ground, at 22:00 LT. (a) 16 km grid increments, (b) 4 km grid increments, and (c) 2 km grid increments.

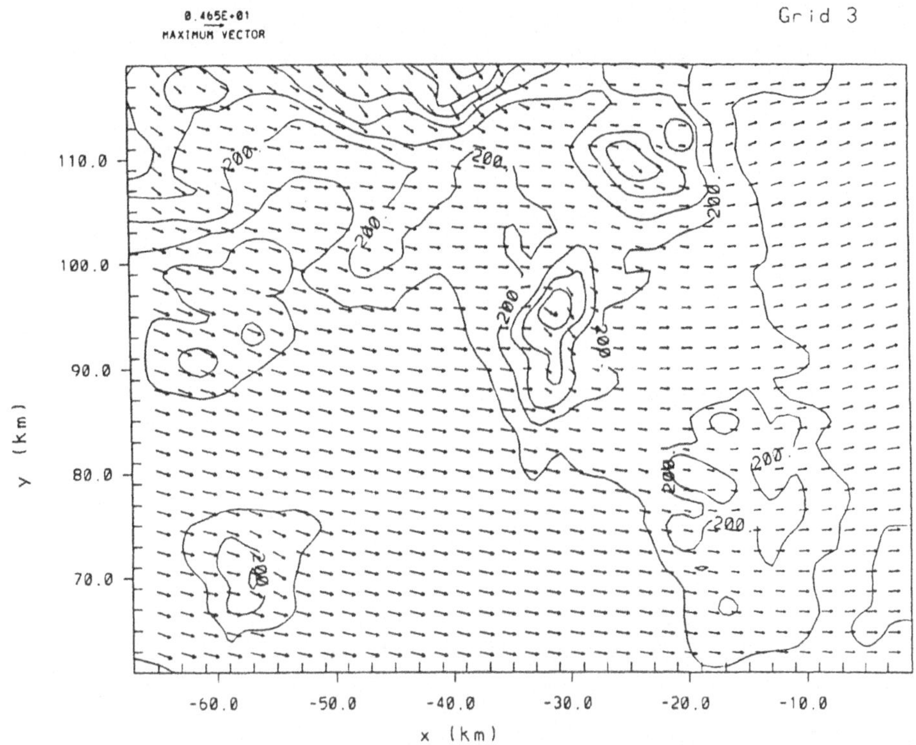

Fig. 10. Vertical cross-section of the potential temperature and winds (v and w components) along the Athens basin at 14:00 LT. The grid increments are 2 km while the contour interval is 1° K.

concentrations of air pollutants (primary, secondary and particulates) reach very high levels. The air quality standards are violated for a significant number of days. Clear-sky conditions favour these processes. Bad dispersion is a result of a combination of synoptic conditions and local physiographic characteristics. In general, during transient and winter seasons, stagnant conditions are more likely to occur. During spring, warm advection over the relatively cool sea and wet land causes very stable atmospheric conditions over the GAA, the mixing layer is very shallow and mesoscale circulations do not fully develop.

Sea breezes are usually thought to be associated with air pollution episodes in Athens. As was shown above, this is not the case. The worst episodes are usually associated with stagnant conditions and/or with warm advection in the lower troposphere. During days with full development of the sea breeze cells in the region, air quality in Athens is regularily poor during the morning and evening hours. It was also found that the synoptic – mesoscale flow interactions play a very important role in the formation of appropriate atmospheric conditions for the formation of air pollution episodes.

In general, most of the models used for sea breeze simulations could provide some general characteristics of the flow over Athens quite accurately (e.g. SSW or SW flow over Athens) because the key features are due to the topography and

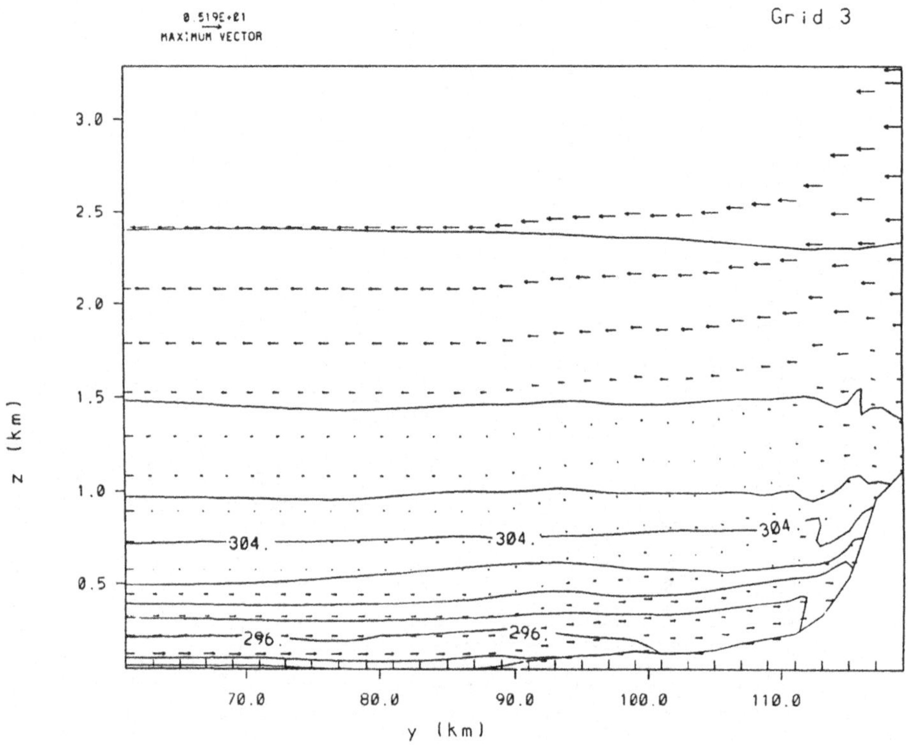

Fig. 11. The horizontal wind field at the 2 km grid increment model-domain, 1100 m above ground, at 14:00 LT.

land-sea distribution. The mesoscale flow over GAA is a combination of sea (land) breezes and upslope (downslope) flows. In the case where simple mesoscale models are used, the different-scale flow interactions cannot be accurately described. They must at least cover a large area (at least a part of NE Peloponnese, the Isthmus of Korinth and the southern part of island of Evoia) with grid increments of less than 4 km in order to resolve the flow in critical areas and describe very interesting phenomena such as recirculation of air pollutants. Atmospheric models with nesting capabilities are more appropriate for such simulations. The use of photochemical models in this area is very problematic because an accurate and detailed emission inventory required does not exist.

5. Acknowledgements

This research has been partially supported by IBM-Hellas, the General Secretariat of Science and Technology of Greece and Colorado State University, Department of Atmospheric Science by NSF Grant # ATM-89-15265. Part of model simulations were completed at the National Center for Atmospheric Research (NCAR) computer facilities. NCAR is supported by the NSF. The authors would like to thank Craig Tremback and Bob Walko for their assistance in the use of the RAMS model. Thanks also to the Hellenic Meteorological Service (EMY) and the Program for Air Quality Monitoring in Athens (PERPA) for providing the necessary meteorological and air quality data.

References

Asimacopoulos, D.N., D. Deligiorgi, C. Drakopoulos, C. Helmis, K. Kokori, D.P. Lalas, D. Sikiotis, and C. Varostsos. 1991. *An experimental study of air pollutant transport over complex terrain.* Atmos. Envir., in press.

Avissar R., and R.A. Pielke, 1989. *A parameterization of heterogeneous land surfaces for atmospheric numerical models and its impact on regional meteorology.* Mon. Wea. Rev., **117**, 2113–2136.

Carapiperis, L. 1951. *On the periodicity of the Etesian in Athens.* Weather, Dec. 1951.

Carapiperis, L. and Catsoulis N. 1977. *Contribution to the study of sea breeze in Athens area during the winter.* Bull. Hell. Meteorol. Soc. **2**, 1–18.

Dikaiakos, J.G., 1983. *The air pollution regime during Halkyon days.* Weather., 1983.

Glendening, J.W., B.L. Ulrickson, and J.A. Businger. 1986. *Mesoscale variability of boundary layer properties in the Los Angeles basin.* Mon. Wea. Rev., **114**, 2537–2549.

Helmis, C.G., D.N. Asimacopoulos, D. Deligiorgi, and D.P. Lalas. 1987. *Observations of the sea breeze front structure near the shoreline.* Boundary-Layer Meteorol., **38**, 395–410.

Kallos, G., and P. Kassomenos. 1991. *Weather conditions during air pollution episodes in Athens, Greece: An overview of the problem.* Preproceedings of the 19th ITM of NATO-CCMS on Air Pollution Modelling and its Application, Ierapetra, Greece, 77–102.

Katsoulis, B.D., 1988a. *Aspects of the Occurrence of Persistent Surface Inversions over Athens Basin, Greece.* Theor. and Appl. Climatol., **39**, 98–107.

Katsoulis, B.D., 1988b.*Some Meteorological Aspects of Air Pollution in Athens, Greece.* Meteorol. Atmos. Phys., **39**, 203–212.

Lalas, D.P., V.R. Veirs, G. Karras and G. Kallos, 1982. *An analysis of SO_2 concentration levels in Athens, Greece.* Atmos. Environ. **16**. 531–544.

McKendry, I.G., 1989. *Numerical simulation of sea breezes over the Auckland region, New Zealand – Air quality implications.* Boundary-Layer Meteorol., **49**, 7–22.

Pielke, R.A., R.T. McNider, M. Segal, and Y. Mahrer. 1983. *The use of a mesoscale numerical model for evaluations of pollutant transport and diffusion in coastal regions and over irregular terrain.* Bull. American Meteorol. Soc. **64**, 243–249.

Pielke, R.A., M. Garstang, C. Lindsey and J. Gusdorf, 1987. *Use of a synoptic classification scheme to define seasons.* Theor. Appl. Climatol., **38**, 57–68.

Pielke, R.A., and R. Avissar, 1990. *Influence of landscape structure on local and regional climate.* Landscape Ecology, **4**, 133–155.

Pielke, R.A., R.A. Stocker, R.W. Arrit and R.T. McNider, 1991. *A procedure to estimate worst-case air quality in complex terrain.* Environmental International, **17**, 559–574.

Pielke, R.A., W.R. Cotton, R.L. Walko, C.J. Tremback, W.A. Lyons, L.D. Grasso, M.E. Nicholls, M.D. Moran, D.A. Wesley, T.J. Lee and J.H. Copeland, 1992. *A comprehensive meteorological modelling system – RAMS.* Submitted to Modeling Atmos. Phys., pp 67.

Prezerakos, N.G., 1986. *Characteristics of the sea breeze in Attica, Greece.* Boundary-Layer Meteorol., **36**, 245–266.

Segal, M., W.E. Schreiber, G. Kallos, J. R. Garatt, A. Rodi, J. Weaver, and R. A. Pielke. 1989a. *The impact of crop areas in northest Colorado on midsummer mesoscale thermal circulations.* Mon. Wea. Rev., **117**, 809–825.

Segal, M., J.R. Garratt, G. Kallos, and R.A. Pielke. 1989b. *The impact of wet soil and canopy temperatures on daytime boundary-layer growth.* J. Atmos. Sci., **46**, 3673–3784.

Segal, M., G. Kallos, J. Brown, and M. Mandel. 1992. *On morning temporal variations of shelter level specific humidity.* J. of Appl. Meteorol., **31**, 74–85.

Steyn, D.G., and G. Kallos. 1992. *A study of dynamics of hodograph rotation in the sea breezes of Attica, Greece.* Boundary-Layer Meteorol., **58**, 215–228.

Ulrickson, B., and C. F. Mass, 1990a*Numerical investigation of mesoscale circulations over the Los*

Angeles basin, Part I: A verification study. Mon. Wea. Rev., **118**, 2138–2161.
Ulrickson, B., and C. F. Mass, 1990b. *Numerical investigation of mesoscale circulations over the Los Angeles Basin, Part II: Synoptic influences and pollutant transport*. Mon. Wea. Rev., **118**, 2162–2184.
Xian, Z. and R.A. Pielke, 1991. *The effects of width of landmasses on the development of sea breezes*. J. Appl. Meteorol., **30**, 1280–1304.
Yu, C.H. and R.A. Pielke, 1986. *Mesoscale air quality under stagnant synoptic cold season conditions in the Lake Powell area*. Atmos. Environ., **20**, 1751–1762.

THE USE OF A MESO-GAMMA SCALE MODEL FOR EVALUATION OF POLLUTION CONCENTRATION OVER AN INDUSTRIAL REGION IN ISRAEL (HADERA)

Y. TOKAR, J. GOLDSTEIN, Z. LEVIN and P. ALPERT
Department of Geophysics and Planetary Sciences, Tel-Aviv University, Israel 69978

(Received October 1991)

Abstract. A model was developed for pollutant dispersion from a point source simulating the Hadera (Israel) power plant stack. The model is based on the NCAR mesoscale meteorological MM4 model that provides the wind fields and coefficients of turbulent diffusion. The model was implemented using an implicit numerical scheme with changing directions. A comparison between the model calculations and an analytical solution for the advection-diffusion equation shows good agreement. Relatively low numerical diffusion of the adopted advection scheme was noted. Results for the hilly region of central Israel are presented for a summer case.

1. Introduction

The problem of environmental pollution is an important one, even for such relatively unpolluted countries as Israel. Measurements show that there is an urgent need to control emissions over urban and industrial areas and to monitor the dispersion of pollutants. In this paper we develop a model for the evaluation of pollution concentrations; the model has promise of providing reliable forecasts of air pollutant dispersion. An attempt has been made to calculate fields of air pollution concentrations for typical meteorological conditions encountered in Israel. The model can predict pollution levels at different heights under a variety of meteorological conditions. The model was implemented for the 108 by 90 km region with the Hadera power plant in its center. Topography was included. The PSU/NCAR mesoscale numerical model MM4 (Anthes *et al.*, 1987) with 31 height levels and fine 2 km resolution was adapted to fit the Israeli conditions.

In the near future it is planned to upgrade the MM4 model using a non-hydrostatic approximation.

2. Model Description

The development of a pollution dispersion model requires the inclusion of atmospheric dynamics, advection and turbulent diffusion as well as microphysical transformations of pollutant species produced by of different processes like nucleation, condensation, photochemical reactions, etc.

As a starting point, the problem of pollution dispersion due to atmospheric dynamics is studied. A three-dimensional region with dimensions of X_{max}=108 km, Y_{max}=90 km and Z_{max}=2 km was considered. It is assumed that pollutants do not diffuse above 2 km. The axes x, y, z are directed correspondingly eastward, northward and vertically upward. In the middle of the region at time $t=0$, a pollutant source is activated with a known constant output Q kg/s. The pollutant is supposed to be passive. The source simulates the Hadera power plant stack having a height of 250 m. Under these conditions the time variations of the pollutant concentration fields are calculated assuming that the meteorological fields are given. The main governing equation for the advection-diffusion process is

$$\frac{\partial c}{\partial t}+\mathbf{V}\nabla c=\frac{\partial}{\partial x}(K_x\frac{\partial c}{\partial x})+\frac{\partial}{\partial y}(K_y\frac{\partial c}{\partial y})+\frac{\partial}{\partial z}(K_z\frac{\partial c}{\partial z})+Q\delta(x-x_o)\delta(y-y_o)\delta(z-z_o) \quad (1)$$

where $c(x,y,z,t)$ is the pollutant concentration at (x,y,z) at time t; $K_x(x,y,z,t)$, $K_y(x,y,z,t)$, $K_z(x,y,z,t)$ are coefficients of turbulent diffusion; Q, x_o, y_o, z_o are source intensity and source coordinates; $\mathbf{V}=(u,v,w)$ is the wind velocity vector and $\delta(x)$ is the Dirac delta function. The last term on the right-hand side of Eq. (1) represents a continuous point source. The coefficients of turbulent diffusion are calculated in the MM4 model using the equations formulated by Smagorinsky *et al.* (1965) and Blackadar (1976,1979).

The transformation

$$t/t* \rightarrow t;$$

$$x/X_{\max} \rightarrow x; \qquad y/Y_{\max} \rightarrow y; \qquad (z-\varphi(x,y))/Z_{\max} \rightarrow z;$$

$$ut^*/X_{\max} \rightarrow u; \qquad vt^*/Y_{\max} \rightarrow v; \qquad wt^*/Z_{\max} \rightarrow w;$$

$$K_x t^*/X_{\max}^2 \rightarrow K_x; \quad K_y t^*/Y_{\max}^2 \rightarrow K_y; \quad K_z t^*/Z_{\max}^2 \rightarrow K_z;$$

$$cX_{\max}Y_{\max}Z_{\max}/(t^*Q) \rightarrow c,$$

was performed in order to nondimensionalize Eq. (1). $\varphi(x,y)$ is the underlying surface elevation above mean sea level. $t^*=1$ sec was chosen as the time scale unit because there is no proper natural time scale in the model. The resulting equation is

$$\frac{\partial c}{\partial t}+u\frac{\partial c}{\partial x}+v\frac{\partial c}{\partial y}+(w-f_x u-f_y v)\frac{\partial c}{\partial z}=$$

$$=\frac{\partial}{\partial x}(K_x\frac{\partial c}{\partial x})+\frac{\partial}{\partial y}(K_y\frac{\partial c}{\partial y})+\frac{\partial}{\partial z}[(K_z+f_x^2K_x+f_y^2K_y)\frac{\partial c}{\partial z}]- \quad (2)$$

$$-[f_x\frac{\partial}{\partial z}(K_x\frac{\partial c}{\partial x})+\frac{\partial}{\partial x}(K_xf_x\frac{\partial c}{\partial z})+f_y\frac{\partial}{\partial z}(K_y\frac{\partial c}{\partial y})+\frac{\partial}{\partial y}(f_yK_y\frac{\partial c}{\partial z})+$$

$$+\delta(x-x_o/X_{\max})\delta(y-y_o/Y_{\max})\delta(z-z_o/Z_{\max}),$$

where

$$f_x = \frac{X_{\max}}{Z_{\max}} \frac{\partial \varphi}{\partial x}, f_y = \frac{Y_{\max}}{Z_{\max}} \frac{\partial \varphi}{\partial y}.$$

Eq. (2) is solved in the unit cube with the following boundary conditions:

1) Side boundaries:

$$\text{for } \mathbf{V}_n > 0; \quad c = c_f, \qquad \text{for } \mathbf{V}_n < 0; \ \frac{\partial c}{\partial n} = 0, \tag{3}$$

where $\mathbf{V}_n$ is the wind velocity normal component to the boundary (directed inward), $\partial c/\partial n$ is the normal derivative of concentration field and c_f is the background pollutant concentration outside of the simulation domain.

2) Underlying surface:

$$-K_z \frac{\partial c}{\partial z} = q_s - V_d c \tag{4}$$

where $q(x, y, t)$ is the pollutant source (taken to be zero in the present problem) and V_d is deposition velocity.

3) Upper boundary:

$$\frac{\partial c}{\partial z} = 0. \tag{5}$$

3. External Parameters

The numerical mesocale model is used to predict the meteorological fields. The model was applied over a horizontal grid of 55 by 46 points at 2 km intervals. The number of vertical levels was 31. The vertical grid was adjusted to increase the resolution near the underlying surface. Radiosonde data from one site (Hadera) were used for model initialization. The following output fields from the MM4 were employed:

1) Wind fields.
2) Fields of horizontal and vertical turbulent exchange coefficients.

4. Numerical Procedure

To numerically solve the problem formulated in Eqs. (2)–(5) an implicit scheme with 6 intermediate time steps was adopted (Marchuk, 1974). The finite-differences operators denoted by Λ_x, Λ_y, Λ_z correspond to the differential operators:

$$D_x = u\frac{\partial}{\partial x} - \frac{\partial}{\partial x}(K_x \frac{\partial}{\partial x}),$$

$$D_y = v\frac{\partial}{\partial y} - \frac{\partial}{\partial y}(K_y \frac{\partial}{\partial y}),$$

$$D_z = (w - uf_x - vf_y)\frac{\partial}{\partial z} - \frac{\partial}{\partial z}[(K_z + f_x^2 + f_y^2 K_y)\frac{\partial}{\partial z}].$$

Λ_{13}, Λ_{31}, Λ_{23}, Λ_{32} are operators corresponding to the terms in the square brackets in Eq. (2).

The numerical scheme (Marchuk, 1974) for the n-th time step then becomes

$$\begin{aligned}
(E + \frac{T}{2}\Lambda_x)c^{n+\frac{1}{6}} &= (E - \frac{T}{2}\Lambda_x - \frac{T}{2}\Lambda_{13})c^n \\
(E + \frac{T}{2}\Lambda_y)c^{n+\frac{2}{6}} &= (E - \frac{T}{2}\Lambda_y - \frac{T}{2}\Lambda_{23})c^{n+\frac{1}{6}} \\
(E + \frac{T}{2}\Lambda_z)c^{n+\frac{3}{6}} &= (E - \frac{T}{2}\Lambda_z)c^{n+\frac{2}{6}} + \frac{T}{2}q \\
(E + \frac{T}{2}\Lambda_z)c^{n+\frac{4}{6}} &= (E - \frac{T}{2}\Lambda_z)c^{n+\frac{3}{6}} + \frac{T}{2}q \qquad (6) \\
(E + \frac{T}{2}\Lambda_y)c^{n+\frac{5}{6}} &= (E - \frac{T}{2}\Lambda_y - \frac{T}{2}\Lambda_{32})c^{n+\frac{4}{6}} \\
(E + \frac{T}{2}\Lambda_x)c^{n+1} &= (E - \frac{T}{2}\Lambda_x - \frac{T}{2}\Lambda_{31})c^{n+\frac{5}{6}},
\end{aligned}$$

E is a the unit operator, τ is the time step value, q is the finite-differences operator representing the source function:

$$q = \delta_{iI}\delta_{jJ}\delta_{kK}/(\Delta x \Delta y \Delta z).$$

Here δ_{lm} is the Kronecker delta, i, j, k are the grid points indices, I, J, K are the coordinates of the source in the grid and $\Delta x, \Delta y, \Delta z$ are the grid intervals at point (I, J, K) in the directions x, y, z respectively.

The procedure (6) has been implemented on a 55 by 46 by 31 grid. The coding was carried out in C language using an IBM RISC/6000 workstation.

5. Results and Discussion

The computations were first performed assuming constant parameters and flat topography, in order to compare with analytical solutions. Fig. 1 shows the contours of $log(c)$ obtained for a simple and analytically solvable two-dimensional model where both the wind velocity and the coefficients of turbulent diffusion are constant ($u = 5$ m/s, v=0, $K_x = K_y = 1000$ m^2/s). Figs. 1,b,c compare these results with the analytical solution of the advection-diffusion equation. The latter is a moving Gaussian (Seinfeld, 1986). The comparison shows quite good agreement. The velocities of pollutant cloud drift obtained numerically and analytically are nearly equal. Maximal height of the numerical concentration distribution diminishes with time no faster than predicted by the analytical solution, which indicates that the

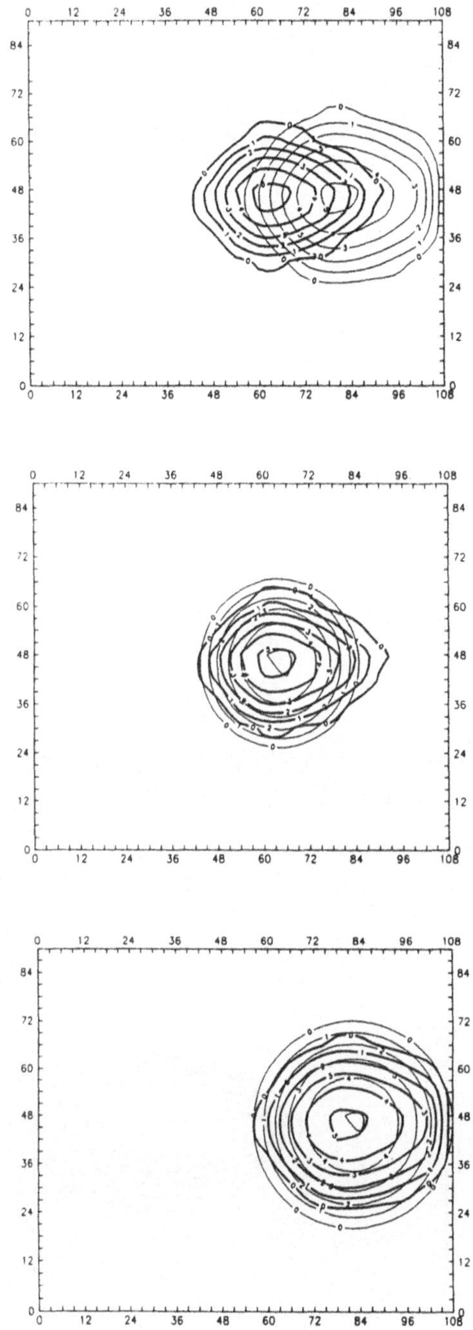

Fig. 1. Comparison between the numerical and the analytical solutions for the case of an instantaneous source at point (46,46). a) Contours of the numerical solution for t=1 h (wide) and t=2 h (thin). b) Contours of the numerical (wide) and the analytical (thin) solutions for t=1 h. c) Same as b), but for t=2 h. Distances are in kilometers.

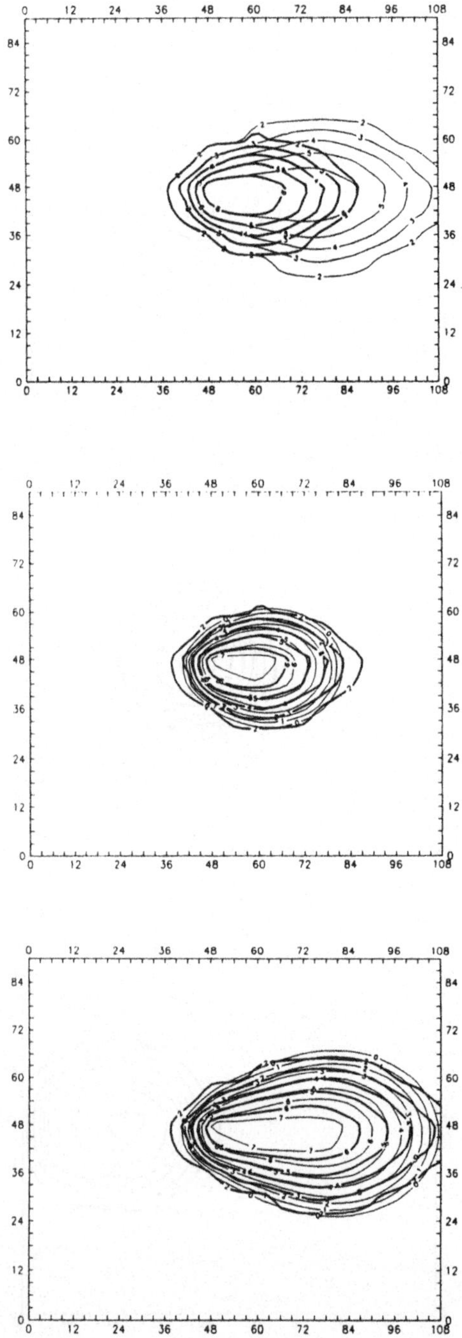

Fig. 2. Comparison between numerical and analytical solutions for the case of a continuous source. a), b) and c) descriptions are the same as for Fig. 1.

Fig. 3. The contours of $log(c)$ in the case of pollutant dispersion from the Hadera power plant at different elevations above sea level and times. a) 100 m of elevation, t=2 h. b) The same as a), but for t=6 h. c) 400 m of elevation, t=2 h. d) The same as c), but for t=6 h. e) 1000 m of elevation, t=2 h. f) The same as e), but for t=6 h. One distance unit corresponds to 2 km.

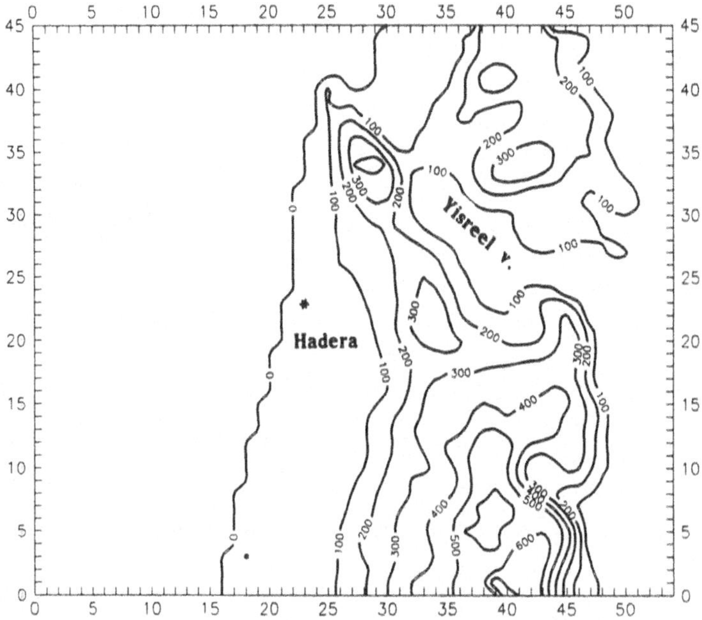

Fig. 4. The topography of the Hadera region. The asterisk denotes the source location. One distance unit corresponds to 2 km.

effect of the so-called "artificial viscosity" in the adopted numerical scheme is relatively small. This result was obtained by using the Smolarkiewicz (1986) numerical scheme for the solution of the advective equation, thus minimizing the artificial viscosity, which generally is considered to present a serious problem. Fig. 2 presents results analogous to those of Fig. 1, but obtained for a continuous source. It should be noted, that since the grid step is 2 km, the source is numerically distributed around its origin into a square area with a 4 km side. This led to a discrepancy between the numerical and analytical solutions in the neighborhood of the source. This problem will be solved in the future by increasing the grid resolution near the source.

As an example of more realistic results, Fig. 3 shows the concentration field evolution at horizontal surfaces with elevations of 100, 400 and 1000 m above sea level. The topography of the domain is presented in Fig. 4. The meteorological data have been obtained using the MM4 model based on the radiosonde data measured at August 1, 1989. This date was typical for summer meteorological conditions in Israel with increased air pollution. Present calculations were performed under model conditions $Q = 1$, $c_f = 0$, $v_d = 0.1$ m/s, for 6 hours of real time. The source is assumed to have constant intensity and the pollutant admixture is supposed to be inert. Using the Pasquill and Smith (1983) formula, the effective height of the Hadera power plant stack was calculated to be approximately 400 m. Fig. 4 shows that the height of the hilly area downstream of the pollutant source is below 400 m. The source is located at the point (23,23) and is denoted by an asterisk on Fig. 4.

Figs. 3a–d demonstrate that in the case of a constant source and westerly wind (characteristic for this area during summer), the pollutant cloud is quickly stabilized and the mainly polluted areas are within a radius of 6 km of the source and the Yisreel Valley. Small quantities of pollutant slowly penetrate the north-eastern part of Israel. Only above 1000 m where the wind was north-easterly, was it observed (Fig. 3e,f) that the pollutant was transferred southwards.

Obviously, the lack of sufficient upper-level meteorological data to input into MM4 is critically affecting the accuracy of the predicted wind and the computed turbulent diffusion coefficients. At the moment, our calculations are based on the one-point radiosonde measurements, performed near Hadera. In the future, the focus will be on developing the microphysical and chemical processes as well as improving the numerical predictions by using additional data, both surface and upper-air.

References

R.A. Anthes, E.Y.Hsie, Y.H.Kuo (1987).*Description of the Penn State/NCAR Mesoscale Model Version 4 (MM4)*. NCAR Technical Note NCAR/TN–282+STR.

Blackadar, A.K. (1976). *Modeling the nocturnal boundary layer.* Preprints of Third Symposium on Atmospheric Turbulence, Diffusion and Air Quality, Raleigh, NC, 19–22 October 1976, Amer. Meteor. Soc., Boston, 46–49.

Blackadar, A.K. (1979). *High resolution models of the planetary boundary layer*. Advances in Environmental Science and Engineering, 1, No. 1, Pfafflin and Ziegler, eds., Gordon & Breach Sci.Pub., N.Y., 50–85.

Marchuk, G. (1974). *Numerical Methods in Weather Prediction*. Academic Press.

Pasquill, F. and F.B. Smith (1983).*Atmospheric Diffusion*. John Wiley & Sons, 437pp.

Seinfeld, J.H. (1986). *Atmospheric Chemistry and Physics of Air Pollution*. John Wiley & Sons, 738pp.

Smagorinsky, J., S. Manabe and J.L. Holloway, Jr. (1965) *Numerical results from a nine-level general circulation model of the atmosphere*. Mon. Wea. Rev., **92**, 727–768.

Smolarkiewicz, P.K. and T.L. Clark (1986). *The multidimensional positive definite advection transport algorithm: further development and applications*. J. Comp. Phys., **67**, 2, 396–438.

PART II

ANALYSIS OF CONCENTRATION FLUCTUATIONS

RECENT DEVELOPMENTS IN THE LAGRANGIAN STOCHASTIC THEORY OF TURBULENT DISPERSION

B. L. SAWFORD
CSIRO Division of Atmospheric Research
Private Bag #1, Mordialloc, VIC 3195, Australia

(Received October 1991)

Abstract. In this paper some fundamental aspects of the Lagrangian stochastic theory of turbulent dispersion are discussed. Because of their similar mathematical form, the one- and two-particle theories are treated in parallel. Particular issues identified and discussed include the lack of uniqueness and universality, the role of Reynolds number and intermittency, the importance of two-particle acceleration correlations in relative dispersion and the imposition of consistency constraints between one- and two-particle models.

1. Introduction

The Lagrangian stochastic theory of turbulent dispersion is enticingly simple in concept but its development for other than the simplest turbulent flows has not been straightforward. Initial extensions of the basic Langevin equation to include effects of inhomogeneity, non-stationarity and non-Gaussian behaviour were made on an *ad hoc* basis (see Sawford (1985) for a review) but more recent work by Thomson (1987, 1990) has begun to place the theory on a more general and rigorous foundation.

In this paper we review recent work aimed at exploring some of the remaining difficulties in the development of a rigorous theory and the impact of some of the assumptions which are made. Formally, the mathematics is the same for both one-particle models (which deal with absolute dispersion in a fixed reference frame and the mean concentration field) and two-particle models (which deal with relative dispersion and the mean-square concentration field). Thus there are advantages in treating both together, although naturally in some respects the one-particle case is simpler and has been developed more fully.

2. Basis for the Theory

Here we briefly review Thomson's (1987) and (1990) development of the general stochastic equation describing turbulent dispersion, commencing with the one-particle case and then generalising to two particles.

2.1. ONE-PARTICLE CASE

The basic assumption underlying the theory is that the velocity of a fluid element, $\mathbf{u}$, is a Markov process which is a continuous function of time. There has apparently been no rigorous demonstration of the conditions under which the Markov assumption can be justified. However, a number of authors (Lin and Reid, 1963; Durbin, 1983; Borgas and Sawford, 1991a) have argued that, since in three-dimensional turbulence the acceleration of a fluid element is significantly correlated only for times much shorter than the lifetime of the energetic eddies, the velocity is Markovian in the large Reynolds number limit. This immediately leads to our first question;

What is the importance of finite Reynolds number corrections to the theory?

Given the velocity of a fluid element, its position, $\mathbf{x}$, can be determined by integrating the deterministic equation

$$dx_i = u_i dt. \tag{1}$$

(Throughout this paper we use subscripts i,j,k... to denote Cartesian tensors with summation over repeated subscripts). Since $\mathbf{u}$ is continuous and Markovian, it follows from (1) that the joint process $(\mathbf{u}, \mathbf{x})$ is also continuous and Markovian. Thus in its most general form the equation describing the evolution of $\mathbf{u}$ is the stochastic differential equation

$$du_i = a_i(\mathbf{x}, \mathbf{u}, t)dt + b_{ij}(\mathbf{x}, \mathbf{u}, t)dW_j(t) \tag{2}$$

where the random term, $d\mathbf{W}(\mathbf{t})$, is the incremental Wiener process, which is Gaussian with zero mean and variance

$$< dW_i(t)dW_j(t+\tau) >= \delta_{ij}\delta(\tau)dtd\tau \tag{3}$$

and the notation $<>$ represents an ensemble average.

According to Kolmogorov's theory of local isotropy (Monin and Yaglom, 1975, pp, 345-377), the Lagrangian structure function has the inertial sub-range form

$$< \mathcal{U}_i \mathcal{U}_j >= \delta_{ij} C_o \bar{\varepsilon} t \qquad (t_k \ll t \ll t_E) \tag{4}$$

where $\mathcal{U}(t)$ = $\mathbf{u}$(t) - $\mathbf{u}$(0), $\bar{\varepsilon}$ is the mean rate of dissipation of turbulence kinetic energy, $t_E = \sigma^2/\bar{\varepsilon}$ is a time scale representative of the energy-containing scales of the turbulence, $t_k = (\nu/\bar{\varepsilon})^{1/2}$ is the Kolmogorov microscale, $\sigma^2 = \frac{1}{3} < u_j u_j >$ is the velocity variance and C_o is a universal constant. To leading order in t as $t \longrightarrow 0$ (but with $t \gg t_k$ as implicity required by the Markov approximation), we have from (2) and (3)

$$b_{ik} b_{jk} t = \delta_{ij} C_o \bar{\varepsilon} t. \tag{5}$$

Since the statistics of $\mathbf{u}$ are determined by $b_{ij}b_{jk}$, we are free to choose any b_{ij} which satisfies (5). The simplest choice is

$$b_{ij} = \delta_{ij}\sqrt{C_o \bar{\varepsilon}} \tag{6}$$

which prompts the next two questions:

How do intermittency corrections to Kolmogorov's theory affect the present theory?

Is C_o, as it appears in the present theory, truly universal and is it influenced by the Reynolds number?

2.2. Two-Particle Case

For the joint motion of a pair of particles, equations formally identical to (1)-(3) arise from the corresponding assumption that $\mathbf{u}^{(1)}$ and $\mathbf{u}^{(2)}$ are jointly Markovian and continuous, where the superscripts $^{(1)}$ and $^{(2)}$ are particle labels. In this case, the vectors in (1) to (3) are six-dimensional, with components i = 4,5,6 representing particle 2. (Thomson, 1990; Borgas and Sawford, 1991b).

In this case the second-order tensors corresponding to those discussed above consist of four subtensors, two of which are one-particle tensors and two of which are two-particle tensors. For example, the velocity structure function is

$$< \mathcal{U}_i \mathcal{U}_j >= \begin{bmatrix} < \mathcal{U}_{i'}^{(1)} \mathcal{U}_{j'}^{(1)} > & < \mathcal{U}_{i'}^{(1)} \mathcal{U}_{j"}^{(2)} > \\ < \mathcal{U}_{i"}^{(2)} \mathcal{U}_{j'}^{(1)} > & < \mathcal{U}_{i"}^{(2)} \mathcal{U}_{j"}^{(2)} > \end{bmatrix} \qquad (7)$$

where the indices i, j run over the components of six-dimensional phase space while the indices $i'j', i", j"$ run over three-dimensional physical space.

Now the two one-particle subtensors can be treated exactly as for the one-particle case set out in (4) through (6). The two-particle subtensors (i.e. the off-diagonal ones in (7)) are a little different since they depend also on the initial magnitude, Δ_o, of the separation, $\Delta(t) = \mathbf{x}^{(2)}(t) - \mathbf{x}^{(1)}(t)$. This dependence introduces a new time scale, $t_o = (\Delta_o^2/\bar{\varepsilon})^{1/3}$, which is the time taken for the initial separation to be 'forgotten' (Batchelor, 1950).

For $t \ll t_o$ the Taylor series expansion of the two-particle structure function gives (Borgas and Sawford, 1991)

$$< \mathcal{U}_i^{(1)}(t)\mathcal{U}_j^{(2)}(t) > \approx < \mathbf{A}_i(\mathbf{x}^{(1)}(0))\mathbf{A}_j(\mathbf{x}^{(2)}(0)) > t^2 \qquad (8)$$

where $< \mathbf{A}_i(\mathbf{x}^{(1)}(0))\mathbf{A}_j(\mathbf{x}^{(2)}(0)) >$ is the Eulerian two-point acceleration covariance. For inertial sub-range separations, $\eta \ll \Delta_o \ll L$, where η is the Kolmogorov microscale and $L = \sigma^3/\bar{\varepsilon}$ is a length scale representative of the energy-containing scales, $< \mathbf{A}_i(\mathbf{x}^{(1)}(0))\mathbf{A}_i(\mathbf{x}^{(2)}(0)) > \approx \kappa\bar{\varepsilon}/t_o$(Monin and Yaglom, 1975 pp. 371) and therefore

$$< \mathcal{U}_i^{(1)}(t)\mathcal{U}_i^{(2)}(t) >= \kappa\bar{\varepsilon}t^2/t_o + o(t^2/t_o). \qquad (9)$$

Thus in the limit $t \longrightarrow 0$ the off-diagonal subtensors in (7) are negligible compared with the one-particle subtensors and therefore to $O(t)$ the six-dimensional structure function (7) is diagonal; that is, to be consistent with Kolmogorov's theory, (5) and (6) should carry over directly in six-dimensional form to the two-particle theory. This is consistent with the often-made assumption (Novikov, 1963; Lin and Reid, 1963; Monin and Yaglom, 1975, p. 546; Gifford, 1982 and 1983; Sawford, 1984) that the contribution of the two-particle acceleration covariance is negligible. However, for later times such that $t \approx t_o$, the right-hand side (RHS) of (9) becomes $\kappa\bar{\varepsilon}t$ which is of the same order as the one-particle structure function! Thus, although the expansion (9) is anticipated to break down at this point, it suggests that the two-particle structure function increases in importance with time and is comparable with the one-particle function for $t_o \ll t \ll t_E$. Since the velocity structure functions are directly related kinematically to the acceleration correlation functions, this raises our next question:

What is the form of the two-particle acceleration correlations and how is this reflected in the Lagrangian stochastic theory?

2.3. The Fokker-Planck Equation

An alternative, Eulerian, description of the Lagrangian equations (1)-(3) is available through the Fokker-Planck equation (Gardiner, 1983, p. 52), which is a diffusion equation for the joint probability density function (pdf) for **u** and **x**, P(**u**,**x**,t) and is

$$\frac{\partial P}{\partial t} + u_i \frac{\partial P}{\partial x_i} = -\frac{\partial}{\partial u_i}(a_i P) + \frac{1}{2} C_o \bar{\varepsilon} \frac{\partial^2 P}{\partial u_i \partial u_i}. \tag{10}$$

Now the distinction between Eulerian and Lagrangian statistics is merely one of sampling or conditioning which should not affect the form of (1)-(3) or equivalently of (10). We wish to use these equations to calculate Lagrangian statistics but they are equally valid for Eulerian velocity statistics and if we assume that the Eulerian statistics are known, then they represent a constraint on the form of the equations. In particular, if the Eulerian velocity pdf is denoted by P_E (**u**,**x**,t), then from the Fokker-Planck equation (10) we have

$$\frac{\partial}{\partial u_i}(a_i P_E) = -\frac{\partial P_E}{\partial t} + \frac{1}{2} C_o \bar{\varepsilon} \frac{\partial^2 P_E}{\partial u_i \partial u_i} - u_i \frac{\partial P_E}{\partial x_i}. \tag{11}$$

This equation does not determine **a** uniquely; in general any purely rotational vector (in **u** space) can be added to $\mathbf{a}P_E$ and the resulting vector will still satisfy (11). This result leads to the related questions:

Under what conditions is the non-uniqueness nontrivial?

Are there additional constraints which can eliminate the non-uniqueness?

Much of the research effort of our group over the past couple of years has been directed towards answering these questions. The task is by no means complete but in the next Sections we give an overview of the progress made.

3. Non-Uniqueness

3.1. ONE-PARTICLE CASE

A unique solution for the fully three-dimensional case has been shown to exist only for the one-particle problem in homogeneous, isotropic turbulence (Borgas and Sawford, 1991b). In general, the non-uniqueness in the solution to (11) is associated with a vector ϕ_i such that

$$\frac{\partial}{\partial u_i}(P_E\phi_i) = 0. \tag{12}$$

Under the conditions of homogeneity and isotropy we have (Batchelor, 1953, p. 42) $P_E\phi_i(\mathbf{u},t) = \psi(|\mathbf{u}|,t)u_i$. On substituting this form into (12) it can be shown that

$$\phi_i(\mathbf{u},t) = c|\mathbf{u}|^{-3}P_E^{-1}u_i \tag{13}$$

from which it follows immediately that the constant c must vanish since **a** must be finite for all **u**. The further assumption of a Gaussian form for the Eulerian velocity pdf leads to the well-known result for decaying isotropic turbulence (Durbin, 1983)

$$du_i = -(\frac{1}{2}C_o\bar{\varepsilon}\sigma^{-2} - \frac{1}{2}\sigma^2\frac{\partial\sigma^{-2}}{\partial t})u_i dt + \sqrt{C_o\bar{\varepsilon}}dW_i. \tag{14}$$

Note that in this case $\mathbf{a}P_E$ is irrotational in **u**-space.

For homogeneous anisotropic turbulence, it is easy to write down a non-trivial rotational term, for example,

$$P_E\phi_i = \varepsilon_{ijk}u_k\frac{\partial P_E}{\partial u_j} \tag{15}$$

where ε_{ijk} is the unit alternating tensor (Batchelor, 1953, p. 38). Thus, there is no unique solution to (11) in this case.

For inhomogeneous turbulence, Sawford and Guest (1988) write down two different solutions (due to Thomson (1987) and Borgas (private communication)) and show numerically for a simple neutral boundary-layer flow that these solutions are distinct. They also demonstrate that different values of C_o effectively apply to different flows.

3.2. TWO-PARTICLE CASE

Thomson (1986, 1990) pointed out that the complications which enter the one-particle theory due to turbulence inhomogeneity are already present in the two-particle theory for homogeneous isotropic turbulence. This is because mathematically the variation of the two-point velocity covariance with the separation of the points has a similar effect in (2), or equivalently (10), as does the variation of the one-point stress tensor with position. In particular, in both cases the spatial gradient term in (10) is non-zero and **a** depends explicity on the position vector **x**. As a result there is no unique two-particle solution to (11) even for isotropic turbulence.

Under the assumption of a Gaussian form for the two-point Eulerian velocity pdf (which cannot be strictly correct but is a useful starting point) Borgas and Sawford (1991b) demonstrate this non-uniqueness explicity by deriving a whole class of solutions quadratic in **u**.

$$a_i = \Gamma_i - (\frac{1}{2}C_o\bar{\varepsilon}\sigma^{-2})u_i + \gamma_{ijk}u_ju_k \tag{16}$$

where we recall that tensors in (16) are six-dimensional. Two simple examples (for incompressible flow) are Thomson's (1990) model for which

$$\Gamma_i = 0; \qquad \gamma_{ijk} = -\frac{1}{2}\lambda_{il}^{-1}\frac{\partial \lambda_{lj}}{\partial x_k} \tag{17}$$

and a new solution with

$$\Gamma_i = \sigma^{-2}\lambda_{ij}^{-1}\lambda^{-1}\frac{\partial \lambda^{-1}}{\partial x_j}; \qquad \gamma_{ijk} = -\frac{1}{2}\lambda_{il}^{-1}\frac{\partial \lambda_{jk}}{\partial x_l} \tag{18}$$

where $\lambda = |\lambda_{ij}|$ and λ_{ij} is the inverse of the six-dimensional Eulerian velocity covariance tensor. Numerical calculations of the mean-square relative velocity and the mean-square separation of the pair of particles show that the two models produce significantly different results in the inertial sub-range.

In general, after imposing the zero-divergence constraint (12) and various symmetry conditions, Borgas and Sawford (1991b) show that the arbitrariness in the quadratic system of solutions reduces to five unknown scalar functions of the separation. Attempts to resolve this arbitrariness by invoking additional constraints are described in the next sub-section.

3.3. Consistency Constraints

A very general requirement which Thomson (1990) noted is not satisfied by his model, (17), is that one-particle statistics calculated from a two-particle model should be the same as those calculated from the corresponding one-particle model. Borgas and Sawford (1991) have used this requirement, which they call two-to-one reduction, to further constrain their class of quadratic models.

There are infinitely many two-to-one reduction constraints and it can be shown that they cannot all be satisfied by models of the restricted form discussed above, that is, quadratic form models with a Gaussian Eulerian pdf. Nevertheless, if we can order the constraints in some way, we would expect the lowest order constraints to be most important and that enforcing them would lead to improved models. Borgas and Sawford (1991b) develop an ordering procedure (which is not unique) by considering the general two-particle pdf

$$P_{\mathcal{J}}(\mathbf{u},\mathbf{x},t) = \int P_L(\mathbf{u},\mathbf{x},t;\mathbf{u}_o,\mathbf{x}_o,0)P_E(\mathbf{u}_o;\mathbf{x}_o,0)\mathcal{J}(\mathbf{x}_o)d^6\mathbf{u}_o d^6\mathbf{x}_o \tag{19}$$

representing the statistics of particle pairs initially distributed in space according to $\mathcal{J}(\mathbf{x}) = S(\mathbf{x}^{(1)})S(\mathbf{x}^{(2)})$, which is suitably normalised. Note that the Eulerian pdf is recovered from the special limit where S is a uniform distribution and the Lagrangian pdf corresponds to the case S($\mathbf{x}$)=$\delta(\mathbf{x})$.

A general formulation of the two-to-one reduction principal is that the marginal one-point distribution obtained from (19) should be a solution of the corresponding one-particle model - for example, (14) for decaying isotropic turbulence. This general principle is effectively unusable because it involves the Lagrangian pdf, but can be cast into Eulerian form by expanding $P_{\mathcal{J}}$ as a Taylor-series expansion in time about t=0. Two-to-one reduction is satisfied identically for the first two terms of this expansion but the second order terms yield two constraints on **a**. For stationary isotropic turbulence the first is

$$< a_i^{(1)}|\mathbf{u}^{(1)} >_E = -\frac{1}{2}C_o\bar{\varepsilon}\sigma^{-2}\mathbf{u}^{(1)}, \tag{20}$$

and the second is satisfied by the kinematic relation

$$< a_i^{(2)}|\mathbf{u}^{(1)} >_E = - < \frac{du_i^{(2)}}{dt}|\mathbf{u}^{(1)} >_E \tag{21}$$

where the averages $<>_E$ are Eulerian averages over the velocity of particle two conditional on the velocity of particle one. Although the constraints (20) and (21) can be interpreted simply in terms of the mean acceleration of the particles, and can be derived directly, the expansion procedure is important in generating a hierarchy of constraints in a systematic way.

In general (20) corresponds to three scalar equations, but only one of these is independent of the symmetry and divergence constraints already applied. It therefore reduces to four the number of unknown scalar functions in the general quadratic form. Both Thomson's model (17) and the model (18) satisfy (20).

The second constraint (21) leads to a new sort of consistency requirement by substituting in the RHS using the Navier Stokes equations to give

$$< a_i^{(2)}|\mathbf{u}^{(1)} >_E = -\frac{1}{\rho} < \frac{\partial p^2}{\partial \Delta_i}|\mathbf{u}^{(1)} >_E + O(Re^{-1}) \tag{22}$$

where Re=$\sigma L/v$ is the turbulence Reynolds number. The pressure gradient term can be determined from the Eulerian velocity pdf and the velocity correlation function (Batchelor, 1953, p.178). For stationary turbulence we should also include an external forcing term in (22) but this can be ignored in the inertial sub-range. Three new scalar constraints on the four remaining unknown functions result from (22). Neither Thomson's model nor (18) satisfies these new constraints. However, since there is still a single arbitrary function remaining, it is relatively easy to build a quadratic form model which satisfies (22). Borgas and Sawford (1991b) have done

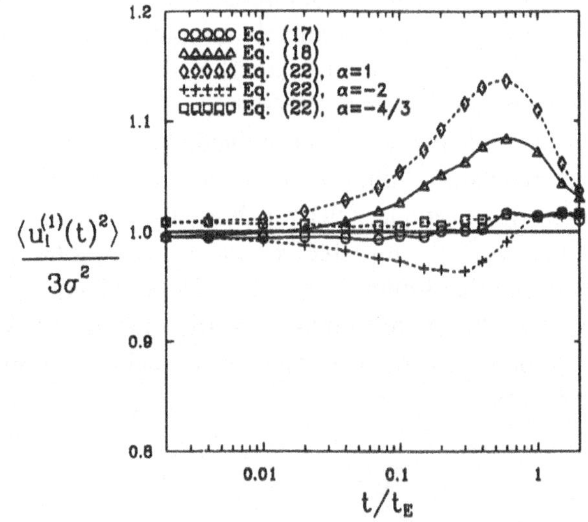

Fig. 1. Performance of various two-particle models in reproducing the correct one-particle velocity variance.

this so as to satisfy inertial sub-range scaling and to involve a single undetermined constant, α say.

The one-particle velocity variance calculated as a function of time is shown in Figure 1 for a range of values of α. Also shown are corresponding results for Thomson's model, which performs well in this regard, reproducing the exact value to within the statistical uncertainty of the calculation, and (18) which performs poorly. The performance of the general quadratic model varies considerably and for some values of α its one-particle velocity variance departs significantly from the correct value. For α=-4/3 it performs as well as Thomson's model while also satisfying the additional Navier Stokes constraint.

Figure 2 shows the mean-square separation of the two particles relative to its initial value and initial trend, $< (\mathcal{X}_i^{(1)} - \mathcal{X}_i^{(2)})^2 >$, where $\mathcal{X}_i = \int_o^t \mathcal{U}_i dt = x_i(t) - x_i(0) - u_i(0)t$, is a function of time for the same range of models. We are particularly interested in times in the inertial sub-range where it is apparent that the various models give quite different relative dispersion predictions. Note though that both Thomson's model and the general model with α =-4/3 give almost identical results. Thus in practical terms Thomson's model captures the relative dispersion characteristics of the 'optimum' more completely constrained model.

It is apparent from (20) and (21) that acceleration statistics play an important role in the structure of these stochastic models. In the next Section we consider more fully the role of acceleration correlations in the relative dispersion process.

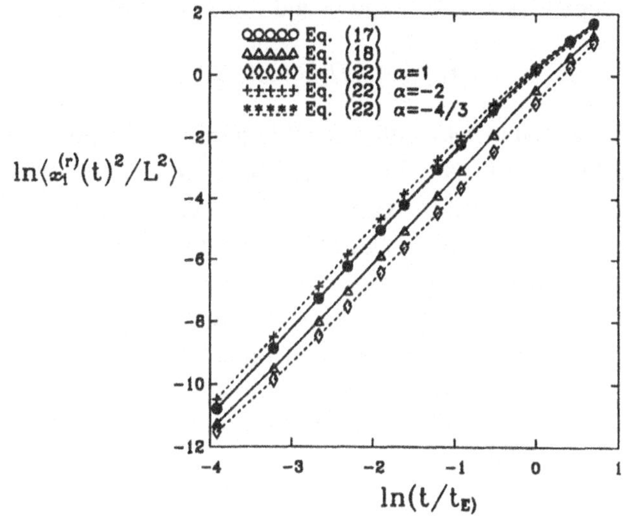

Fig. 2. Relative dispersion predictions of various models.

4. Two-Particle Accelerations

Acceleration covariances are important because they are directly related kinematically to the velocity structure function and hence to the displacement statistics, that is to the dispersion. The conventional approach (Monin and Yaglom, 1975, p. 546; Novikov, 1963, 1989) is to argue that because the accelerations are localised in space (i.e. the Eulerian acceleration covariance decays rapidly with separation) as well as in time, the two-particle acceleration covariance is negligible in the inertial sub-range or, equivalently, that the relative acceleration is stationary. A consequence of this assumption is that if we consider velocities relative to the initial velocity, i.e. $\mathcal{U}_i(t) = u_i(t) - u_i(0)$, and corresponding displacements, $\mathcal{X}_i(t)$, pairwise-relative statistics are related simply to one-particle statistics and the two-particle statistics are redundant. This is not the case for the class of models considered here, since for all these models the calculated relative dispersion is less than that obtained by neglecting the two-particle acceleration covariance. Thomson (1990) addressed this point and argued that although the two-particle acceleration covariance is much smaller than the corresponding one-particle function, its influence persists over a longer time so that it makes a comparable contribution to pair velocity and displacement statistics in the inertial sub-range. Borgas and Sawford (1991a) use dimensional and kinematic arguments to show that there is no basis for the neglect of the two-particle acceleration covariance. The crucial point in their argument is that the two-particle acceleration covariance is non-stationary since the correlation due to the initial proximity of the particles decays as the mean-square separation increases even for zero lag. Thus within the inertial sub-range the covariance depends on both t_1 and t_2 and $\bar{\varepsilon}$ and is therefore of the form

$$< A_i^{(1)}(t_1)A_i^{(2)}(t_2) >= \bar{\varepsilon}t_1^{-1}\mathcal{R}_2(t_1/t_2) + o(\bar{\varepsilon}t_1^{-1}) \tag{23}$$

which must also be symmetric with respect to particle labelling. With only weak restrictions on the form of $\mathcal{R}_2$, Borgas and Sawford (1991a) show that (23) is kinematically consistent with the corresponding inertial sub-range forms for the velocity structure function and displacement covariance respectively. Note that if stationarity is assumed, then integration of the corresponding form for the acceleration covariance, which then is of $O(\bar{\varepsilon}|t_1 - t_2|^{-1})$, leads incorrectly to a logarithmic term in the velocity structure function, a dilemma which can only be resolved by taking the acceleration covariance to be negligible to this order.

The form (23) leads to nontrivial inertial sub-range forms for the velocity and displacement covariances:

$$< \mathcal{U}_i^{(1)}(t)\mathcal{U}_i^{(2)}(t) >\approx 2C_1\bar{\varepsilon}t \qquad (t_o \ll t \ll t_E) \tag{24}$$

$$< \mathcal{X}_i^{(1)}(t)\mathcal{X}_i^{(2)}(t) >\approx (C_1 - \frac{1}{3}C_2)\bar{\varepsilon}t^3 \qquad (t_o \ll t \ll t_E) \tag{25}$$

where $C_1 = \int_0^1 \chi^{-1}\mathcal{R}_2(\chi)d\chi$ and $C_2 = \int_o^1 \mathcal{R}_2(\chi)d\chi$ are two new universal constants. (24) and (25) are of the same order in t as the corresponding one-particle results. However for $t \ll t_o$ the two-particle covariances are negligible (see (8) and Section 2).

Although it is not possible to determine the function $\mathcal{R}_2$ by dimensional reasoning, Borgas and Sawford (1991a) derive some powerful constraints on it and the new inertial sub-range constants by considering the correlation matrix for a cluster of equivalent particles and particle pairs. The most important of these constraints are

$$|\mathcal{R}_2(\xi)| \leq \xi^{1/2}\mathcal{R}_2(1) \tag{26}$$

$$0 \leq \frac{2}{3}C_1/C_o \leq 1 \tag{27}$$

$$0 \leq (C_1 - \frac{1}{3}C_2)/C_o \leq 1. \tag{28}$$

These constraints strengthen the usual Schwarz inequalities.

Borgas and Sawford (1991a) proposed a simple algebraic form which satisfies all of these constraints,

$$\mathcal{R}_2(\xi) = \alpha\xi^\mu/(1 + \xi^{2\mu-1}) \tag{29}$$

where α and μ are positive constants with $\mu > 1$.

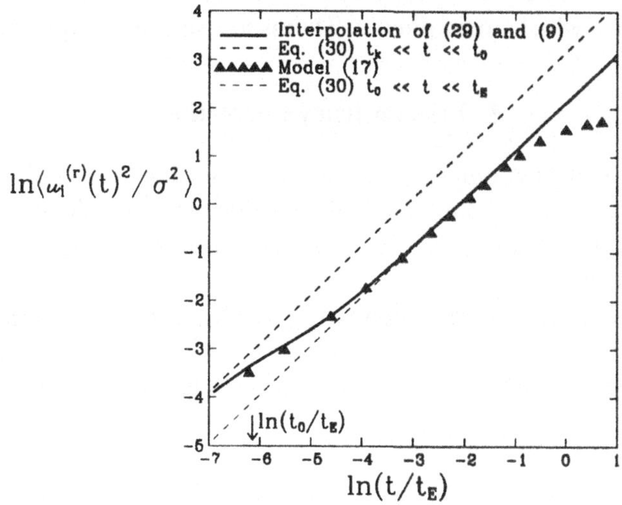

Fig. 3. The relative velocity structure function

The two-particle covariances (24) and (25) make a nontrivial contribution to the corresponding pairwise-relative quantities, resulting in a 'two-stage' inertial subrange behaviour. Thus we have for relative velocity differences, $\mathcal{U}_i^{(r)} = \mathcal{U}_i^{(1)} - \mathcal{U}_i^{(2)}$ and displacement differences, $\mathcal{X}_i^{(r)} = \mathcal{X}_i^{(1)} - \mathcal{X}_i^{(2)}$,

$$< \mathcal{U}_i^{(r)}(t)^2 >= \begin{cases} 6C_0\varepsilon t & (t_k \ll t \ll t_0) \\ (6C_0 - 4C_1)\bar{\varepsilon}t & (t_0 \ll t \ll t_E) \end{cases} \tag{30}$$

and

$$< \mathcal{X}_i^{(r)}(t)^2 >= \begin{cases} 2C_0\bar{\varepsilon}t^3 & (t_k \ll t \ll t_0) \\ (2C_0 - 2C_1 + \frac{2}{3}C_2)\bar{\varepsilon}t^3 & (t_0 \ll t \ll t_E) \end{cases} \tag{31}$$

Figure 3 illustrates this behaviour for the velocity structure function for Thomson's (1991) model with $C_o = 4$. Other details of the calculation are given in Borgas and Sawford (1991a). The upper dashed line shows the behaviour (30) and (31) for $t \ll t_o$ for which the two-particle contribution is negligible. For $t \gg t_o$ the numerical solution approaches a second straight line, the lower dashed line, which corresponds to the numerical values $C_1 = 3.9$ and $C_2 = 2.1$ in (30) and (31).

Eventually, for $t \approx t_E$, the simulation deviates from this second asymptotic regime. Thus the stochastic model implicitly accounts for both the small-time and intermediate-time behaviour and for the transition between these two ranges. The solid line in Figure 3 represents a form for $\mathcal{R}_2$ obtained by Borgas and Sawford (1991a) by interpolating between (29) with μ=3/2 and the small-time limit $\mathcal{R}_2 \approx \kappa\bar{\varepsilon}/t_o$. This interpolation form was fitted to the velocity statistics, but it also represents the displacement statistics very well without further adjustment. It apparently captures the essential ingredients of the acceleration covariance in the inertial sub-range which are implicit in Thomson's stochastic equations.

5. Intermittency Corrections

Eulerian aspects of intermittency corrections to Kolmogorov's (1941) similarity theory have received much attention both experimentally and theoretically, but the Lagrangian consequences have only recently been addressed by Novikov (1990), Pope and Chen (1990) and Borgas (1991).

In an Eulerian framework intermittency is characterised in terms of a local volume average of dissipation, ε_r, where r is the linear dimension of the volume (Monin and Yaglom, 1975, p. 591). The refined similarity hypothesis then gives

$$< |\varepsilon_r|^q > \approx \bar{\varepsilon}^q (r/L)^{(q-1)(D_q-1)} \tag{32}$$

for moments of the local dissipation, and the velocity structure functions are of the form

$$< |u_r|^q > \approx (\bar{\varepsilon} L)^{q/3} (r/L)^{1+(q/3-1)D_{q/3}} \tag{33}$$

where $|u_r|$ is a velocity difference over the distance r within the inertial sub-range. Values of the moment-exponent D differing from unity reflect the intermittency corrections and D is related to the fractal dimension of the region of space for which the dissipation is non-zero (Mandelbrot, 1976).

Many models have been proposed for the distribution of the local average dissipation in order to predict these intermittency corrections but most have been shown either empirically or theoretically to be unsatisfactory. However, the multifractal formalism of Frisch and Parisi (1985) (a mixture of fractals of many dimensions) provides a framework for describing any experimental scaling results (of the form (32), say); i.e. the measured D's can be interpreted as defining a specific multifractal model empirically (Borgas, 1991). It is also possible to parameterise the experimental results using cascade models to generate an appropriate multifractal distribution of the local dissipation (Meneveau and Sreenivasan, 1987).

Borgas (1991) has used the multifractal formalism to connect intermittency corrections in a Lagrangian reference system to those empirically determined (and modelled) for Eulerian statistics. He did this by using the ergodic hypothesis (for stationary homogeneous turbulence) to assert that Eulerian and Lagrangian averages are equivalent. Thus corresponding to (32), for averages of the dissipation over small increments of time, τ, along a particle trajectory he obtained

$$< \varepsilon_\tau^q > \approx \bar{\varepsilon}^q (\tau/t_E)^{(q-1)(D_q-1)} \tag{34}$$

where the Lagrangian exponent, D_q, is determined explicity by the Eulerian exponent D_q. For example, Figure 4 shows some experimental data for the Eulerian

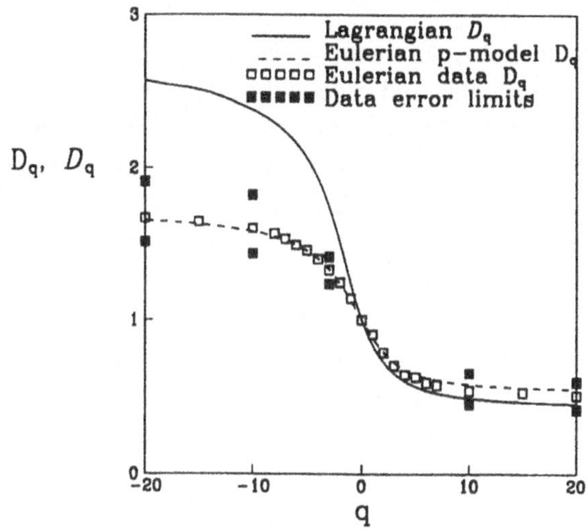

Fig. 4. Eulerian and Lagrangian intermittency moment-exponents

moment-exponent fitted with a cascade model, the so-called p-model of Meneveau and Sreenivasan (1987) and the corresponding Lagrangian moment-exponent calculated by Borgas.

Borgas and Sawford (1991c) have used these results to incorporate intermittency into a Lagrangian stochastic model. The generalisation is achieved by replacing $\bar{\varepsilon}$ by its instantaneous value along the particle trajectory; i.e. for stationary isotropic turbulence, the particle velocity is described by the equation

$$du_i = -(\frac{1}{2}C_o\varepsilon(t)\sigma^{-2})u_i dt + \sqrt{C_0\varepsilon(t)}dW_i(t). \qquad (35)$$

Pope and Chen (1990) use a similar approach for non-stationary turbulence (i.e. based on (14) but use a log-normally distributed $\varepsilon(t)$, which is itself described by an Ito equation.

(35) is best interpreted as a conditional stochastic differential equation; i.e. given a realisation of ε(t), then the velocity is Markovian and continuous and satisfies (35).

In order to implement (35) numerically, that is to generate trajectories, Borgas and Sawford (1991) use a cascade model which is essentially a random version of the p-model and which is known as the ρ-model to generate Lagrangian time series of the locally averaged dissipation, ε_T. This model provides a very good representation of the Lagrangian moment-exponent shown in Figure 4. Trajectories generated with the intermittent model are not significantly different in shape from those calculated from the usual Langevin model.

6. Reynolds Number Effects

So far we have dealt exclusively with the large-Reynolds number limit. However, most laboratory experiments and especially direct numerical simultations are at relatively low Reynolds number. In order to assess the effect on Lagrangian statistics and dispersion in particular, Sawford (1991) proposed a simple extension of the Langevin equation for stationary isotropic turbulence. He modelled the acceleration as a Markov process thus introducing a second time scale which is related to the Kolmogorov time scale. Both time scales can be specified precisely by matching the small-time behaviour and the inertial sub-range behaviour of the velocity structure function to Kolmogorov's similarity theory in the dissipation sub-range and the inertial sub-range respectively. The two time scales define a turbulence Reynolds number which is thus a parameter of the theory.

Sawford's stochastic equation for the acceleration is

$$T_L^{(\infty)} d\mathbf{A}(t) + (1 + Re^{*1/2})\mathbf{A}(t)dt + \frac{Re^{*1/2}}{T_L^{(\infty)}} \int_o^t A(t')dt'dt = \qquad (36)$$

$$= \sqrt{\frac{2\sigma^2}{T_L^{(\infty)}} Re^*(1 + Re^{*-1/2})} d\mathbf{W}(t)$$

where $T_L^{(\infty)} = 2\sigma^2/(C_o\bar{\varepsilon})$ is the Lagrangian integral time scale at infinite Reynolds number, the second Lagrangian time scale is $C_o t_k/2a_o$, a_o is the universal constant (for large Reynolds number) associated with the magnitude of the acceleration variance and

$$Re^* = (16a_o^2/C_o^4)t_E^2/t_k^2 = (16a_o^2/C_o^4)Re. \qquad (37)$$

Unconditional statistics can then be obtained by averaging over many realisations of $\varepsilon(t)$, the moments of which are given by the multifractal form (34), as well as over the white noise. For example, the velocity structure function can be calculated using this two-stage averaging process as

$$< \mathcal{U}_i(t)\mathcal{U}_j(t) >= 2\sigma^2\delta_{ij}[1- < exp(\frac{-1}{2}C_o\varepsilon_t\sigma^{-2}t) >_\varepsilon] \qquad (38)$$

which can be evaluated in terms of the Lagrangian moment exponents by expanding the exponential and using (34). At leading order in time, the Kolmogorov result (4) is retrieved since the second-order structure function is linear in ε at this order. The extension (35) ensures that the well-mixed condition of Thomson (1987) is satisfied and Borgas' (1991) theory ensures consistency with measured Eulerian intermittency corrections.

In Figure 5, the structure function (38) is compared with the non-intermittent result, which can be obtained from (38) by carrying the averaging operator inside the exponential function. The difference is small. The mean-square displacement

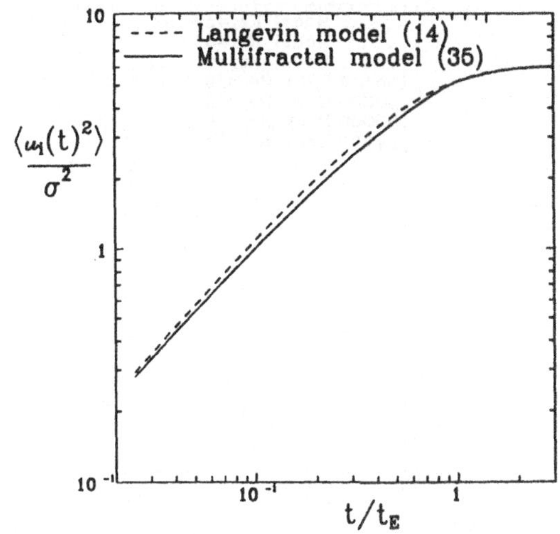

Fig. 5. Effect of intermittency on the Lagrangian velocity structure function

of particles (i.e. the dispersion) is practically identical for the two cases and so is not shown here.

Note that the separation of the Lagrangian scales depends on the magnitude of the two universal constants and may therefore be significantly different from the separation of the corresponding Eulerian scales. This difference is reflected in the numerical factor relating the two Reynolds numbers and is important at low Reynolds numbers. The Lagrangian integral time scale for finite Reynolds number is given by

$$T_L = T_L^{(\infty)}(1 + Re^{*-1/2}) = \frac{2\sigma^2}{C_o \bar{\varepsilon}}(1 + Re^{*-1/2}). \tag{39}$$

Sawford tested this model by applying it to the low Reynolds number direct numerical simulations (DNS) of turbulence by Yeung and Pope (1989). He estimated the parameters a_o (as a function of Reynolds number at low Reynolds number) and $C_o \approx 7$ by fits to the DNS results for a_o itself and to the Lagrangian integral time scale respectively. The model then provides an excellent representation of the DNS results for second-order statistical quantities such as the velocity structure function, the acceleration correlation function and dispersion.

Figure 6 shows the Lagrangian velocity structure function plotted in such a way that if the inertial sub-range form (4) held, then the plot would show a plateau of height C_o. Agreement with the DNS data is remarkable, especially considering that the value $C_o = 7$ was estimated from a 'bulk' property of the statistics, the integral time scale, and here we are dealing with small-scale statistics. However, there is no sign at the Reynolds number of these simulations, $Re_\lambda = (15Re)^{1/2} = 38 - 93$, of either a collapse of the data under Kolmogorov scaling or of an approach to a plateau

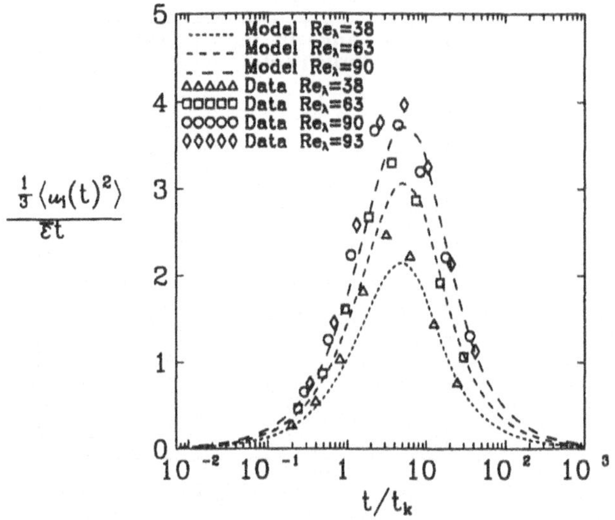

Fig. 6. Comparison of model (36) predictions for the Lagrangian velocity structure function with data of Yeung and Pope (1989)

at $C_o = 7$. Clearly the apparent value C_o estimated as the maximum value in these plots is considerably less than 7 at these Reynolds numbers. Model calculations at higher Reynolds number show that C_o^* exceeds about 90% of its asymptotic limit only for $Re_\lambda \geq 10^3$, a value not approached in laboratory experiments or numerical simulations. In contrast, Eulerian statistics from the DNS results (in particular the energy spectrum) when plotted in an analogous way do collapse at small scales and show a much weaker dependence on Reynolds number.

Sawford also examined the effect of Reynolds number on the dispersion of marked particles and showed that the primary influence is through the Lagrangian integral time scale. At the Reynolds numbers of the DNS results using the Lagrangian integral time scale (which may not be readily measured in practice) to non-dimensionalise the dependence of the dispersion on time accounts for all but at most 6% of the dependence on Reynolds number.

This finding has an important consequence for attempts to estimate a numerical value for C_o by fitting the predictions of a Langevin model to laboratory measurements of dispersion (Sawford and Guest, 1988; Pope and Chen, 1990; Anand and Pope, 1985). This corresponds closely to determining an effective value C_o' by 'fitting' (39) by its large Reynolds number limit to obtain

$$C_o' = C_o(1 + Re^{*-1/2})^{-1}. \tag{40}$$

Sawford concludes that an unachievably high (at present) Reynolds number is required in order to accurately estimate C_o from laboratory or numerical data in this way.

7. Conclusions

Over the past few years our group has focussed on a number of important questions underlying the Lagrangian stochastic theory of turbulent dispersion. We have demonstrated explicity that the theory yields a unique model only for the case of one-particle motion in isotropic turbulence and that for all other cases, including two-particle motion in isotropic turbulence, the non-uniqueness is nontrivial.

In the latter case we have explored the properties of a class of quadratic form models in conjunction with a Gaussian Eulerian velocity pdf and have attempted to further constrain these models by requiring a two-particle model to reproduce appropriate one-particle statistics. Although it is impossible for models in this class to satisfy all such two-to-one constraints, expansion methods provide a means of ordering the constraints so that the lower order ones can be identified and enforced. This procedure leads to the incorporation of constraints involving explicit consideration of the Navier Stokes equations. A potentially important practical result of this work is that although Thomson's (1990) model does not satisfy these constraints, its performance in reproducing one-particle statistics is as good as the 'optimum' more fully constrained model. Furthermore it predicts two-particle statistics almost identical to those from the more general and complex model.

Also in the context of two-particle models we have shown that there is no kinematic or dimensional requirement for the two-particle acceleration covariance to be neglected (compared with the one-particle covariance) and that therefore there is a nontrivial two-particle contribution to the relative velocity structure function and to relative dispersion in the inertial sub-range. The inertial sub-range behaviour of Thomson's (1990) model (and the other two-particle models discussed in Section 4) can be explained quantitatively in terms of a single analytical functional form for the two-particle acceleration covariance which is consistent with kinematic, dimensional and symmetry constraints, including some new inequality constraints.

We have developed procedures for relating Lagrangian intermittency corrections for Kolmogorov's local similarity theory to measured Eulerian corrections using a multifractal formalism. These Lagrangian results have been used to incorporate intermittency effects into Lagrangian stochastic models in a way which is consistent both internally and with Eulerian measurements. These intermittency corrections are shown to be unimportant for second-order statistical quantities such as the velocity structure function and the dispersion, but are significant for higher order moments. Particle trajectories do not show significant effects of intermittency and in particular are not fractal.

Finally, we have modelled the effect of finite Reynolds number by treating the acceleration as a Markov process. Second-order statistics from direct numerical simulations are represented very well by this approach. Our model suggests that the separation of viscous and energy containing scales of motion is effectively much less for Lagrangian than for Eulerian quantities at low Reynolds number.The effect of Reynolds number on dispersion is accounted for almost completely by the dependence of the Lagrangian integral time scale on Reynolds number. Attempts to estimate the universal inertial sub-range constant C_o fitting predictions of the Langevin equation (or its extensions) to low Reynolds number dispersion data results in significant underestimation of this constant, typically by up to a factor of two.

Acknowledgement

It is a pleasure to acknowledge the major contribution made by Dr M. S. Borgas to the work outlined in this paper.

References

Anand, M. S. and Pope, S. B. (1983). 'Diffusion behind a Line Source in Grid Turbulence', in L.J.S. Bradbury, F. Durst, B.E. Launder, F.W. Schmidt and J.H. Whitelaw (eds), *Turbulent Shear Flows 4*, Springer, pp. 46-52.

Batchelor, G. K. (1950). 'The Application of the Similarity Theory of Turbulence to Atmospheric Diffusion', *Quart. J. Roy. Meteor. Soc.* **76**, 133-146.

Batchelor, G. K. (1953). *The Theory of Homogeneous Turbulence*, CUP, Cambridge,

Borgas, M. S. (1991). 'The Multifractal Lagrangian Nature of Turbulence'. *Proc. Roy. Soc. London* A (Submitted).

Borgas, M. S. and Sawford, B. L. (1991a). 'The small-Scale Structure of Acceleration Correlations and its Role in the Statistical Theory of Turbulent Dispersion', *J. Fluid Mech.***228**, 295-320.

Borgas, M. S. and Sawford, B. L. (1991b). 'Stochastic Models for Two-Particle Dispersion in Isotropic, Homogeneous and Stationary Turbulence', (In preparation).

Borgas, M. S. and Sawford , B. L. (1991c). 'Stochastic Equations with Multifractal Random Increments for Modelling Turbulent Dispersion', *Phys. Fluids* (To be submitted)

Durbin, P. A. (1983). 'Stochastic Differential Equations and Turbulent Dispersion', NASA Reference Publication 1103, NASA Lewis Research Center, Cleveland, Ohio.

Frisch, U. and Parisi, G. (1985). 'On the Singularity Structure of Fully developed Turbulence', in M. Ghil, R. Benzi and G. Parisi (eds), *Turbulence and Predictability in Geophysical Fluid Dynamics and Climate Dynamics*, North-Holland, Amsterdam, pp. 84-88.

Gardiner, C. W. (1983), *Handbook of Stochastic Processes for Physics, Chemistry and the Natural Sciences*, Springer-Verlag, Berlin.

Gifford, F. A. (1982). 'Horizontal Diffusion in the Atmosphere: A Lagrangian-Dynamical Theory' *Atmos. Environ.* **16**, 505-512.

Gifford, F. A. (1983). Discussion of 'Horizontal Diffusion in the Atmosphere: A Lagrangian-Dynamical Theory', *Atmos. Environ* **17**, 196-197.

Lin, C. C. and Reid, W. H. (1963). 'Turbulent Flow, Theoretical Aspects', in S. Flügge (ed), Encyclopedia of Physics, Springer-Verlag, Berlin, **VIII/2**, 438-523.

Mandelbrot, B. (1976). 'Intermittent Turbulence and Fractal Dimension: Kurtosis and the Spectral Exponent 5/3 + B' in R. Teman (ed), *Turbulence and Navier Stokes Equations, Lecture Notes in Mathematics*, Springer-Verlag, Berlin, pp. 121-145.

Meneveau C. and Sreenivasan, K. R. (1987). 'Simple Multifractal Cascade Model for Fully Developed Turbulence', *Phys. Rev. Lett.* **59**, 1424-1427.

Monin, A. S. and Yaglom, A. M. (1975). '*Statistical Fluid Mechanics: Mechanics of Turbulence*, Vol. 2, MIT Press, Cambridge MA.

Novikov, E. A. (1963). 'Random Force Method in Turbulence Theory', *Sov. Phys.* JETP **17**, 1449-1454.

Novikov, E. A. (1989). 'Two-Particle Description of Turbulence, Markov Property, and Intermittency' *Phys. Fluids* **A1**, 326-330.

Novikov, E. A. (1990). 'The effects of Intermittency on Statistical Characteristics of Turbulence and Scale Similarity of Breakdown Coefficients'. *Phys. Fluids* **A2**, 814-820.

Pope, S. B. and Chen, Y. L. (1989). ' The Velocity-Dissipation Probability Density Function Model for Turbulent Flows'. *Phys. Fluids* **A2**, 1437-1449.

Sawford, B. L. (1984). 'The Basis for, and some limitations of, the Langevin Equation in Atmospheric Relative Dispersion Modelling', *Atmos. Environ.* **18**, 2405-2411.

Sawford, B. L. (1985). 'Lagrangian Statistical Simulation of Concentration Mean and Fluctuating Fields'. *J. Climate Appl. Meteor.* **24**, 1152-1166.

Sawford, B. L. and Guest, F. M. (1988). 'Uniqueness and University in Lagrangian Stochastic Models of Turbulent Diffusion', 8th Symposium on Turbulence and Diffusion, Am. Meteor. Soc., Boston, MA, pp. 96-99.

Sawford, B. L. (1991). 'Reynolds Number Effects in Lagrangian Stochastic Models of Turbulent Dispersion', *Phys. Fluids* **A3**, 1577-1586.

Thomson, D. J. (1986). 'On the relative Disperson of Two Particles in Homogeneous Stationary Turbulence and the Implications for the Size of Concentration Fluctuations at Large Times'. *Quart. J. Roy. Meteor. Soc.* **112**, 890-894.

Thomson, D. J.(1987). 'Criteria for the Selection of Stochastic Models of Particle Trajectories in Turbulent Flows'. *J. Fluid Mech.* **180**, 529-556.

Thomson, D. J. (1990). 'A stochastic Model for the Motion of Particle Pairs in Isotropic High-Reynolds Number Turbulence, and its Application to the Problem of Concentration Variance', *J. Fluid Mech.* **210**, 113-153.

Yeung, P. K. and Pope, S. B. (1989). 'Lagrangian Statistics from Direct Numerical Simulations of Isotropic Turbulence', *J. Fluid Mech.* **207**, 531-586.

A THREE-DIMENSIONAL MODEL FOR CALCULATING THE CONCENTRATION DISTRIBUTION IN INHOMOGENEOUS TURBULENCE

H. KAPLAN and N. DINAR
Israel Institute For Biological Research
P. O. B. 19 Ness-Ziona, Israel

(Received October 1991)

Abstract. A new approach for calculating the concentration distribution in inhomogeneous turbulence is suggested. The model is a 3-D model, constrained to describe incompressible flow. The model requires a knowledge of the covariance matrix of the Eulerian velocities and the two-point third moments. The model is applied for three types of turbulent field: homogeneous isotropic turbulence, constant flux neutral boundary layer and free convective turbulence. The required Eulerian moments are calculated using the 'eddy model' of the turbulent field. Concentration moments are calculated and results are compared to experimental data. Other model predictions which have no experimental support can be compared to measurements when available.

1. Introduction

Recently, there has been much interest in methods for calculating the probability distribution function of concentration fluctuations. Those methods are based on an Eulerian approach - high order closure models (Sykes *et al.*,1984), p.d.f. methods (Pope, 1985), or a Lagrangian statistical approach (Sawford,1982; Thomson, 1986; Durbin, 1980 and others). The Lagrangian methods are based on the statistics of fluid particle trajectories, and most of these techniques are capable of calculating the mean and variance of concentration fluctuations only. While there is a consensus among all these models with respect to the behaviour of the averaged concentration, there are still differences of opinion regarding the behaviour of the variance even in idealized flows like stationary homogeneous turbulence. Another problem of the stochastic models for relative diffusion is that they are all formulated in one-dimensional space. For the calculation of concentration averages where only a one- particle stochastic model is needed, motions of different components in space can be decoupled. However, when one calculates the variance and uses the statistics of particle pair motion, this decoupling is no longer justified. The reason is that in incompressible flow the three components of the relative velocity are related through the continuity equation. Kaplan and Dinar (1988b) suggested a 3-D model for calculating the variance of concentration fluctuations in 3-D space and in homogeneous turbulence, taking into account the incompressibility constraint. They have compared their calculations with data of Warhaft (1984), who measured concentration fluctuations in grid turbulence (Kaplan and Dinar, 1989a). The model results are found to be in good agreement with the experimental data. However such

Boundary-Layer Meteorology **62:** 217–245, 1993.

a model cannot deal with the more realistic cases such as constant flux boundary layer, where data of many field experiments are available [see for example: Hanna (1984), Sawford *et al.* (1985), Dinar et al. (1988)], or the free convective turbulence (Willis and Deardorff, 1976,1978). In this work we extend the 3-D model to the case of inhomogeneous turbulence. We calculate the average concentration and the behaviour of fluctuation intensity in the case of a ground-level source in a neutral boundary layer and of an elevated source in the free convective turbulence. The calculated results are compared with measurements.

2. A Stochastic Model for the Diffusion of N Particles

2.1. The Lagrangian Equations of Motion

In analogy to the one-particle equations of motion we assume that the motion of N particles in a turbulent field is a Markov process, described by a stochastic differential equation of the form:

$$du_i^\alpha = a_i^\alpha(\mathbf{r}_1,\ldots,\mathbf{r}_N,\mathbf{u}_1,\ldots,\mathbf{u}_N,t)dt + \sum_j b_{ij}^{\alpha\beta}(\mathbf{r}_1,\ldots,\mathbf{r}_N,\mathbf{u}_1,\ldots,\mathbf{u}_N,t)d\xi_j^\beta \qquad (2.1)$$

$$dr_i^\alpha = (u_i^\alpha + U^\alpha(\mathbf{r}_i))dt\,.$$

Latin subscripts stand for the particle, Greek subscripts denote the components of the vector in a Cartesian reference system, and the summation convention is applied for Greek indices (i.e. for terms in which an index appears twice). a_i^α,$b_{ij}^{\alpha\beta}$ are functions of time, velocities and locations of particles, and $d\xi_j^\beta$ represent the increments of a Wiener process. $\mathbf{U}(\mathbf{r})$ is the ensemble average Eulerian velocity field which is assumed to be known. With the initial condition that at time t the particles are at locations $\mathbf{r}_1,\ldots,\mathbf{r}_N$ with velocities $\mathbf{u}_1,\ldots,\mathbf{u}_N$ respectively, equation (2.1) can be solved for the probability $P(\mathbf{r}_1,\ldots,\mathbf{r}_N,t;\mathbf{r}'_1,\ldots,\mathbf{r}'_N,t')$ that at time t' the particles are at locations $\mathbf{r}'_1,\ldots,\mathbf{r}'_N$ respectively. We denote by $g(\mathbf{r}_1,\ldots,\mathbf{r}_N,\mathbf{u}_1,\ldots,\mathbf{r}_N,t)$ the density function of the phase space distribution of the N particles in the ensemble of flows. If the particle trajectories are a solution of equation (2.1), then $g(\mathbf{r}_1,\ldots,\mathbf{r}_N,\mathbf{u}_1,\ldots,\mathbf{u}_N,t)$ is a solution of the Kolmogorov forward equation:

$$\frac{\partial g}{\partial t} + \sum_i [U^\alpha(\mathbf{r}_i)\frac{\partial g}{\partial r_i^\alpha} + \frac{\partial}{\partial r_i^\alpha}(u_i^\alpha g) + \frac{\partial}{\partial u_i^\alpha}(a_i^\alpha g) - \sum_j \frac{\partial^2}{\partial u_i^\alpha \partial u_j^\beta}(B_{ij}^{\alpha\beta} g)] = 0 \qquad (2.2)$$

where $B_{ij}^{\alpha\beta} = 0.5\sum_k b_{ik}^{\alpha\tau} b_{jk}^{\tau\beta}$. We denote by $g_E(\mathbf{r}_1,\ldots,\mathbf{r}_N,\mathbf{u}_1,\ldots,\mathbf{u}_N,t)$ the Eulerian distribution function of the fluid elements at points $\mathbf{r}_1,\ldots,\mathbf{r}_N$. If the initial conditions are such that at time t the locations of the N particles are derived from a homogeneous distribution function and their velocities are derived from the Eulerian distribution function at the sites of the points, then for an incompressible flow, we expect the distribution function $g(\mathbf{r}_1,\ldots,\mathbf{r}_N,\mathbf{u}_1,\ldots,\mathbf{u}_N,t)$ to remain equal to $g_E(\mathbf{r}_1,\ldots,\mathbf{r}_N,\mathbf{u}_1,\ldots,\mathbf{u}_N,t)$. This is a generalization of the well-mixed principle suggested by Thomson(1987) for the motion of N particles.

2.2. MODELING a_i^α AND $B_{ij}^{\alpha\beta}$

The equality between g and g_E leads, using (2.2), to a functional relation between the parameters a_i^α and $B_{ij}^{\alpha\beta}$ of the stochastic model. This relation can be written in the form:

$$a_i^\alpha g_E = \sum_j \frac{\partial}{\partial u_j^\beta}(B_{ij}^{\alpha\beta} g_E) + \phi_i^\alpha \tag{2.3}$$

where ϕ_i^α satisfies:

$$\sum_i \frac{\partial \phi_i^\alpha}{\partial u_i^\alpha} = -\frac{\partial g_E}{\partial t} - \sum_i [U^\alpha(\mathbf{r}_i)\frac{\partial g_E}{\partial r_i^\alpha} + \frac{\partial}{\partial r_i^\alpha}(u_i^\alpha g_E)]. \tag{2.4}$$

Additional assumptions that we make are the following:

1) $a_i^\alpha(\mathbf{r}_1,\ldots,\mathbf{r}_N,\mathbf{u}_1,\ldots,\mathbf{u}_N)$ depends only on the velocity and location of the ith particle.

$$a_i^\alpha(\mathbf{r}_1,\ldots,\mathbf{r}_N,\mathbf{u}_1,\ldots,\mathbf{u}_N) = a_i^\alpha(\mathbf{r}_i,\mathbf{u}_i). \tag{2.5}$$

This property of a_i^α is a result of a requirement that the one-particle statistics derived from the model do not depend on the location of other particles. In other words, the trajectory of a particle with given initial conditions does not depend on the initial location of the other particles.

$$B_{ii}^{\alpha\beta} = \frac{[\sigma^{\alpha\alpha}]^2}{\tau^\alpha}\delta^{\alpha\beta} \prec \epsilon^{\alpha\alpha}. \tag{2.6}$$

This choice assures that the small time behaviour of particles from an instantaneous source is correct [see Thomson(1987)]. If $g_E(\mathbf{r}_1,\ldots,\mathbf{r}_N,\mathbf{u}_1,\ldots,\mathbf{u}_N)$ is known then $a_i^\alpha(\mathbf{r}_i,\mathbf{u}_i)$, $B_{ij}^{\alpha\beta}$ can be determined uniquely. Let us denote the one particle equilibrium function by $G(\mathbf{r}_i,\mathbf{u}_i)$:

$$G(\mathbf{r}_i,\mathbf{u}_i) = \int\ldots\int g_E(\mathbf{r}_1,\ldots,\mathbf{r}_N,\mathbf{u}_1,\ldots,\mathbf{u}_N) \prod_{k=1,k\neq i}^{N} dr_k \prod_{k=1,k\neq i}^{N} du_k \tag{2.7}$$

From (2.3) it follows that:

$$a_i^\alpha G = B_{ii}^{\alpha\alpha}\frac{\partial G}{\partial u_i^\alpha} + \phi = \frac{\sigma_{ii}^{\alpha\alpha}}{\tau^\alpha(\mathbf{r}_i)}\frac{\partial G}{\partial u_i^\alpha} + \phi \tag{2.8}$$

where ϕ satisfies:

$$\frac{\partial \phi}{\partial u_i^\alpha} = -\frac{\partial G}{\partial t} - \frac{\partial}{\partial r_i^\alpha}(u_i^\alpha G)\,. \tag{2.9}$$

Those equations are used to evaluate $a_i^\alpha(\mathbf{r}_i, \mathbf{u}_i)$. Once $a_i^\alpha(\mathbf{r}_i, \mathbf{u}_i)$ is determined, equation (2.3) is used to evaluate $B_{ij}^{\alpha\beta}$, assuming g_E is known.

2.3. The Moments Approximation

The above algorithm for evaluating a_i^α, $B_{ij}^{\alpha\beta}$ requires knowledge of the function $g_E(\mathbf{r}_1, \dots, \mathbf{r}_N, \mathbf{u}_1, \dots, \mathbf{u}_N)$. Usually g_E is not known and only limited information is available about its low order moments. In this section we suggest an approximation for a_i^α,$B_{ij}^{\alpha\beta}$ which is constrained to conserve all the known properties of g_E. The approximation is based on expanding $a_i^\alpha(\mathbf{r}_i, \mathbf{u}_i)$ as a power series in u_i^α.

$$a_i^\alpha(\mathbf{r}_i, \mathbf{u}_i) = C_0^\alpha(\mathbf{r}_i) + C_1^{\alpha\beta}(\mathbf{r}_i)u_i^\beta + C_2^{\alpha\beta\gamma}(\mathbf{r}_i)u_i^\beta u_i^\gamma + \dots \tag{2.10}$$

If we denotes by $\hat{g} = \int_0^\infty g e^{i\mathbf{u}\boldsymbol{\theta}}\, d^3u$ the characteristic function of g, then substituting this expansion in (2.2), yields the equation for $\hat{g}$:

$$\frac{\partial \hat{g}}{\partial t} + \sum_i [U^\alpha(\mathbf{r}_i)\frac{\partial \hat{g}}{\partial r_i^\alpha} - i\frac{\partial^2 \hat{g}}{\partial r_i^\alpha \partial \theta_i^\alpha} - i\theta_i^\alpha [C_0^\alpha(\mathbf{r}_i)\hat{g} - iC_1^{\alpha\beta}(\mathbf{r}_i)\frac{\partial \hat{g}}{\partial \theta_i^\beta}\,. \tag{2.11}$$

$$-C_2^{\alpha\beta\gamma}(\mathbf{r}_i)\frac{\partial^2 \hat{g}}{\partial \theta_i^\beta \partial \theta_i^\gamma} + \dots] + \sum_j B_{ij}^{\alpha\beta}\theta_i^\alpha \theta_j^\beta \hat{g}] = 0$$

By evaluating derivatives of (2.11) with respect to θ_i at $\boldsymbol{\theta} = 0$ and using the fact that the first moments of g_E are zero, one get an expression which connects the coefficients of expansion (2.10) to the moments of g_E. For stationary turbulence those equations are:

$$\frac{\partial}{\partial r_k^\alpha}\mu_{kk}^{(2)\delta\alpha} - C_0^\delta(\mathbf{r}_k) - C_2^{\delta\beta\gamma}(\mathbf{r}_k)\mu_{kk}^{(2)\beta\gamma} = 0$$

$$\frac{\partial}{\partial r_k^\alpha}\mu_{lkk}^{(3)\delta\epsilon\alpha} + \frac{\partial}{\partial r_l^\alpha}\mu_{lkl}^{(3)\delta\epsilon\alpha} + U^\alpha(\mathbf{r}_k)\frac{\partial}{\partial r_k^\alpha}\mu_{kk}^{(2)\epsilon\delta} + U^\alpha(\mathbf{r}_l)\frac{\partial}{\partial r_l^\alpha}\mu_{ll}^{(2)\epsilon\delta} - \tag{2.12}$$

$$-\mu_{lk}^{(2)\delta\beta}C_1^{\epsilon\beta}(\mathbf{r}_k) - \mu_{kl}^{(2)\epsilon\beta}C_1^{\delta\beta}(\mathbf{r}_l) -$$

$$\mu_{lkl}^{(3)\gamma\epsilon\beta}C_2^{\delta\beta\gamma}(\mathbf{r}_l) - \mu_{kkl}^{(3)\beta\gamma\delta}C_2^{\epsilon\beta\gamma}(\mathbf{r}_k) + 2B_{kl}^{\delta\epsilon} = 0\,.$$

When deriving these equations, we have used the fact that because of the fluid incompressibility:

$$\frac{\partial}{\partial r_{i_1}^{\alpha_1}}\mu_{i_1,\dots,i_M}^{(M)\alpha_1,\dots,\alpha_M} = 0 \qquad \text{for all} \quad i_1 \neq i_2 \neq \dots \neq i_M\,. \tag{2.13}$$

In the special case of one particle statistics l=k, (2.12) yields:

$$\frac{\partial}{\partial r_k^\alpha}\mu_{kk}^{(2)\delta\alpha} - C_0^\delta(\mathbf{r}_k) - C_2^{\delta\beta\gamma}(\mathbf{r}_k)\mu_{kk}^{(2)\beta\gamma} = 0 \tag{2.14}$$

$$2\frac{\partial}{\partial r_k^\alpha}\mu_{kkk}^{(3)\delta\epsilon\alpha} + 2U^\alpha(\mathbf{r}_k)\frac{\partial}{\partial r_k^\alpha}\mu_{kk}^{(2)\epsilon\delta} - \mu_{kk}^{(2)\delta\beta}C_1^{\epsilon\beta}(\mathbf{r}_k) - \mu_{kk}^{(2)\epsilon\beta}C_1^{\delta\beta}(\mathbf{r}_k)-$$

$$-\mu_{kkk}^{(3)\gamma\epsilon\beta}C_2^{\delta\beta\gamma}(\mathbf{r}_k) - \mu_{kkk}^{(3)\beta\gamma\delta}C_2^{\epsilon\beta\gamma}(\mathbf{r}_k) + 2B_{kk}^{\delta\epsilon} = 0\,.$$

The decorrelation of the velocity fluctuations at a fixed point in space is due to energy dissipation at this point. Therefore the time scale of the correlation will be $\frac{[\sigma^{\alpha\alpha}]^2}{\epsilon^{\alpha\alpha}}$. Using dimensional considerations it follows that:

$$C_1^{\alpha\beta}(\mathbf{r}_k) = \frac{\partial a^\alpha}{\partial u^\beta}|_{r_k} = \delta^{\alpha\beta}\frac{[\sigma^{\alpha\alpha}]^2}{\epsilon^{\alpha\alpha}}\,. \tag{2.15}$$

The approximation we suggest is based on the information we have about g_E. If only three lower order moments are known, then we truncate the series (2.10) after the third term, and use (2.14),(2.15) to evaluate $C_0^\delta(\mathbf{r}_k), C_2^{\delta\beta\gamma}(\mathbf{r}_k)$. Once those coefficients are evaluated, $B_{kl}^{\delta\epsilon}$ are determined from equation (2.12). The calculated $a_i^\alpha, B_{ij}^{\alpha\beta}$ can be used to determine $P(\mathbf{r}_1,\ldots,\mathbf{r}_N,t;\mathbf{r}'_1,\ldots,\mathbf{r}'_N,t')$ with equation (2.1). This procedure ensures that the equilibrium function of the process, g, has the same low moments as g_E. On the other hand, its higher moments are not necessarily identical to those of g_E. If information on higher moments of g_E is supplied, then more terms in the expansion of a_i^α should be added and equations for higher moments should be derived from (2.15). The procedure described above is used to calculate the parameters of the Markov process in such a way that the 'well mixed' principle is preserved [see Thomson (1987)]. The model is thus designed in a way which is consistent with inertial subrange theory and satisfies the criteria suggested by Thomson to distinguish good models from bad. In particular it describes motion in incompressible flow and prevents fluctuations contributed by changes in the fluid density, which appear in all one-dimensional models [see Kaplan and Dinar (1988c)]. It is reasonable to believe that the more information that is supplied about the Eulerian distribution function g_E, the more accurate is the model. However we assume that for the purpose of concentration fluctuation calculations, knowledge of the second moment of g_E is enough to give results in satisfactory accord with measurements. This assumption can be examined only by comparing predictions with observations.

3. Examples

In order to evaluate the parameters for the Markov process (2.1), one needs information about the following:

1) The Lagrangian time scale $\tau^{\alpha}(\mathbf{r}_i)$;

2) The second and third moments of the one-particle distribution function. Usually these parameters are only partially known. In several cases one can use symmetry or models to supply the missing information. In the following, three such examples are described:

(i) The idealized case of homogeneous isotropic turbulence, in which symmetry considerations can be used.

(ii) The constant flux layer, in which the correlation matrix is modeled.

(iii) Free convective turbulence for which the second and third moments are needed.

3.1. Homogeneous Isotropic Turbulence

3.1.1. Modeling the covariance matrix and the third moments

The Eulerian correlation matrix $R^{\alpha\beta}(\mathbf{r}_i, \mathbf{r}_j)$ in homogeneous isotropic turbulence can be simply written using symmetry considerations and the incompressibility constraint [see Batchelor (1956)]. It can be expressed in terms of scalar functions of $|\mathbf{r}_{ij}|$, which is the distance between the locations of particles i and j:

$$R^{\alpha\beta}(\mathbf{r}_i, \mathbf{r}_j) = \sigma^2[f(|\mathbf{r}_{ij}|) + 0.5|\mathbf{r}_{ij}|f'(|\mathbf{r}_{ij}|)\delta^{\alpha\beta} - 0.5f'(|\mathbf{r}_{ij}|)\frac{r_{ij}^{\alpha} r_{ij}^{\beta}}{|\mathbf{r}_{ij}|^2}] \tag{3.1}$$

where $f(0) = 1$. For i = j, $R^{\alpha\beta}(\mathbf{r}_i, \mathbf{r}_j) = \sigma^2\delta^{\alpha\beta}$. The third moments are zero except $\mu_{lkk}^{(3)\alpha\beta\gamma}$ and $\mu_{lkl}^{(3)\alpha\beta\gamma}$. We define:

$$T^{\epsilon\gamma}(\mathbf{r}_{kl}) = \frac{\partial}{\partial r_k^{\alpha}}\mu_{lkk}^{(3)\epsilon\gamma\alpha} + \frac{\partial}{\partial r_l^{\alpha}}\mu_{lkl}^{(3)\epsilon\gamma\alpha} \,. \tag{3.2}$$

This is the inertial term in the energy decay equation given by Batchelor (1956).

$$\frac{\partial}{\partial t}R^{\epsilon\gamma}(\mathbf{r}) - T^{\epsilon\gamma}(\mathbf{r}) - 2\nu\,\nabla^2\,R^{\epsilon\gamma}(\mathbf{r}) = 0 \,. \tag{3.3}$$

The third term is the energy dissipation by viscosity forces. For $\mathbf{r}$ outside the dissipation range, the third term is negligible and the energy decay is balanced by the inertial term. The average velocity $\mathbf{U}$ is constant.

3.1.2. Determination of the constants for the stochastic process

The Lagrangian time for isotropic homogeneous turbulence is $\tau^{\alpha\beta} = T_L\delta^{\alpha\beta}$. Therefore the equations for the parameters of the stochastic model are:

$$\begin{gathered} C_0^{\delta}(\mathbf{r}_k) - C_2^{\delta\beta\gamma}\sigma^2\delta^{\beta\gamma} = 0 \\ \frac{2}{T_L}R_{kl}^{\delta\epsilon} - B_{kl}^{\delta\epsilon} = 0 \,. \end{gathered} \tag{3.4}$$

The Markov process describing the diffusion will be:

$$dr_i^\alpha = u_i^\alpha dt \tag{3.5}$$

$$du_i^\alpha = \frac{-u_i^\alpha dt}{T_L} + \Theta^\alpha(\mathbf{r}_i)$$

where $\Theta^\alpha(\mathbf{r}_i) = b_{ij}^{\alpha\beta} d\xi_j^\beta$ is a random field with covariance given by:

$$< \Theta^\alpha(\mathbf{r}_i)\Theta^\beta(\mathbf{r}_j) >= \begin{cases} R^{\alpha\beta}(r_{ij})\frac{2}{T_L}dt & \alpha \neq \beta \\ (R^{\alpha\beta}(r_{ij})\frac{2}{T_L} - \frac{\partial}{\partial t}R^{\alpha\beta}(r_{ij})\delta_{ij})dt & \alpha = \beta \end{cases} \tag{3.6}$$

3.1.3. Results

The above expressions were used by Kaplan and Dinar (1988b, 1989a) to calculate the concentration fluctuations and correlations between concentrations due to two sources and found to be in good agreement with experiments. Also (see Kaplan and Dinar, 1989b), the model predicts the correct behaviour of the root mean square particle separation (proportional to $t^{1.5}$), as shown in Batchelor (1952).

4. Diffusion in Inhomogeneous Turbulence

Modeling the covariance matrix and the third moments in the case of inhomogeneous turbulence is more complicated. Very limited information is available about the structure function and usually measurements are taken along the principle axes [see for example: Townsend(1956); Deardorff and Willis (1985)]. In this section we suggest an eddy model which enables us to calculate the structure function and the third moments as functions of space. The resulting calculated moments are constrained to be consistent with the measured quantities and with the symmetry of the problem. We adopt an approach similar to Townsend(1956). The assumptions of the eddy model are the following: a) The turbulent field is composed of self-similar eddies with various sizes. b) The velocity of each eddy fulfills the incompressibility constraint $\nabla \mathrm{v} = 0$. c) The shape of eddies and its distribution in space is consistent with the symmetry of the problem and with the boundary conditions. d) The eddy sizes vary in range from λ_0 to L, where λ_0 is the Kolmogorov length scale and L is a typical length of the flow. The distribution of eddy sizes is proportional to dl/l^3. e) The amplitude of each eddy is a function of its size and is determined in such a way that the resulting calculated structure function is consistent with measurements. Once the eddies are determined, the covariance matrix and third moments are calculated. The procedure assures that the calculated moments are consistent with measurements, symmetry and the incompressibility constraint. Two cases of inhomogeneous turbulence are studied: The constant flux layer and the free convective boundary layer.

4.1. Diffusion In The Constant Flux Boundary Layer

4.1.1. Determination Of The Structure Function And Third Moments

The spatial distribution of eddy centers in this case is homogeneous in the (x,y) plane but not in the vertical direction. The presence of the wall prevents eddies from having scales in the vertical direction larger than the distance of the eddy center from the wall. We therefore assume that for $z << L$, eddies are attached to the wall and the vertical velocity of each eddy should be zero at the surface. The velocity of an eddy of size l is:

$$u^z = \frac{z}{l}[1-(\frac{y-\eta}{l})^2)exp(-\frac{1}{2}(\frac{x-\xi}{l})^2-\frac{1}{2}(\frac{y-\eta}{l})^2-\frac{1}{2}(\frac{z}{l})^2]S_v$$

$$u^x = \frac{u^z}{K}S_v \tag{4.1}$$

$$u^y = -\frac{y-\eta}{l}(1-(\frac{z}{l})^2+\frac{(x-\xi)z}{Kl^2})exp[-\frac{1}{2}(\frac{x-\xi}{l})^2-\frac{1}{2}(\frac{y-\eta}{l})^2$$

$$-\frac{1}{2}(\frac{z}{l})^2]S_v$$

where S_v is an amplitude, K=0.4 is the ratio of σ_z^2 to the constant flux and (ξ,η) is the eddy location in the (x,y) plane. This choice is similar to the double-cylinder eddy suggested by Townsend(1956) and is changed in such a way that $\nabla \mathbf{v} = 0$ to ensure that the fluid is incompressible. The calculated covariance matrix $R_{ij}^{\alpha\beta}$ is:

$$R_{ij}^{\alpha\beta} = 2\int\int\int u^\alpha(x_i-\xi,y_i-\eta,z_i)u^\beta(x_j-\xi,y_j-\eta,z_j)\frac{d\xi d\eta dl}{A^2(\lambda_0^{-2}-L^{-2})l^3} \tag{4.2}$$

where A^2 is the area of the turbulent media. The expressions for its components are:

$$R^{zz} \prec \frac{1}{B^2}-\frac{4\delta y^2}{B^4}+\frac{8\delta y^4}{3B^6} \tag{4.3}$$

$$\delta x = \frac{1}{2}\frac{x_2-x_1}{\sqrt{z_1z_2}} \quad \delta y = \frac{1}{2}\frac{y_2-y_1}{\sqrt{z_1z_2}} \quad B = (\frac{1}{2}\frac{z_1^2+z_2^2}{z_1z_2}+\delta x^2+\delta y^2)^{0.5}$$

$$R^{yy} \prec [\frac{1}{4}E_1((\frac{B\sqrt{z_1z_2}}{L})^2)+\frac{1}{2B^2}(\delta x^2-B^2-\frac{1}{2}(\sqrt{\frac{z_1}{z_2}}-\frac{z_2}{z_1})\frac{\delta x}{K}+\frac{1}{4K^2})+$$

$$+\frac{1}{2B^4}(\delta y^2(\frac{z_1^2+z_2^2}{z_1z_2}-\frac{1}{2K^2})+\frac{\delta x}{K}(\sqrt{\frac{z_1}{z_2}}-\sqrt{\frac{z_2}{z_1}})(\delta y^2-\frac{1}{2})+\frac{1}{2}-\frac{\delta x^2}{2K^2})+$$

$$+\frac{\delta y^2}{B^6}(\frac{\delta x^2}{K^2}+\frac{\delta x}{K}(\sqrt{\frac{z_1}{z_2}}-\sqrt{\frac{z_2}{z_1}})-1]\frac{1}{\frac{E_1((\frac{z_1}{L})^2)}{4}-\frac{1}{4}+\frac{1}{8K^2}} \tag{4.4}$$

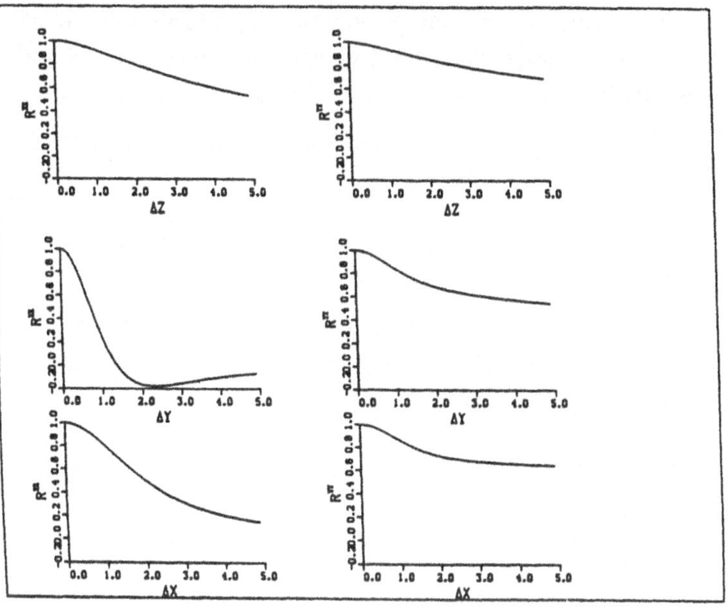

Fig. 1. The calculated correlation matrix elements as functions of principal axes

$$R^{yz} \prec [-\frac{\delta y}{4B^2} + \frac{1}{2B^4}[(\frac{z_1}{z_2} + \sqrt{\frac{z_1}{z_2}}\frac{\delta x}{K})\frac{\delta y}{2} - \delta y^3] + \frac{\delta y^3}{B^6}(\frac{z_1}{z_2} + \sqrt{\frac{z_1}{z_2}}\frac{\delta x}{K})]\sqrt{\frac{z_2}{z_1}} \tag{4.5}$$

where $E_1(z) = \int_z^\infty \frac{exp(-t)}{t} dt$ is the exponential integral. Those expressions are calculated assuming $z << L$ and are valid only in this region. In Fig. 1 the behaviour of the correlation functions along the principle axes is presented. This behaviour is very similar to the measured correlations (Grant *et al.*, 1958). The calculated third moments in this region are:

$$\mu^{yyy} = \mu^{zzy} = \mu^{xxy} = 0$$

$$(4.6)$$

$$\mu^{zzz} = K^2; \quad \mu^{xxx} = K; \quad \mu^{zzx} = K^2; \quad \mu^{xxz} = const.$$

4.1.2. The diffusion equation

The Lagrangian time scale $\tau^\alpha(\mathbf{r}_i)$ in the neutral boundary layer is given by

$$\tau_L^x = \tau_L^y = \tau_L^z = \frac{z}{2\sigma_w} \equiv \tau(z) \tag{4.7}$$

where σ_w is the standard deviation of the vertical wind fluctuations. The motion of particles is determined also by the average wind. In the boundary layer this wind depends on the coordinates and therefore one can not use a moving coordinate system as was done in the case of homogeneous isotropic turbulence. In the constant flux layer $\mathbf{U}(z)$ can be determined by dimensional considerations and is given by:[see Townsend (1956)]

$$U^x(z) = (\frac{\sqrt{\tau_0}}{k}) ln(\frac{z}{z_0}) \tag{4.8}$$

$$U^y = U^z = 0$$

where τ_0 is the constant flux, z_0 is the friction length and k is the Von Karman constant, $k \approx 0.35$. Using these expressions and the calculated moments, the evaluated parameters for the stochastic process are:

$$C_0^{\alpha}(\mathbf{r}_i) = 0$$

$$C_1^{\alpha\beta}(\mathbf{r}_i) = \frac{1}{\tau_L^{\alpha}(\mathbf{r}_i)} \delta^{\alpha\beta} \tag{4.9}$$

$$C_2^{\alpha\beta\gamma}(\mathbf{r}_i) = 0$$

and the resulting Markov process is:

$$(a) \quad d\mathbf{r}_i = (\mathbf{U}(z) + \mathbf{u}_i) dt$$

$$(b) \quad du_i^{\alpha} = \frac{-u_i^{\alpha} dt}{\tau(z_i)} + \Theta_i^{\alpha} \tag{4.10}$$

where the covariance of the random field Θ_i^{α} is given by:

$$< \Theta_i^{\alpha} \Theta_j^{\beta} >= R^{\alpha\beta}(\mathbf{r}_i, \mathbf{r}_j)(\frac{1}{\tau(z_i)} + \frac{1}{\tau(z_j)}). \tag{4.11}$$

4.2. The Free Convective Boundary-Layer

4.2.1. Determination of the covariance matrix and the third moments

The free convective flow is a flow arising in a nonuniform heated fluid in a gravitational field. Archimedean forces are created which cause an upward buoyancy of warm volumes of fluid and downward sinking of cooler volumes. We assume that the variation of density due to pressure fluctuations is small and can be ignored, so that the incompressibility condition is still valid. However, density fluctuations due to temperature variation are not negligible. We confine ourselves to the regime $z >> z_0$ in which the presence of friction forces can be neglected but the structure of velocity fluctuations is affected by the presence of the ground. In the absence of prefered directions in the (x,y) plane, the velocity fluctuations should be isotropic and homogeneous in this plane. On the other hand, unisotropy in the vertical direction is caused by convective forces, as well as by the presence of the ground.

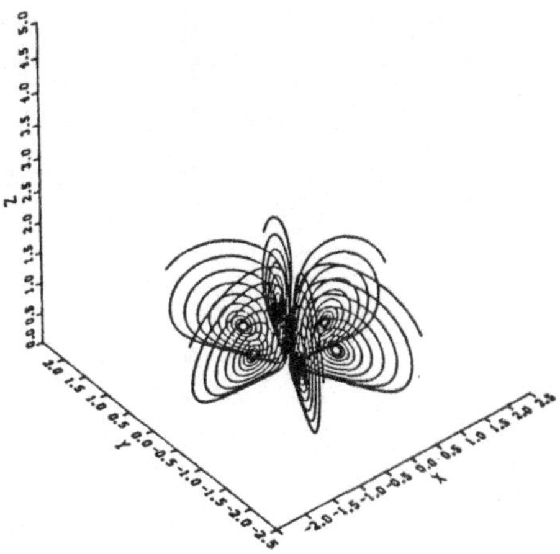

Fig. 2. The streamlines of an eddy defined by equation (4.12)

An eddy which is consistent with these symmetry considerations and with the incompressibility constraint can be of the form:

$$
\begin{aligned}
u_z^l &= \frac{z}{l}(2 - \frac{r^2}{l^2})exp(\frac{r^2 + z^2}{2l^2})S_v(l) \\
u_x^l &= -\frac{x - \xi}{l}(1 - \frac{z^2}{l^2})exp(\frac{r^2 + z^2}{2l^2})S_v(l) \\
u_y^l &= -\frac{y - \eta}{l}(1 - \frac{z^2}{l^2})exp(\frac{r^2 + z^2}{2l^2})S_v(l) \\
r^2 &= (x - \xi)^2 + (y - \eta)^2
\end{aligned}
\tag{4.12}
$$

where(ξ, η) is the eddy center. In Fig. 2 a structure of this eddy is presented.

As in the case of the constant flux layer, we assume that the eddies are distributed homogeneousely in the (x,y) plane, and are attached to the wall. We assume that the amplitude of the eddy is proportional to the one-third power of its size. The covariance matrix of the turbulence is given by:

$$R_{ij}^{\alpha\beta} = 2\int\int\int u^{\alpha}(x_i - \xi, y_i - \eta, z_i)u^{\beta}(x_j - \xi, y_j - \eta, z_j)\frac{d\xi d\eta dl}{A^2(\lambda_0^{-2} - L^{-2})l^3} \tag{4.13}$$

where A^2 is the area of the turbulent media. Using (4.12) the resulting terms of the covariance matrix are:

$$R^{\alpha\beta}ij \prec (z_i z_j)^{\frac{1}{3}}\{\frac{1}{4}(3B^{\frac{2}{3}}\Psi(\frac{B^2 z_i z_j}{L^2}) - (\frac{z_i}{z_j} + \frac{z_j}{z_i})C_{\frac{2}{3}} + C_{\frac{5}{3}})-$$

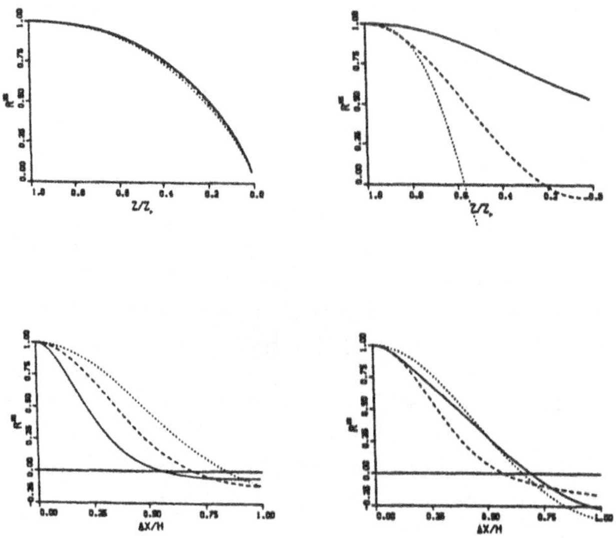

Fig. 3. The correlation function of the free convective turbulence along the principal axes.

$$-\frac{1}{2}\delta\alpha^2[C_{\frac{2}{3}} - (\frac{z_i}{z_j} + \frac{z_j}{z_i})C_{\frac{5}{3}} + C_{\frac{8}{3}}]\} \quad ; \quad \alpha = x, y \tag{4.14}$$

$$R_{ij}^{xy} \prec -\frac{1}{2}(z_i z_j)^{\frac{1}{3}}\delta x \delta y (C_{\frac{2}{3}} - (\frac{z_i}{z_j} + \frac{z_j}{z_i})C_{\frac{5}{3}} + C_{\frac{8}{3}})$$

$$R_{ij}^{zz} \prec (z_i z_j)^{\frac{1}{3}}(C_{\frac{2}{3}} - 2\delta r^2 C_{\frac{5}{3}} + 0.5\delta r^4 C_{\frac{8}{3}})$$

$$R_{ij}^{z\alpha} \prec -\frac{1}{2}\delta\alpha(z_i z_j)^{\frac{1}{3}}\sqrt{\frac{z_i}{z_j}}(2C_{\frac{2}{3}} - (2\frac{z_j}{z_i} + \delta r^2)C_{\frac{5}{3}} + \frac{z_j}{z_i}\delta r^2 C_{\frac{8}{3}}); \quad \alpha = x, y$$

where:

$$\Psi(\xi) = \frac{1}{3}\int_{\xi^2}^{\infty}\frac{exp(-\tau)d\tau}{\tau^{\frac{4}{3}}}$$

$$C_\nu(\xi) = \frac{\int_{\xi^2}^{\infty} exp(-\tau)\tau^{1-\nu}d\tau}{B^{2\nu}}$$

$$\delta x = \frac{1}{2}\frac{x_2 - x_1}{\sqrt{z_1 z_2}}; \quad \delta y = \frac{1}{2}\frac{y_2 - y_1}{\sqrt{z_1 z_2}}; \quad \delta r^2 = \delta x^2 + \delta y^2; \quad B^2 = \frac{1}{2}\frac{z_1^2 + z_2^2}{z_1 z_2} + \delta r^2.$$

In Fig. 3 the behaviour of the correlation functions along the principal axes are presented. This behaviour is very similar to that measured by Deardorff and Willis (1985) in the water tank experiments. In contrast to the constant flux layer turbulence, in the case of free convective turbulence the third moments play a very important role in the diffusion process. The calculated third moments using the above eddy model are:

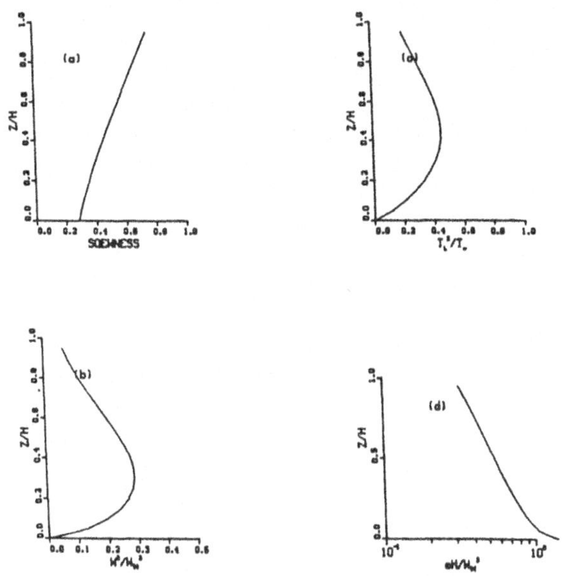

Fig. 4. The parameters of the turbulent field defined by the eddy model (eq. 4.12), as functions of height: a) The skewness; b) The vertical variance of wind fluctuations; c) ϵ^{zz} (defined by eq. 4.18); d) The Lagrangian time scale T_L.

$$\mu^{xxx} = \mu^{yyy} = \mu^{xxy} = \mu^{yyx} = 0. \tag{4.15}$$

Expressions for the other moments are given in Appendix A. In Fig. 4 we present the vertical behaviour of the vertical velocity variance and the skewness of the one-point velocity distribution. Comparison of those results with the measured quantities in laboratory experiments (Deardorff and Willis, 1985) and in the atmosphere (Lenschow *et al.*, 1980) show very good agreement. For $z << H$ the model describes the correct behaviour of the variance ($\prec z^{\frac{2}{3}}$) and of $\mu^{zzz}(\prec z)$ which yield constant skewness [see: Hunt *et al.* (1988),Kaimal *et al.* (1976)]. Fig. 5 presents one realization of the turbulent field. The updraft and downdraft streams are well described by the model.

4.2.2. The diffusion equation

The fluctuations in the convective boundary layer are not isotropic and therefore different Lagrangian time scales are taken for the vertical and horizontal motions. The horizontal Lagrangian time scale [see: Hanna (1981)] is:

$$T_L^x = T_L^y = \frac{0.15}{\sigma_u} \tag{4.16}$$

where σ_u is the horizontal standard deviation of the wind fluctuations. For the vertical motion we assume

$$T_L^z = \frac{\sigma_z^2}{\epsilon(z)} \tag{4.17}$$

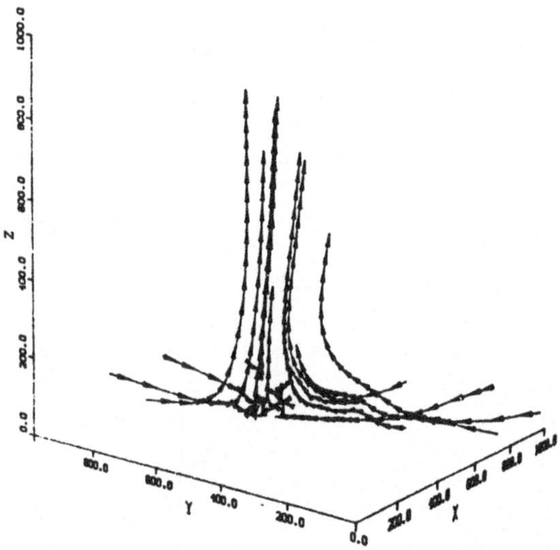

Fig. 5. One realization of the turbulent field defined by the eddy model. The updraft and downdraft are simulated.

where $\epsilon(z)$ is the energy dissipation and σ_z is the standard deviation of the vertical wind fluctuations. For $\epsilon(z)$ we used the functional form suggested by Luhar and Britter(1989):

$$\epsilon(z) = \frac{w_*^3}{H}(1 - (\frac{z}{H})^{\frac{1}{3}}) \tag{4.18}$$

where w_* is the convective velocity. Using these expressions and the calculated moments, the evaluated parameters for the stochastic process are:

$$\begin{aligned}
C_2^{yyz}(z) &= \frac{1}{2\mu^{yzz}} \frac{\partial \mu^{yyz}}{\partial z} \\
C_2^{xxz}(z) &= \frac{1}{2\mu^{xxz}} \frac{\partial \mu^{xxz}}{\partial z} = C_2^{yyz}(z) \\
C_2^{zzz}(z) &= \frac{1}{\mu^{zzz}} (\frac{\partial \mu^{zzz}}{\partial z} - \frac{\partial \mu^{xxz}}{\partial z} - \frac{\partial \mu^{yyz}}{\partial z}) \qquad (4.19) \\
C_0^z(z) &= \frac{\partial \mu^{zz}}{\partial z} - C_2^{zzz}(z)\mu^{zz} - 2C_2^{xxz}(z)\mu^{xx}.
\end{aligned}$$

The equations for the particle's motion are:

$$\begin{aligned}
du_i^z &= [C_0^z(z) - \frac{u_i^z}{T_L^z(z)} - \sum_\epsilon C_2^{\epsilon\epsilon z}(u_i^\epsilon)^2]dt + \Theta^z(\mathbf{r}_i) \\
du_i^\alpha &= -\frac{u_i^\alpha}{T_L^\alpha} + \Theta^\alpha(\mathbf{r}_i); \quad \alpha = x, y
\end{aligned} \tag{4.20}$$

where Θ is a random field with a covariance matrix given by:

$$< \Theta^{\alpha}(\mathbf{r}_i)\Theta^{\beta}(\mathbf{r}_j) >= R^{\alpha\beta}(\mathbf{r}_i\mathbf{r}_j)(\frac{1}{T_L^{\alpha}(\mathbf{r}_i)} + \frac{1}{T_L^{\beta}(\mathbf{r}_j)})dt+$$

$$[\frac{\partial\mu_{ijj}^{\alpha\beta\epsilon}}{\partial r_i^{\epsilon}} + \frac{\partial\mu_{jij}^{\alpha\beta\epsilon}}{\partial r_j^{\epsilon}} - \mu_{jij}^{\epsilon\beta\nu}C_2^{\alpha\epsilon\nu}(\mathbf{r}_i) - \mu_{iij}^{\epsilon\nu\alpha}C_2^{\beta\epsilon\nu}(\mathbf{r}_j)]dt\,.$$

5. Numerical Procedure

5.1. Particle Pair Trajectories

Particle pair trajectories are calculated by solving equation (2.1), given the particle locations $\mathbf{r}_1, \mathbf{r}_2$ the calculated covariance matrix $R^{\alpha\beta}(\mathbf{r}_i, \mathbf{r}_j)$, and the local Lagrangian time scale $\tau(\mathbf{r}_i)$. Then a set of 6 random numbers is derived from a multivariable normal distribution function which has the calculated covariance matrix. This set Θ_i^{α} is used in the diffusion equation to calculate the velocity at the new time step. The location of the particles is changed according to (2.1), where the averaged velocity $\mathbf{U}(\mathbf{r}_i)$ is calculated at the particle location. The initial conditions are:

$$\begin{aligned} &r_1^x = 0; \quad r_1^y = 0; \quad r_1^z = h \\ &r_2^x = 0 + \delta; \quad r_2^y = 0 + \delta; \quad r_2^z = h + \delta \end{aligned} \qquad (5.1)$$

where h represents the detection height. In this work the value of h was 1 m and $\delta = 10^{-4}H$ for the ground-level source in the constant flux layer. In the convective boundary layer, we calculate the concentration moments for three elevated sources, h=0.067H, 0.24H, 0.49H, where H is the inversion height.

5.2. Calculation Of The Concentration Moments

The Nth moment of the concentration at location $\mathbf{r}$ is given by:

$$C^N(\mathbf{r}, t) = \int\int\int Q(\mathbf{r}'_1, s'_1)\ldots Q(\mathbf{r}'_N, s'_N)Pd^3\mathbf{r}'_1\ldots d^3\mathbf{r}'_N d^3 s'_1 \ldots d^3 s'_N \qquad (5.2)$$

where $P(\mathbf{r}'_1, \ldots, \mathbf{r}'_N, s'_1 \ldots, s'_N; \mathbf{r}, \ldots, \mathbf{r}, t, \ldots, t)$ is the probability that the N particles will coincide at point $\mathbf{r}$ at time t, $t > Max(s'_j)$ $(i \leq j \leq N)$ (forward diffusion probability function). In incompressible flow one can use in equation (5.2) the backward diffusion probability function, which is the probability that N particles which at time t are at location $\mathbf{r}$ came from points $(x'_1, y'_1, z'_1) \ldots (x'_N, y'_N, z'_N)$ at time $s'_1 \ldots s'_N (s'_i < t)$ [see Egbert and Baker (1984)] . $Q(\mathbf{r}', s')$ is the spatial and temporal source distribution at time $s' < t$. For an instantaneous source $Q(\mathbf{r}', s') = S(\mathbf{r}')\delta(s' - s_0), s_0 < t$, and $S(\mathbf{r}')$ is the source's spatial distribution. For a stationary continuous source, $Q(\mathbf{r}', s') = S(\mathbf{r}')\frac{\delta(x'-x)}{U(\mathbf{r}')}$. If in addition the turbulence is stationary, the backward diffusion probability function can be calculated by following the trajectories of particles forward in time [see Corrsin (1952), Durbin (1980)]. The procedure for calculating C^N is as follows: We solve equation (2.1) with the initial conditions (5.1). For a continuous source, the lateral and vertical position of the particles (y_i', z_i') i=1,2 at the time they cross the point x are calculated. Then we assign to particle 1 the concentration $\bar{C}_1 = \frac{S(y_1', z_1')}{U^x(z_1')}$ and to particle 2 the concentration $\bar{C}_2 = \frac{S(y_2', z_2')}{U^x(z_2')}$. For an instantaneous source we calculate the positions of particles (x_i', y_i', z_i') i=1,2 at time t. Then we assign to particle 1 the concentration $\bar{C}_1 = S(x_1', y_1', z_1')$ and to particle 2 the concentration $\bar{C}_2 = S(x_2', y_2', z_2')$. We repeat the procedure M times. The averaged concentration is defined by:

$$\bar{C} = 0.5(\bar{C}_1 + \bar{C}_2) \tag{5.3}$$

$$\bar{C}_1 = \frac{1}{M} \sum_{n=1}^{M} C_{in}; \quad i = 1,2$$

where C_{in} is the value of C_i in the nth realization. The second moment of the concentration is given by:

$$\bar{C^2} = \frac{1}{M} \sum_{n=1}^{M} C_{in} C_{2n} \tag{5.4}$$

The fluctuation intensity S is:

$$S = \sqrt{\frac{\bar{C^2} - \bar{C}_1 \bar{C}_2}{\bar{C}_1 \bar{C}_2}}\,. \tag{5.5}$$

In all our calculations we have used M=50000.

6. Results

6.1. The Constant Flux Layer

6.1.1. Comparison of the theory with other theoretical and experimental results
Although the model is designed to calculate higher moments of the concentration fluctuation distribution function, we shall use it first to calculate the average concentration of a line source in the constant flux neutral boundary layer. This is done in order to compare our model with other theoretical and experimental results. The average concentration of a line source was investigated by many authors using methods of K-theory approximation for the diffusion equation (Sutton, 1953), similarity theory (Batchelor, 1964), random walk computation (Corrsin, 1952) and others. Those methods are reviewed in detail in Pasquill and Smith (1983). The procedure suggested by van Ulden (1978) and Horst (1979) was found to fit best the Prairie-Grass experiments. In Fig. 6 we compare our results with

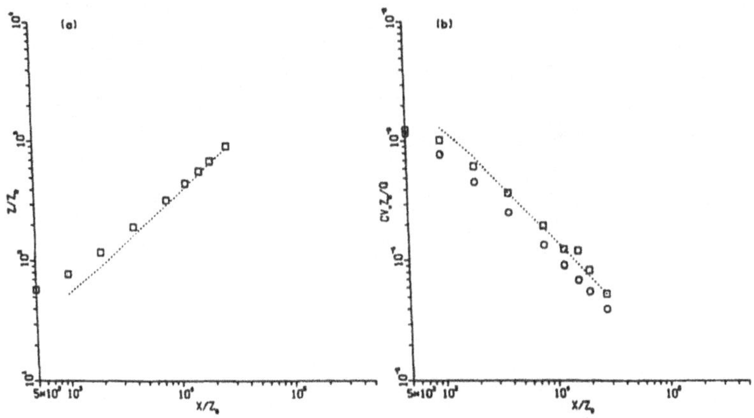

Fig. 6. a) The center of mass of the vertical distribution, $\bar{Z}$ as a function of downwind distance from a line source. Lengths are scaled by z_0 - the friction length.□ - our simulation results. ....- results of Horst-van Ulden b) The average concentration of a line source as a function of downwind distance from the source. The concentration is scaled by $Q/(v_* z_0)$ where Q is the emission rate, v_* the friction velocity and z_0 the friction length. □ - our simulations results. ....- results of Horst-van Ulden o - fit to the Horst-van Ulden approximation $1/[1.37U(0.6\bar{Z})\bar{Z}]$ using $\bar{Z}$ calculated by our simulations.

their calculations for the downwind dependence of the average concentration, and the center of mass of the vertical concentration distribution $\bar{Z}$. The dotted line in Fig.6 represents the Horst-van Ulden predictions and □ our model calculations. The circles present comparisons of the average concentration with the approximation $\bar{C} = \frac{1}{1.37\bar{Z}U(0.6\bar{Z})}$, using $\bar{Z}$ calculated by our model. Results are presented for the dimensionless parameters $\frac{x}{z_0}$ and $\frac{Cv_* z_0}{Q}$, where v_* is the friction velocity, z_0 the friction length and Q the source emission rate. It can be seen that good agreement is achieved. The average concentration was compared also to the wind tunnel experiments of Fackrell and Robins (1982). From their data for a point source we have calculated the cross-wind integrated concentration, which is equal to the average concentration of an infinite line source. In Fig. 7 their data are compared with our predictions for a ground-level source and for an elevated source of height $\frac{H}{z_0} = 2000$. A good agreement is achieved between our theoretical model and the wind tunnel measurements.

6.1.2. Comparison of the model calculations with field experiments

The calculated average concentrations at height z = 1 m are presented in Fig. 8. The source was distributed uniformly on a circular cross-section with 0.25 m diameter, centred at a height of 1 m. The friction length in the calculation was 0.02 m. Those parameters were chosen to be the same as in the experiment of Dinar *et al.* (1988) in order to compare the calculations to measurements. That experiment was carried out over flat terrain. The stability of the atmosphere was near neutral and a mean wind of about 5 m/sec was measured at 2 m height. The tracer used was fog oil smoke which was released through a buffer tube in order to assure

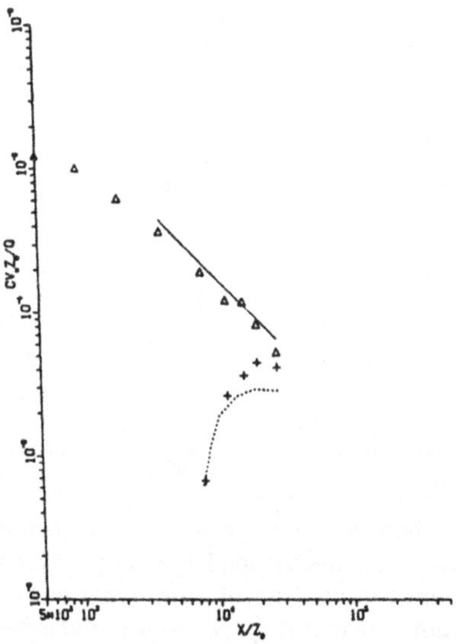

Fig. 7. The average concentration as a function of downwind distance from a line source. —— - Fackrell and Robins results for a ground-level source. . . . . . - Fackrell and Robins results for an elevated source. △ - our model results for a ground-level source. + - our model results for an elevated source.

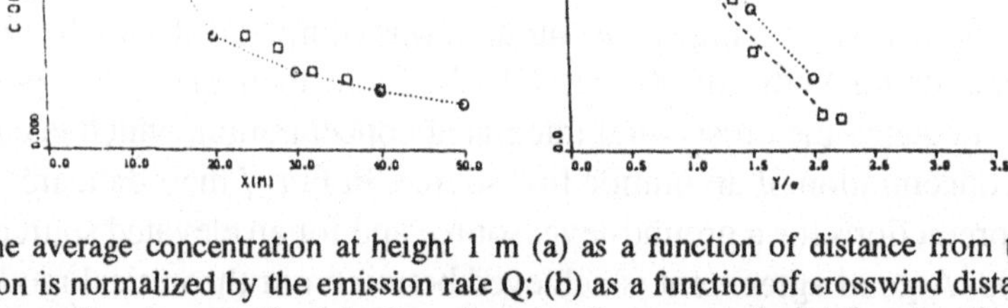

Fig. 8. The average concentration at height 1 m (a) as a function of distance from the source. Concentration is normalized by the emission rate Q; (b) as a function of crosswind distance scaled by σ_y. Concentration is normalized by its value at y=0. o · · · o calculated, using the model; □ measurement (Dinar *et al.* (1988)); - - - - Gaussian fit to the measurements.

constant flow rate. The detectors, based on optical techniques, are described in detail in Dinar *et al.* (1988). The concentration was measured with a resolution of 300Hz. In Fig. 8a the behaviour of the average concentration as a function of downwind distance is presented (dotted line) and compared with experimental results. The agreement between measured and calculated values is very good. In contrast to homogenous turbulence, equation (3.16) cannot be solved analytically and the vertical distribution is not Gaussian. The agreement between measured and calculated values is an indicator that the one-particle statistics are modeled

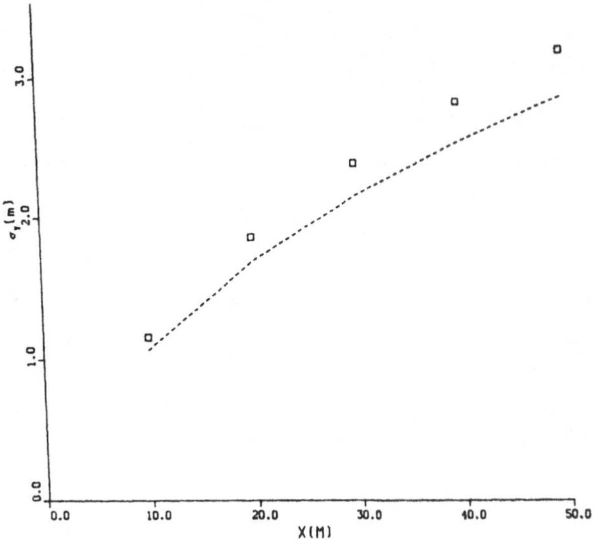

Fig. 9. Downwind behaviour of the lateral standard deviation σ_y: □ calculation of the model; - - - - - equation (6.2)

correctly. In Fig. 8b the cross-wind behaviour of the average concentration is presented. The average concentration is found to be a Gaussian function of cross-wind distance. This is a result of the model itself. The equations of motion of the particle in the lateral direction are:

$$dy = v\,dt$$
$$dv = \frac{-v\,dt}{\tau(z)} + \sqrt{\frac{2}{\tau(z)}}\sigma_v\,d\xi \tag{6.1}$$

where $d\xi$ is a Wiener process. The solution of this equation is Gaussian. Experimental results which are also presented in Fig. 8a (o · · · o), exhibit also a Gaussian shape. A Gaussian fit to the experimental data is the dashed line in the figure. In contrast to the case of homogeneous turbulence in which the Lagrangian time scale is constant, the downwind dependence of the standard deviation (Fig.9) cannot be described by the formula:

$$\sigma_y(x) = \sqrt{2}\sigma_v\tau\sqrt{exp(\frac{-x}{u\tau}) + \frac{x}{u\tau} - 1} \tag{6.2}$$

as in homogeneous turbulence. A comparison of the simulated $\sigma_y(x)$ to (6.2) is given in Fig. 9. It is shown that the simulated value exhibits a larger diffusion rate than that predicted by (6.2) based on the local Lagrangian time scale. This larger rate is found because the effective Lagrangian time scale in the constant flux layer increases with the expansion of the cloud. In Fig. 10a the behaviour of fluctuation

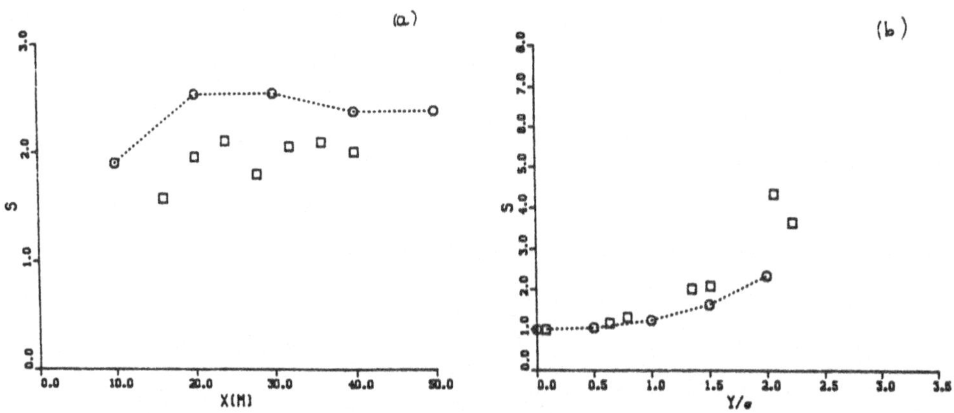

Fig. 10. Fluctuation intensity at height 1 m (a) as a function of distance from the source; (b) as a function of crosswind distance scaled by σ_y. (Notations are as in Fig. 3)

intensity as a function of downwind distance from the source is presented. It can be seen that in the range at which the calculations were carried out, the fluctuation intensity is almost constant. This behaviour (found also in the experimental data) is different from that expected in the case of homogeneous isotropic turbulence. In the last [see Kaplan and Dinar (1988b)] we found that for $t > \tau$ the fluctuation intensity decays with downwind distance from the source. The local Lagrangian time scale in this work was $\tau = 1.4$ sec. In Fig. 10a one can see that even at a distance of 50 m where the travel time is x/u=10 sec, this decay is not observed and the fluctuation intensity is high. Experimental results of Dinar *et al.* (1988) exhibit the same downwind behaviour. This behaviour was observed also in the wind tunnel experiments of Fackrell and Robins (1982). However, we could not compare our results quantitatively because their measurements were done for $x/z_0 > 8000$, while our calculations were for $x/z_0 < 4000$. In Fig. 10b the cross-wind behaviour of the fluctuations intensity is presented. The cross-wind distance is scaled by σ_y. Experimental data of Dinar *et al.* (1988) are presented as well. The agreement is good, and the scaling is found to be in accordance with other measurements [see Hanna (1984)]. It should be mentioned that although the model predicts quite well the behaviour of fluctuation intensity, it overestimates its value (see Fig. 10a). This overestimation may be attributed to inaccurate modelling of the lateral correlations (see Fig. 1). It was found that those correlations decay slower than the measured ones. This causes a more intensive lateral meandering of the cloud than in reality, thus increasing the fluctuation intensity. Better modelling of the correlations may improve the results.

6.2. Free Convective Turbulence

6.2.1. *Comparison of the theory with other theoretical and experimental results for the mean concentration*

Due to the large size of eddies comparable with the inversion height H, K-theory models are not suitable to describe the diffusion in the convective boundary layer. Therefore in the last decade people have tried statistical models which incorporate all the complex features and nonhomogeneity of the turbulence. [See for example, Misra(1982), Venkatram (1983), Bernstein and Berkowicz(1984), de Baas *et al.* (1986), Sawford and Guest(1987), Luhar and Britter(1989)]. These models are 1-D models,and were used to describe the average cross-wind concentration. The model of Luhar and Britter(1989) was also consistent with the well-mixed condition of Thomson (1987), and predicts well the water tank experiments of Willis and Deardorff (1976,1978). The first test to the model that we suggested, was to compare its results for the mean concentration with those experiments. This comparison is presented in Fig. 11 which shows that good agreement is achieved. In addition, as our model is a 3-D model, downwind concentration from a point source can be calculated. Those calculations were carried out for three source heights: 0.067H, 0.24H, 0.49H. Results are presented in Fig. 12. The lateral spread of the plume which is defined by

$$\sigma_y = \sqrt{\int\int C(x,y,z)y^2\,dz\,dy} \tag{6.3}$$

is described in Fig. 14. σ_y depends on source height and decreases with height (in the range $z < 0.5h$ where the calculations were performed). The dependence on time is very similar to that measured by Willis and Deardorff (1978) although the value of σ_y in our calculations is lower by a factor of 0.7. The reason is that the value of σ_v, the lateral wind standard deviation used by our model, is also lower than that measured in the experiments by the same factor. In contrast to the case of homogeneous turbulence, it is not straightforward that the concentration downwind from a line source C_l is given by:

$$C_l = C_p(0)\sqrt{2\pi}\sigma_y \tag{6.4}$$

where $C_p(0)$ is the concentration downwind from a point source at the plume center. This equality was assumed by Deardorff when evaluating concentration fluctuation intensity from his measurements (Deardorff, 1988). We therefore evaluate C_l from the calculation of C_p (Fig. 12) and σ_y (Fig. 14) using equation (6.4). Results are presented in Fig. 13. Comparing those results with the direct calculation of C_l (Fig. 11), it can be seen that (6. 4) still holds in the convective boundary layer. In Fig. 14 we present the vertical spread of the cloud σ_z as a function of time, and the agreement with the water tank measurements is very good.

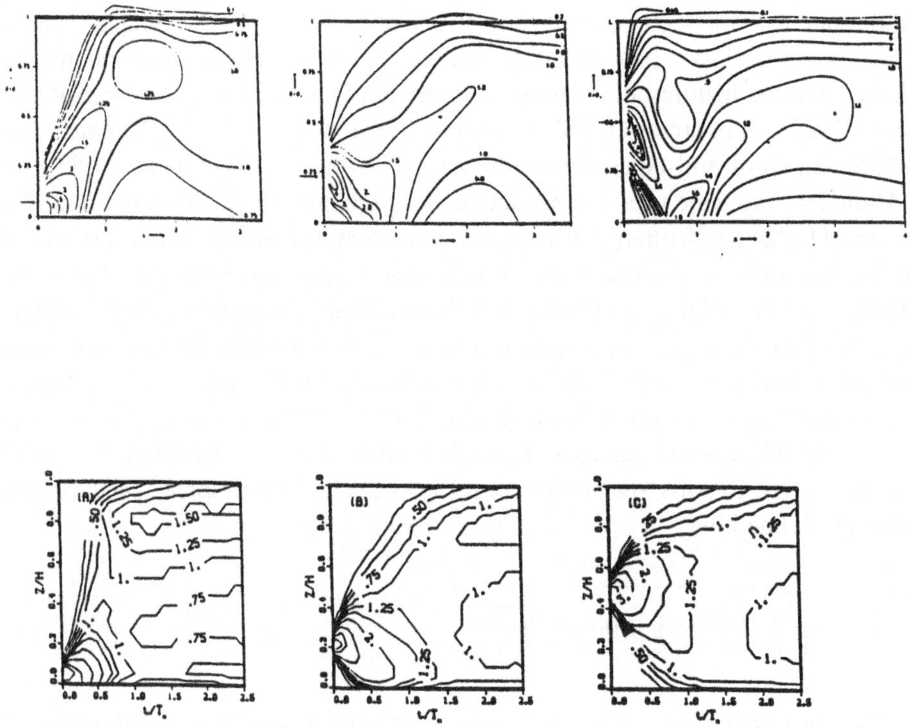

Fig. 11. The calculated average concentration of a line source for three source heights: a) h=0.067H, b) h= 0.24H, c) h= 0.49H. Bottom: model results. Top: Measurements in water tank (Willis and Deardorff, 1976,1978)

6.2.2. Concentration fluctuations

The calculated fluctuation intensity $\sigma_c/\bar{C}$ is presented in Fig. 15 as a function of source height and of the detector height. It is seen that after $t > 1.5T_L$, the fluctuations do not depend on the source height. The value of the fluctuation intensity is about 1 and it decays slowly with time. At $t > 1.5T_L$ the fluctuations do not depend on the detector height at the range at which the calculations were carried, i.e., $z < 0.5H$. On the other hand for $t < T_L$, fluctuation intensity depend strongly on the source height, and as expected it is lower when the detector is located at the source height, and higher when the detection height is far from the source height. Those fluctuations are contributed mainly by the meandering of the plume center in the vertical direction. The numerical model was not accurate

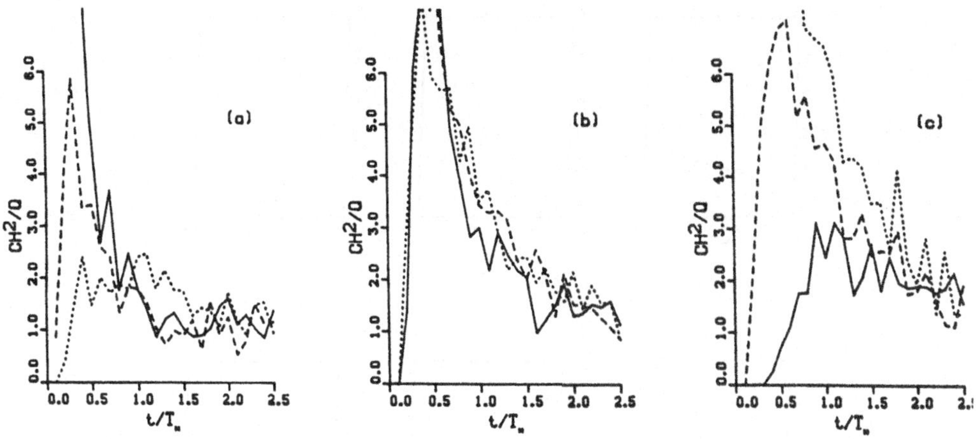

Fig. 12. The mean concentrations downwind from a point source as a function of time. a) Detector height: 0.067H, b) Detector height: 0.24H, c) Detector height: 0.49H; (solid line: source height .067H; dashed line: source height .24H; dotted line source height .49H.)

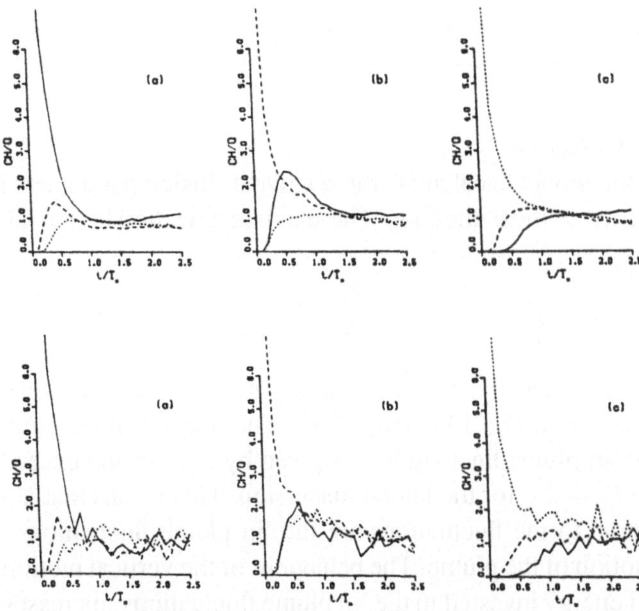

Fig. 13. The mean concentrations downwind from a line source as a function of time. Top: direct calculations; Bottom: reconstruction using eq. (6.4). (Notations as in Fig. 12.)

enough to describe the fluctuation intensity downwind from a point source, and therefore comparison with the experimental results of Deardorff(1988) was not possible. It should be emphasised that the fluctuation intensity at the center of the plume downwind from a point source, is not equal to that downwind from a line source.

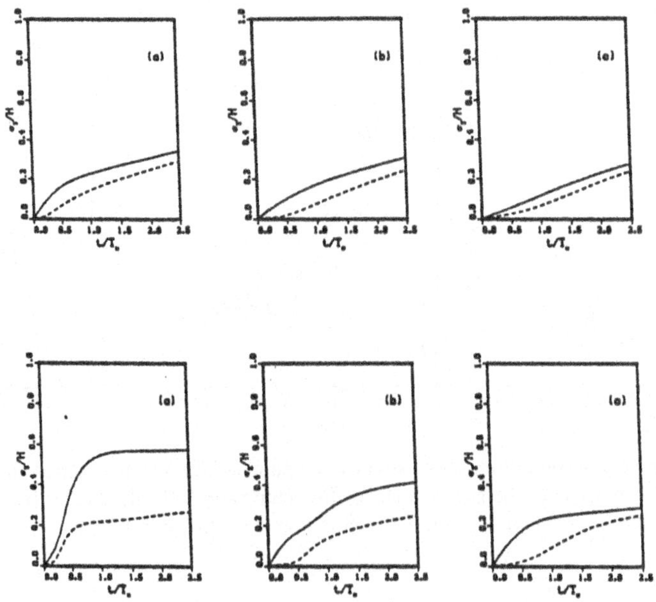

Fig. 14. The lateral spread σ_y and the vertical spread σ_z as functions of time (solid line). The lateral spread σ_y^R and the vertical spread σ_z^R of the relative motion (dashed line). (a,b,c as in Fig.12)

6.2.3. Relative diffusion

We have used the model to calculate the relative diffusion parameters in the lateral direction and in the vertical direction. The standard deviation for the relative motion is defined by:

$$\sigma_R^\alpha = \sqrt{< (r_i^\alpha - r_j^\alpha)^2 >}; \quad \alpha = x, y \tag{6.5}$$

where $<>$ denotes an average over all pairs and all realizations. Results of calculations are presented in Fig. 14 (dashed line). The fraction of the turbulent energy invested in the 'in plume fluctuations' is given by α_R^2/α^2 and exceeds a value of 0.63-0.75 at $t/T_* = 2.5$ for the lateral dispersion. This means that at this time the main contribution to the fluctuations are the 'in plume fluctuations' and not the meandering motion of the plume. The behaviour of the vertical motion is different. The fraction of energy invested in the 'in plume fluctuations' increases with source height. At $t/T_* = 2.5$ this fraction is 0.21, 0.37, 0.77 for the source at heights 0.067H, 0.24H, 0.49H respectively.

7. Comparison with Other Models

Concentration fluctuations using random walk models were studied by many authors. Corrsin (1952) used the one-particle statistics to evaluate the variance of concentration fluctuations. Durbin (1980) have shown that in order to model correctly the concentration variance, one should use the particles' pair probability

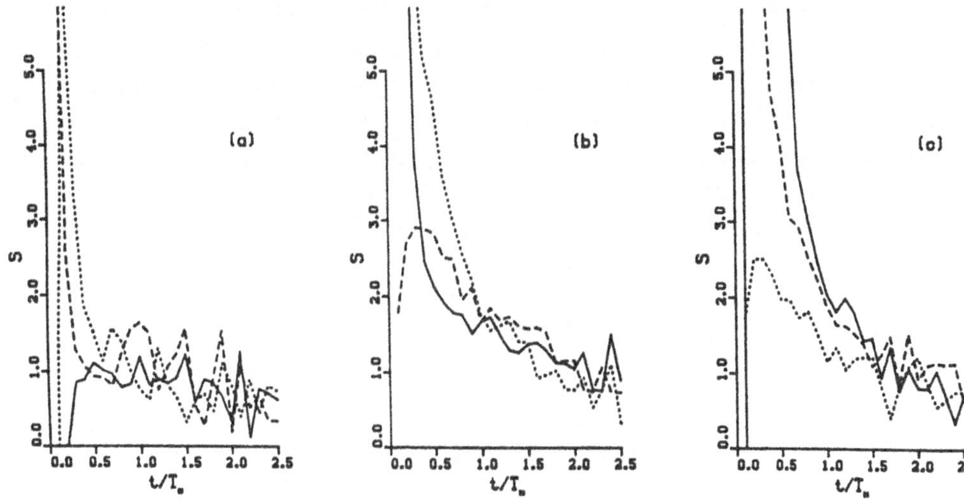

Fig. 15. Fluctuation intensity downwind from a line source as a function of time. (Notation as in Fig. 12.)

function. The random walk two-particle models of the form (2.1) have been used in the literature with different modeling of a_i^α, $B_{ij}^{\alpha\beta}$. Novikov (1963), Gifford (1982), and Lee and Stone (1983) have used a diagonal matrix for $B^{\alpha\beta}$. In their models, both a_i^α and $B_{ii}^{\alpha\alpha}$ were functions of the ith particle coordinates or velocities only, (in fact $B_{ii}^{\alpha\alpha}$ was taken as a scalar matrix). The correlation between the particles' pair was considered only in the initial conditions. In Lee and Stone (1983) the initial conditions were chosen in such a way that the two particles initial velocity correlations will be equal to the Eulerian velocities correlations. In addition, the particles' pair motions become independent after a time of the order of the Lagrangian time scale. The motion of the two particles remains independent even if they have approached each other at some time after their release. Those models do not satisfy the well-mixed condition suggested by Thomson (1987). Durbin (1980) has suggested a one-dimensional model, in which the motion of pair of particles is correlated along their path. Durbin's model is not consistent with the well-mixed condition. It also does not fulfill the incompressibility condition (Egbert and Baker, 1984) and suffers from the fact that the one-particle probability density function depends on the initial conditions of the second particle, which seems unreasonable. Thomson (1990) suggests a model in which the spatial correlation between the pair of particles is included in a_i^α, i. e. in the part of the correlation which is correlated in time. $B_{ij}^{\alpha\beta}$ in his model was taken as a scalar matrix. His model is compatible with the well-mixed condition. However, the well-mixed distribution that was used by him was multinormal with zero odd moments. It is known that in isotropic decaying turbulence the second moment decay in the inertial range is balanced by the third moment (see Batchelor, 1956). Choosing zero third order moments violates this balance. Also in the Thomson model the one-particle motion depends on the initial conditions of the other particle. Kaplan and Dinar (1988a,b) have suggested a different approach to the isotropic homogeneous turbulence. In their model the spatial correlations of the particle velocities are included in the part of the acceleration which is time independent. The equilibrium function of this model is compatible with the well-mixed condition and with the decay equations of those correlations, as derived from the Navier-Stokes equations (Batchelor 1956). The one-particle probability distribution function is modeled correctly. This model was extended also to the N-particle statistics which enable one to calculate higher moments of the distribution (see Kaplan and Dinar, 1988). In this work we use the same approach for inhomogeneous turbulence. As an example we applied the theory for the constant-flux neutral boundary layer, and to free-convective turbulence.

8. Summary

In this work we suggest a 3-D model for the calculation of the concentration distribution in inhomogenous, turbulent, incompressible flow. The model requires knowledge of the covariance matrix of velocity. This information can be supplied either by measurements or by modelling. In this work we implement the model for two different cases: The constant flux neutral boundary layer and the free convective boundary layer. In the first case the model was used to calculate averaged concentration and fluctuation intensities of a ground-level source in the constant flux neutral boundary layer. The covariance matrix of velocities was modeled using an eddy description of turbulence, as suggested by Townsend (1956). The averaged concentration was found to be in good agreement with measurements. The model also predicts well the spatial behaviour of fluctuation intensity. The model overestimates the magnitude of fluctuation intensity and this may be attributed to the crudity of the correlation matrix model. The behaviour of fluctuation intensity in the constant flux layer is found to be different from that of homogeneous isotropic turbulence. In the case of free convective turbulence we compare the model's results with the water tank experiments of Deardorff(1976,1978). The model predicts correctly the behaviour of downwind mean concentration from an elevated point source, as well as the lateral and vertical spread. The model is consistent with the incompressibility of the flow, a fact that ensures that it is free from fluctuations contributed by changes in the fluid density. This result cannot be achieved with one-dimensional models.

Appendix

The Calculated Third Moments For Free Convective Turbulence

Using the eddy model [equation (4.13)], the third moments are given by:

$$\mu_{ijk}^{\alpha\beta\gamma} = \int\int\int u^{\alpha}(x_i - \xi, y_i - \eta, z_i) u^{\beta}(x_j - \xi, y_j - \eta, z_j) u^{\gamma}(x_k - \xi, y_k - \eta, z_k) d\xi d\eta \frac{dl}{l}$$

The calculated moments for use in equation (4.9) are:

$$\mu_{iij}^{zzz} \prec z_i \sum_{k=0}^{3} A_k(\mathbf{r}_i, \mathbf{r}_j) C_k$$

where

$$C_k = \frac{\Gamma_{k+1}(\bar{B}^2 \frac{z_i z_j}{H^2})}{\bar{B}^{2k+2}}.$$

$\Gamma_k(\xi)$ is the incomplete gamma function of order k and $\bar{B} = \frac{z_i}{z_j} + \frac{z_j}{2z_i} + \frac{4\delta r^2}{3}$;

$$\delta r^2 = \delta x^2 + \delta y^2 + \delta z^2$$

$$A_0 = \frac{32}{9}\,;\, A_1 = -\delta r^2 \frac{64}{9}\,;\, A_2 = \delta r^4 \frac{64}{27} + \delta x^2 \delta y^2 \frac{128}{27}\,;$$

$$A_3 = -\delta r^6 \frac{256}{729} - (\delta x^4 \delta y^2 + \delta y^4 \delta x^2) \frac{256}{243};$$

$$\mu_{iji}^{\alpha\alpha z} \prec [\frac{2}{9} E_1(\frac{\bar{B}^2 z_i z_j}{H^2}) + \sum_{k=1}^{5} D_k(\mathbf{r}_i, \mathbf{r}_j) C_k] z_i$$

where

$$E_1(\xi) = \int_{\xi}^{\infty} \frac{exp(-t)}{t} dt$$

$$D_1 = -\frac{28}{27}\delta\alpha^2 - \frac{4}{27}(\delta r^2 - \delta\alpha^2)$$

$$D_2 = \frac{32}{81}\delta\alpha^4 + \frac{32}{81}\delta\alpha^2(\delta r^2 - \delta\alpha^2) - \frac{2}{9}(\frac{z_i}{z_j} + \frac{z_j}{z_i})$$

$$D_3 = -(\frac{z_i}{z_j} + \frac{z_j}{z_i}) D_1 + \frac{2}{9}$$

$$D_4 = -(\frac{z_i}{z_j} + \frac{z_j}{z_i}) D_2 + D_1$$

$$D_5 = \frac{32}{81}\delta\alpha^4 + \frac{32}{81}\delta\alpha^2(\delta r^2 - \delta\alpha^2).$$

References

Batchelor, G.K. (1952). *Diffusion in a field of homogeneous turbulence II:The relative motion of particles*. Proc. Camb. Phil. Soc. **48**, 345-362.

Batchelor, G.K. (1956). *The theory of homogeneous turbulence*. Cambridge University Press.

Batchelor, G. K. (1964). *Diffusion from sources in a turbulent boundary layer*. Archiv. Mechaniki Stoswanej.,**3**, 661.

Bernstein J. H., Berkowicz R. (1984), *Monte-Carlo simulation of plume dispersion in the convective boundary layer.*, Atmos. Environ., **18**, 701-712.

Corrsin, S. (1952). *Heat transfer in isotropic turbulence*. J. Appl. Phys. 23, **1**, 113-117.

Deardorff J. W. Willis G. E. (1985), *Further results from a laboratory model of the convective planetary boundary layer*, Bound. Lay. Met., **32**, 205-236.

Deardorff J. W. Willis G. E. (1988), *Concentration fluctuations within laboratory convectively mixed layer*, in "Lectures on Air Pollution Modeling" ed. by A. Venkatram and J. C. Wyngaard AMS Boston, 366-370.

De Baas A. F., Van-Dop H., Nieuwstadt F. T. M. (1986), *An application of the Langevin equation for inhomogeneous conditions to dispersion in the convective boundary layer*. Atmos. Environ., **18**, 701-712.

Dinar N., Kaplan H. and Kleiman M.(1988). *Characterization of concentration fluctuations of a surface plume in a neutral boundary layer* Bound. Lay. Met. **45**, 157-175.

Durbin, P.A. (1980). *A Stochastic model of two-particle dispersion and concentration fluctuations in homogeneous turbulence*. J. Fluid Mech. **100**, 279-302.

Egbert, G.D. and Baker, M.B. (1984). *Comments on the effect of Gaussian particle-pair distribution functions in the statistical theory of concentration fluctuations in homogeneous turbulence. B.L. Sawford 1983, 339-353*. Q.J.R. Met. Soc. **110**, 1195-1199.

Fackrell, J.E. and Robins, A.G. (1982). *Concentration fluctuations and fluxes in plumes from point source in a turbulent boundary layer*. J. Fluid Mech. **117**, 1-26.

Grant, H.L. (1958). *The large eddies of turbulent motion*. J. Fluid Mech. **4**, 149-190.

Gifford, F. A. (1982).*Horizontal diffusion in the atmosphere: A Lagrangian dynamical theory*. Atmos. Environ., **16**, 505-512.

Hanna, S.R. (1981). *Turbulent energy and Lagrangian time scale in the planetary boundary layer*. A.M.S. 5th Symp. on turbulence diffusion and air pollution, Atlanta,Ga, 61-62

Hanna, S. R. (1984). *The exponential probability density function and concentration fluctuations in smoke plumes* Bound. Lay. Met. **29** ,361-375

Horst, T. W., (1979). *Lagrangian similarity modelling of vertical diffusion from a ground level source*. J. App. Met., **18**, 733-740.

Hunt J. C. R., Kaimal J. C., Gaynor J. E. (1988) *Eddy structure in the convective boundary layer-new measurements and new concepts.*, Q. J. R. Met. Soc., **114**, 827-858.

Kaimal, J.S., Wyngaard, J.C., Hangen, D.A., Cotè, O.R., Izumi, Y., Caughey, J.J. Reading, C.J. (1976). *Turbulent structure in the convective boundary layer*. J. Atm. Sci. **33**, 2152-2169.

Kaplan, H. and Dinar, N. (1988a). *A stochastic model for dispersion and concentration distribution in homogeneous turbulence*. J. Fluid. Mech. **190**, 121-140

Kaplan, H. and Dinar, N., (1988b). *A three dimensional stochastic model for concentration fluctuation statistics in isotropic homogeneous turbulence*. Journal of Computational Physics, **79**, No.2, 317-335.

Kaplan, H. and Dinar, N. (1988c). *Comments on the paper: On the relative dispersion of two particles in homogeneous stationary turbulence and the implication for the size of concentration fluctuations at large time. By D.J. Thompson (1986), Q.J.R. Met. Soc. 12, 890-894*. Q.J.R. Met. Soc. **114**, 545-550.

Kaplan, H. and Dinar, N. (1989a). *The interference of two passive scalars in a homogeneous isotropic turbulent field*. J. Fluid. Mech. **203**, 273-287.

Kaplan H., Dinar N. (1989b). *Diffusion of an instantaneous cluster of particles in homogeneous turbulence*. Atmos. Envi., **23**, 1459-1463.

Lee, J. T. and Stone, G. L. (1983). *The use of Eulerian initial conditions in a Lagrangian model of turbulent diffusion*. Atmos. Environ., **17**, 2477-2481.

Lenschow, D. H., Wyngaard, J. C., Pennel, W. T. (1980) *Mean field and second moment budgets in a baroclinic, convective boundary layer.*, J. Atmos. Sci., **37**, 1313-1326.

Luhar A. K. and Britter R. E. (1989), *A random walk model for dispersion in inhomogeneous turbulence in a convective boundary layer.*, Atmos. Environ., **23**, No. 9, 1911-1924.

Misra P. K. (1982). *Dispersion of non-buoyant particles inside a convective boundary layer*. J. Atmos. Sci., **41**, 3162-3169.

Novikov, E. A. (1963). *Random force method in turbulence theory.* Soviet Physics JEPT, **17**, 1449-1454.

Pasquill, F. and Smith, F. B. (1983). *Atmospheric Diffusion*, third edition, Ellis Horwood Chichester.

Pope, S.B. (1985). *Pdf methods for turbulent reactive flows*. Prog. Energy Combust. Sci. **11**, 119-192.

Sawford, B.L. (1982). *Lagrangian Monte-Carlo simulation of a turbulent motion of a pair of particles.* Q.J.R. Met. Soc. **108**, 207-213.

Sawford, B. L., Frost C. C. and Allen T. C.(1985) *Atmospheric boundary layer measurements of concentration statistics from isolated and multiple sources* Bound. Lay. Met. **31** ,249-268.

Sawford B. L., Guest F. M., (1987), *Lagrangian stochastic analysis of flux gradient relationship in the convective boundary layer*. J. Atmos. Sci., **44**, No. 8, 1952-1165.

Sutton, O. G. (1953). *Micrometeorology*, McGraw-Hill, New-York.

Sykes, R.I., Lewellen, W.S. and Parker, S.F. (1984). *A turbulent transport model for concentration fluctuations and fluxes*. J. Fluid. Mech. **139**, 193-218

Thomson, D.J. (1986). *On the relative dispersion of two particles in homogeneous stationary turbulence and the implication for the size of concentration fluctuations a large times*. Q.J.R. Met. Soc. **12**, 890-894.

Thomson, D.J. (1987). *Criteria for the selection of stochastic models of particle trajectories in turbulent flows*. J. Fluid Mech. **180**, 529-556.

Thomson, D.J. (1990). *A stochastic model for the motion of particle pairs in isotropic high Reynold number turbulence, and its application to the problem of concentration variance*. J. Fluid Mech. **210**, 113-153

Townsend, A.A. (1956). *The structure of turbulent shear flow*. Cambridge Univ. Press.

van Ulden, A. P. (1978). *Simple estimates for vertical diffusion from sources near the ground.* Atmos. Environ.,**12**, 2125-2129.

Venkatram A. (1983). *On dispersion in the convective boundary layer*, Atmos. Environ., **17**, 529-533.

Warhaft, Z.(1984) *The interference of thermal fields for line sources in grid turbulence* J. Fluid. Mech. **144** 363-381.

Willis G. E. Deardorff J. W. (1976), *A laboratory study of diffusion into the convective planetary boundary layer*, Q. J. R. Met. Soc., **102**, 427-445.

Willis G. E. Deardorff J. W. (1978), *A laboratory study of dispersion from an elevated source within a modeled convective planetary boundary layer*, Atmos. Environ.,**12**, 1305-1311.

FLUCTUATIONS OF LINE INTEGRATED CONCENTRATIONS ACROSS PLUME DIFFUSION IN GRID GENERATED TURBULENCE AND IN SHEAR FLOWS

M. POREH[(1)], A. HADAD[(1)] and J. E. CERMAK[(2)]
(1) Technion- Israel Institute of Technology
(2) Colorado State University

(Received October 1991)

Abstract. The fluctuations of the instantaneous values of line integrated concentrations across plumes from point sources diffusing in turbulent shear flows, and in grid generated turbulence, have been studied experimentally using a fast response system which measured the attenuation of the intensity of an infrared beam crossing the plume. Analysis of the measurements show that the dimensionless statistical properties of the fluctuations at different distances from the source at each flow are approximately similar, in the sense that they depend primarily on the relative off-center location of the line of integration and almost independent of the distance from the source and the nature of the turbulence in the flows, as long as the characteristic length of the mean plume is not large compared to the size of the large eddies. The characteristic time of the fluctuations, on the other hand, was found to grow with the distance from the source and the autocorrelations of the fluctuations, particularly in the case of a plume diffusing in grid generated turbulence, were it found to be proportional to the lateral size of the mean plume. A $-5/3$ decay law of the power spectrum of the fluctuations was observed in the low frequency range which corresponds to the scale of the large eddies. The decay of the fluctuations caused by smaller eddies was much faster, as expected.

1. Introduction

Many effects of aerosols or gaseous contaminants diffusing in turbulent flows are determined by the values of their instantaneous concentrations or short time averages at a point or along a specific path. The chance of igniting an inflammable gas and adverse health effects of a highly toxic gas are related to the instantaneous level of the local concentration of the gas and the visibility of an object viewed through aerosol plumes is determined by the instantaneous value of the integral of the concentration along the line of vision.

As the instantaneous concentrations in turbulent flow fluctuate strongly in time and space, the statistical properties of their fluctuations are of both theoretical and practical interest. In particular, one is interested in estimates of the probability that instantaneous values exceed critical values, in the duration of such events and in the intermittency of the concentration field.

Most researchers have studied the fluctuations of the concentrations at a point (Gifford, 1959; Hanna, 1984a and 1984b; Deardorff and Willis, 1984; Sawford and Stapountzis, 1986; Lewellen and Sykes,1986; Sawford, 1987; Dinar *et al.*, 1988; Hanna and Insley, 1989; Mylne and Mason, 1991). The authors have been interested in the fluctuations of obscuration by smoke plumes and have thus studied the fluctuations of line-integrated concentrations across plumes diffusing from ground

Boundary-Layer Meteorology **62:** 247–267, 1993.

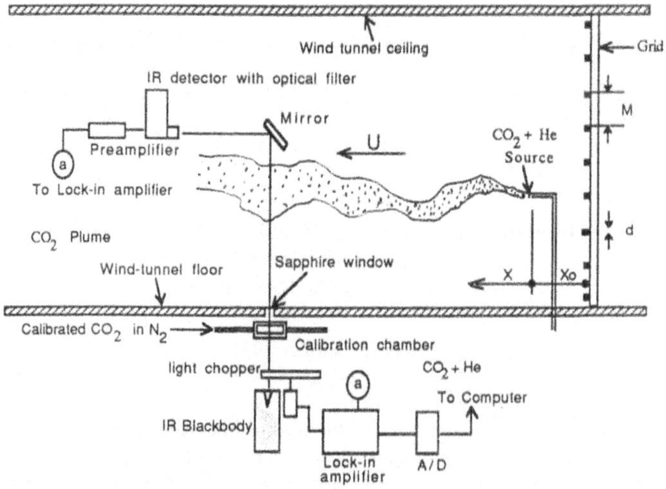

Fig. 1. Schematic description of the experimental system

level sources in a simulated adiabatic atmospheric surface layers over a smooth surface up to distances of two boundary-layer (BL) heights from the sources (Poreh and Cermak, 1987 and 1990; Poreh *et al.*, 1989, 1990 and 1991) and over a rough surface, up to distances of nine BL heights (Poreh and Cermak, 1991a), and across a plume diffusing in grid generated turbulence (GGT) (Poreh and Cermak, 1991b; Kistler and Vrebalovich, 1966).

The present paper describes the experimental system developed for measuring line-integrated concentration fluctuations, presents an overview of the above studies, the experimental results and their analysis.

2. The Experimental System and Procedures

The experiments were conducted at the Meteorological and the Industrial Wind Tunnels at Colorado State University and the Environment Wind Tunnel at the Technion.

A schematic view of an experimental configuration for the case of a plume diffusing in GGT is shown in Fig. 1. The mesh size of the grid was M = 7.62 cm, and its porosity was p = 0.64. Upstream of the grid, a 5 cm thick honeycomb (mesh size 0.3 cm) was placed to attenuate any large-scale eddy motion that might have been generated upstream of the grid. In studying plumes diffusing in the surface layer, the grid was removed, spires were placed at the entrance to the wind-tunnel, roughness elements were placed on the floor, to produce desired boundary-layer characteristics, and the source was placed on the floor.

The tracer gas was a neutrally buoyant mixture of carbon dioxide (62.5%) and helium (37.5%). The discharge of the gas mixture was measured using rotameters, which had been calibrated using bubble flow-meters. It was adjusted for each experiment so that the exit velocity of the gas from the source tube (d=3-5 mm, depending on the distance from the source to the point of measurement) was matched with the mean wind speed at the mean height of the source. The values of the mean wind tunnel speeds in the experiments were in the range U=2.4-3.0 m/sec.

The IR/CO_2 sensor was always located at the same position in the wind tunnel. The position of the grid and the source were changed from one experiment to the other in order to vary the distance x between the source and the location of vertical IR beam. The distance between the grids and the sources in the GGT experiments was constant, $x_o/M = 20$ (see Fig. 1).

A Blackbody emitting IR radiation into the wind tunnel through a 4 mm diameter sapphire window was placed underneath the floor of the wind tunnel. The beam crossed the plume and was deflected by a mirror toward a side looking dewar in which a LN cooled Indium-Antimonide Photovoltaic detector was placed. The beam crossed a narrow band ($4.257 + 0.04\mu m$) optical filter, which passes only IR radiation in the absorption band of CO_2, and a calibration chamber was placed. In addition, the IR beam was chopped by an optical light chopper with a chopping frequency of 2000 Hz. The response of the sensor was determined by measuring the attenuation of its output by certified mixtures of CO_2 in nitrogen that filled the calibration chamber. The output of the detector was fed via a preamplifier into a Lock-in-Amplifier, which also sensed the chopping frequency of the IR beam and produced a modulated DC signal proportional to the intensity of the IR beam. The signal was sampled at a rate of 600 Kz and was filtered by a 24 db/octave filter with a cut-off frequency of 300 Hz. In addition, an $RC = 1ms^{-1}$ filter had been installed by the manufacturer of the lock-in amplifier, so that the cutoff frequency of the system was $f_c = 160$ Hz.

The overall ratio of the noise to the mean value of the signal at the centerline of the plume was usually small, O(0.5%) at $x/M = 20$ and O(1%) at $x/M = 102$. However, the relative noise level was high at the far edges of the plume, where the mean signal is very small. Because the IC signal is highly intermittent at the edges of the plume, the appearance of the noise during periods where IC is almost zero was bothersome, although its contribution to the measuremennts of the mean and rms values of IC is negligible. Unfortunately, the noise could not be separated from the signal and a special method for reducing its effect on the measurements of the intermittency was developed.

3. Presentation and analysis of the results

3.1. Integrated concentrations accross a plume in GGT

The measurements of IC will be presented using the dimensionless variable

$$IC^* = IC\,U\,M/Q \tag{1}$$

where Q is the strength of the source.

The instantaneous value IC^*, will be divided into a mean value ICM^* and a fluctuating value ic^*, so that

$$IC^*(t) = ICM^* + ic^*(t). \tag{2}$$

The rms of ic^* will be denoted by $ic^{*'}$.

The lateral distributions of $ICM^*(y)$, at the different locations downwind of the source, were found to be closely Gaussian, namely $ICM^*(y) = ICM^*(0)e^{-y^2/2\sigma^2}$, where σ is the lateral length parameter of the mean plume.

Because the mean plume is expected to be symmetric, it follows from the continuity equation, assuming that the longitudinal turbulent flux is negligible, that

$$ICM^*(0)\sigma/M = (2\pi)^{-1/2} \tag{3}$$

The deviations of the measurements from equation (3) were small.

Figure 2 shows typical measurements of IC^* at different off-center locations at x/M=20. Each graph shows only 1/6th of the 16000 points sampled during each run. The abscissa in this figure shows the dimensionless time $T^* = tU/M$. The upper graph in the figure shows measurements near the centerline of the mean plume, the central graph shows measurements at y/σ =1.61 and the lower graph shows measurements at the far edge of the mean plume, $y/\sigma = 2.33$. The mean and rms values of IC^* at each location are given at the right-hand side of the figure. One sees from the figure that $IC^*(t)$ at the far edge of the plume is highly intermittent and has relatively large peaks, as large as 15 times its mean value. It is also observed that the mean value of IC^* at $y/\sigma = 2.33$ is only 7% of its value near the centerline of the mean plume (note that $\exp[-(2.33^2/2)] = 0.07$), whereas the ratios of the rms of the fluctuations and the peaks of the signal to the mean value at $y/\sigma = 0$ are approximately 1:3.

The measurements of IC^* at $x/M = 36$ and 102, were quite similar to those at $x/M = 20$, except that the values of IC^* were much smaller and the time scale of the fluctuations appeared to grow with this distance.

Figure 3 and 4 show the measured lateral distributions of $ic^{*'}(y/\sigma)/ic^{*'}(0)$ and $ic^{*'}(y/\sigma)ICM^*(y/\sigma)$, respectively, at different distances from the source. Both graphs suggest that the relative concentration fluctuations at different distances are similar. Figure 4 clearly shows that large values of the rms/mean ratio exist at the edges of the mean plume. The relatively small scatter renders a high degree of reliability to the data.

The Probability Distribution Function (PDF) of IC^*/ICM^*, which is defined as the probability that the instantaneous value of IC^*/ICM^* is larger than a given ratio a, and which will be denoted by $P(\mathbf{a})$, has also been calculated. Before presenting the results, the general properties of the $P(\mathbf{a})$ will be briefly discussed.

As IC is always positive, $P(0) = 1$ and $P(\infty) = 0$. When IC is intermittent, namely larger than 0 for only part of the time, say γ, P(a) is not continuous at a=0 and γ, which is termed the intermittency, is equal to $P(0^+)$.

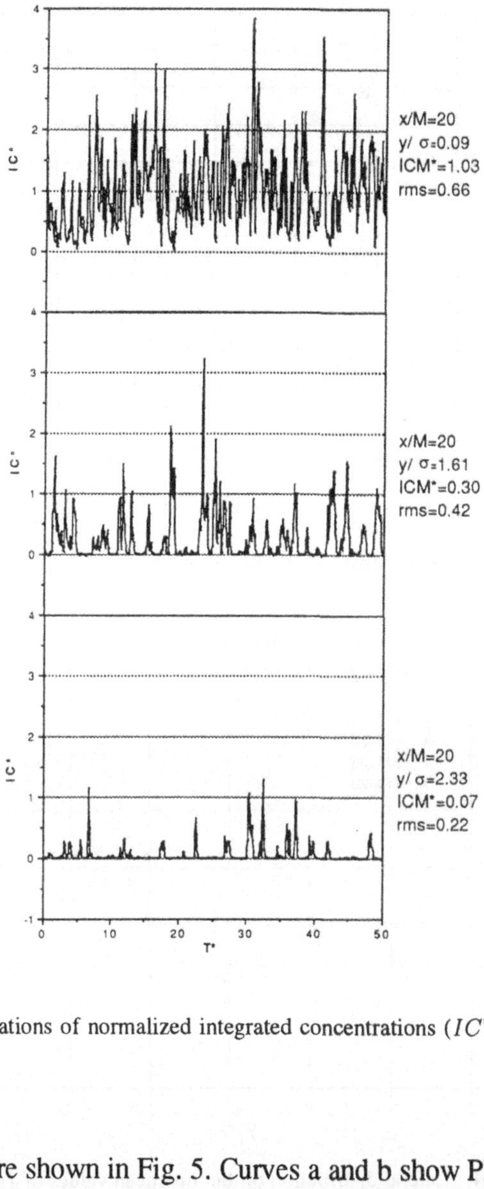

Fig. 2. Typical fluctuations of normalized integrated concentrations (IC^*) at different off-center locations at x/M=20.

Typical PDFs are shown in Fig. 5. Curves a and b show PDFs for intermittent signals, $\gamma = 0.46$ and 0.8, respectively. The intermittency, (see Fig.5), is given by the value of the PDF at 0^+. Curve c is typical for measured PDF near the centerline of the plume, where the minimum values of the signal are slightly above zero and $\gamma = 1$. Curve d is typical of signals with relatively small rms/mean values. Finally, the straight lines: P=1 for $0 < \mathbf{a} < 1$, and P=0 for $1 < \mathbf{a}$, which are denoted by e, are the limiting case of the PDF for infinitely small fluctuations.

Figure 6 shows a typical curve of P near $\mathbf{a} = 0$ and depicts the effect of noise on the measurement of $P(\mathbf{a})$ in this region. Values of P<1 are measured at a<0, due to the negative values of the measured signal, and a similar increase of $P(\mathbf{a})$ for small positive values of **a** is identified. The effect is limited to the region

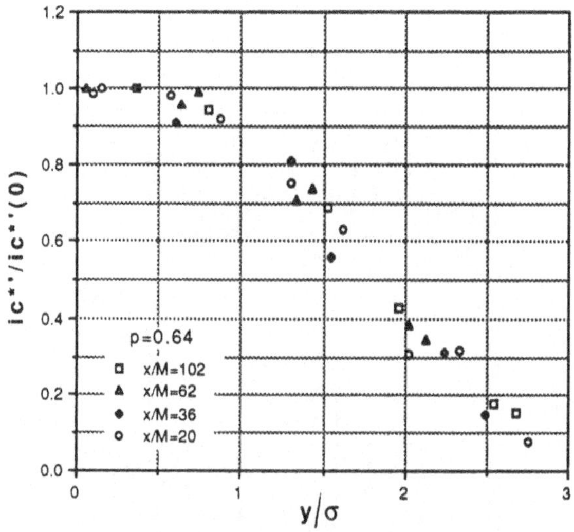

Fig. 3. The lateral distributions of the $ic^{*'}/ic^{*'}(0)$ at different x/M

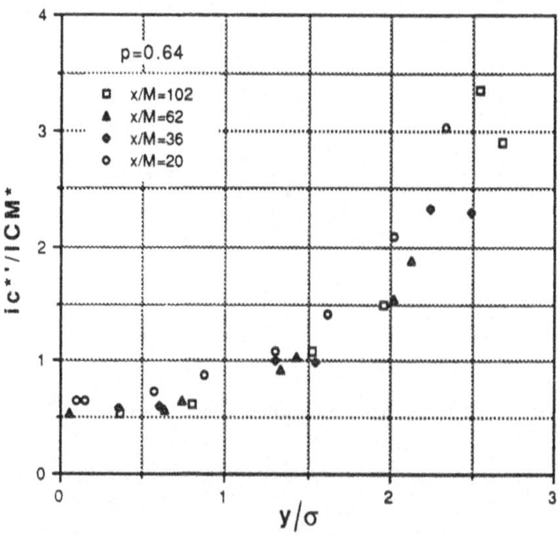

Fig. 4. Lateral distribution of the rms/mean values of IC^*

$-3n' < \mathbf{a} < 3n'$, where n' is the rms value of the noise in equivalent $ic^{*'}/ICM^*$ units, which in this figure is of the order of 0.02. If the noise is symmetrical with respect to 0, the apparent intermittency γ_{ap} becomes $\gamma_{ap} = \gamma + (1-\gamma)/2$. The error due to the noise is, thus, small for $\gamma = O(1)$, but it approaches 0.5 for small values of γ. In Fig. 6, the apparent intermittency is $\gamma_{ap} = 0.9$, whereas the true intermittency is approximately $\gamma = 0.8$.

When the noise is symmetrical, the correct distribution of $P(\mathbf{a})$ can be calculated from the apparent one, P_{ap}, by the equations: $P(\mathbf{a}) = 1$ for $\mathbf{a} \leq 0$ and $P(\mathbf{a}) =$

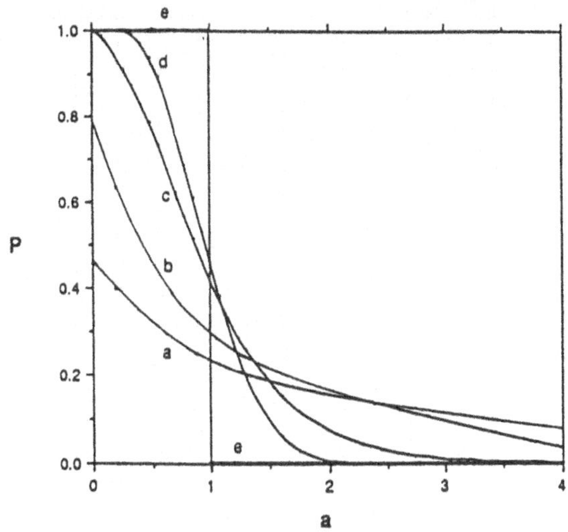

Fig. 5. Typical shapes of the probability density function P(a)

$P_{ap}(a)[1 - P_{ap}(-a)]$ for **a**$>$ 0. The accuracy of this procedure for correcting measurements at the edge of the plume, where γ is always small and n' is relatively large, is decreased by several effects: the Least Significant Bit (LSB) of the 12 bit analog-to-digital conversion in the experiments is not always small compared to the noise level. Drift of instruments could also affect the measurement of the zero value. Thus, the exact zero might not have been accurately determined and the symmetry of the noise could have been distorted by the A/D conversion.

An alternate method for determining P from P_{ap} is to extrapolate the measured curve from the region **a**$>$ $3n'$ to **a**=0, as shown in Fig. 6. We have not corrected the graphs of the PDFs which are presented below for the effect of the noise. However, we have corrected the values of the intermittency γ, using the first method for cases where γ was close to 1 and using the second one for cases where γ was relatively small or whenever there was a suspicion that a drift might have been present.

In general the probability $P(\mathbf{a})$ is expected to be a function of both y/σ and x/M. Figures 7 through 10 show measured PDF. The measurements suggest that the PDFs at different distances up to x/M=O(100) are similar, in the sense that they are determined primarily by y/σ. However, a close inspection of the data shows that the shape of the PDFs, particularly near $y/\sigma = 0$, changes very slowly with distance and has a decreased level of fluctuations at larger distances.

The measured values of the intermittency γ at different locations are plotted in Fig. 11 versus y/σ. These values have been corrected for the effect of noise, as outlined earlier. Again, one finds that the distributions of γ at different distances are approximately similar. The figure also shows that intermittency starts at approximately y/σ=0.8.

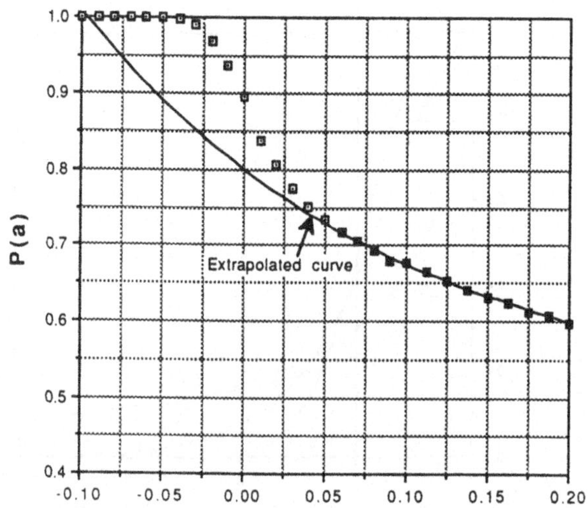

Fig. 6. The effect of noise on the measurement of the intermittency and the probability that IC^*/ICM^* >a near a=0

To get an insight into the time dependence characteristics of IC^*, the autocorrelations and power spectra of the fluctuations were calculated.

In Fig. 12 the autocorrelations are plotted versus $\tau U/\sigma$. The figure clearly shows that the time scale of the fluctuations at all distances is proportional to σ/U, namely to the time that a volume of fluid which has a longitudinal dimension of σ passes a stationary point, and suggests that the longitudinal length parameter of parcels of fluid with correlated concentrations is of the order of σ. This information is of both basic and practical importance as it enables one to estimate the duration of relatively high values of IC. For example, R becomes smaller than 0.1 for $\tau/(\sigma/U) > 1.4$, it may be concluded that episodes with elevated concentrations will last, on the average, for a period of only a fraction of $1.4\sigma/U$. As σ increases with the distance, such episodes will last longer at larger distances downwind of the source.

The spectral density distributions of the IC fluctuations, S(n), at different distances downwind of the source along the centerline of the plume were calculated and presented in Fig. 13 using the dimensionless parameters: $S^* = S(n)U/[(ic^{*'})^2\sigma]$ and $n^* = n\sigma/U$. Note that the integral of S^*n^* is equal to 1. As seen in this figure, identical non-dimensional spectral density distributions were obtained, substantiating the previous conclusion that the time scales of the IC fluctuations at different distances are proportional to σ/U. (We have omitted from the figure the data measured at $x/M = 20$, which showed slightly higher energy at low frequencies.)

An approximate -5/3 power-law variation of S^* is observed in the approximate range $0.5 < n\sigma/U < 1.5$, namely fluctuations with time scales of the order of σ/U. The -5/3 line in the figure is given by $S^* = a_c(n\sigma/U)^{-5/3}$, where $a_c = O(0.12)$. This range is followed by a range with a -11/3=(-5/3 -2), power-law variation, up

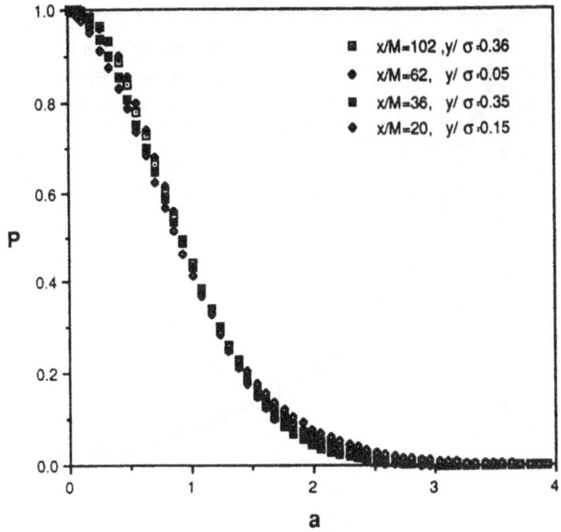

Fig. 7. Typical distributions of P(a) near the centerline of the plume.

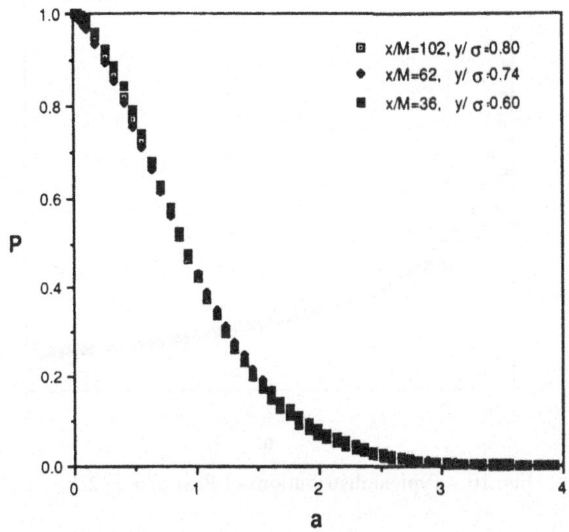

Fig. 8. Typical distributions of P(a) around $y/\sigma = 0.7$

to approximately $n\sigma/U = 5$.

In Fig. 14 we have plotted the distribution of n^*S^* versus n^* at two stations. The data suggest that most of the contributions to the variance of the fluctuations come from the range $0.2 < n\sigma/U < 1.5$.

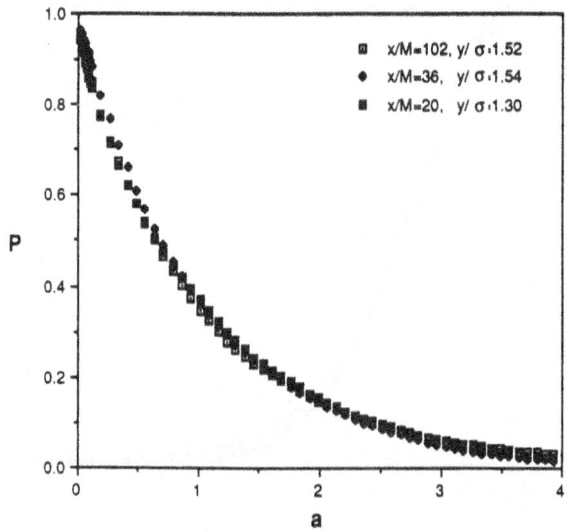

Fig. 9. Typical distributions of P(a) around $y/\sigma = 1.5$

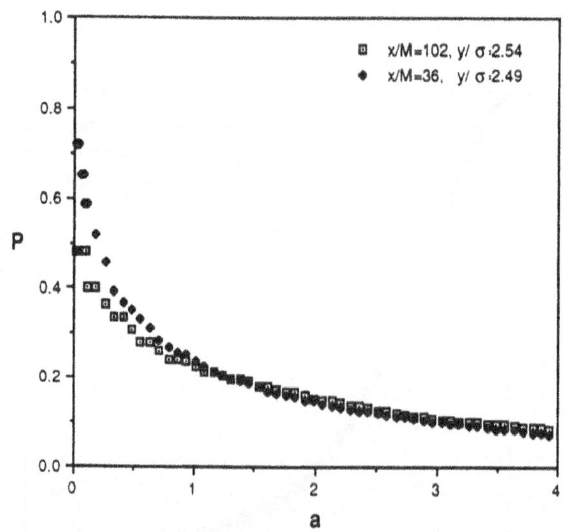

Fig. 10. Typical distributions of P(a) $y/\sigma = 2.5$

3.2. VERTICAL INTEGRATED CONCENTRATIONS ACROSS A PLUME DIFFUSING IN A BOUNDARY LAYER DEVELOPING OVER A SMOOTH SURFACE

The Vertical Integrated Concentrations (VIC), defined as:

$$VIC(x, y, t) = \int_o^\infty C(x, y, z, t) dz, \tag{4}$$

were measured across plumes diffusing in a boundary layer over a relatively smooth surface, in the region $x/\delta = 0.3 - 2.5$. The thickness of the boundary layer was

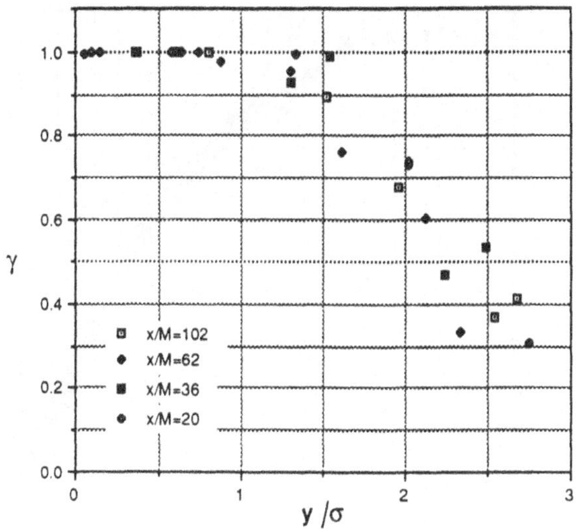

Fig. 11. Lateral distribution of the intermittency (p=0.64)

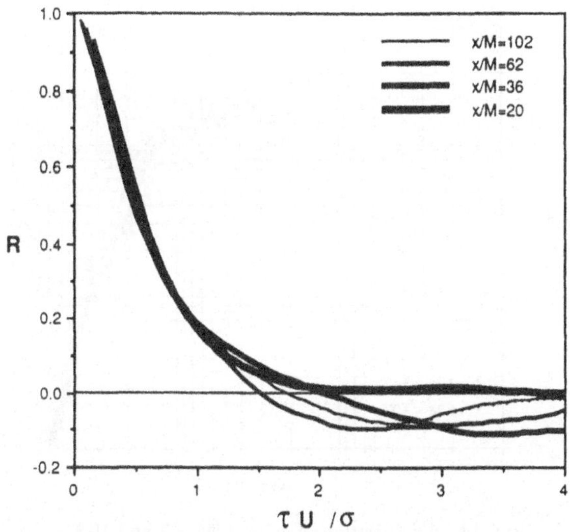

Fig. 12. Plot of the autocorrelation of IC^* versus $\tau U/\sigma$ near the centerline of the plume.

δ =1.0 m and the velocity profiles near the wall region followed a power law with an exponent n=0.16.

The lateral distributions of VIC were found to be Gaussian and the values of lateral length scale $\sigma = \sigma_y$ varied between 2.8 to 12.9 cm. The lateral distributions of the mean ground level concentrations C, measured using a propane tracer and an FID detector, gave almost identical values of σ at the same distances.

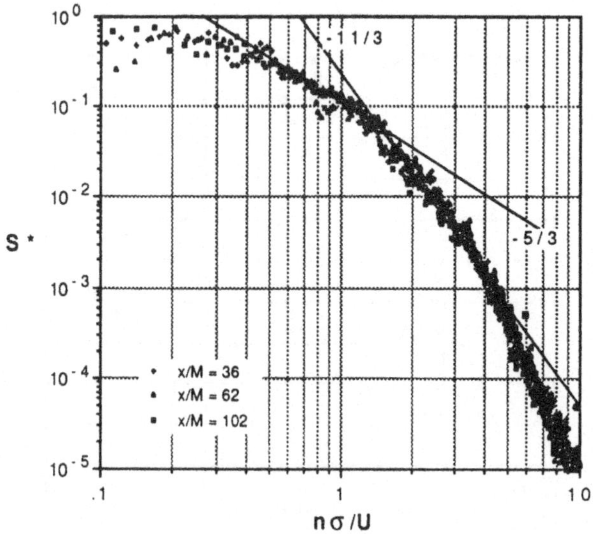

Fig. 13. Dimensionless spectral energy density of ic^* at x/M=36, 62 and 102.

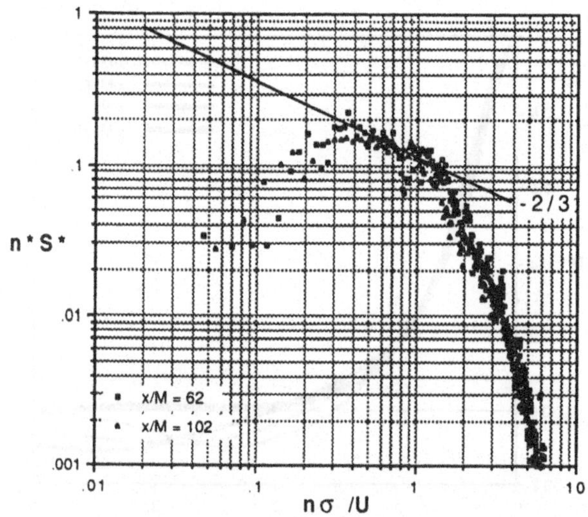

Fig. 14. Distribution of n^* at x/M=62 and 102

The fluctuations of VIC^* in this BL, where VIC^* is defined similarly to IC^* as

$$VIC^* = VIC\delta U/Q \tag{5}$$

were also found to be approximately similar, namely a function of y/σ and only slightly dependent on x/δ. Surprisingly, as shown in Figs. 15 and 16, which compares the PDFs of IC^*/ICM^* and of $VIC^*/VICM^*$, they are quite similar to those of the integrated concentrations across the plume diffusing in the GGT. Of

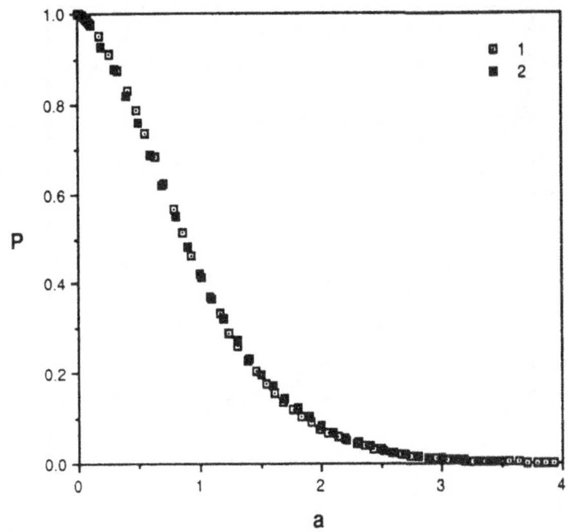

Fig. 15. The PDF of 1) IC^*/ICM^*, $x/M = 20, y/\sigma = 0.15$. 2) $VIC^*/VICM; \delta = 1.0m$; $x/\delta = 0.5$, $y/\sigma = 0.21$.

course, in both cases, the shape of the PDFs was slightly dependent on the distance, and the similarity of the two cases is only approximate.

3.3. VERTICAL INTEGRATED CONCENTRATIONS ACROSS A PLUME IN A BOUNDARY LAYER DEVELOPING OVER A ROUGH SURFACE

The similarity of the integrated concentration fluctuations in the above two cases was rather surprising, as fully explained in the next section, and it was decided to examine the fluctuations of VIC^* across a plume diffusing in a boundary layer developing over a rougher surface than the one described earlier, up to distances of $x/\delta = 9$.

The velocity profile power law in this BL was n = 0.2. Downwind of the roughness elements, however, a layer with very low velocities existed. Typical measurements of VIC^* in this BL are shown in Figs. 17 and 18. A marked difference between these and the previously presented ones is that the VIC signal at the edge of the plume in the $\delta = 0.25m$ BL is hardly intermittent.

Figure 19 shows the lateral distribution of the rms/mean values of VIC^* at different locations downwind of the source. One clearly sees that the relative magnitudes of the fluctuations decrease considerably with the distance from the source and that their values are smaller than those in the previous BL, see Fig.4. The distributions of $vic^{*'}/vic^{*'}(0)$ at different distances from the source in the two cases were, however, approximately similar.

Figures 20 through 22 show the measured variation with the distance from the source of the PDFs of $VIC^*/VICM$ at approximately equal values of y/σ. Comparison of these figures with Figures 7-10 shows the effect of the roughness

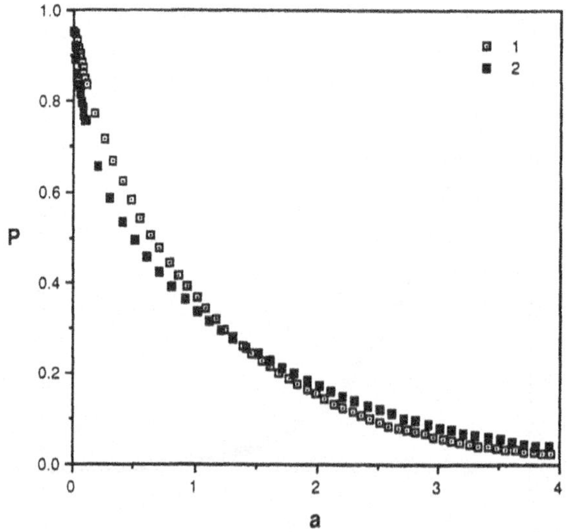

Fig. 16. The PDF of 1) IC^*/ICM^*, $x/M = 20, y/\sigma = 1.30$, 2) $VIC^*/VICM$; $\delta = 1.0m$; $x/\delta = 0.5$, $y/\sigma = 1.36$.

and the shear on the statistical properties of the VIC fluctuations across the plume as it diffuses downwind in the $\delta = 0.25$ boundary layer: a continuous reduction of the fluctuation level, disappearance of the intermittency and, of course, a lack of similarity of the VIC fluctuations at different distances.

Autocorrelations of VIC^* near the centerline of the plume at different distances are plotted in Fig. 23 versus $\tau U/\delta$. It is seen from this figure that values of R>0.2 were observed along the centerline of the plume up to approximately $\tau U/\delta = O(1)$, or $\tau U/\sigma > 3(\sigma/\delta$ in these experiments varied in the range 0.2 - 0.35), suggesting that puffs with relative high concentrations of the tracer are continuously stretched by shearing of the mean flow near the ground.

4. Discussion and Conclusions

Measurements have been made of line-integrated concentrations (IC) across a plume diffusing in grid-generated turbulence (GGT) and vertical integrated concentrations (VIC) across plumes diffusing in a $\delta = 1.0m$ BL over a smooth surface and in a $\delta = 0.25m$ BL developing over a relatively rough surface, using a fast response IR/CO_2 system.

In general, it is expected that the statistical properties of the fluctuations of IC would be a function of the distance from the source, of the dimensionless off-center location y/σ and of dimensionless parameters describing the turbulence in the flow and the relative size of the turbulent eddies. Analysis of the measured data in the GGT experiments showed, however, that the effect of the distance and of the turbulent field on the dimensionless fluctuations was very small. The local

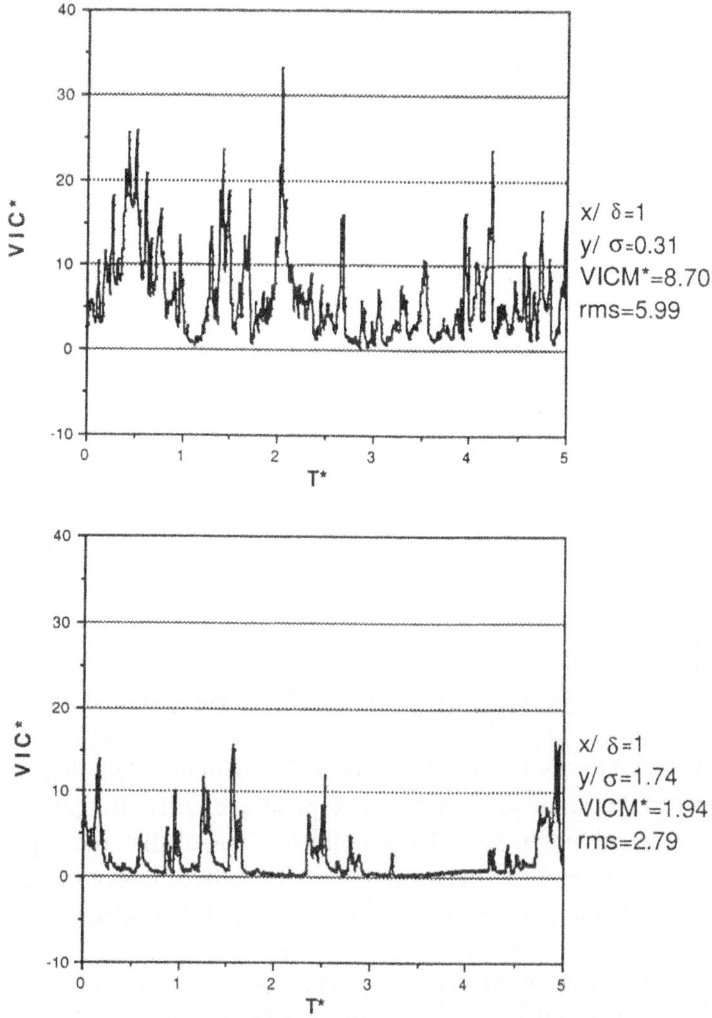

Fig. 17. Typical VIC^* fluctuations at $x/\delta = 1$ ($\delta = 1.0m, y/\sigma = 0.31$ and $y/\sigma = 1.74$)

variance of fluctuations and the intermittency γ, in the above mentioned range of distances, were mainly a function of the relative off-center location y/σ. Similarly, the probability distribution functions of IC^*/ICM^* measured in the $\delta = 1.0m$ BL over the smooth surface in the range 0.2 <x/M<2.2, were found to depend primarily on the off-center location, although, in this case too, a very mild effect of the distance was noticed.

The results are somewhat surprising, but particularly surprising is the observed similarity between the statistical properties of the fluctuations of IC^* across a plume diffusing in GGT and of the fluctuations of VIC across a ground-level

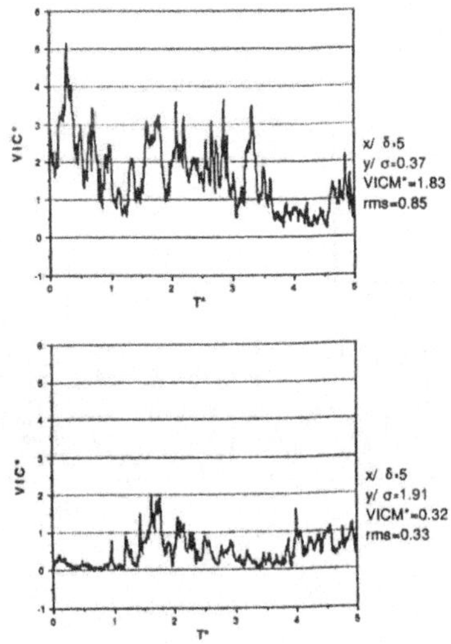

Fig. 18. Typical VIC^* fluctuations at x/δ=5 ($\delta = 1.0m, y/\sigma = 0.37$ and $y/\sigma = 1.91$)

plume diffusing in that boundary layer. One is inclined to conclude from these observations that this similarity might be universal. However, such a conclusion would be incorrect.

When analyzing the dynamics of plumes diffusing in turbulent flow, one distinguishes between the instantaneous plume (typical lateral length scale σ_i) and the mean plume (typical length scale σ). The fluctuations of the concentrations at a point are viewed as a result of the meandering of the instantaneous plume and the concentration fluctuations within the instantaneous plume (in-plume fluctuations). A similarity of the concentration fluctuations within a certain range implies that the ratio σ_i/σ remains constant at that range. Theoretical considerations (Poreh and Cermak, 1991b) suggest that σ_i and σ grow at different rates at different distances from the source, and that the relative role of the meandering reduces with the distance: very close to the source, at least for ideally small sources, the fluctuations are primarily due to meandering caused by relatively large eddies. At large distances, on the other hand, when the size of the instantaneous source becomes larger than that of the turbulent eddies in the flow, the effect of meandering should decrease, the ratio σ_i/σ approaches one and the relative magnitude of the fluctuations decrease.

In the present investigations we have used finite size sources and have started the measurements in the GGT experiments at $x/M = 20$, so that, apparently, the initial region was not observed, except for mild deviations of the measurements at $x/M = 20$ from the rest of the data. We have terminated our measurement at $x/M = 100$, as the turbulence downwind of the grid decayed very fast and the growth of the mean plume beyond that location was very slow. We have also been guided by practical interests and have focused the study of the VIC fluctuations

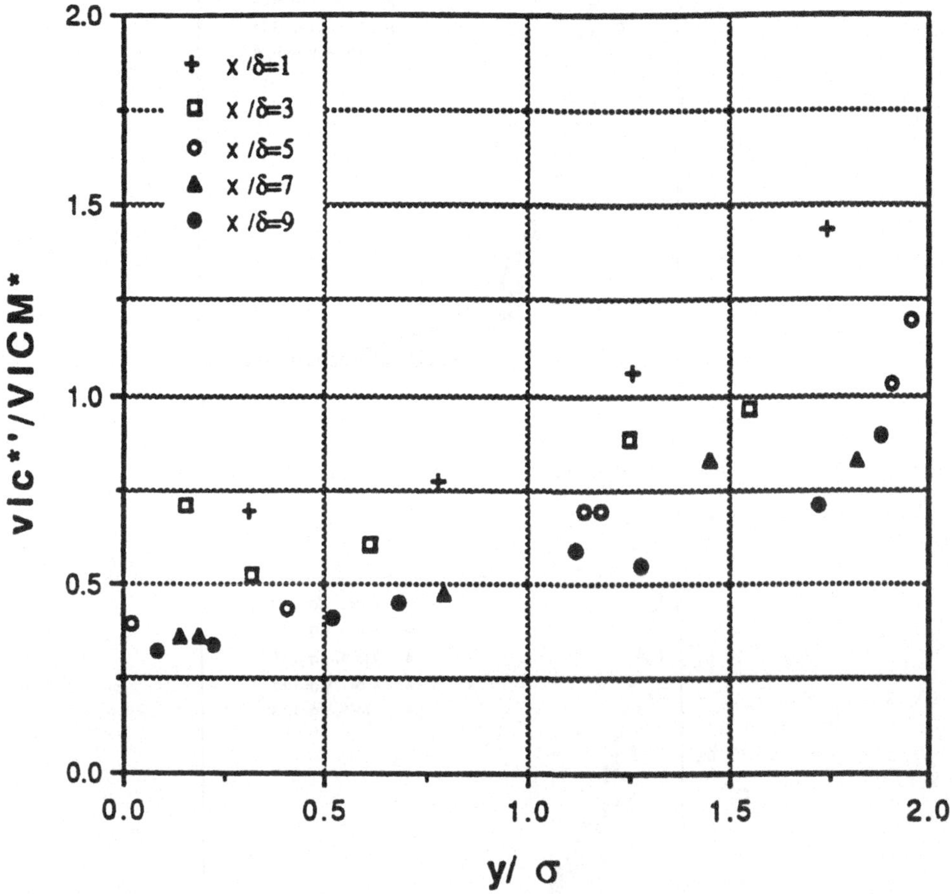

Fig. 19. The lateral distribution of the rms/mean values of VIC^* at different distances downwind of the source in the $\delta = 0.25m$ BL.

across the plume in the $\delta = 1.0m$ BL to the region $x/\delta < 2$ (approximately 600 m - 1500 m from the source in the atmospheric surface layer). Thus, we have not observed the expected effect of the distance and the shear on the fluctuations of IC and VIC, which is apparently rather slow.

The measurements of Vertical Integrated Concentrations (VIC) across a plume diffusing in a boundary layer ($\delta = 0.25$) over a slightly rougher surface in the range $1.0 \leq x/\delta \leq 9.0$ confirm the expected effect of the shear. The VIC fluctuations were not similar any more and exhibited a reduction of the intermittency and fluctuations with the distance. A more pronounced effect is expected in urban areas with a large relative roughness length. However, a larger scale simulation is

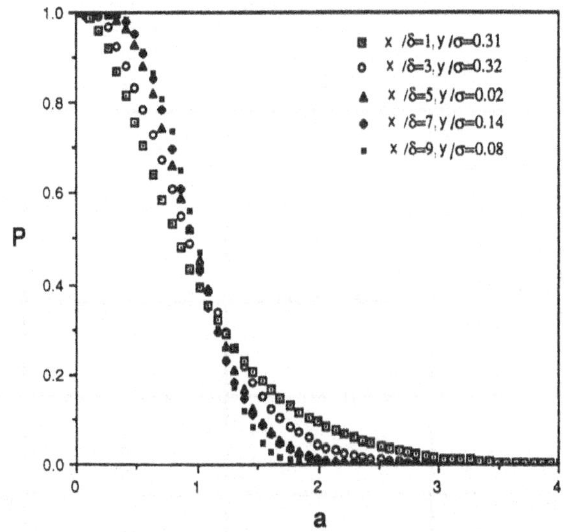

Fig. 20. The effect of the distance on the PDFs near the centerline of the plume in the $\delta = 0.25m$ BL.

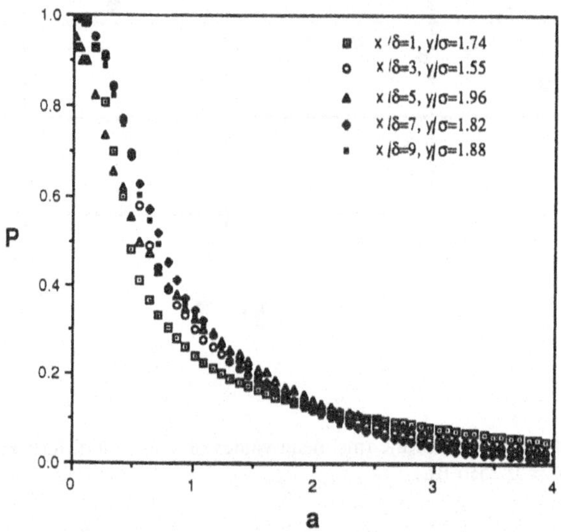

Fig. 21. The effect of the distance on the PDFs around $y/\sigma = 1.8$ in the $\delta = 0.25m$ BL.

required to obtain a more quantitative estimate of the dispersion in such areas.

The time dependence of the integrated concentration fluctuations in the GGT was studied by measuring autocorrelations and spectral density distributions of the fluctuations of IC at different distances from the source. Both were found to scale as σ/U.

The observed similar dimensionless autocorrelation curves suggest that fluid parcels of high concentrations pass a given point within a time period of less than

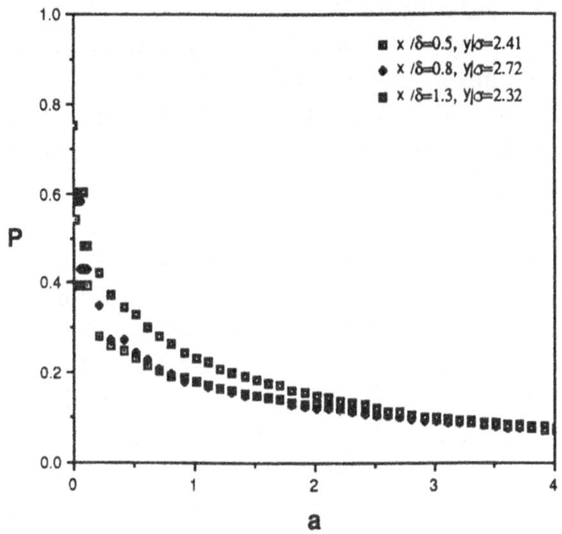

Fig. 22. The effect of the distance on the PDFs around $y/\sigma = 2.5$ in the $\delta = 0.25m$ BL.

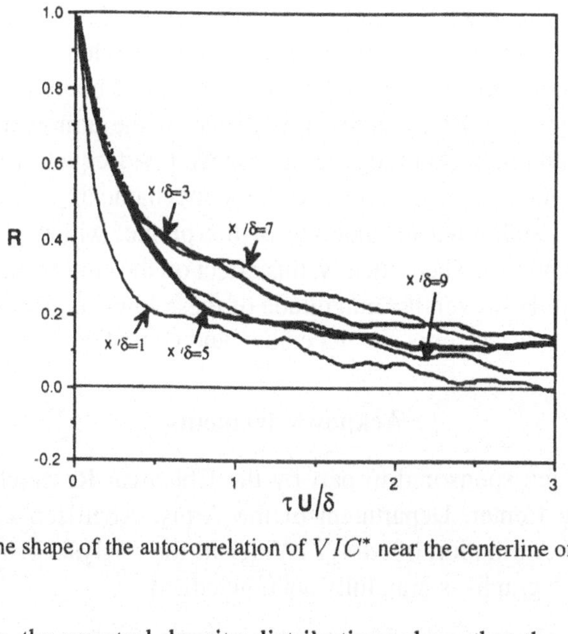

Fig. 23. The shape of the autocorrelation of VIC^* near the centerline of the plume

$2\sigma/U$. Similarly, the spectral density distributions show that the major contribution to the variance of the integrated concentration fluctuations comes from the frequency range $0.2 < n\sigma/U < 1.5$, which represents eddies with scales of 0.7σ to 5σ.

These results suggest that the instantaneous plume disintegrates into regions of relatively high concentrations separated by regions of low concentrations. The lateral distance between the centroids of close regions with elevated concentrations is not negligible, relative to σ_i, so that the autocorrelations at $tU/\sigma > 2$ are very small. Preliminninary analysis of flow visualization of an instanatneous plume in a boundary-layer flow using digital image processing techniques used by Wu *et al.* (1960) qualitatively supports this conclusion.

The integrated concentration spectral density has been shown to exhibit an inertial subrange shape in the range $0.5 < n\sigma/U < 1.5$. Namely

$$S^* = a_c(n\sigma/U)^{-5/3} \tag{6}$$

with a_c=0.12. It is interesting to note that Hanna and Insley (1989) have found the same power law with the same value in the inertial subrange of measured point concentration fluctuations in various field experiments. It is thus not surprising that the integrated concentrations exhibit a similar distribution in this range which corresponds to eddies with a length scale of the size of the plume, σ. In the case of point concentration fluctuations, however, the inertial subrange behavior extends to small eddies, whereas in the case of integrated concentrations, the spectrum density function of the in-plume fluctuation (eddies smaller than σ) is attenuate by integration, and assumes a - 11/3 (=-5/3 - 2) low, up to eddies of the order of 0.2 σ.

Clearly, the results of the present investigation provide a better understanding of the nature of integrated concentration fluctuations as well as a guide for many practical applications. Further study of the nature of integrated concentration fluctuations across plumes in an urban area is recommended.

Finally, it is appropriate to stress that the wind-tunnel boundary layer simulates only the shear-generated mechanical turbulence of the atmosphere. Very often, large-scale topographical inhomogeneities induce low frequency variations in the wind direction, which are not coupled with the mechanical turbulence. Such variations will cause additional meandering of the plume, which is not depicted in the present experiments. Theoretically, this effect on the concentration fluctuations can be calculated. However, the magnitude of these low-frequency changes in the atmospheric boundary layer is site dependent and cannot be easily predicted.

Acknowledgements

The work has been sponsored in part by the Chemical Research, Development and Engineering Center, Department of the Army, Aberdeen Proving Ground, Maryland. The contribution of Mrs. Z. Vider in the analysis of the data and the preparation of the graphs is gratefully acknowledged.

References

Bradshaw, P. (1964). *Wind-Tunnel Screens: Flow Instability and Its Effects on Aerofoil Boundary*, J. Roy. Aero. Soc., **68**, p. 198.

Deardorff J. W. and G. E. Willis (1984). *Ground level Concentration Fluctuations from a Buoyant and a non-Buoyant Source Within a Laboratory Convectively Mixed Layer*, Atmospheric Environment, **18**, No. 7, pp. 1297-1307.

Dinar N., H. Kaplan and M. Kleiman (1988). *Characterization of Concentration Fluctuations of a Surface Plume in a Neutral Boundary Layer.* Boundary-Layer Meteorology, **45**, pp. 157-175.

Gifford, F. G. (1959). *Statistical Properties of a Fluctuation Plume Dispersion Model*, Adv. Geophys., **6**, pp. 117-138

Hanna, S.R. (1984a). *Concentration Fluctuations in a Smoke Plume*, Atmospheric Environment, **18**, pp. 1091-1106.

Hanna, S. R. (1984b). *The Exponential Probability Density Function and Concentration Fluctuations in Smoke Plumes*, Boundary-Layer Meteorology, **29**, pp. 361-375.

Hanna, S. R. and E. M. Insley (1989). *Time Series Analyses of Concentration and Wind Fluctuations*, Boundary-Layer Meteorology, **47**, pp. 131-147.

Hinze (1975). Turbulence, 2nd Edition, McGraw Hill.

Kistler, A. L. and T. Vrebalovich (1966). Journal of Fluids Mechanics, **26**, pp. 37.

Lewellen, W. S. and R. I. Sykes (1986). *Analysis of Concentration Fluctuations from Lidar Observations of Atmospheric Plumes*, J. of Climatic and Applied Meteorology, **25**, pp. 1145-1154.

Morgan, P. G. (1960). *The stability of Flow Through Porous Screen*, J. Roy. Aero. Soc., **64**, pp. 359-362.

Mylne, K. R. and P. J. Mason (1991). *Concentration Fluctuation Measurements in a Dispersing Plume at a Range of Up to 1000 m*, Q.J.R. Meteorol. Soc., **117**, pp. 177-206.

Poreh, M. and J. E. Cermak (1987). *Experimental Study of Aerosol Plume Dynamics*, III: Wind Tunnel Simulation of Vertical integrated Concentration fluctuations, CSU Report, CER87-88MP-JEC4.

Poreh, M., A. Hadad and J. E. Cermak (1989). *Analysis of the Fluctuations of Obscuration Through a Ground Level Aerosol Source*, Proceeding of the 1989 CRDEC Conference on Obscuration and Aerosol Research, Aberdeen Proving Ground, MA.

Poreh, M. and J. E. Cermak (1990). *Small Scale Modeling of Line Integrated Concentration Fluctuations*, J. of Wind Engineering and Industrial Aerodynamics, **36**, pp. 665-673.

Poreh, M., A. Hadad and J. E. Cermak (1990). *Fluctuations of Visibility through Ground Level Aerosol Plume*, Preprint Volume Ninth Symposium on Turbulence and Diffusion, Roskilde, Denmark.

Poreh, M., A. Hadad and J. E. Cermak (1991). *Dynamics of Integrated Concentration Fluctuations Across Plumes Diffusing in Grid-Generated Turbulence*, Preprint Volume CRDEC Conference on Obscuration and Aerosol Sciences.

Poreh M., J. E. Cermak (1991a). *Wind Tunel Measurements of Line Integrated Concentrations*, Atmospheric Environment, **25A**, No. 7, pp. 1181-1187.

Poreh M. and J. E. Cermak (1991b). *Fluctuations of Integrated Concentrations Across a CO_2 Plume Diffusing in Grid Generated Turbulence* Eighth International Conference on Wind Engineering, London, Ontario, Canada.

Sawford, B. L. and H. Stapountzis (1986). *Concentration Fluctuations According to Fluctuations Plume Models in One and Two Dimensions*, Boundary-Layer Meterology, **37**, pp. 89-105.

Sawford, B. L. (1987). *Conditional Concentration Statistics for Surface Plumes in the Atmospheric Boundary Layer*, Boundary-Layer Meteorology, **38**, pp. 209-223.

Wu, G., K. Higuchi and R. N, Meroney (1990). *Applications of Digital Image Processing in Wind Engineering*, 8th International Conference on Wind Engineering, London, Ontario.

THE STRUCTURE AND MAGNITUDE OF CONCENTRATION FLUCTUATIONS

P.C. CHATWIN and PAUL J. SULLIVAN
Dept. of Applied and Computational Mathematics, University of Sheffield, Shefield, S10 3TN, U.K.
and
University of Western Ontario, London, Ont.,Canada

(Received October 1991)

Abstract. This paper is concerned with the science of turbulent diffusion and not, except incidentally, with its numerous practical applications. It discusses some recent research, particularly that by the authors and their collaborators. Among the topics considered are (i) the intermittency factor, (ii) the relationship between the mean of the concentration and its variance, and (iii) the interpretation of data. The principal aim of the paper is to draw attention to some outstanding basic questions which would seem promising targets for future research. Without progress on these questions (and others), regulatory models of air quality will continue - inevitably - to be unreliable and hardly worth using.

1. Introduction

The increased importance accorded by the public worldwide, and their governments, to air quality, and to the assessment of hazards associated with the accidental release of dangerous gases into the atmosphere, is resulting in a larger demand for mathematical models for regulatory (and associated) purposes. The number of such models - and their developers - is increasing; so moreover, are quasi-political pressures for the international "harmonisation" of such models. Although less advanced, a similar process is taking place in respect of water pollution.

From this point of view, it is unfortunate that the science of turbulent diffusion, which underpins these important practical problems, is still not understood to enough depth to allow (in general) such models to be well founded and reliable, as well as practically useful. While undeniable, this fact is undoubtedly surprising to most non-experts; nor, regrettably, has it inhibited many people from producing and selling software which, however attractively packaged, cannot fulfil the purposes for which it was purchased since it is based on inadequate, often wrong, science.

Therefore it continues to be important and timely, to conduct research into turbulent diffusion. (Indeed, proper recognition of the real situation ought to lead to vastly increased financial support for such work, but that is another story!) In this paper we summarise some of our recent results on the fundamental science, and consider what we believe to be promising developments. Throughout, the emphasis is on a better understanding of the underlying physics and mathematics. We have given complementary, more practical viewpoints elsewhere (e.g., Chatwin and Sullivan, 1990b; Chatwin, 1991).

Boundary-Layer Meteorology **62:** 269–280, 1993.

2. Basic Background

We denote by $\Gamma = \Gamma(x,t)$ the concentration (in arbitrary units) of a dispersing contaminant, and we shall suppose throughout this paper that the dependence of Γ on x and t is determined by advection by the ambient fluid, with velocity fluid $Y = Y(x,t)$, and by molecular diffusion, with diffusivity κ. In particular we ignore chemical changes. The equation governing Γ is

$$\frac{\partial \Gamma}{\partial t} + Y \cdot \nabla \Gamma = \kappa \nabla^2 \Gamma \,. \tag{1}$$

Since the ambient flow is turbulent, Y is a random vector field (satisfying the Navier-Stokes equations); so therefore is Γ. The correct mathematical description of Γ (and Y) must therefore be a statistical one. (This simple truth is largely ignored by the present generation of model developers.) A statistical description requires a definition of the underlying ensemble (Chatwin & Sullivan, 1979; Sullivan, 1990). Since Γ is a random variable it has, in any ensemble, a probability density function $p(\theta; x, t)$ defined in the normal way by

$$p(\theta; x, t) = \frac{d}{d\theta}[prob\{\Gamma(x,t) \leq \theta\}] \,. \tag{2}$$

The equation governing the evolution of p can be derived from (1) (Chatwin, 1990) and is

$$\frac{\partial p}{\partial t} + \nabla \cdot E\{Y\delta[\Gamma(x,t) - \theta]\} = \kappa \nabla^2 p - \kappa \frac{\partial^2}{\partial \theta^2} E\{\nabla \Gamma)^2 \delta[\Gamma(x,t) - \theta]\} \tag{3}$$

where the symbol E denotes "expected value" in the technical statistical sense.

Equations for some of the simplest and most commonly used statistical properties of Γ can be derived from (3). The mean concentration $\mu(x,t)$, where

$$\mu(x,t) = \int_0^\infty \theta p(\theta; x, t) d\theta = E\{\Gamma(x,t\}, \tag{4}$$

satisfies

$$\frac{\partial \mu}{\partial t} + U \cdot \nabla \mu + \nabla \cdot E\{uc\} = \kappa \nabla^2 \mu, \tag{5}$$

in (5), $U = U(x,t) = E\{Y(x,t)\}$ is the mean velocity field, and $u = u(x,t)$, $c = c(x,t)$ are the "fluctuations" defined by

$$u = Y - U, \quad c = \Gamma - \mu \,. \tag{6}$$

The variance $\sigma^2(x,t)$ of the concentration, often termed the mean-square fluctuation, satisfies

$$\sigma^2(x,t) = \int_o^\infty (\theta - \mu)^2 p(\theta; x, t) d\theta = E\{\Gamma^2(x,t)\} - \mu^2(x,t), \tag{7}$$

and its governing equation is:

$$\frac{\partial \sigma^2}{\partial t} + U \cdot \nabla \sigma^2 + \nabla \cdot E\{uc^2\} + 2 \nabla \mu \cdot E\{uc\} = \kappa \nabla^2 \sigma^2 - 2\kappa E\{(\nabla c)^2\}. \quad (8)$$

Details of the derivation of the standard equations (5) and (8) from (3) are given in Chatwin (1989). Justification for the use of the standard statistical notations μ and σ^2 (rather than symbols using overbars and dashes that are still more conventional in turbulence and turbulent diffusion research) is given by Chatwin (1990).

Except in very rare cases (e.g., Chatwin & Sullivan, 1979), exact results cannot be obtained from (5) and (8), let alone from (3). This is because of the "closure problem" which, in (3) is evident in the last (i.e., second) terms on each side of the equation. The expected values in these terms are defined by equations like (4) and (7), but with the crucial difference that the relevant probability density functions are in neither case $p(\theta; x, t)$, but are more complicated. For example, the last term on the left-hand side of (3) involves the joint probability density function of velocity and concentration. Therefore equation (3) for p is not "closed", and neither, consequently, are the equations for μ and σ^2.

Many attempts have been made to solve, or avoid, the closure problem among the earliest of which are Gaussian plume models, and models using the long-discredited concept of eddy diffusivity. These models, and later ones, such as most of those considered in the papers by Hanna and Klug in this volume, are usually of the mean concentration μ. They do not (except in very rare instances) consider the statistics of the deviations between μ and what is observed, that is of the concentration fluctuations; indeed they are not capable of doing so. This is despite the fact that the magnitude of these deviations is known to be at least of the order of μ itself and sometimes much larger. An interesting development in recent years which does account for the fluctuations is the use of random walk models (see Sullivan, 1971; Allen, 1982). Latest work in this technique is described in papers in this volume by van Dop, by Kaplan & Dinar, and by Sawford. It is now known that such models, as currently structured, have inevitable inconsistencies (Thomson, 1990). While the consequences of these appear to be numerically small, the inconsistencies are fundamental and cannot be removed without a radical change in model structure. Real mastery of the closure problem will probably occur only when a future generation of computers is large enough and fast enough to allow full 3D and unsteady solutions of (1) to be directly calculated to adequate accuracy, which requires, in particular, a satisfactory resolution of all length scales down to the conduction cut-off length ($\approx 10^{-4} m$), and of all time scales. In addition, enough solutions must be generated for each ensemble of velocity fields to permit direct estimation of the statistical properties of $\Gamma(x, t)$ to within acceptable limits which will, of course, require a large enough sample size. What evidence there is (Thomson, 1990) suggests that the number of solutions will need to be at least 10^4 (and perhaps larger than 10^5) for the estimation of statistical properties as simple as $\sigma^2(x, t)$. Such power is unlikely to be available soon.

In a series of papers (Chatwin & Sullivan, 1979, 1980, 1989a, 1990a) we have adopted a different approach, using physical reasoning to extend simple results for idealised cases to real situations. One success of this approach (Chatwin & Sullivan, 1979) was the demonstration that the magnitude of σ^2 depends significantly on source size and geometry, and much more so than μ. Small sources generate large fluctuations. In the present paper we focus on our later results and some of their implications.

3. The Intermittency Factor

A statistical property of $\Gamma(x,t)$ that has not yet been considered but which is prominent in many research papers is the intermittency factor $\gamma = \gamma(x,t)$, conventionally defined by

$$\gamma(x,t) = prob\{\Gamma(x,t) > 0\} . \tag{9}$$

(Some workers - and not without linguistic justification - use a complementary definition in which the right-hand side of (9) is 1-γ.) But, because of molecular diffusivity, solutions of the basic equation (1) for Γ have $\Gamma(x,t) > 0$ everywhere for all times after release of the contaminant. It then follows from (9) that γ(x,t)=1 everywhere after release and, therefore, that the definition (9) is theoretically meaningless.

The only logical way in which (9) can be made meaningful is to reject (1) and, in particular, to insist that its replacement does not permit instantaneous diffusion of matter, i.e. diffusion with "infinite velocity". Since molecular velocities are not infinite, such amendments to (1) are physically reasonable (and were, indeed, considered by Russian scientists over 35 years ago). However, there is no experimental evidence whatsoever that (1) is not an entirely satisfactory description of the evolution of $\Gamma(x,t)$ on the continuum scale. (It is pertinent to note that if (1) is to be rejected because it predicts the instantaneous diffusion of matter then so, logically, should be the Navier-Stokes equations because they predict the instantaneous diffusion of vorticity). Rejecting (1) would therefore be too drastic a resolution of the dilemma on present evidence, and this possibility will not be considered further here.

It follows that it is the definition (9) that must be discarded. Before considering replacements, it is important to note that reported values of γ less than 1 (but greater than 0) must occur because of instrumentation characteristics (almost all inevitable) or because of signal processing strategy such as thresholding. In other words, reported values of γ between 0 and 1 have no relevance at all to turbulent diffusion.

However the concept of intermittency of the velocity field is so useful that it is important to seek a new definition of $\gamma(x,t)$ that

(i) represents all relevant properties of the velocity field Y, and

(ii) is meanginful.

In particular, (ii) implies that a new definition of γ should be one that can, in principle at least, be a legitimate goal of mathematical modellers.

Such a definition was proposed by Chatwin & Sullivan (1989a), in a paper on which this section is largely based. It is convenient in what follows to use a zero subscript. e.g. γ_o, μ_o..., to denote properties in a hypothetical ensemble of releases in which the velocity field and geometry are identical to those in the real situation, but in which there is no molecular diffusion i.e. $\kappa = 0$. Suppose that, at $t = 0$, there is a release of contaminant of uniform concentration θ_1. In the hypothetical ensemble of releases there is no molecular diffusion; it follows that $p_o(\theta; x, t)$, the probability density function of concentration in this hypothetical ensemble, must have the form

$$p_o(\theta; x, t) = \gamma_o(x, t)\delta(\theta - \theta_1) + \{1 - \gamma_o(x, t)\}\delta(\theta) . \tag{10}$$

Equation (10) indicates that in this hypothetical ensemble the only values of Γ_o that occur are θ_1 and 0; moreover, without molecular diffusion, the logical objection to (9) disappears, so that $\gamma_o = \gamma_o(x, t)$ in (10) is defined by (9) but with Γ_o replacing Γ. Use of (4) and (10) gives

$$\mu_o(x, t) = \theta_1\gamma_o(x, t) \rightarrow \gamma_o(x, t) = \frac{\mu_o(x, t)}{\theta_1} . \tag{11}$$

In (10) and (11), the properties γ_o and μ_o are determined by the velocity field and geometry in the real situation. It is very likely, and commonly supposed, that $\mu(x, t)$, the real mean concentration, is insensitively dependent on κ, i.e. that $\mu(x, t) \approx \mu_o(x, t)$. We therefore propose that the definition (9) be replaced by $\gamma(x, t) = \gamma_o(x, t)$, where $\gamma_o(x, t)$ is as in (11), and that, in practice, γ be estimated from data by

$$\gamma(x, t) \approx \frac{\mu(x, t)}{\theta_1} . \tag{12}$$

One of the merits of (12) is that μ and θ_1 are two properties of the real concentration field that are most straightforward to measure reliably.

Because measurements of γ between 0 and 1 that are claimed to be obtained using the conventional definition (9) cannot satisfy (9), and, instead, reflect only characteristics of the instrumentation and signal processing strategy, little, if any, support is provided for the proposed new definition (12) by the fact that most graphs of γ reported in the research literature are at least qualitatively similar to the corresponding graphs of μ, and sometimes very close. Similar remarks apply to the agreement between profiles of γ and μ obtained by numerical simulation - see, for example, Figure 3 of Kaplan and Dinar (1988). Nevertheless the new definition of γ does appear to have the required properties. Further developments are discussed in our cited paper and also in Chatwin & Sullivan (1989b).

4. Relationship Between μ And σ

The hypothetical ensemble, identical to the real one except that κ=0 has been successfully exploited in another way by Chatwin & Sullivan (1990a). Use of (7) and (10) gives

$$\sigma_o^2(x,t) = \theta_1^2\gamma_o(x,t) - \theta_1^2\gamma_o^2(x,t), \tag{13}$$

and elimination of θ_1 from (11) and (13) gives

$$\sigma_o^2 = \mu_o(\theta_1 - \mu_o), \tag{14}$$

for all x and t. As in the previous section, this relationship incorporates the real velocity field completely (except insofar as this enhances the effects of κ in the real case by, for example, stretching).

Because molecular diffusion is a "weak" process compared with advection, we thought that (14) could be adapted to the real case by relatively simple changes. In particular we proposed that for dispersion in self-similar situations, such as jets and wakes (in which all data available to us had been measured), we allowed for molecular diffusion by

- (i) replacing the source (and maximum) concentration θ_1 by $\alpha\mu_*$, where μ_* is the maximum value of μ at any cross section (and so depends on position downstream of the source) and, in view of the application to self-similar flows, α is a constant;
- (ii) introducing a constant of proportionality β.

We therefore proposed the following relationship between σ and μ for the real situation:

$$\sigma^2 = \beta\mu(\alpha\mu_* - \mu). \tag{15}$$

In brief, our reasoning was that the term involving α allowed for the reduction of the maximum concentration by κ, and that the constant β represented the effects of dissipation of σ^2. We also had in mind that instrument smoothing could cause measured values of σ^2 to be less than the real values, and we recognised that this could affect the value of β.

The agreement between (15) and the data available to us was remarkably good; full details are given in Chatwin and Sullivan (1990a). In all cases examined, data from the self-similar dispersion region obeyed (15) to within normal experimental errors, bearing in mind also the uncertainty in the measurements of σ^2 due to statistical noise. With one exception, the value of the constant α was between 1 and 2 (but depended on the particular flow), and it can be shown easily that such values ensure that the maximum of σ^2 occurs at an off-axis location, different from that at which $\mu = \mu_*$. This phenomenon is of course well-known. In the one exception (Nakamura, et al., 1987), the maxima of μ and σ^2 coincided and the value of α

was 3; nevertheless these measurements also satisfied (15). We discuss further measurements from this group later. The values of the constant β also varied from flow to flow but satisfied $0 < \beta \leq 1$ in all cases.

We also discussed extensions of the ideas to other statistical properties (higher moments and the probability density function). Although there were substantially fewer measurements of such properties available to us, the comparisons that were made were reasonably encouraging given, especially, the greatly enhanced statistical noise.

Many workers have been interested in the asymptotic (far downstream of the source) value of σ/μ, the concentrations intensity. Although we are not convinced that this is an important measure from the point of view of basic understanding, it is interesting to record that (15) gives, on the axis,

$$\frac{\sigma}{\mu} = \sqrt{\{\beta(\alpha - 1)\}}, \tag{16}$$

which is a constant whose value varied from 0.1 to (approximate) 1 for the data that we examined.

The success of the comparison of (15) with data from self-similar regimes naturally led to attempts to apply it more generally. Preliminary ideas were discussed in Chatwin et al., (1990), and led to a model (Moseley, 1991; Moseley and Sullivan, 1991) which showed good agreement with data from grid turbulence for all distances dowstream of the grid including those prior to the establishment of the self-similar dispersion regime. Figure 1, taken from Moseley and Sullivan (1991), is typical; further comparisons are given in Moseley (1991).

In brief, the extended model retains (15) but recognizes that α and β cannot be constant throughout the dispersion regime. On the basis of a simple hypothesis - in essence a closure hypothesis - evolution equations for α and β as functions, for example, of downstream distance are obtained and solved. These equations (coupled ordinary differential equations) incorporate source size and geometry (not explicity included in our first model for the self-similar regime) and involve the growth rate of the dispersing contaminant plume. The solutions of these evolution equations tend to the constant values in the first version as downstream distance tends to infinity.

The generic problem in turbulent diffusion is arguably the one arising from instantaneous release of a finite quantity of contaminant, and therefore intrinsically both (statistically) non-stationary and inhomogeneous. In view of the success of the model based on (15), it seems likely that further extensions to these more difficult (but more realistic-in practice at least) situations would be worth attempting. Unfortunately, but understandably, very few measurements are available.

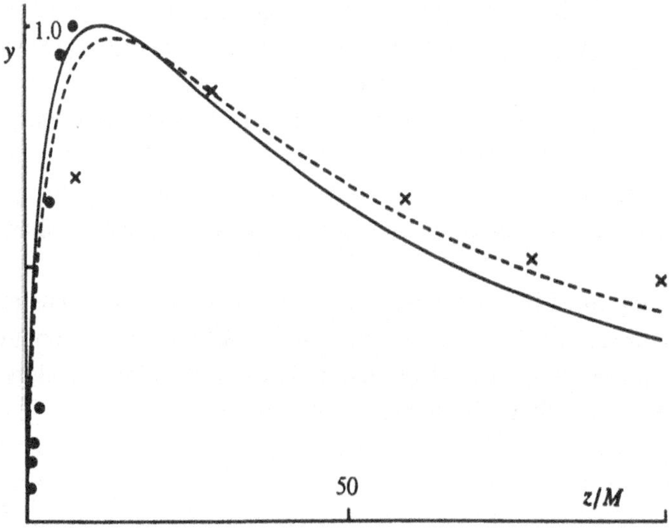

Fig. 1. Comparison of extension of (15) from Moseley & Sullivan (1991) with data of Warhaft (1984) from grid turbulence. The ordinate y is the centre-line value of σ^2 normalized by its absolute maximum, z is downstream distance and M is the grid mesh length. The solid circles and crosses are data from the 0.025mm and 0.127mm sources respectively, and the solid and dashed curves are the corresponding predictions from the extended theory.

5. Some Unanswered Problems

Research in turbulent diffusion is so difficult that any success in achieving increased understanding requires the intimate interplay of theory with experiment; that is one of its most attractive features. But recent developments seem to us to have consequences for research and future research priorities that are more important than seems to have been generally realized.

The three most important such developments are perhaps:

(i) Increasing power and availability of computers.

(ii) Improved measurements techniques.

(iii) Growing public demand for the control and monitoring of air and water quality, including the assessment of potential dangerous accidents involving the release of harmful substances into the atmosphere or natural water bodies.

In themselves these developments are welcome, but that is not necessarily true of their immediate consequences.

The sophistication of present data collection and acquisition system has led to the generation of enormous quantities of data, orders of magnitude greater than some data analysts and theoreticians (including at least one of us!) have been accustomed to. Consequently much data are not being examined or analysed or interpreted to the extent which is merited, or which the experimenter would wish. For us, the implication is not only that data analysts should show more foresight but that, given the inevitable resource limitations, all experiments must now be planned with the data analysis regarded as an integral part. Otherwise much data will continue to be ignored. It is obvious that the experiments should be designed with a clear purpose in mind, particularly the use(s) to which the data analysis will be put by the theoreticians on the project. Less obvious perhaps is the need to be clear about the underlying ensemble; many potential valuable experimental projects have been corrupted, sometimes beyond redemption, by arbitrary changes in the ensemble. One example is provided by investigations on the effects of buildings, where the demands of sponsors have often led to so many (apparently random) changes in the building(s) configuration during the experimental series being made that no quantitative results of value can be obtained for any ensemble. Even when this problem does not arise, it is necessary to assess in advance whether the uncertainty in the estimated statistical properties, due to limited sample size or length of record, will be acceptably small. In many cases of importance, the cost of obtaining such acceptably small uncertainty in full-scale (or field) trials will be too great. It will therefore be necessary to continue to use wind or water tunnels, and the demands of sponsors for "answers" to increasingly more sophisticated questions (such as effects of atmospheric stability or buildings) then require increased research into the capability of these facilities to model full-scale conditions.

Associated with the points above is the treatment of raw data before it is "validated" for transmission to the analysts. Such points as the signal noise and its deconvolution, thresholding strategy (if any), and baseline drift, are so crucial to the interpretation of experiments that they need more attention than has been customary (Mole, 1989, 1990a,b).

One of the key theoretical problems which such experimental phenomena influence is the behaviour of $p(\theta; x, t)$ as $\theta \rightarrow 0+$. The theoretical arguments used earlier to discredit the conventional definition (9) of the intermittency factor would seem to suggest $p(\theta; x, t)$ should tend to zero as $\theta \rightarrow 0+$, but this is not observed in experiments. An obvious explanation is that noise and the other factors mentioned in the previous paragraph make it extremely difficult, if not impossible, to measure very small concentrations with reliability or discrimination.

An associated, but different, problem connected with instrumentation is that, even with vastly improved modern measurement techniques, it is not likely that the small-scale dynamics of the fluctuating concentration field can be accurately resolved in all three space dimensions and in time. (It was noted earlier that significant dynamics occurs at length scales down to $O(10^{-4}m)$ in the atmosphere). Instrument smoothing is therefore inevitable, but its degree and type will depend on the characteristics of the instrumentation system. Figure 2, due to Sakai et al.,(1990), appears categorically to show that instrument smoothing can be significant. The measured value of the constant β in the relationship (15), but not that of α, is shown to depend strongly on probe size d_o. Details are given in Table I. It will be noted that (15) describes the data well in all four cases but that as d_o decreases (increased resolution), the measured value of β, i.e. of σ^2, increases.

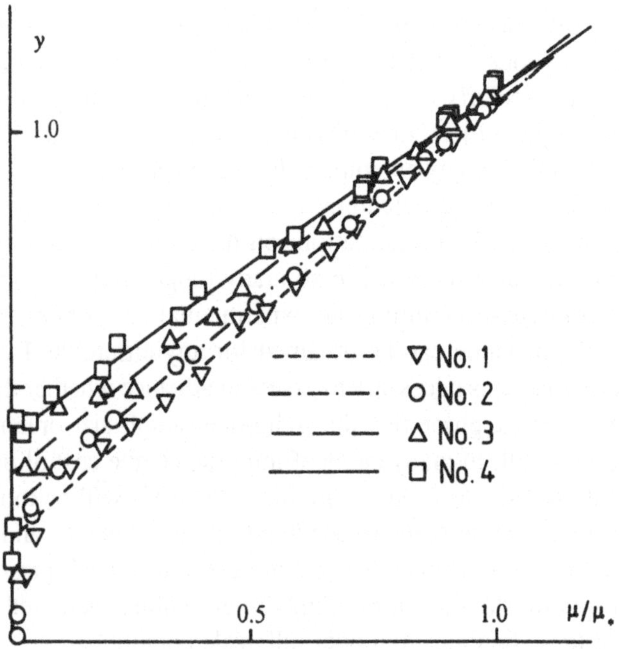

Fig. 2. The influence of probe size on the constants α and β in (15). Data from Sakai et al. (1990). The ordinate y is $(\mu^2 + \sigma^2)(\mu\mu_*)$ and rearrangement of (15) gives $y = \alpha\beta + 1(1 - \beta)(\mu/\mu_*)$. Numerical values of α and β are given in Table 1.

Probe no.	d_o/mm	α	β
1	0.54	1.31	0.16
2	0.30	1.25	0.22
3	0.13	1.33	0.25
4	0.10	1.25	0.33

Although some theoretical investigations of instrument effects have been undertaken by Mole (loc.cit.) and others, we believe that increased understanding of turbulent diffusion in general, and concentration fluctuations in particular, is being severely inhibited because insufficient attention is being placed on them by the research community. As examples, we record our opinion that there is a strong case (supported by data like that in Figure 2) for measurements of concentration fluctuations to be carried out by two (or more) transducer systems operating simultaneously, and that attempts should be made to see whether calibration procedures are valid for the actual experimental conditions.

Except indirectly, we have deliberately not discussed purely theoretical problems in this section.

Acknowledgements

We are grateful to the UK Ministry of Defense, the Commission of the European Communities, and the Natural Sciences and Engineering Research Council of Canada for their financial support. We would like to thank the organisers of the OHOLO conference for their invitation to present the views expressed in this paper, and for their generosity.

References

Allen, C.M. 1982 *Numerical simulation of contaminant dispersion in estuary flows*. Proc. Roy. Soc. Lond.**A381**, 179-194.

Chatwin, P.C. 1989 *Scalar transport in turbulent shear flows*. Lecture Series 1989-03 (Turbulent Shear Flows), von Karman Institute for Fluid Dynamics, Rhode-St-Genèse. Belgium.

Chatwin, P.C. 1990 *Statistical methods for assessing hazards due to dispersing gases*. Environmetrics **1**,143-162.

Chatwin, P.C. 1991 *New research on the role of concentration fluctuation in useful models of the consequences of accidental releases of dangerous gases*. Proc. Int. Conf. & Workshop on Modeling and Mitigating the Consequences of Accidental Releases of Dangerous Materials. (New Orleans, LA; published by AIChE, New York), 327-339.

Chatwin, P.C. & Sullivan, P.J. 1979. *The relative diffusion of a cloud of passive contaminant in incompressible turbulent flow*. J. Fluid Mech.**91**, 337-355.

Chatwin, P.C. & Sullivan, P.J. 1980. *Some turbulent diffusion invariants*. J. Fluid Mech. **97**, 405-416.

Chatwin, P.C. & Sullivan, P.J. 1989a. *The intermittency factor of scalars in turbulence* Phys. Fluids **A1**, 761-763.

Chatwin, P.C. & Sullivan, P.J. 1989b. *The intermittency factor of dispersing scalars in turbulent shear flows. Some applications of a new definition*. Proc. 7th Symp. on Turb. Shear Flows (Stanford Univ., CA), 29.4.1-29.4.6.

Chatwin, P.C. & Sullivan, P.J. 1990a. *A simple and unifying physical interpretation of scalar fluctuation measurements from many turbulent shear flows*. J. Fluid Mech. **212**, 533-556.

Chatwin, P.C. & Sullivan, P.J. 1990b. Vol. **1**, No. 2 of Environmetrics (special editors).

Chatwin, P.C., Sullivan, P.J. and Yip, H. 1990 *Dilution and marked fluid particle analysis*. Proc. Int. Conf. on Phys. Modelling of Transport & Dispersion (in conjunction with *The Garbis H. Keulegan Symp.*) (MIT, Cambridge MA, edited by E. Eric Adams & George E. Hecker), 6B.3-6B.8.

Kaplan, H. & Dinar, N. 1988. *A stochastic model for dispersion and concentration distribution in homogeneous turbulence*. J. Fluid. Mech.**190**, 121-140.

Mole, N. 1989. *Estimating Statistics of concentration fluctuations from measurements*. Proc. 7th Symp. on Turb. Shear Flows (Stanford Univ., CA), 29.5.1-29.5.6.

Mole, N. 1990a. *A model of instrument smoothing and thresholding in measurements of turbulent dispersion*. Atmos. Envir. **24A**, 1313-1323.

Mole, N. 1990b. *Some interactions between turbulent dispersion and statistics*. Environmetrics **1**, 179-194.

Moseley, D.J. 1991. *A closure hypothesis for contaminant fluctuations in turbulent flow*. M. Sc. thesis, Faculty of Graduate Studies, Univ. of Western Ontario, London, Canada.

Moseley D.J. & Sullivan, P.J. 1991. *A simple closure hypothesis for the prediction of contaminant concentration fluctuations in turbulent flows*. (private communication of draft MS).

Nakamura, I., Sakai, Y. & Miyata, M. 1987 *Diffusion of matter by a non-buoyant plume in grid-generated turbulence*. J. Fluid Mech. **178**, 379-403.

Sakai, Y. Nakamura, I., Tsunoda, H. & Shengian, L. 1990. Private communication (and lecture presented at Euromech 253, Brunel Univ., Uxbridge, UK. August 1989).

Sullivan, P.J. 1971. *Longitudinal dispersion within a two-dimensional turbulent shear flow*. J. Fluid Mech.**49**, 551-576.

Sullivan, P.J. 1990. *Physical modeling of contaminant diffusion in environmental flows*. Environmetrics **1**, 163-177.

Thomson, D.J. 1990. *A stochastic model for the motion of particle pairs in isotropic high-Reynolds-number turbulence, and its application to the problem of concentration variance*. J. Fluid Mech.**210**, 113-153.

Warhaft, Z. 1984. *The interference of thermal fields from line sources in grid turbulence*. J. Fluid Mech.**144**, 363-387.

DISPERSION IN SHEARED GAUSSIAN HOMOGENEOUS TURBULENCE

J.D. WILSON[(1)], T.K. FLESCH[(1)] and G.E. SWATERS[(2)]
University of Alberta, Edmonton, Alberta, Canada T6G 2H4.
(1) Geography Dept., (2) Mathematics Dept.

(Received October 1991)

Abstract. By integrating the Fokker-Planck equation corresponding to a Lagrangian stochastic trajectory model, which is consistent with the selection criterion of Thomson (1987), an analytical solution is given for the joint probability density function $p(x_i, u_i, t)$ for the position (x_i) and velocity (u_i) at time t of a neutral particle released into linearly-sheared, homogeneous turbulence. The solution is compared with dispersion experiments conforming to the restrictions of the model and with a short-range experiment performed in highly inhomogeneous turbulence within and above a model crop canopy. When the turbulence intensity, wind shear and covariance are strong, the present solution is better than simpler solutions (Taylor, 1921; Durbin, 1983) and as good as any numerical Lagrangian stochastic model yet reported.

1. Introduction

Counter-gradient fluxes are often observed in crop canopies (Denmead and Bradley, 1985), a consequence of non-diffusive contributions from nearby sources ("near field" effects). To relate scalar concentrations to source distributions in canopies, Raupach (1989) applied Taylor's (1921) Lagrangian solution to calculate the "near field" contributions. His necessary (and justified) assumption of local homogeneity highlights the paucity of analytical Lagrangian solutions more flexible than Taylor's.

Durbin (1983) gave a Gaussian solution for cross-stream diffusion in uniformly-sheared homogeneous turbulence. The cross-stream velocity was modeled by the Langevin equation, while the along-stream velocity fluctuation was omitted. We have extended that solution by including the along-stream fluctuation and its covariance with cross-stream velocity (these additions were recognised as straightforward by Durbin).

2. Solution for Shear Dispersion

Consideration of scalar dispersion in uniformly sheared homogeneous turbulence goes back at least as far as Corrsin (1952). The diffusion solution has been provided (e.g. Okubo and Karweit, 1969) but of course fails in the near field.

Our solution is two-dimensional. Motion along the (x, z) axes takes place with instantaneous total velocities $(u(t), w(t))$ that have mean values $(U(z), 0)$. The fluctuating velocities $(u', w') = (u - U, w)$ may have distinct time scales τ_u, τ_w

Boundary-Layer Meteorology **62:** 281–290, 1993.

and velocity scales σ_u, σ_w, and have covariance $-u_*^2$. The mean velocity field is $U(z) = U_o(1 + \alpha z)$, and the Eulerian velocity pdf is:

$$g_a = \frac{1}{2\pi\sigma} exp(-\frac{(u-U)^2\sigma_w^2 + w^2\sigma_u^2 + 2(u-U)wu_*^2}{2\sigma^2})$$

where $\sigma^2 = \sigma_u^2\sigma_w^2 - u_*^4$.

We assume modified Langevin equations for the increments in velocity[1]:

$$du = -\frac{(u-U)}{\tau_u}dt + b_{uu}d\xi_u + b_{uw}d\xi_w$$

$$dw = -\frac{w}{\tau_w}dt + b_{wu}d\xi_u + b_{ww}d\xi_w$$

where the $d\xi_i$ are Gaussian, with vanishing mean and variance dt, and have vanishing expectation $< d\xi_i(t_1)d\xi_j(t_2) >$ for distinct (i, j) and/or distinct (t_1, t_2). Shear stress is forced to arise through the random accelerations, and our model is therefore inconsistent with Kolmogorov's theory of local isotropy (the covariance $< dudw >$ does not vanish for time intervals dt lying in the inertial subrange). We are not interested in times so short.

The joint position-velocity probability density function $p(x, z, u, w, t)$ evolves according to the Fokker-Planck equation corresponding to our Langevin model:

$$\frac{\partial p}{\partial t} = -\frac{\partial}{\partial x}[u\,p] - \frac{\partial}{\partial z}[w\,p] - \frac{\partial}{\partial u}[-\frac{(u-U)}{\tau_u}p] -$$

$$-\frac{\partial}{\partial w}[-\frac{w}{\tau_w}p] + B_{uu}\frac{\partial^2 p}{\partial u^2} + (B_{uw} + B_{wu})\frac{\partial^2 p}{\partial u \partial w} + B_{ww}\frac{\partial^2 p}{\partial w^2}.$$

By enforcing Thomson's (1987) well-mixed constraint (i.e. insisting that g_a be a steady-state solution to this Fokker-Planck equation) we obtain for the model coefficients $B_{ij} = (1/2)b_{ik}b_{jk}$ the prescription:

$$B_{uu} = \frac{\sigma_u^2}{\tau_u} - u_*^2\frac{\partial U}{\partial z}, \qquad B_{ww} = \frac{\sigma_w^2}{\tau_w},$$

$$B_{uw} + B_{wu} = -u_*^2(\frac{1}{\tau_u} + \frac{1}{\tau_w}) + \sigma_w^2\frac{\partial U}{\partial z}.$$

Solution of the Fokker-Planck equation for the time evolution of the mean state has been performed along the lines suggested by Risken (1985). For particles

[1] We also solved the Fokker-Planck equation corresponding to an alternative treatment of the alongstream velocity $du' = -\frac{u'}{\tau_u}dt + b_{uu}d\xi_u + b_{uw}d\xi_w$, $dx = (U(z) + u')dt$ for which case the b's are independent of the mean shear. Predicted mean concentrations were as good as those given here, but the predicted alongstream fluxes were less satisfactory.

released at the origin with a random velocity from the Eulerian pdf, the initial value of the joint pdf is

$$p = \delta(x,0)\delta(z,0)\frac{1}{2\pi\sigma} exp[-\frac{(u-U_o)^2\sigma_w^2 + w^2\sigma_u^2 + 2(u-U_o)wu_*^2}{2\sigma^2}]$$

the solution is:

$$p = \frac{1}{(\sqrt{2\pi})^4\sqrt{detM}} exp[-\frac{1}{2}(y_i - m_i)m_{ij}^{-1}(y_j - m_j)].$$

The y_i denote coordinates in phase space (x, u, z, w); the m_i expected values for the particle coordinates at time t ($m_z = m_w = 0, m_x = U_o t, m_u = U_o$); and m_{ij} and m_{ij}^{-1} are the elements of the covariance matrix M (given in the Appendix) and its inverse.

An alternative expression of the solution is:

$$p = \frac{1}{(\sqrt{2\pi})^4\sqrt{m_{xx}m_{zz} - m_{xz}^2}\sqrt{2\Psi}\sqrt{2X}}\cdot$$

$$\cdot exp[-\frac{m_{xx}z^2 + m_{zz}(x - U_o t)^2 - 2m_{xz}z(x - U_o t)}{2(m_{xx}m_{zz} - m_{xz}^2)}] exp[-\frac{(u+\Lambda)^2}{4\Psi} - \frac{(w+\Omega)^2}{4X}]$$

where the velocity moments are:

$$\Lambda(x,z,t) = \frac{(x - U_o t)(m_{xz}m_{uz} - m_{zz}m_{ux}) + z(m_{xz}m_{ux} - m_{xx}m_{uz})}{m_{xx}m_{zz} - m_{xz}^2} - U_o$$

$$\Psi(t) = \frac{m_{xx}m_{uu}(m_{xx}m_{zz} - m_{xz}^2) - m_{ux}^2(m_{xx}m_{zz} - m_{xz}^2) - (m_{xz}m_{ux} - m_{xx}m_{uz})^2}{2m_{xx}(m_{xx}m_{zz} - m_{xz}^2)}$$

$$\Omega(x,z,u,t) = \frac{(x - U_o t)(m_{xz}m_{wz} - m_{zz}m_{wx}) + z(m_{xz}m_{wx} - m_{xx}m_{wz})}{m_{xx}m_{zz} - m_{xz}^2} + \nu(u + \Lambda)$$

$$X(t) = [\frac{m_{xx}m_{ww} - m_{wx}^2}{2m_{xx}} - \frac{(m_{xz}m_{wx} - m_{xx}m_{wz})^2}{2m_{xx}(m_{xx}m_{zz} - m_{xz}^2)}] -$$

$$-\nu[\frac{(m_{xz}m_{wx} - m_{xx}m_{wz})(m_{xz}m_{ux} - m_{xx}m_{uz}) + (m_{xx}m_{zz} - m_{xz}^2)(m_{ux}m_{wx} - m_{xx}m}{m_{xx}(m_{xx}m_{zz} - m_{xz}^2)}$$

where:

$$\nu(t) =$$

$$= \frac{(m_{xz}m_{wx} - m_{xx}m_{wz})(m_{xz}m_{ux} - m_{xx}m_{uz}) + (m_{ux}m_{wx} - m_{xx}m_{uw})(m_{xx}m_{zz} - m_x^2}{(m_{xx}m_{zz} - m_{xz}^2)(m_{xx}m_{uu} - m_{ux}^2) - (m_{xz}m_{ux} - m_{xx}m_{uz})^2}$$

Mean concentration is obtained from the joint position-velocity pdf by integrating out the velocity dependence:

$$C(x,z,t) = \frac{1}{2\pi\sqrt{m_{xx}m_{zz} - m_{xz}^2}} e^{-\frac{m_{xx}z^2 + m_{zz}(x-U_o t)^2 - 2m_{xz}z(x-U_o t)}{2(m_{xx}m_{zz}-m_{xz}^2)}} .$$

By calculating $\int C dx$, one correctly recovers the Gaussian distribution in z with variance $m_{zz}(t)$. Similarly, $\int C dz$ yields a Gaussian for the distribution about the mass-weighted mean streamwise position

$$< x >= \int_{-\infty}^{\infty}\int_{-\infty}^{\infty} xC(x,z,t)dxdz = U_o t$$

with variance $m_{xx}(t)$. In the eventuality that α=0, $u_* = 0$, (no mean shear and no covariance), $m_{xz}(t) = 0$ and $C(x,z,t)$ reduces to a product of independent Gaussian distributions about the drifting point $z = 0$, $x = U_o t$. This is a Taylor solution with independent diffusion in z and x. Should $\sigma_u \to 0$, the Gaussian in x reduces to δ $(x - U_o t)$, and Taylor's solution (strictly speaking, the Gaussian distribution with Taylor's solution for the variance) is again recovered.

For large t, the dominant term in the streamwise position-variance $m_{xx}(t)$ about the centre of mass is $(2/3)\alpha^2 U_o^2 \sigma_w^2 \tau t^3$ (see Appendix). Streamwise spread is dominated not by alongwind "diffusion" (which would involve the streamwise velocity variance σ_u^2) but by the joint action of vertical turbulent convection and differential advection in the mean shear. The earliest demonstrations that the alongwind variance of a puff released into an unbounded, linearly-sheared atmosphere increases with t^3 were given by Saffman (1962) and Smith (1965).

In the next section solutions for a continuous source are obtained by integrating in time the instantaneous-source solution for unit release at all previous release times. Similarly, fluxes for a steady source are built from the instantaneous flux densities, eg. the instantaneous along-stream turbulent flux density is

$$F_x = \int_{-\infty}^{\infty}\int_{-\infty}^{\infty} [u - U(z)]p(x,z,u,w,t)dwdu.$$

3. Comparison with Observations

Measurements of the heat plume from a continuous line source in uniformly-sheared, approximately homogeneous turbulence have been reported by Tavoularis and Corr- sin (1981; TC), Stapountzis and Britter (1989; SB), Karnik and Tavoularis (1989); KT), and Chung and Kyong (1989); CK). But the data do not provide a useful test of the present solution with respect to Durbin's, due to factors such as (i) flow disturbance by the source wire (CK); (ii) unreported source strength (TC); (iii) use of scaling that cannot be "undone" and forces the model and observed concentrations to overlap (KT); (iv) horizontal inhomogeneity; (v) very small turbulence intensity σ_u/U; and (vi) incompletely reported flow statistics. All solutions we have examined (ours, Durbin's, or our implementation of the numerical model of SB) predict displacement of the plume centreline (location of maximum temperature) in the direction of lower mean velocity. The opposite displacement was observed by SB and KT near the source (buoyancy was a possible factor). If the observed asymmetry is not due to buoyancy, it is also not due to any mechanism encapsulated in our model or that of SB, and the simulation reported by SB is puzzling.

Flow properties for simulating the KT experiments, derived from KT and companion papers, were: $\sigma_u = 0.06$, $\sigma_w = 0.04$, $U_o = 1$, $\alpha = 0.016$, $u_* = 0.021$, $\tau = 13$, where numbers are dimensionless on the mean centreline velocity U_o (varying from run to run) and the constant "mesh length" M=0.0254m. The small turbulence intensity implies little advantage of our model relative to Durbin's. Figure 1 compares our solutions with the KT data, at scaled downstream distances (measured from the source) of x/M=0.25, 10, 80. KT gave their data in the form of a dimensionless temperature perturbation χ formed by scaling the observed perturbation temperature on the maximum observed perturbation temperature for that cross-section of the flow. The cross-stream position was scaled on the local half-width of the plume ("x_2/w"). Unfortunately that presentation discarded much information (centreline temperature, plume width), leaving a shape largely constrained by the scaling. Differences between the different models are invisible. However at x/M=80 there is a fourfold difference between the present and the Taylor (cross-stream dispersion only) solution for (unscaled) centreline perturbation temperature.

Legg et al. (1988; LRC) performed dispersion experiments in the vertically-inhomogeneous flow about an artificial crop in a boundary-layer wind tunnel. Flow statistics were well-specified and adequately uniform alongstream, and all information needed to predict actual temperature rise in the tunnel can be gleaned from LRC and related work cited below. The features of the LRC experiment useful to us are the large turbulence intensity (σ_u/U at the source was 0.67) and shear stress, which might allow discrimination between models. The objection that homogeneous flow models should not be compared with data from inhomogeneous flows can be subdued by comparing model and observation near the source (where the plume has sampled a narrow range on the inhomogeneous axis). But a short flight does not exclude the influence of other factors not accounted for in the present model, eg. velocity skewness.

Details of the experiments are found in Legg et al. (1986), Raupach et al. (1986), and Raupach et al. (1987). A heating wire was stretched across the flow at height z_s=51 mm, within a canopy of height h_c=60 mm. At the source height U=2.6 ms^{-1}, $< u'w' >= -0.8\ m^2s^{-2}$, $\sigma_u = 1.7\ ms^{-1}$, and $\sigma_w = 1.3\ ms^{-1}$. We formed dimensionless variables using a friction velocity $u_{*s} = 0.9\ ms^{-1}$ based on the shear stress at the source height, and on LRC's estimate for the

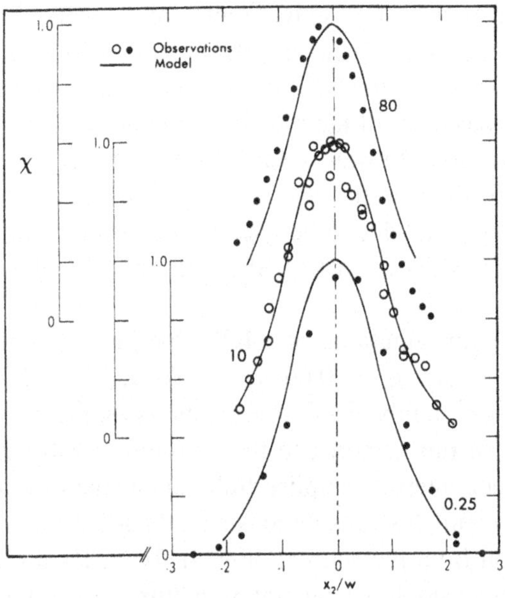

Fig. 1. Comparison of solution with the observations of Karnik and Tavoularis (1989) of the normalised mean temperature rise downstream of a heater-wire in uniformly-sheared homogeneous turbulence. x_2/w is the cross-stream coordinate scaled on the plume half-width. Comparisons at distances from the source (scaled on mesh length) x/M=0.25, 10, 80.

Lagrangian timescale within the crop (derived from the dispersion data), namely $\tau = 0.3h_c/u_{*c}$=0.018 s, where $u_{*c} = 1.0ms^{-1}$ is the friction velocity based on the shear stress measured immediately above the crop. The dimensionless variables are: $U_o = 2.8, \ \alpha = 0.44, \ \sigma_u = 1.9, \ \sigma_w = 1.4, \ u_* = 1, \ \tau = 1.$ We examine LRC's results at the closest point of observation downwind from the source, in their terminology x_L=0.023m, or in our dimensionless notation x=1.44. Advection time from the source to this station is 0.5 (i.e. half a timescale). Even at this short fetch, the plume covers roughly a height range of 20 to 80 mm. Inspection of the velocity statistics given by the experimenters shows that plume depth is too great to justify a claim of homogeneity. Down at 20 mm the plume has encountered a mean velocity larger (by about 50%) than given by our linear profile, and much attenuated turbulent velocities.

It is necessary to multiply our dimensionless concentration and turbulent along-wind flux by $U_o z_s/(\tau u_{*s}^2)$=9.0 to obtain the normalised mean temperature θ/θ_* and flux $< u'\theta' > /u_*\theta_*$ presented by LRC (their Figure 16a). Figure 2a compares our solution, Durbin's, and Taylor's (1-dimensional) against the data. Though it accounts for shear, Durbin's solution performs little better than Taylor's. Its over-prediction at low heights may be because, omitting u', it is more sensitive to underestimation of true windspeed by our shear profile.

Figure 2b shows improvements (relative to the 1-dimensional Taylor solution) as first streamwise diffusion is accounted for (using the present solution with

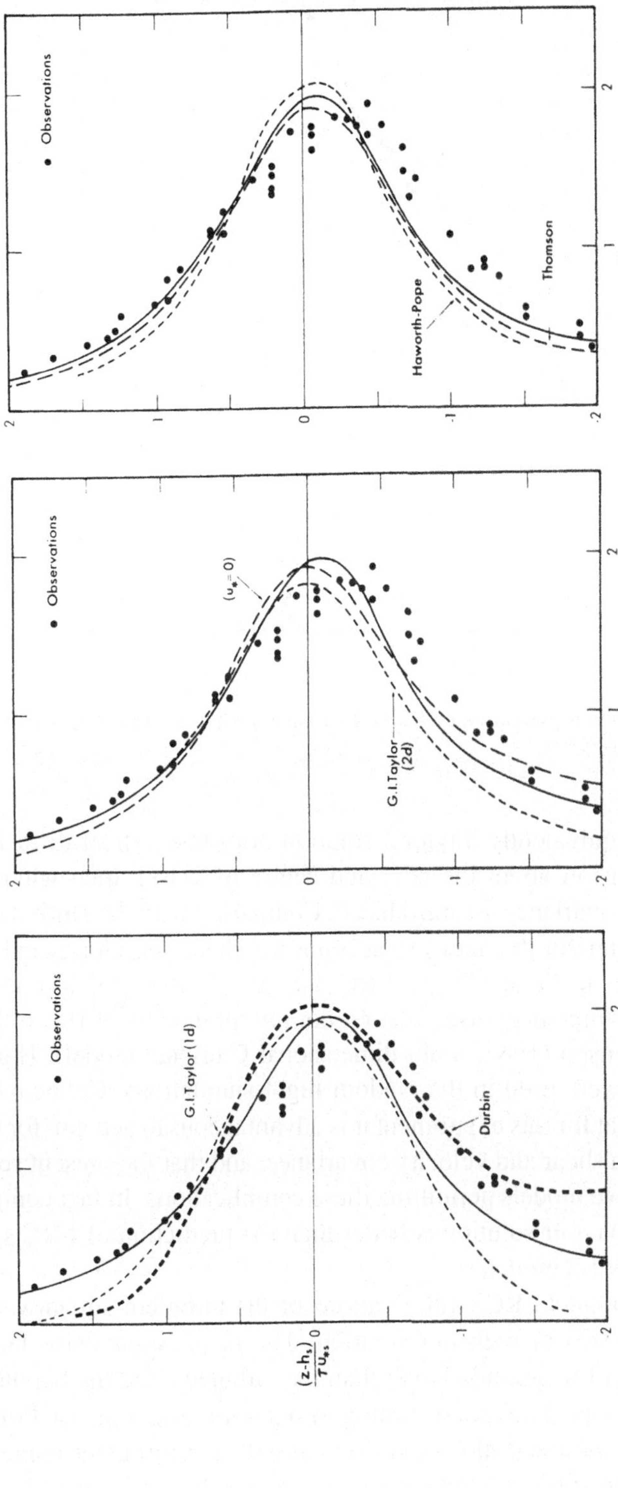

Fig. 2. Comparison of models with observations of Legg et al. (1986) of dispersion from a line source placed in strong velocity shear near the top of a wind-tunnel crop. Height axis spans z=20 mm to z=82 mm, and $(h_C - h_S)/(\tau u_{*s})$=0.58. Solid lines: present solution with shear and covariance.

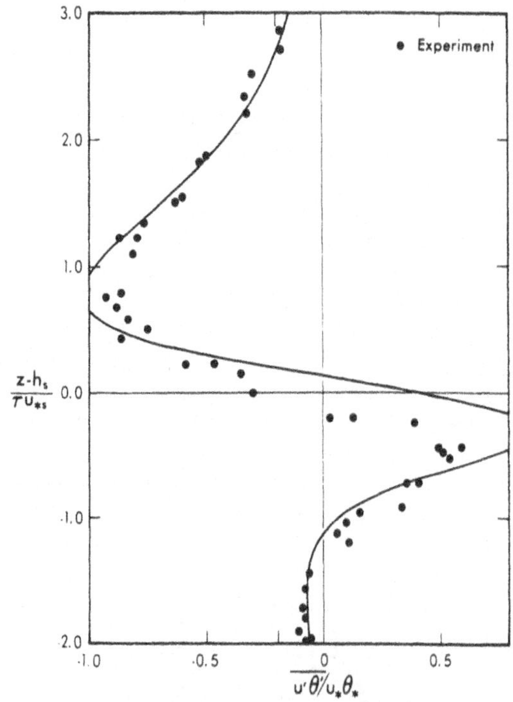

Fig. 3. Comparison of the present solution for the turbulent streamwise heat flux with the observations of Legg et al. (1986).

$\alpha = u_* = 0$; equivalently Taylor's solution both for vertical diffusion and for streamwise diffusion about the advected centre of mass), then wind shear, and finally velocity covariance are introduced. Comparing with the Durbin solution we see that it is better (in this case) to account for shear and alongwind turbulence than shear alone. In Figure 2c the LRC data are compared with our model and with numerical simulations using the random flight models of Haworth and Pope (1986) and Thomson (1987; multi-dimensional Gaussian model). Homogeneous flow properties were used in the random flight simulations. Collectively Figures 2a-2c indicate that for this experiment it is advantageous to account for streamwise dispersion, mean shear and velocity covariance, and that the present solution is as good as alternative models permitting these complications. In fact comparing with LRC's Figure 16a, our solution is better than the prediction of LRC's numerical, height-dependent LS model.

Figure 3 compares LRC's observations of the turbulent streamwise heat flux density at x_L=0.023 m with our solution. The mean streamwise flux densities UC are an order of magnitude larger than the turbulent, and the height-integrated lateral flux at this position is (according to our solution) composed of 116% due to advection by the mean flow and -16% due to the turbulent component. Our solution overestimates the height at which the turbulent flux changes sign, as did the numerical, height-dependent LS model of LRC.

4. Conclusion

Although the Langevin model we assumed is open to criticism, the solution it implies matches the observed near-field of a source in high-intensity turbulence better than simpler solutions (Taylor, 1921; Durbin, 1983) and as well as any numerical Lagrangian stochastic model yet reported. Models (like Raupach, 1989) which for simplicity use an assumption of local homogeneity, could in principle be improved by adopting this extended Gaussian solution.

5. Acknowledgements

The authors acknowledge support from the Natural Sciences and Engineering Research Council of Canada and the Atmospheric Environment Service of Environment Canada.

Appendix

Determination of the moments $m_{xz}(t) =< xz > - < x >< z >$ (etc) is laborious but straightforward. We give the results for the case $\tau_u = \tau_w = \tau$. In none of the experiments that we considered was there any benefit from distinguishing the timescales.

$$m_{zz} = 2\sigma_w^2\tau(t - \tau(1 - e^{-t/\tau}))$$

$$m_{xx} = 2\sigma_u^2\tau(t - \tau(1 - e^{-t/\tau}))+$$

$$+4\alpha U_o u_*^2\tau^2 t - 2(\alpha U_o u_*^2\tau + \alpha^2 U_o^2\sigma_w^2\tau^2)t^2 + (2/3)\alpha^2 U_o^2\sigma_w^2\tau t^3 +$$

$$+(4\alpha U_o u_*^2\tau^3 - 8\alpha^2 U_o^2\sigma_w^2\tau^4)e^{-t/\tau} - 8\alpha^2 U_o^2\sigma_w^2\tau^3 te^{-t/\tau}$$

$$-2\alpha^2 U_o^2\sigma_w^2\tau^2 t^2 e^{-t/\tau} + 8\alpha^2 U_o^2\sigma_w^2\tau^4 - 4\alpha U_o u_*^2\tau^3$$

$$m_{ww} = \sigma_w^2, \quad m_{uu} = \sigma_u^2, \quad m_{wz} = \sigma_w^2\tau(1 - e^{-t/\tau}),$$

$$m_{ux} = \sigma_u^2\tau(1 - e^{-t/\tau}) - 2\alpha U_o\tau(u_*^2 + \alpha U_o\tau\sigma_w^2)t + \alpha^2 U_o^2\sigma_w^2\tau t^2 +$$

$$+2\alpha^2 U_o^2\sigma_w^2\tau^2 te^{-t/\tau} + \alpha^2 U_o^2\sigma_w^2\tau t^2 e^{-t/\tau} + 2\alpha U_o u_*^2\tau^2(1 - e^{-t/\tau})$$

$$m_{wx} = (\alpha U_o\sigma_w^2\tau^2 - u_*^2\tau)(1 - e^{-t/\tau}) - \alpha U_o\sigma_w^2\tau te^{-t/\tau}$$

$$m_{uz} = -(3\alpha U_o\sigma_w^2\tau^2 + u_*^2\tau)(1 - e^{-t/\tau}) + \alpha U_o\sigma_w^2\tau te^{-t/\tau} + 2\alpha U_o\sigma_w^2\tau t$$

$$m_{xz} = -2(u_*^2\tau + \alpha U_o\sigma_w^2\tau^2)(t - \tau(1 - e^{-t/\tau})) + \alpha U_o\sigma_w^2\tau t^2$$

$$m_{uw} = -u_*^2 + \alpha U_o \sigma_w^2 \tau (1 - e^{-t/\tau}).$$

On first sight it may confuse the reader that m_{uw} does not tend to -u_x^2 for large t. The explanation is that

$$m_{uw} =< uw > - < u >< w >=< uw >$$

since $< w >= 0$. Therefore

$$m_{uw} =< (u - U)w > + < Uw >=< u'w > + < U_o(1 + \alpha z)w >=$$

$$=< u'w' > + \alpha U_o m_{wz},$$

i.e. the term $< Uw >$ contributing to m_{uw} does not vanish at large t.

References

Chung M.K. and N.H. Kyong, 1989. *Measurement of turbulent dipersion behind a fine cylindrical heat source in a weakly sheared flow*, J. Fluid Mech. **205**, 171-193.

Corrsin, S., 1952. *Heat transfer in isotropic turbulence*, J. Appl. Physics **23**, 113.

Denmead, O.T., and E.F. Bradley, 1985. *Flux-gradient relationships in a foresr canopy. (In the Forest-Atmosphere Interaction, B.A. Hutchison and B.B. Hicks, editors), D. Reidel Publ. Co.*, ISBN 90-277-1936-5.

Durbin, P.A. 1983. *Stochastic differential equations and turbulent dispersion.* NASA reference publication 1103.

Haworth, D.C., and S.B. Pope, 1986. *A generalized Langevin equation for turbulent flows*, Phys. Fluids.**29**, 387-405.

Karnik, U., and S. Tavoularis, 1989. *Measurements of heat diffusion from a continuous line source in a uniformly sheared turbulent flow*, J. Fluid Mech. **202**, 233-261.

Legg, B.J., M.R. Raupach, and P.A. Coppin, 1986. *Experiments on scalar dispersion within a model plant canopy, part III: An elevated line source*, Boundary-Layer Meteorol., **35**, 277-302.

Okubo, A., and M.J. Karweit, 1969. *Diffusion from a continuous source in a uniform shear flow*, Limnol. Ocean. **14**, 514-520.

Raupach, M.R., 1989. *A practical Lagrangian scheme for relating scalar concentrations to source distributions in vegetation canopies*, Quart. J.R. Meteorol. Soc., **115**, 609-632.

Raupach, M.R., P.A. Coppin, and B.J. Legg, 1986. *Experiments on scalar dispersion within a model plant canopy, part I: The turbulence structure*, Boundary-Layer Meteorol., **35**, 21-52.

Raupach, M.R., P.A. Coppin, and B.J. Legg, 1987. *Erratum with respect to "Experiments on scalar dispersion within a model plant canopy, part I: The turbulence structure, Boundary-Layer Meteorol.,* ***35****, 21-52."* Boundary-Layer Meteorol., **39**, 423-424.

Risken, H., 1985. *Methods for solving the Fokker Planck Equation. (In Stochastic Processes Applied to Physics, eds. L. Pesquera and M.A. Rodriguez).* World Sci. Pub. Co. ISBN 9971-978-20-2.

Saffman, P.G., 1962. *The effect of wind shear on horizontal spread from an instantaneous ground source*, Quart. J.R. Meteorol. Soc., **88**, 382-393.

Smith, F.B., 1965. *The role of wind shear in horizontal diffusion of ambient particles*, Quart. J.R. Meteorol. Soc., **91**, 318-329.

Stapountzis, H., and R.E. Britter, 1989. *Turbulent diffusion behind a heated line source in a nearly homogeneous turbulent shear flow. In turbulent shear flows* **6**. *Selected papers from the 6th Int'l Symposium on Turbulent Shear Flows*, Toulouse, Springer-Verlag.

Tavoularis, S., and S. Corrsin, 1981. *Experiments in a nearly homogeneous shear flow with a uniform mean temperature gradient*, J. Fluid. Mech. **104**, 311.

Taylor, G.I., 1921. *Diffusion by continuous movements*, Proc. London Math. Soc. Ser. 2, **20**, 196-212.

Thomson, D.J., 1987. *Criteria for the selection of stochastic models of particle trajectories in turbulent flows*, J. Fluid Mech., **180**, 529-556.

FRACTAL ASPECTS OF INTEGRATED CONCENTRATION FLUCTUATIONS

A. HADAD, M. STIASSNIE, M. POREH[(1)] and J.E. CERMAK[(2)]
(1) Technion- Israel Institute of Technology, Dept. of Civil Engineering, Haifa, 32000, Israel
(2) Colorado State University, Ft. Collins, Colorado, 80523, U.S.A.

(Received October 1991)

Abstract. Time series of vertically integrated concentrations (VIC) across neutrally buoyant plumes are used to study the fractal and multifractal characteristics of passive scalar fluctuations in turbulent flow fields. Here, the multifractal analysis is based on a novel definition of the singularity spectrum-$F(\alpha)$ of the time records. Approximations for quantities such as the fractal dimension and the spectral exponent are derived as functions of $F(\alpha)$ and are compared with the experimental results. Among other things, we show that VIC records are characterized by two typical subdomains. One domain, which is related to integrated concentration fluctuations, is a subfractal process; whereas the second one, which is directly related to the concentration fluctuations, is a fractal process.

1. Introduction

The phenomenon of turbulent fluctuations of passive scalars, such as concentrations, is important in various theoretical and practical problems. In air pollution problems, one is interested in the concentration fluctuations at some points in the plume, whereas in visibility obscuration problems, one is usually interested in the integrated concentrations across the plume.

The topic of concentration fluctuations is widely discussed in the scientific literature, where different approaches and models for predicting various measures which characterize the dispersion fields are suggested. These include, among others, closure models, random walk models, statistical models, and similarity models. Each of these approaches has some support by experimental measurements of concentrations, or combined measurements of concentrations and velocities.

Relatively little effort has so far been made to study integrated concentrations. Bowers and Black (1985), for example, took measurements of cross-wind integrated concentrations from plumes in the atmospheric boundary layer. The energy spectrum of these measurements was calculated by Hanna and Insley (1989) who found a 5/3 power law behavior between some bounds of wave numbers. In the present paper we refer to Poreh *et al.* (1990), who took measurements of VIC across neutrally buoyant plumes diffusing in a wind tunnel boundary layer and grid turbulence.

In the last decade investigators started to apply fractal analysis techniques in order to get new insight and understanding of turbulent fluctuations in general, and of passive scalar fluctuations in particular. So far, this analysis has been applied to various turbulent problems, such as cloud structure, flame surfaces, tracer plumes,

Boundary-Layer Meteorology **62:** 291–302, 1993.

etc. Sreenivasan (1991), gives a good survey of fractals and multifractals in fluid turbulence.

Special attention was given to the multifractal nature of the energy dissipation rate (Meneveau and Sreenivasan, 1987) and of the scalar dissipation rate (Prasad et al, 1988) in various turbulent flows, using the formalism of the singularity spectrum-$f(\alpha)$, (Halsey *et al.*, (1986)).

The present work focuses on the fractal and multifractal facets of measured time records. In section 2 we introduce a somewhat new approach for the analysis of the multifractal nature of time records. Relations between the derived singularity spectrum and other properties of the time records, such as fractal dimension and the spectral exponent are presented.

2. Theory

Turbulence is multifractal in nature, i.e., measured quantities are well described by fractal power laws (Mandelbrot, 1974; Frisch and Parisi, 1983). The bulk of existing work with multifractals is focused on the derivation of the singularity spectrum for the energy dissipation rate or the scalar dissipation rate, in various fully developed flows (Meneveau and Sreenivasan, 1987; Prasad *et al.*, 1988). Such an analysis deals with measures which are in some way related to the geometry of the measured records, but not with the geometry itself. However, the geometry of the records is multifractal. Thus, we suggest that the singularity spectrum of the measured records is the appropriate way for a multifractal analysis. Furthermore, we show that this technique is a generalization of the more common approach.

2.1. DEFINITIONS

A different formulation for the derivation of the singularity spectrum was recently proposed by Stiassnie (1991). The main steps of this formulation are summarized below. We start with a definition of a set of points which is a sampling of some continuous, bounded, single-valued function-$y(x)$, in the domain [0,1]. We divide the x-axis in the domain [0,1] into n segments of size size $\delta, n = \delta^{-1}$, and denote $x_i = i\delta$; i=0,1,....,n (Fig. 1). The number of δ-boxes which is needed to cover $y(x)$ in the domain between x_i and $x_i + \delta$ is $\Delta(x_i, \delta)/\delta$, where

$$\Delta(x_i, \delta) = y_{max}(x, \delta) - y_{min}(x, \delta), \quad x_i \leq x \leq x_i + \delta \tag{1}$$

thus, the total number of boxes-$N(\delta)$ which is needed to completely cover $y(x)$ is $\Sigma_{i=o}^{n-1} \Delta(x_i, \delta)/\delta$. The box-counting (fractal) dimension-D_b of $y(x)$ is defined and exists only when the measure M is finite:

$$M = lim_{\delta \to 0} \delta^{D_b - 1} \Sigma_{i=o}^{n-1} \Delta(x_i, \delta). \tag{2}$$

The local behavior of the function $y(x)$, in terms of the Lipschitz-Holder (L-H) exponent-α, is represented by

$$\Delta(x_i, \delta) \propto \delta^{\alpha} \tag{3}$$

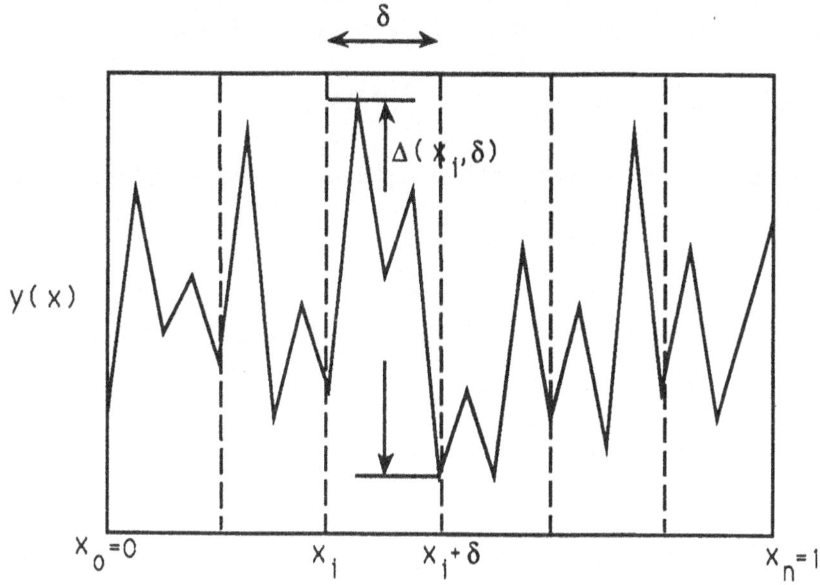

Fig. 1. Illustration of $\Delta(x_i, \delta)$.

where α determines the singularity strength, or the rate by which the derivative tends to infinity. For turbulent data we would expect $y(x)$ to have different values of α at different locations. Actually, $y(x)$ is characterized by a union of an infinite number of subsets each related to a typical singularity-α and supported by a dust on [0,1] with a fractal dimension- $F(\alpha)$. The $F(\alpha)$ curve is a 'singularity spectrum'. The number of boxes needed to cover a subset with L-H exponent between α and $\alpha + d\alpha$ is

$$N(\alpha, \delta) \propto \delta^{-F(\alpha)} d\alpha. \tag{4}$$

2.2. SINGULARITY SPECTRUM

The extension of the measure concept for multifractals is given by the q-measure of the set:

$$M_q = lim_{\delta \to 0} \delta^{\tau(q)} \Sigma_{i=o}^{n-1} \mu_i^q \tag{5}$$

where $\tau(q)$ are the 'mass exponents' and μ_i which denotes the relative weight in the i-th segment, is defined as

$$\mu_i = \Delta(x_i, \delta) / \Sigma_{i=o}^{n-1} \Delta(x_i, \delta). \tag{6}$$

From eqs. (2), (3) and (6) we get an expression for the relative weight in a segment as a function of its L-H exponent and of the box-counting dimension:

$$\mu(\alpha, \delta) \propto \delta^{\alpha + D_b - 1}. \tag{7}$$

Using eqs. (4) and (7) we can write the q-measure in the form

$$M_q \propto lim_{\delta \to 0} \int_0^\infty d\alpha \delta^{-F(\alpha)+(\alpha+D_b-1)q+\tau(q)}. \tag{8}$$

The integral in eq. (8) is calculated by the steepest descent approximation. In the limit of small δ this integral is dominated by

$$M_q \propto \delta^{-F(\alpha)+(\alpha+D_b-1)q+\tau(q)}; \quad \alpha = \alpha(q) \tag{9}$$

where $F'(\alpha) = q$ and $F"(\alpha) < 0$. The measure M_q in eq. (9) is finite only when

$$F(\alpha) = (\alpha + D_b - 1)q + \tau(q) \qquad (a) \tag{10}$$

taking the derivative of eq. (10.a) with respect to q we find

$$\alpha = 1 - D_b - d\tau(q)/dq. \qquad (b)$$

Eqs. (10.a) and (10.b) define the singularity spectrum $F(\alpha)$, when the mass exponents curve $\tau(q)$ and the fractal dimension D_b of the set are known.

In the literature on turbulence the multifractal analysis is usually related to the dissipation rate of energy or scalars, i.e., the relative weight in a segment is given by

$$\mu_i = \frac{\int_{xi}^{x_i+\delta} (dy/dx)^2 dx}{\int_0^1 (dy/dx)^2 dx}. \tag{11}$$

In terms of the formalism of the singularity spectrum, the common notation is

$$f(\alpha) = \alpha q + \tau(q) \qquad (a) \tag{12}$$

$$\alpha = -d\tau(q)/dq \qquad (b)$$

where α and q are the same as before, and $\tau(q)$ is slightly different due to the different definition of the relative weight in a segment. Note that if one defines a new process Y=$\int_0^x (dy/dx)^2 dx$, it can be shown that D_b of this process is unity and that eqs. (10.a,b) yield eqs. (12.a,b) which renders $f(\alpha) = F(\alpha)$.

2.3. THE FRACTAL DIMENSION

From eqs. (2),(3) and (4) one finds that

$$M \propto lim_{\delta \to 0} \delta^{D_b - 1} \int_0^\infty d\alpha \delta^{-F(\alpha)} \delta^\alpha. \tag{13}$$

Using the method of steepest descent and the fact that M has a finite value, one gets

$$D_b = 1 - \alpha_1 + F(\alpha_1); \quad dF(\alpha)/d\alpha_{|\alpha=\alpha_1} = 1 \tag{14}$$

where α_q is used for $\alpha(q)$.

2.4. THE SPECTRAL EXPONENT

Here, we are looking for the value of the spectral exponent-β in the inertial-subrange where the spectrum is expected to follow a power law of the form $k^{-\beta}$. We estimate the spectral exponent by applying the notion of the structure function:

$$< \Delta^p(x_i, \delta) > \;\propto\; \delta^{\zeta p} \tag{15}$$

where ζp is the scaling exponent and <> stands for an average over all segments. Since the number of segments in the domain is δ^{-1}, with p=2 we obtain from eqs. (3), (4) and (15)

$$< \Delta^2(x_i, \delta) > \;\propto\; lim_{\delta \to 0} \int_0^\infty d\alpha \frac{\delta^{-F(\alpha)}}{\delta^{-1}} (\delta^\alpha)^2 \tag{16}$$

which yields

$$< \Delta^2(x_i, \delta) > \;\propto\; \delta^{-F(\alpha_2)+2\alpha_2+1}; \quad dF(\alpha)/d\alpha_{|\alpha=\alpha_2} = 2. \tag{17}$$

Thus, the spectral exponent in wave number space satisfies

$$\beta = 2 + 2\alpha_2 - F(\alpha_2). \tag{18}$$

Using a parabolic approximation for F(α), eqs. (18) and (14) yield

$$\beta = 3 - 2D_b + 2F(\alpha_1). \tag{19}$$

Since the fractal dimension has values in the range $1 \leq D_b \leq 2$, and since $F(\alpha_1)_{max} = 1$, we conclude that the spectral exponent is bounded by $-1 + 2F(\alpha_1) \leq \beta \leq 3$. Exponents lower than the lower limit represent nonphysical processes, and exponents higher than the higher limit represent a non fractal process. For example, $3 \leq \beta \leq 4$ gives a process with dimension one; but the first derivative of this process is a fractal. Such a process is called subfractal.

3. Experimental work

We apply the theoretical procedure outlined in above to measured records of Vertically Integrated Concentrations (VIC) across buoyant plumes from a continuous point source. The VIC is defined as

$$VIC(x, y, t) = \int_0^\infty C(x, y, z, t)dz. \tag{20}$$

The experimental system used for measuring VIC works on the principle of absorption of IR radiation across a carbon-dioxide plume (Poreh and Cermak, 1987).

Using this experimental system we took measurements of VIC at different locations in the down and cross-wind directions. Here we concentrate on analyzing

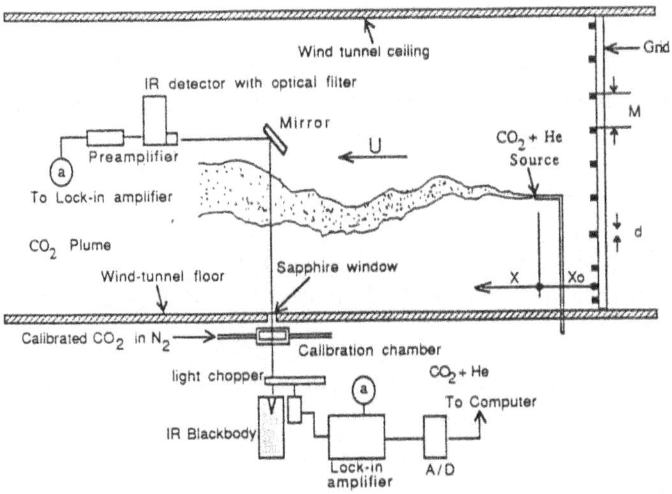

Fig. 2. Schematic description of the experimental setup for measuring VIC. Grid turbulence configuration. M=7.62cm; d=1.51cm (porosity=0.64); U=2.4m/s.

one single typical time record of VIC fluctuations across a plume diffusing in a grid-generated turbulence configuration (Fig. 2).

A time series of 256 samples which are part of this record is shown in Fig. 3. This measurement series was taken at a down-wind distance x/M=62.3 from the tracer source, and in the center-line of the plume. The complete record consists of 8000 measurements sampled at a rate of 300Hz, which yields a measurement duration of about 840M/U (U=2.4m/s) and measurement resolution, in terms of the down-wind displacement between samples, of about 8mm.

A significant length-scale of the dispersing plume is the standard deviation- σ_y of the average cross-wind plume, which for the above down-wind distance was found to be 55.2 mm. We shall use the value of σ_y as a reference length-scale in our calculations.

4. Data Analysis

The above mentioned VIC time record was processed using different statistical, spectral, and fractal tests, which are discussed below. Starting with spectral analysis, Fig. 4 shows a spectral density curve of the VIC time record in terms of non-dimensional scalar 'energy density'- S(m)=S(m)UN$\Delta t/(\sigma^2_{VIC}\sigma_y)$ vs. cycles $m^* = m\sigma_y/(UN \Delta t)$, where N denotes the number of points in the record, Δt the time step between two sampling points, σ_{VIC} the standard deviation of the VIC record at the point of measurement,σ_y the standard deviation of the cross-wind average plume, and U the average wind in the tunnel. From the spectral analysis it appears that the VIC process is characterized by two typical subdomains. In terms of

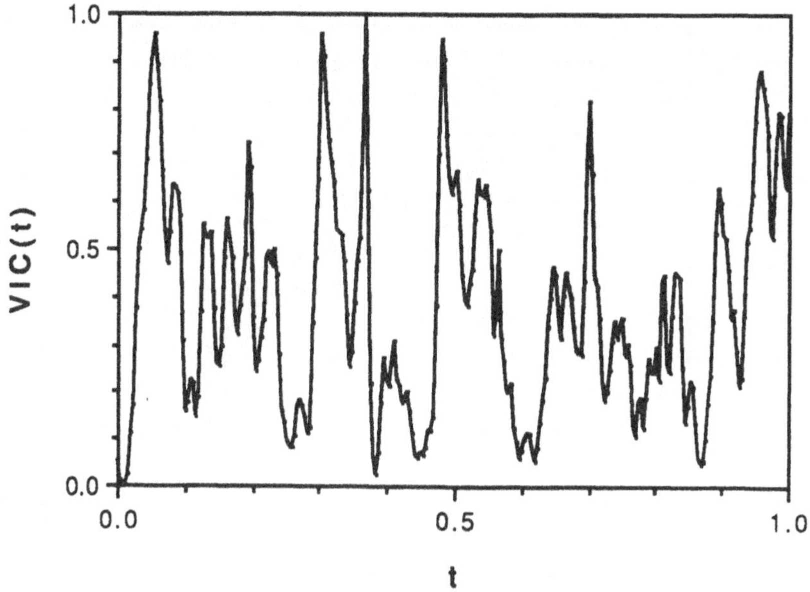

Fig. 3. A typical section of the VIC time record from grid-generated turbulence. x/M=62.3; y/σ_y=0; σ_y=55.2mm

wave lengths the transition between the domains is around σ_y. In range II (see Fig. 4) of length-scales between σ_y and about $5\sigma_y$ (i.e., of order of magnitude of the width of the average plume) we find a spectral exponent of 5/3, similar to what is usually observed in spectra of velocities and concentrations in the inertial subrange. In range I (see Fig. 4) of length-scales between $0.3\sigma_y$ (which corresponds to the Nyquist frequency of the measurement resolution and is longer than both the IR beam diameter and the smallest eddy in the field) and σ_y, we find a spectral exponent of 11/3. From this behavior we conclude that, from the point of view of the cascade, our measurements are indifferent to the integration for scales greater than the integration length, and therefore range II reflects concentration fluctuations. Hence we find a 5/3 law above length-scales around σ_y, whereas below that value, the integration yields a (5/3+2) law.

Spectral power law decay is an indication of a singular or fractal nature. If our VIC record had been a monofractal set, we would expect it to have a fractal dimension $D = (5 - \beta)/2$, (see eq. (19)), which yields values of 2/3 in the range I, and 5/3 in the range II. We would also expect it to have a singularity or Lipschitz-Holder (L-H) exponent $\alpha = 2 - D$, which yields values of 4/3 in range I, and 1/3 in range II. But, since our VIC records are embedded in two-dimensional space, its fractal dimension can vary between 1 and 2, and hence its L-H exponent varies between 0 and 1 respectively. We thus conclude that the process in range I is subfractal, whereas the process in range II is fractal. In other words, we claim that integrated concentration fluctuations are subfractal processes, whereas concentration fluctuations yield a fractal process. Therefore the discussion of the

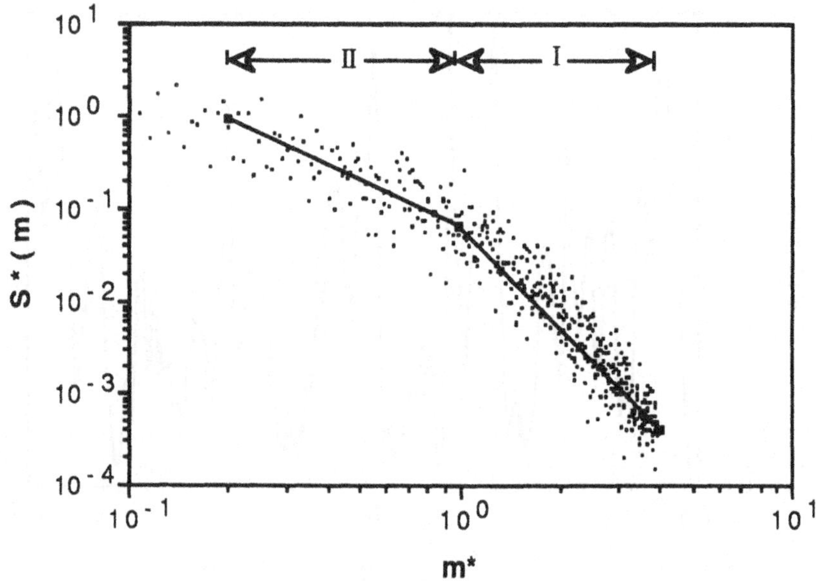

Fig. 4. Spectral density curve of the VIC time record. Non-dimensional scalar 'energy density' vs. cycles. (.) An average of 7 subsets of 1024 sampling points each. (—) Theoretical estimate. Horizontal axis in terms of wave lengths is estimated by σ_y/m^*.

fractal and multifractal nature of the VIC record will be limited to range II, which maintains its fractal nature. However, turbulent processes are multifractal in nature. This should be reflected in the singularity spectrum-$F(\alpha)$ of the record. Since the $F(\alpha)$ curve depends on the box-dimension-D_b of the record, we apply the box-counting algorithm to calculate D_b first.

According to eq. (2) D_b is equal to the slope of the curve $logN(\delta)$ vs. $\log(\delta^{-1})$ in the appropriate wavelength range. Fig. 5 shows, in terms of non-dimensional variables $N = \Sigma_{i=0}^{n-1} \Delta(x_i, \delta)/(\delta^* \sigma_{VIC})$ vs. $\delta^* = \delta U/\sigma_y$, the number of boxes of size δ needed to cover our VIC record in range II. In terms of wave lengths we find self-similarity in a range between σ_y and $15\sigma_y$. The slope of the curve, i.e., the box-dimension D_b, in this range was found to be 1.62 ± 0.03.

Using eqs. (10a) and (10b) with the above value of D_b we have calculated the singularity spectrum $F(\alpha)$. In Fig. 6 we show the singularity spectrum of the given VIC record in range II. As we could expect, α varies in the range 0 to 1, with α_{min}=0.105 and $\alpha_{max} = 0.917$. The dimension F, on the other hand, is limited by the value of 1; the value of α where $F = 1$ (denoted by α_0) is the most probable value of α along the record. Here we find α_o=0.422. The value of the box-dimension is recalculated using eq. (14). From Fig. 6 we find α_1=0.343 and $F(\alpha_1)$=0.963, which yields $D_b = 1.620$. This serves as a check for the robustness of the numerical procedure. The value of the spectral exponent, independent of the spectral analysis, is calculated using eq. (18). From Fig. 6 we find α_2=0.285 and $F(\alpha_2)$=0.877, which yields $\beta = 1.693$. We find that this value is very close to 1.667 obtained from the classical Kolmogrov's 5/3 law.

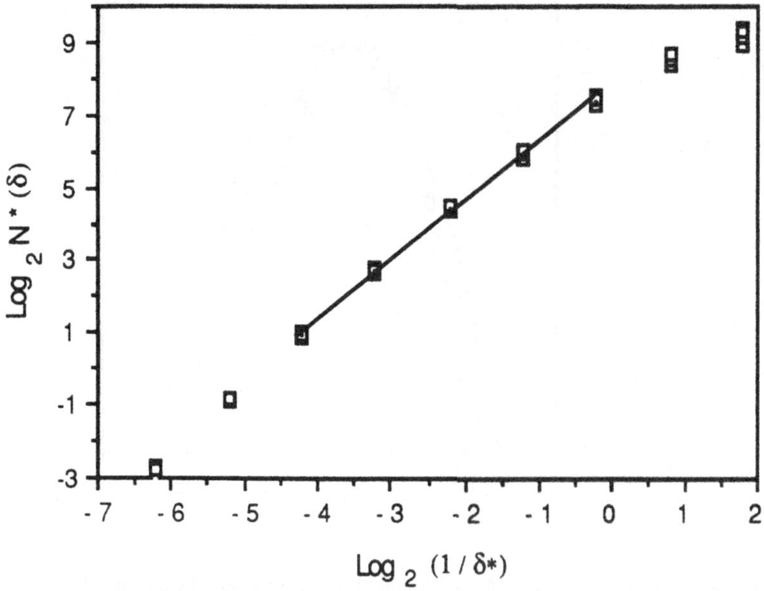

Fig. 5. Box-counting algorithm. Non-dimensional number of boxes of size δ needed to cover the given VIC record in range II: (□)7 subsets of 1024 sampling points each. (—) The average box-dimension D_b of all subsets is 1.62±0.03. Horizontal axis in terms of wave lengths is estimated by $\delta^* \sigma_y$.

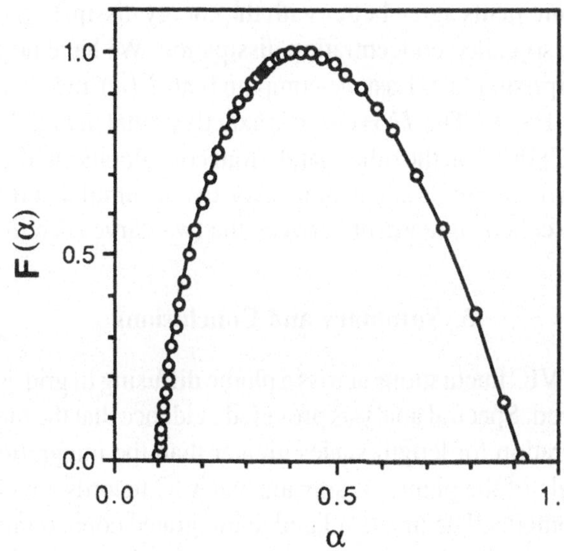

Fig. 6. The singularity spectrum $F(\alpha)$ of the given VIC record in range II.

At this stage we compare our data with other available results, which are mainly related to the dissipation rate. In Fig. 7 we present $f(\alpha)$ for our VIC record and compare it with results from velocity time series of fully developed flows, spatial

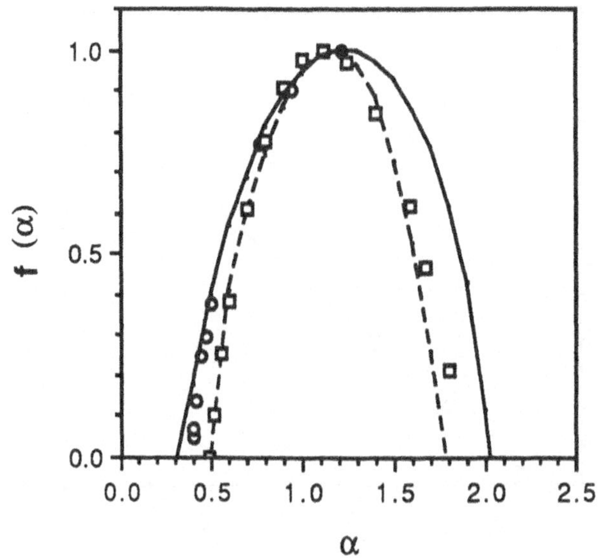

Fig. 7. Singularity spectrum f(α) of the dissipation rate from different experiments; (- - -) Velocity time series (Meneveau and Sreenivasan, 1987); (——) Spatial instantaneous pictures of concentrations (Prasad *et al.*, 1988); (o) Temperature time series (Prasad *et al.*, 1988); (□) Our VIC record in range II.

instantaneous pictures of concentrations, and temperature time series. Our results for the VIC measurements agree better with the energy dissipation rate results than with those for the so-called concentration dissipation. We have no explanation for this somewhat surprising fact. Last we compare both $F(\alpha)$ and $f(\alpha)$ from our VIC record in range II (Fig. 8). The $f(\alpha)$ curve is exactly as that in Fig. 7, i.e. its measure is based on $[d(VIC)/dt]^2$. On the other hand, from considerations already mentioned in section 2.2, $F(\alpha)$ in Fig. 8 is calculated for the accumulated dissipation along the record. The excellent agreement between the two curves is rather reassuring.

5. Summary and Conclusions

Measurements of VIC fluctuations across a plume diffusing in grid-generated turbulence were analyzed. Spectral analysis provided evidence that the measurements are invariant to integration for length-scales greater than the integration length-scale, i.e. the length-scale of the plume. It appears that VIC records are characterized by two typical subdomains. The first is related to integrated concentration fluctuations and is indicative of a subfractal process, whereas the second subdomain is related to the concentration fluctuations and is indicative of a fractal process.

Multifractal analysis indicates that VIC records, in the range where the VIC reflects concentration fluctuations, are multifractal and singular everywhere. These conclusions are substantiated by the singularity spectrum-$F(\alpha)$ of the record. The $F(\alpha)$ spectrum is related to the scaling properties of the "statistical moments".

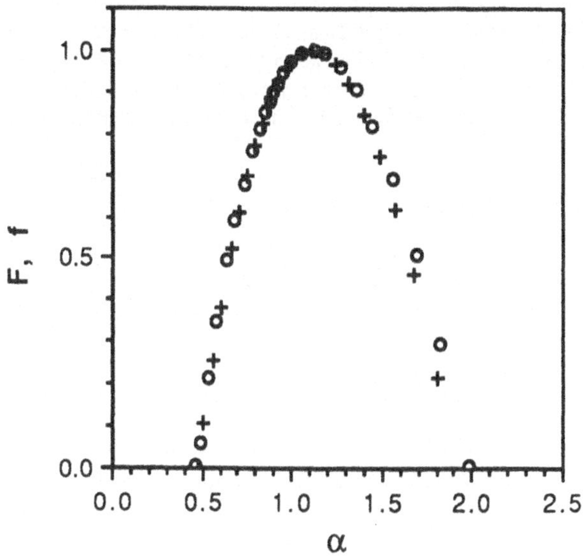

Fig. 8. Comparison between F(α)(o) vs. f(α)(+) in range II.

The first two "moments" are related to the fractal box-dimension and to the spectral exponent respectively. For the given VIC record, in the range which reflects concentration fluctuations, we found that $D_b = 1.62 \pm 0.03$ and $\beta = 1.69$. These values are in agreement with corresponding results obtained by the box-counting algorithm and spectral analysis.

Comparison between our findings and others' work shows that the $f(\alpha)$ curve for VIC is in better agreement with the $f(\alpha)$ curve for velocity fluctuations, than with that of passive scalar fluctuations.

In the future we hope to compare $F(\alpha)$ curves of velocity and VIC fluctuations. Such a comparison will give us a better understanding of the scaling nature of turbulent fluctuations in general and of passive scalar fluctuations in particular.

6. Acknowledgements

The work has been sponsored in part by the Chemical Research, Development and Engineering Center, Department of the Army, Aberdeen Proving Ground, Maryland.

The contribution of Mrs. Z. Vider in the analysis of the data and the preparation of the graphs is gratefully acknowledged.

References

Bowers J.F. and Black R.B. (1985). *Test report-product improved M3A3 (M3A3E2) smoke generator (mobile applications)*. U.S. Army Dugway Proving Ground, Dugway, UT 84022-5000.

Frisch U. and Parisi G. (1985). *On the singularity structure of fully developed turbulence*. Turbulence and predictability in Geophysical Fluid Dynamics and Climate Dynamics, New York, 84-88.

Halsey T.C., Jensen M.H. Kadanoff L.P., Procaccia I. and Shraiman B.I. (1986). *Fractal measures and their singularities: The characterization of strange sets*, Phys. Rev. **A 33**, 1141-51.

Hanna S. R. and Insley E.M. (1989). *Time series analysis of concentration and wind fluctuations.* Bound. Layer Meteor. **47** 131-147.

Mandelbrot B.B. (1974). *Intermittent turbulence in self-similar cascades: Divergence of high moments and dimensions of the carrier*. J. Fluid Mech. **62**, 331-358.

Meneveau C. and Sreenivasan K.R. (1987). *The multifractal spectrum of the dissipation field in turbulent flows*. Nucl. Phys. B. **2**, 49-76.

Poreh M. and Cermak J.E. (1987). *Experimental study of aerosol plume dynamics-III, Wind tunnel simulation of Vertical Integrated Concentration fluctuations*, CSU Project No. 5-3 2571 CER87-88MP-JEC4.

Poreh M., Hadad A. and Cermak J.E. (1990). *Fluctuations of line integrated concentrations across plume diffusing in grid generated turbulence*, CSU Project No. 5-3 8765 CER91-92MP-AH-JEC2.

Prasad R.R., Meneveau C. and Sreenivasan K.R. (1988). *Multifractal nature of the dissipation field of passive scalars in fully developed turbulent flows*. Amer. Phys. Soc. **61**, 74-77.

Sreenivasan K.R. (1991). *Fractals and multifractals in fluid turbulence*. Annu. Rev. Fluid. Mech. **23**, 539-600.

Stiassnie M. (1991). *The multifractal structure of the ocean surface*. Proceedings of the Nonlinear Water Waves Workshop. Bristol 22-25 Oct. 1991.

MODELLING NO CONVERSION BY USING PROBABILITY DENSITY FUNCTIONS

PETER BANGE
NV KEMA Arnhem, The Netherlands.

(Received October 1991)

Abstract. An expression for concentration fluctuations in a smoke plume is derived from airborne measurements of NO_X. A linear relation between the standard deviation of the fluctuations around a Gaussian concentration profile and the average gradient in the concentrations is assumed. With this relation the probability density function of expected NO_2 concentrations at 3 km from a source of NO_X is modelled under the assumption of photostatic equilibrium, and is compared with measurements. A parametrisation for the concentration fluctuations of std(C') = 26(+/-7)$*d\bar{c}/dr$ is proposed (r in metres). Calculated NO_2 distributions are in reasonable agreement with the measurements and the average NO_2 concentration appeared not to be affected by the concentration fluctuations in the NO_X concentration. The spatial resolution of all measurements was 40 m.

1. Introduction

Nitrogen oxide (NO) emitted from a tall stack into ambient, ozone-containing air, oxidizes into NO_2. The rate at which NO_2 is formed depends on the rates of the reactions involved and on the rate at which the ozone-containing air is mixed into the plume of effluents.

In Bange et al. (1991) it is shown that even at short distances from the source the NO_2 concentrations are already near the photochemical equilibrium values provided one uses instantaneous dispersion parameters instead of time-averaged values. When time-averaged dispersion parameters are used, meandering effects will lead to an overestimation of the mixing of O_3 into the plume. It was proposed that small deviations from the photostatic equilibrium in the instantaneous plume are due to space-averaging effects, caused by the assumption that the instantaneous plume has a Gaussian shape, which is not the case. In this paper an attempt is made to parameterise the concentration fluctuations in the instantaneous plume.

The concentration fluctuations are defined as the difference between measured values of NO_x and the mean value.

$$[NOx]' = [NOx] - [\overline{NO}x]. \tag{1}$$

For each place in the plume a distribution of $[NO_X]'$ values will be found. We describe this distribution with its standard deviation and neglect higher order moments.

As a first approximation a linear dependence of this standard deviation of $[NO_X]'$ on either the distance from the plume centre line (r), the mean concentration ($[\overline{NO}_X]$), or the gradient in the mean concentration (d$[\overline{NO}x]$/dr) is assumed.

Boundary-Layer Meteorology **62:** 303–311, 1993.

TABLE I
Meteorological and instantaneous plume parameters. N is the cloud cover, P is the Pasquill stability class, u(h) the windspeed at plume height h, σ the calculated and the measured plume width k_3 is the first order reaction rate (see Eq. 7) and n the number of crossings.

fl.	N (oct)	P	k_3 (s^{-1})	u(h) (m/s)	dist (km)	h (m)	σ(m) meas	σ(m) calc.	n
310	0-1	C	0.008	10	1.0	250	64	63	7
					3.0	270	240	118	6
327	2-7	D	0.007	10	.5	350	78	42	7
					3.0	400	365	118	5

Knowing these fluctuations for inert NOx, one may calculate (i) the expected fluctuations in the reaction product NO_2, and (ii) the effect of concentration fluctuations on the average NO_2 concentration.

2. Measurements

Profiles of NO_X, NO and O_3 were measured while crossing through a power plant plume in the south of the Netherlands. Crossings were performed at various distances from the stack in an aircraft. Flight circumstances and number of crossings per distance are listed in Table I. Data samples were taken every 0.5s and stored on disk. The flight speed of the aircraft during the measurements was about 80 m/s.

3. Data Processing

The data are processed as follows:

1. Measured data of NOx are deconvoluted to account for the effect of monitor averaging. With this procedure the spatial resolution of the measurements is enhanced to about 40 m.
2. The background concentrations are subtracted and all non-plume data are removed.
3. The plume width σ, centrepoints (point of gravity) and integrated concentration (ppb.m) are calculated for each crossing.
4. Data are sorted into groups with equal distance from the source and the average plume width is calculated within each group.
5. Within each group the crossing with the highest integrated concentration is assumed to be flown through the centre of the plume; for the other crossings the distance at which the centre is missed is estimated from the integrated concentration and the average plume width. The average plume is assumed to be circular.

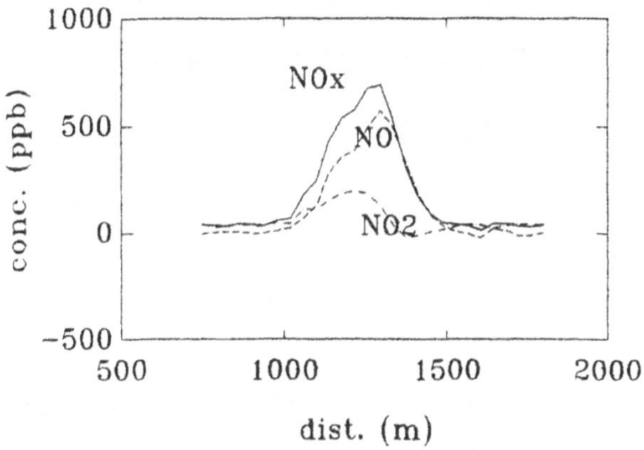

Fig. 1. Negative NO_2 concentrations resulting from small differences in lagtime of the NO_X and the NO monitor at 1 km from the source (flight 310).

6. For each crossing the average Gaussian plume is calculated using the average plume width, the assumed distance at which the centre is missed and the centerpoint (point of gravity) of the crossing.

7. With these a list can be made for each group of crossings:
   a The average (Gaussian) concentration ($\bar{C}$)
   b The average (Gaussian) gradient ($d\bar{C}/dr$)
   c The distance from the centre of the plume (r)
   d The difference between the measured concentration and the average (Gaussian) concentration: C'

8. The list is sorted relative to the explaining variable ($\bar{C}, d\bar{C}/dr$ or r). Then, for each successive five values for C' the standard deviation of C' is calculated. In this way a value of std(C') as a function of $\bar{C}$, $d\bar{C}/dr$ or r is found.

9. Finally a linear least square fit between one of the three explaining parameters and std(C') is calculated.

NO_2 concentrations were derived by subtracting the deconvoluted NO measurements from the deconvoluted NO_X measurements. It appeared that due to a small difference in lagtimes, negative NO_2 concentrations resulted. Errors in the derived NO_2 concentration depend on the gradients in NO_X and NO (see figure 1). This lagtime difference could not be corrected because it was smaller than the sampling interval of 0.5 s. Therefore only the 3 km data are used to compare modelled NO_2 concentrations with measured data. At this distance concentration gradients are smaller, and errors due to differences in lagtimes will be less. Still an additional scattering of the NO_2 data has to be expected.

4. Modelling

The reactions involved in the NO oxidizing process are:

$$[NO] + [O_3] - k_1 \rightarrow [NO_2] + [O_2] \tag{2}$$

and the reverse reaction during daytime:

$$[NO_2] + [O_2] = uv - light\ - k_3 \rightarrow [NO] + [O_3] \tag{3}$$

where square brackets denote concentrations in ppb, and k_1,k_3 the reaction rates. The mixing may be described as a first approximation with the well known Gaussian plume equation. Assuming radial symmetry we get:

$$C = Q/(u \cdot \sigma^2 \cdot \pi) * exp(-.5 * (r^2/\sigma^2)) \tag{4}$$

in which C is the averaged concentration of an inert component, u the averaged windspeed in m/s, x the off-wind distance from the source, Q the stack emission and r the distance from the centreline of the plume. The plume is assumed not to touch the ground.

For σ one generally assumes the following dependence on the downwind distance from the source:

$$\sigma = a.x^b \tag{5}$$

with a and b functions of atmospheric stability and averaging time. From the definition of the reaction constants k_1 and k_3 one can derive that

$$d[NO_2]/dt = -k_1[NO][O_3] \tag{6}$$

and

$$d[NO]/dt = -k_3[NO_2]. \tag{7}$$

When reactions (2) and (3) balance we get:

$$[NO][O_3]/[NO_2] = k_3/k_1. \tag{8}$$

Assuming the total of oxidant to be constant

$$[Ox] = [NO_2] + [O_3] \tag{9}$$

and the mean concentration of the inert NO_X $(= NO + NO_2)$ to be described by (4) one can calculate $[NO_2]$ for all x and r, depending on the values of k_1 and k_3. (Carmichael and Peters, 1974).

Knowing the background concentrations of NO_X, NO and O_3 the background of $O_X(= NO_X - NO + O_3)$ can be calculated. Assuming that 5% of the NO_X is emitted from the stack as NO_2 a profile of Ox can be calculated with:

$$Ox(r) = Ox(background) + 0.05 * NOx(r); \tag{10}$$

k_3 is calculated using the empirical formula derived by Dickerson et al. (1982):

$$k_3 = 0.0167 * exp(-.575/cos(\theta)) \quad (sec^{-1}) \tag{11}$$

with θ the solar angle. For k_1 we use (Hamson and Garving 1978):

$$k_1 = 15.33/T * exp(-1450/T)\,(ppb^{-1}sec^{-1}) \tag{12}$$

with T the temperature in K. Then, under the assumption of photostatic equilibrium (8), a profile of NO_2 can be calculated from the NOx profile, the Ox profile, k_1 and k_3.

When we perform this operation for (i) the Gaussian NO_X profile, (ii) this same NO_X profile with the standard deviation of the fluctuations added and (iii) with the standard deviation subtracted, three NO_2 profiles will be obtained. These three profiles represent the 16, 50 and 84% percentiles of the NO_2 distribution. This procedure to calculate the NO_2 distribution is essential. Because of the non-linearity of the reactions involved, the higher order moments of the NO_2 distribution are not necessarily zero, even if the NO_X distribution is correctly described by its first and second moment. If the calculated distribution of NO_2 is skewed, this is likely to show: the 84% percentile may lie higher above the median than the 16% under it or viceversa.

5. Results and Discussion

Averaged instantaneous plume widths can be found in Table I. Also listed are instantaneous plume widths according to $\sigma = a.x^b$ with a = 1.23 and b=0.57 according to Bange et al. (1991). The higher measured plume width will be partly due to the fact that bimodal plumes (plumes showing two clear maxima) were omitted from analysis in their derivation of instantaneous plume dispersion parameters.

Figure 2 shows a plot of the averaged (Gaussian) concentration $[\overline{NOx}]$ and the measured fluctuations $[NOx]'$ as a function of the gradient in the averaged concentration (d($[\overline{NOx}]/dr$), at 1 km from the source. Also shown are the actual averaged concentration and its standard deviation calculated from successive groups of 5 successive values of the measured concentration. At this distance, the Gaussian plume seems to be a reasonable estimate of the averaged concentration. Assuming:

$$\bar{C} = A * d\bar{C}/dr + B \tag{13}$$

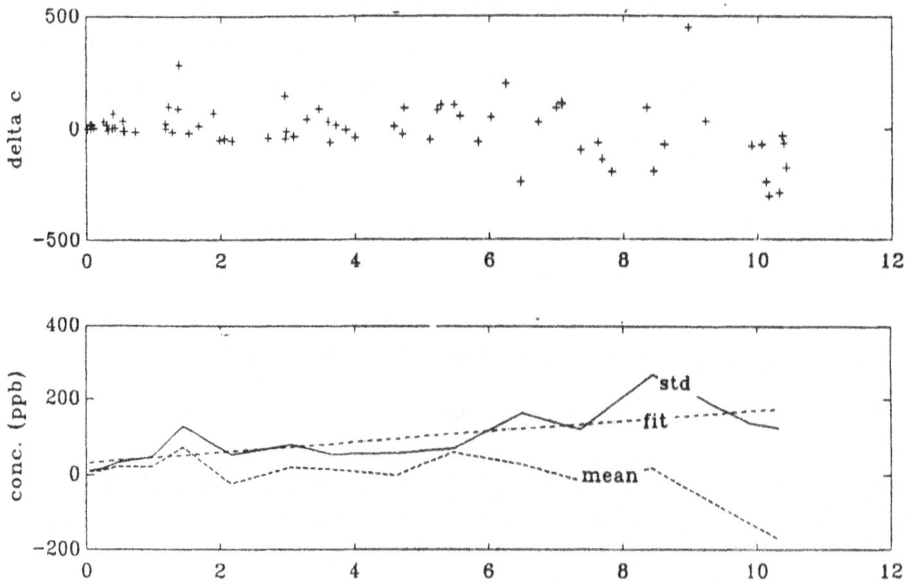

Fig. 2. (a) Deviations of the NO_X concentration from a Gaussian profile as a function of the averaged gradient of NO_X, measured during flight 310 at 1 km from the source.
(b) Five-point averages and five-point standard deviations of the values shown in 2a. Also ahown is a straight line fit through the standard deviations.

we found: A=14(+/-4) meter and B=31 (+/-20) ppb. Values nearer than 50 m to the centre of the plume are omitted from the analysis because of the uncertainty in the height at which the plume is crossed. At this distance a small change in this height leads to large errors in either [$\overline{NO}$x], d[$\overline{NO}x$]/dr or r. The value B represents fluctuations in the background NO_X concentration.

In Table II the results of both flights are collected. Listed are the least square fit coefficients with errors for the three different explaining parameters $[\overline{NO}x]$, $d[\overline{NO}x]/dr$ and r. It is clear that all the calculated A and B values depend on the distance from the source and that their error increases considerably at larger distances. Considering the error in the fit, the dependence on $\bar{C}$ is the most reliable. Physical considerations however favour the dependence on the gradient. Turbulence will only induce concentration fluctuations when gradients exist. In a homogeneously mixed atmosphere, turbulence can not induce fluctuations.

Figure 3 shows modelled and measured NO_2 concentrations for flight 310 at a distance of 3 km from the source as a function of distance from the plume centreline. The dashed curves are the NO_2 profiles calculated with the concentration fluctuations added to and subtracted from the Gaussian NO_X profile. The fluctuations are described according to Table II, as a function of the gradient of NO_X. The average NO_2 concentration is underestimated by 10%. The fluctuations in NO_2 are underpredicted too, as about 50% of the measured NO_2 data lie outside the 16 and 84% percentiles. It can be seen that fluctuations hardly affect the average

TABLE II
Parameterisation of the concentration fluctuations. r, C and dC/dr denote the explanatory parameters. See table I for the distances from the source. A fit is calculated using the following: $std(C') = A(+/-dA) * R + B(+/-db)$ etc.

Flight	r				C				d			C/dr
	a	dA	B	dB	A	dA	B	dB	A	dA	B	dB
310	-0.64	0.13	170	21	0.20	0.04	35	15	14	4	31	20
	-0.03	0.02	38	7	0.11	0.06	23	4	21	22	24	6
327	-1.10	0.17	271	32	0.25	0.04	29	14	19	3	11	12
	0.01	0.01	28	3	0.13	0.05	15	20	48	18	14	3
Mean	-0.43	0.05	127	10	0.17	0.02	26	7	26	7	20	6

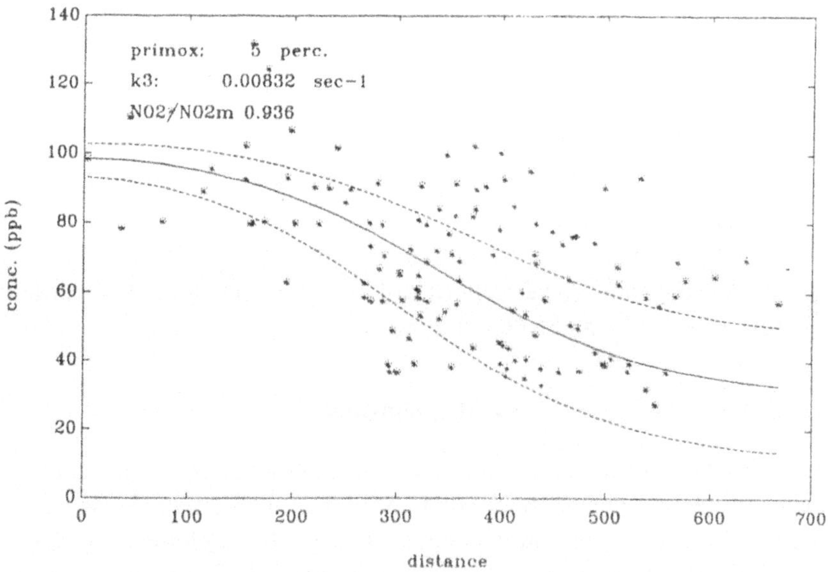

Fig. 3. Measured (*) and modelled NO_2 concentrations for flight 310 at 3 km from the source. On the horizontal axis the cross-wind distance from the centre of the plume. The dashed NO_2 curves are calculated from NO_X with the standard deviations of the concentration fluctuations of NO_X added and subtracted. 50% of the measurements should lie between these two curves. The NO_X concentration fluctuations are parameterised as a function of the gradient of NO_X.

NO_2 concentrations. Otherwise one of the two dashed curves would lie closer to the median than the other.

Figure 4 shows the same plot for flight 327. The NO_2 concentration is overpredicted by 15%, mostly caused by a bad prediction of the expected NO_2 concentration in the core of the plume. Again the fluctuations in NO_2 are underpredicted, with 60% of the data lying outside the 16 and 84% percentiles. The overprediction of the average NO_2 may be caused by the influence of clouds, which was not taken

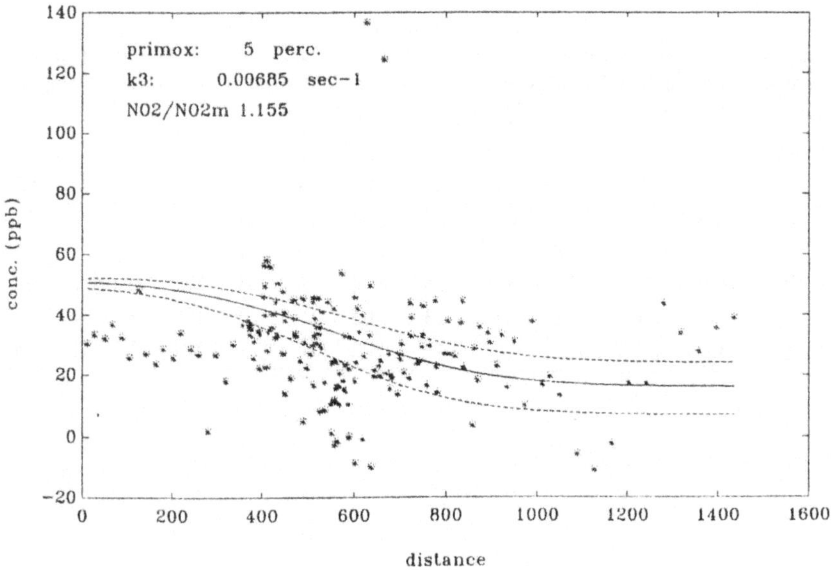

Fig. 4. As figure 3, but for flight 327, at 3 km from the source.

into account while calculating k_3. In both cases the underprediction of the $[NO_2]'$ may be explained by additional scattering of the NO_2 data due to small differences in lagtimes of the NO_X and NO monitors.

6. Conclusions

At short distances from the source, the concentration fluctuations found in an instantaneous smoke plume show good correlation with either the distance from the plume axis r, the concentration predicted by a Gaussian plume C or the gradient in this C (dC/dr). At larger distances from the source this correlation is less. As a first approximation one may use for the first 5 km of plume dispersion:

$$std(C') = 26 * d\bar{C}/dr \tag{14}$$

with r in m.

Distributions of NO_2 calculated from NO_X distributions under the assumption of photochemical equilibrium underestimated the measured fluctuations of NO_2. This however may be due to extra scattering introduced in the NO_2 data by small lagtime differences in the NO_X and NO monitors.

The average NO_2 concentration did not seem to be affected by the fluctuations of NO_X.

Acknowledgements

This study was supported financially by the Joint Dutch Electricity Utilities.

References

Bange P., Janssen L.H.J.M., Nieuwstadt F.T.M., Visser, H. and Erbrink J.J. (1991). *Improvement of the modelling of daytime nitrogen oxide oxidation in plumes by using instantaneous plume dispersion parameters.* Atmos. Envir., **25A**, pp 2321-2328.

Carmichael G. R. and Peters L. K. (1991). *Application of the mixing-reaction in series model to $NO_X - O_3$ plume chemistry.* Atmos. Envir., **15**, pp 1069-1074.

Dickerson R. R., Stedman D. K. and Delanay A. C. (1982). *Direct measurements of ozone and nitrogen dioxide photolysis rates in the troposphere.* Journ. Geogrph. Res., **87**, pp 4933-4946.

Hamson R. F. and Garvin D. (1978). *Reaction rates and photochemical data for atmospheric chemistry.* NBS spec. publ. (U.S. department of commerce), No 513.

PART III

EXPERIMENTAL TECHNIQUES AND METHODS OF DATA ACQUISITION

FINAL RESULTS OF THE CONDORS CONVECTIVE DIFFUSION EXPERIMENT

GARY A. BRIGGS
National Oceanic and Atmospheric Administration
Air Resources Laboratory
Research Triangle Park, North Carolina 27711, USA.[*]

(Received October 1991)

Abstract. The Convective Diffusion Observed by Remote Sensors (CONDORS) field experiment conducted at the Boulder Atmospheric Observatory used innovative techniques to obtain three-dimensional mappings of plume concentration fields, χ/Q, of oil fog detected by lidar and "chaff" detected by Doppler radar. It included extensive meteorological measurements and, in 1983, tracer gases measured at a single sampling arc. Final results from ten hours of elevated and surface release data are summarized here. Many intercomparisons were made. Oil fog χ/Q measured 40m above the arc are mostly in good agreement with SF_6 values, except in a few instances with large spacial inhomogeneities over short distances. After a correction scheme was applied to compensate for the effect of its settling speed, chaff $\int \chi dy/Q$ agreed well with those of oil except in two cases of oil fog "hot spots". Mass or frequency distribution vs. azimuth or elevation angle comparisons were made for chaff, oil, and wind, with mostly good agreements. Spacial standard deviations, σ_y and σ_z, of chaff and oil agree overall and are consistent at short range with velocity standard deviations σ_v and $\sigma_w \approx 0.6w^*$ (the convective scale velocity), as measured at $z > 100$m. Surface release σ_y is enhanced up to 60% at small x, consistent with the Prairie Grass measurements and with larger σ_v and reduced wind speed measured near the surface. Decreased σ_y at small dimensionless average times is also noted. Finally, convectively scaled $\int \chi dy$, C_y, were plotted versus dimensionless x and z for oil, chaff, and corrected chaff for each 30-60 min period. Aggregated CONDORS C_y fields compare well with laboratory tank and LES numerical simulations; surface-released oil fog compares expecially well with the tank experiments. However, large deviations from the norm occurred in individual averaging periods; these deviations correlated strongly with anomalies in measured w distributions.

1. Introduction

The traditional view of Gaussian-like plumes moving downwind and diffusing vertically from a mode at a fixed height was severely challenged by experimental results in the late 1970's. Using a 1.2 square m. water tank with a heated bottom to simulate convectively-induced mixing, Willis and Deadorff (WD) (1976) found that plumes of neutrally buoyant oil droplets released near the surface lifted off after a short time (about five minutes at atmospheric scales); they then lofted into the *upper* half of the mixing layer before becoming more-or-less uniformly mixed in the vertical at long times (about 1/2 hour or more at atmospheric scales). The same behavior of surface releases was also discovered by Lamb (1978,1979) using a large-eddy simulation (LES) numerical model for convective boundary layers (CBLs). His results for elevated releases were also surprising. The concentration modes descended almost to the surface, causing up to 2.9 times the surface impact

[*] On assignment to the US Environmental Protection Agency, Atmospheric Research and Exposure Assessment Laboratory, RTP, NC.

Boundary-Layer Meteorology **62:** 315–328, 1993.

predicted by traditional Gaussian models. Lambs's results inspired further laboratory work in the convective tank, this time using elevated releases (WD, 1978, 1981). Once again, the numerical modeling and laboratory tank results essentially agreed.

These results caused controversy among diffusion modelers because of the doubt they cast on time-honored Gaussian plume models for convective conditions, which prevail much of the time (e.g., during most daylight hours except during high winds, heavy overcasts, or snowcover). Furthermore, convective conditions tend to produce the highest ground concentrations from large sources. Much more about CBL diffusion has been learned in the last decade using the above tools. Experiments have been performed simulating buoyant sources, fumigation, and concentration fluctuations (WD, 1983; Deardoff and Willis, 1982 and 1984). These have revealed a few further surprises, such as the delayed downward mixing of large buoyancy plumes which loft to the top of the mixing layer; this is due to residual buoyancy which resists downward mixing (Briggs, 1985; WD, 1987; Willis and Hukari, 1984).

Although these laboratory and numerical results have been very enlightening, it is hard to have total faith in these results without a degree of confirmation from field experiments. Such experiments are expensive and would be hard to carry out for some of the above situations, due to lack of controlled conditions and large stochastic variability in the field. However, it would be helpful to have at least more basic results, like convective diffusion from non-buoyant sources, tested via fields experiments. This was the basic purpose of the experiment carried out in 1982 and 1983 at the Boulder Atmospheric Observatory (BAO). The best previously existing field measurements for testing WD's results were from the 1956 "Prairie Grass" experiment (Barad, 1958). These were very limited in that only surface concentrations were measured, from only surface releases, to distances no more than 800m. The meteorological information was exceptionally complete, so that the convective scaling parameters (Deardorff, 1970) used in the tank and numerical studies could be reliably estimated; without these parameters, it would be difficult to apply laboratory scale results to full scale and difficult to generalize. Both Briggs and McDonald (1978) and Nieuwstadt (1980) used the aircraft soudings to determine the mixing depths, z_i, and wind and temperature tower profiles in conjunction with similarity theory to estimate buoyancy flux, H^*, needed to determine the convective scaling velocity, w^*=$(H^* z_i)^{1/3}$. They both found good agreement between Prairie Grass $\int \chi dy$ (crosswind-integrated concentration) versus x and WD's tank results, but only a few points at the largest x and the lowest wind speeds had large enough dimensionless x $(X = (x/U)w^*/z_i)$ to show the drastic reduction of surface χ during the plume "liftoff" stage (nonetheless, these few points probably caused the accelerated growth of Pasquill's "A" and "B" curves for vertical diffusion, based as they were on the Gaussian plume assumption and surface $\int \chi dy$; see Briggs, 1988).

The BAO experiment was called CONDORS (<u>Con</u>vective <u>D</u>iffusion <u>O</u>bserved with Remote Sensors). It went well beyond previous basic diffusion experiments in that it attempted to provide three-dimensional measurements of concentration fields in convective conditions to heights exceedings z_i, which ranged up to 1600m. This required refinement of processing techniques, particularly for the lidar returns, to achieve quantitative estimates of χ/Q, where Q is source strength, rather than simply qualitative pictures of plume shapes. To achieve the principal goal, comparisons with the results of laboratory and numerical experiments, the releases were accompanied by a very full complement of meteorological measurements (much inspired by the classic "Prairie Grass" experiment of 1956).

The results given here are a summary from a much larger paper with too many details and figures for a conference presentation (Briggs, 1992).

2. Scope and Quality of Data

The CONDORS experiment has been described many times, so details will not be repeated here (Moninger et al., 1983; Eberhard et al., 1985; Briggs et al., 1986; Kaimal et al., 1986; Eberhard et al., 1988; Briggs, 1989). Briefly, oil fog and "chaff" (aluminized mylar filaments) were released either from a carriage near the top of the 300m BAO meteorological tower or from a few meters above the surface about 100m west of the tower. The oil fog plume was detected by a lidar about 4km north of the tower, making complete vertical scans at 4 to 6 azimuths repeated about every 3 1/2 min. The chaff plume was detected by a Doppler radar 3.5 km east of the tower, making horizontal sweeps in about 0.7° vertical increments repeated about every 2 1/4 min. Thus, 30 to 60 min average χ fields were obtained; the averaging periods were often limited by substantial changes in one of the meteorological parameters, spoiling any approximation to steady state conditions. Very extensive meteorological measurements were obtained from the sonic anemometers, temperature, and humidity sensors at eight tower levels, an acoustic sounder, and frequent radiosonde releases. The 2 hour releases made in September 1982 were considered "trial runs" for testing the remote sensing techniques; they produced some usable measurements, but it was decided to reposition the lidar, to use higher molecular weight oil (to reduce evaporation), to produce more oil fog when possible (to overcome background haze), and to use chaff only as an elevated release in 1983 (the chaff had a set settling speed of 0.3 m/s). The main runs in 1983 also included gaseous tracer releases, primarily SF_6, and a single sampling arc at $x \sim 1200$m west of the tower to provide some "ground truth" on χ/Q to compare with values inferred from the remote sensors; these could not "see" within 30 to 70m above the ground, and because neither tracer was conservative (oil fog partially evaporated and some chaff deposited), $U \int\int \chi dydz$ was used in place of Q for normalizing χ.

Table I gives a brief summary of the averaging periods actually processed, 16 periods totaling about 10 hours (periods 21-25 are from 1982)(z_s is the source height). Altogether, about 24 hours of measurements were taken, but processing costs much exceeded acquisition costs, and many potential averaging periods were

TABLE I
Principal variables, CONDORS averaging periods.

Pd#	Dur, min	oElev. □ Surf. Oil	Chaff	SF6	z_i, m	z_s/z_i Elev.	U, m/s	w^* m/s	U/w^*
21	29	o	o		520	0.45	5.8	1.5	4.0
22*	35	o	o		730	0.32	6.2	1.4	4.4
23	40	o	o		960	0.17	2.8	1.5	1.8
24	44	□	□		980	—	2.4	1.8	1.3
25*	42	□	□		1260	—	1.6	1.8	0.87
31	30	□	oX	o	1600	0.17	3.1	2.0	1.6
32	30	□	o	o	1240	0.19	1.9	2.0	0.95
33*	60	□	oX	o	1400	0.17	2.6	2.0	1.3
34	50	□	o	o	1100	0.25	1.9	1.9	1.0
35	40	o	o	o	880	0.32	2.5	1.6	1.5
36*	30	o	o	o	880	0.32	2.6	1.7	1.5
37*	30	o	o	o	880	0.32	3.3	1.6	2.0
38	40	o	o	o	640	0.41	4.4	1.4	3.2
39*	40	o	o	o	780	0.34	4.6	1.5	3.1
40	30	□	oX		900	0.26	2.1	1.8	1.2
41*	40	□	oX		870	0.27	1.6	1.9	0.84

* Same run as previous period. X Processing incomplete.

spoiled by breaks in acquisition, large changes in wind speed or direction, rapid growth of z_i, or sudden drops in H^* due to a passing cloud. It can be seen that most periods where quite convective, with $U/w* < 2$; $U/w* < 6$ is considered convectively-dominated turbulence. When $U/w^* < 1$, operational problems occured when the oil fog occasionally wafted upwind.

Most tower measurements were quite typical of very convective boundary layers. On the average, wind speed was constant within a few percent of U (averaged over the 150, 200, 250 and 300m levels) from 50m on up. Wind direction shear was insignificant except in a few individual periods. The vertical and lateral turbulent velocity standard deviations, σ_w and σ_v, both approached $0.6w^*$ at the upper tower levels; σ_w/w^* grew roughly as $z^{1/3}$ near the ground, as expected, while σ_v/w^* increased to about 0.8 near the ground. Such an increase was not observed in the 1973 Minnesota boundary-layer experiment (Izumi and Caughey, 1976), but it seems consistent with the horizontal divergence that must occur as convective downdrafts impact the surface. It is also consistent with the enhanced lateral dispersion we observed for surface releases. Skewness of vertical velocity was normally positive, consistent with the frequently-observed larger area occupied by downdrafts than updrafts in bottom-heated convective turbulence. However, two periods, 38 and 39, had a reversed skewness; this coincided with anomalous plume behaviors. The average vertical velocity, $\bar{w}$, departed significantly from zero in nearly half the periods, with $|\bar{w}| > 0.2m/s$ and $|\bar{w}|/w^* > 0.15$. These were nearly equally divided between negative and positive values, but a single 60 min averaging period, No. 33 with $\bar{w}$ = -0.7m/s (!), more than doublexd the $|\bar{w}|$ of any other periodi, biased the overall $\bar{w}$ slightly negative.

A number of cross checks were made between oil, chaff, and SF_6 distributions; some comparisons were also made with wind direction and elevation distributions, separated into 5° bins. Oil fog χ/Q, extrapolated to the surface from about 40m, and SF_6 χ/Q versus azimuth were compared for the five periods with co-releases, Nos. 35-39. Four of these periods showed good to excellent agreement, with most oil and SF_6 values differing by less than 10% of the peak χ/Q at all azimuths except in the northern portion of the arc; there, the nearest lidar scan missed the arc by up to 30% of x at the arc (agreement was excellent in the 250° to 305° azimuth sector, where one lidar scan lined up with the arc to within 4% of x). Period 38 compared poorly, with the oil fog peak 35% less and 8° to the north of the SF_6 peak, due west of the tower; however, the oil fog $\int \chi dy$ versus x itself strongly peaked at this distance, indicating that strong longitudinal gradients existed in this case. Overall, it appears that oil fog χ normalized by $U \int \int \chi dydz$ provided a very satisfactory estimate of passive trace χ/Q.

The coarser lateral resolution ($\sim$ 160m) and 250m longitudinal averaging of the chaff made it unsuitable for this type of comparison, but it was compared with oil and SF_6 in the form of plots of $\int \chi dy/Q$ versus x. Chaff always impacted the ground from elevated sources sooner than co-released oil fog, which was expected in light of its settling speed, w_s. Several methods were tried to correct for the effect of w_s; the best one was called the "profile projection correction" (PPC) method. A profile of chaff $\int \chi dy$ vs. z just before surface impact, at x_i, was raised by w_s/U times x_i and linearly projected to all $x > x_i$ assuming reflections at z= 0 and z_i; the projected original (unraised) profile was then substracted from this to generate a whole field of corrections in x and z to add to the measured field. Unreaslistic "shocks" appeared at intermediate distances due to the simplistic reflection assumptions, but in 3 of 4 cases the initial surface impact of PPC chaff was in excellent agreement with that of oil fog (period 38 again compared rather poorly). The corrected chaff surface $\int \chi dy/Q$ averaged within 30% of the oil fog value at the largest oil fog x, and always approached the uniform-mixing asymptote at large x ($C_y \equiv U z_i \int \chi dy/Q = 1$). It was concluded that the PPC-altered chaff provides a very good estimate of passive tracer χ fields up to the point of maximum surface impact and a satisfactory estimate of surface concentrations at larger distances, improving to excellent as $C_y = 1$ is approached.

Comparisons were also made of oil and chaff $\int \chi dy$ from elevated sources and wind elevation distributions, in terms of mass or frequency per radian, at small x, before surface impact occured. Two cases, periods 38 and 39, showed almost exact agreement between oil, chaff, and tower wind distributions versus elevation angle. The agreement was probably helped by relatively large chaff x ($\sim$400m) in these cases, improving the angular vertical resolution, while dimensionless $X < 0.2$

(i.e., travel time was small compared to the Lagrangian time scale). Agreements were also good in the other four cases, except for chaff in Nos. 32 and 37 (but x was only $\approx$130m, while 250m longitudinal averaging had been applied). Periods 34-36 showed secondary peaks in chaff just below z_i suggesting enhancement by reflection.

Comparisons of oil and chaff $\int \chi dz$ versus azimuth and wind azimuth distributions showed tracer distributions narrower than wind distributions, especially at larger X, as expected from statistical theory. Distributions were not often close to Gaussian; skews and secondary peaks were rather common. However, oil fog and chaff distributions were in good agreement in all cases at $x > 1500m$ ($X > 0.6$), and in particularly excellent agreement for the two cases with $X > 1.9$, in spite of different release heights (Nos. 32 and 34).

3. Results for σ_y and σ_z

The diffusion coefficients σ_y and σ_z were determined in a straightforward manner, as the standard deviations of lateral or vertical distributions of material about the mean in the y and z directions. The results for σ_z are simpler because there was little evidence of averaging time or release height effects. All the data points approximate.

$$\sigma_z/z_i = 0.6X/(1+5X^2)^{1/2} \tag{1}$$

where $X = (x/U)(w^*/z_i)$. The scatter is especially small at $X > 1$, where the data approximate the uniform mixing asymptote $\sigma_z/z_i = 12^{-1/2}$. The scatter is typically $\pm$40% at small X, where the Eq. 1 asymptote is $\sigma_z = 0.6w^*(x/U)$. Elevated-source σ_z fit this very well, consistent with the small-z statistical theory prediction $\sigma_z = \sigma_w t, t = x/U$, and $\sigma_w = 0.6w^*$.

Surface-source σ_z appear to be slightly less, but may be underestimated at small x because the sensors cannot see close to the ground. It can also be inferred from near-surface $\int \chi dy$ and the Gaussian plume assumption, which the oil fog measurements and WD's tank results show to be approximately valid at $X < 0.5$, before plume lift-off occurs. Here, our oil fog measurements show $C_y \approx 0.9X^{-3/2}$, in agreement with Nieuwstadt's (1980) analysis of the Prairie Grass and WD (1976) data. From this, it can be inferred that $\sigma_z \approx 0.9z_iX^{3/2} = 0.9H^{*1/2}(x/U)^{3/2}$ at $X < 0.3$ for surface releases, in agreement with Yaglom's (1972) dimensional analysis prediction. At $X > 0.3$, Eq. 1 is valid for all release heights.

In contrast to σ_z, CONDORS $\sigma_y/(w^*t)$ showed a definite trend to smaller values when the Eulerian dimensionless averaging time, $\tau = t_aU/z_i$, was < 6 (t_a represents averaging time). This is because not enough convective eddies, with diameters $\sim z_i$, pass by the source to give a good ensemble average for horizontal dispersion. (Thus, if $U = 1.5$ m/s and z_i=1800m, $t_a \geq$2 hr. is required for σ_y not to be affected by t_a!) For our $\tau > 6$ cases, most points fit within $\pm$ 30% of the small-x statistical theory prediction with $\sigma_v = 0.6w^* : \sigma_y = 0.6w^*(x/U) = 0.6z_iX$. A slight downard turn was suggested at $X > 1$, but we had few points beyond $X = 2$. To provide more definite recommendations, I developed an extended plot to $X = 18$ by supplementing CONDORS data with power plant plume σ_y, using the MPPSP data cited in Briggs (1985) and the EPRI data from Bull Run and Kincaid (S.R. Hanna of Sigma Research, Inc. 1990, personal communication). The power plant data had to be screened to remove cases where σ_y was likely to be enhanced by buoyancy-induced growth: $F^* = F/(Uw^{*2}z_i) < 0.1$, where F is the standard buoyancy flux parameter used in plume rise models (Briggs, 1985). Also I required $\tau > 6$ and $U/w^* < 6$ (convective-dominated turbulence). The resulting plot, Fig. 1, shows a good merger of points from the various field measurements, but with about 1.5 times as large scatter in the power plant measurements (still, far less scatter than in Hanna's (1986) analysis of the EPRI data with no screening). $X > 1$ points fit,as well as anything, $\sigma_y = 0.6z_iX^{2/3}$. Figure 1 shows a rather abrupt transition between this and the small-x asymptote. A good fit overall to the elevated-release cases is

$$\sigma_y/(w^*x/U) = 0.6(1 + X^2)^{-1/6} \tag{2}$$

To extend the plot in Fig. 1 to small X for surface releases (solid symbols), I added the convective cases ($U/w* < 6$) from the classic Prairie Grass experiment (U was estimated from $\bar{u}$ at 16m by comparing similar measurements in the CONDORS and Minnesota 1973 data sets; this yielded a good fit to $U = 1.4\bar{u}^{0.9}$, m/s units). Again, there is a good merger with CONDORS data points in the region of overlap. (Fig. 1 really demonstrates how well convective scaling works over a wide range of data, with $x = 50$ to 50,000m and σ_y=8 to 9500m). Obviously, at small X the surface-release σ_y are considerably larger than the elevated-release σ_y. This can be attributed to two factors: 1)CONDORS tower measurements showed σ_v about 30% larger in the lowest several percent of the CBL, approximating $0.6w^*$ only above $0.1z_i$; 2) travel time to a given x is longer for a plume confined near the surface, since its $\bar{u} < U$. As the surface-release plume grows in thickness, its σ_y becomes nearly the same as for elevated releases at $X = 0.4$, near the "lift-off" distance. There are only a dozen CONDORS points and a single Prairie Grass point beyond this X, but they suggest some undershooting of elevated-source σ_y. This, we expect, would disappear at very large X, after multiple reflections from $z = 0$ and z_i.

To account for all of these surface-release effects theoretically, we would have to develope σ_v/w^* and $\bar{u}/U$ as functions of σ_z/z_i, roughness length, and Obukhov length, resulting, no doubt, in a very complicated expression. Considering the good agreement of the above two data sets, when convectively scaled, it seems adequate for now to suggest a simple empirical correction to Eq. 2, plotted in Fig. 1 as dashed line:

$$\sigma_y/(w^*x/U) = 0.6(1 + X^2)^{-1/6} + (0.4 - X)/(1 + 6X^2 + 200Z_s^2). \tag{3}$$

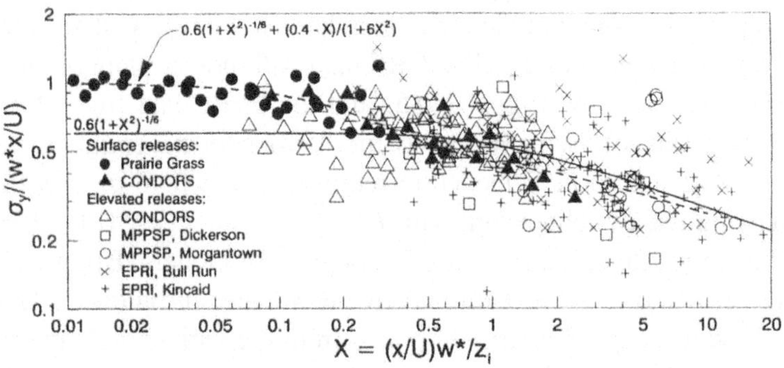

Fig. 1. CONDORS, Prairie Grass, and power plant plume $\sigma_y/(w^*x/U)$ versus X. All data points with $\tau > 6$, $U/w^* < 6$, and $F^* < 0.1$ (to minimize buoyancy effects). Lines are Eqs. 2 and 3 ($Z_s = 0$).

I included the Z_s term in an attempt to universalize Eq.3, althought it is based on scant information, With it, the final term of Eq.3 has less than 10% effect on σ_y when $Z_s > 0.18$, which practically includes all CONDORS elevated releases (our smallest Z_s was 0.17). For Z_s=0.067, it gives 35% and 16% σ_y enhancement at $X = 0$ and 0.2, respectively; this is in rough agreement with the relative values of σ_y, compared with more elevated sources, in WD's tank experiments, the only data of which I am aware with a Z_s in the intermediate range between "elevated" and "surface" release heights. (Tank and LES σ_y in convective scaling are generally 20 to 30% less than field values, probably because of the lack of mesoscale eddies and lack of inhomogeneous surface heatingr; thus, I choose to use only their relative values).

All of the above results are for $\tau = t_a U/z_i > 6$. The CONDORS averaging periods ranged down to τ = 2.8, and the trend in averaging time effects of σ_y was fit by

$$\sigma_y \propto [1 - exp(-0.3\tau)]. \qquad (4)$$

This gives a plume spread $\propto w^* t_a/z_i$ at very small t_a, consistent with a wind direction change $\propto w^*/U$ in an eddy passage time $\propto z_i/U$. However, this accounts for only the meandering conponent of diffusion, whereas relative diffusion still gives $\sigma_y > 0$ at $t_a = 0$. A convenient way to modify Eq. 4 to account for relative diffusion, consistent with theoretical expectation that $\sigma_y \propto H^{*1/2}(x/u)^{3/2}$ at t_a = 0 and small X, would be to set $\sigma_y \propto [1 - exp(-(0.09\tau^2 + b^2X)^{1/2})]$. For elevated sources, this, with Eq. 2, would give $\sigma_y = 0.6bH^{*1/2}(x/U)^{3/2}$ at $t_a = 0$ and small X. It gives vectorial addition of relative and meandering diffusion, it gives credit for increased eddy sampling as a short-release plume diffuses in the longitudinal direction, and it gives relative diffusion approaching total diffusion at very large X (within 10% at $X = 5.3/b^2$). Based on the recently published LES results of Nieuwstadt(1992), I estimate that b is of the order of 0.5, but the relative diffusion contribution to σ_y appears to be small for CONDORS averaging times.

4. Results for $\int \chi dy$ in X and Z

The primary goal of CONDORS was to see if the radically non-Gaussian vertical diffusion behaviors observed in the tank and LES simulations proved true in the field. This is why we obtained three-dimensional measurements. The comparison is best made by plotting, in convective scaling, crosswind integrated concentrations, $\int \chi dy$, versus x and z, as was done by Willis and Deardorff (WD), Lamb and many others. Thus, the appropriate variables are $C_y = z_i U \int \chi dy/Q$, which approaches unity in the uniform mixed regime, $X = (w/U)w^*/z_i$, and $Z = z/z_i$. To make the comparisons much easier, I interpolated previously published results to standard contour levels (C_y=0.1, 0.2, 0.4, 0.6, 0.8, 1, 1.2, 1.4, 1.7, 2, 2.5, 3, 4, 5, 7, and 10). I also squeezed or expanded their scales so that the X-units would approximately equal the Z-units. The full complement of 47 such figures are presented in Briggs (1992), but examples for surface and elevated oil fog releases is given here in Fig. 2.

The comparison of the eight surface-release CONDORS C_y mappings, which were predominantly for oil fog, with WD's tank results for Z_s =0.067 and LES model results of Lamb and of Nieuwstadt and de Valk (1987) was unequivocal. While any of the simulations did far better than a Gaussian plume model, the field results came closest to matching the laboratory tank results. The 8-period oil fog average, which extended only to $X = 0.53$ when dropout occurred, looked nearly identical to WD's C_y versus X and Z plot, except for some wiggles in the high-up contours and for slightly larger C_y at the surface near the source, attributable to the lower CONDORS release height (Z_s ~0.01, when 5 to 10m of plume rise is included). Surface C_y came very close to matching Nieuwstadt's (1980) $C_y = 0.9X^{-3/2}$ fit to Prairie Grass measurements, and the mode of maximum C_y just began to lift off the surface at $X = 0.5$.

An examination of the minimum in C_y at the surface, C_{ymin}, for all eight periods is presented in Table II. Chaff, with no correction for settling speed, is used for periods 24 and 25 because the 1982 oil fog measurements extended only to $X \approx 0.6$, compared to $X \approx 3$ for chaff (the additional oil fog generated and heavier oil used in 1983 made a big improvement in lidar acquisition distance). It can be seen that the chaff plumes strongly lofted, with $C_{ymin} < 0.3$. Medians are used advisedly, because this is a small data set with large scatter, and medians are far more robust than averages. In determining the median for each X, I have assumed that the period 31 value is lower than the median, considering the very strong lofting observed for this case at its maximum X. Looking at each period as a separate occurrence, we find a median $C_{ymin} \approx 0.4$ and a median X_{min} = 1.45.

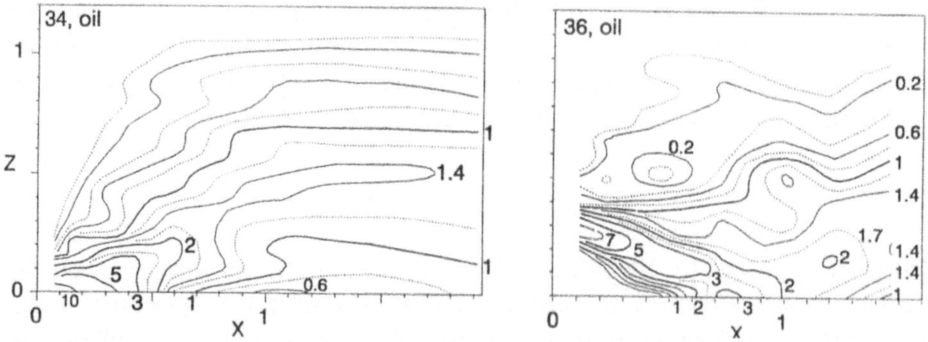

Fig. 2. Examples of $C_y = z_i U \int \chi dy/Q$ versus $X = (x/U)w^*/z_i$ and $Z = z/z_i$ plots from CONDORS, to compare with similar plots published for convective diffusion simulations.

TABLE II
Surface source minimum C_y and X of occurrence.

Pd#	C_{ymin}	X_{min}	Average C_y in ranges of X: 1.0 - 1.2	1.2 - 1.4	1.4 - 1.7	1.7 - 2.0
24[a]	0.29	1.36	0.41	0.29	0.70	0.78
25[a]	0.25	1.61	0.61	0.40	0.25	0.30
31	$\leq 0.11^b$	$\geq 0.85^b$	-	-	-	-
32	$\leq 1.05^b$	$\geq 2.28^b$	1.70	1.74	1.64	1.49
33	$\leq 1.35^b$	$\geq 1.54^b$	1.79	1.49	1.35	
34	0.55	1.08	0.56	0.64	0.73	0.84
40	0.35	1.04	0.35	0.36	0.52	0.74
41	0.42	2.33	0.67	0.64	0.76	0.64
Median	0.385	1.45	0.58	0.52	0.72	0.76

a=Chaff; b= Minimum C_y is at maximum X of measurement

However, if we examine surface C_y in several ranges of X, taking the median in each range to effectively create a much better ensemble average ($\tau \approx 30$ instead of the individual period $\tau = 2.8$ to 6.6), we find $C_{ymin} \approx 0.5$ at $X_{min} \approx 1.3$. These are the appropriate numbers for comparison with WD's results, which had effectively $\tau = 28$; extrapolated to the surface, they showed $C_{ymin} \approx 0.5$ at $X_{min} \approx 1.55$. Thus, the agreement is excellent except for about 20% larger X_{min} for the tank measurements.

As suggested by Table II, the field results for surface-release C_y showed great variability. Periods 32, 33, and 41 all had rather strong negative $\bar{w}$ near the release point, -0.3 to -0.7 m/s at the upper tower levels. This coincided with very slow liftoff of the mode in these cases (nonetheless, the period 41 plume lofted very strongly in the $X = 1.7$ to 4 range). A few other periods, especially 24, 25, and 31, showed very rapid lofting, beginning around $X = 0.2$. In spite of such variability,

TABLE III
Elevated source C_{ymax} vs. predictions.

	Z_s	X_{max}/Z_s	$C_{ymax}Z_s$	Ratio of $C_{ymax}Z_s$ to: Gaussian	$0.484(1+2Z_s)$
Simulations					
Lamb	0.75	2.07	1.39	1.92	1.15
Lamb	0.50	2.30	1.00	2.00	1.03
N&dV	0.49	1.61	0.83	1.68	0.87
W&D	0.49	1.63	0.88	1.78	0.92
Lamb	0.25	2.80	0.58	1.19	0.79
W&D	0.24	2.08	0.72	1.49	1.00
			Avg.	1.67	0.96
Chaff(PPC)					
Pd 21	0.45	2.01	0.86	1.77	0.94
38	0.41	3.98	0.45	0.93	0.51
39	0.34	2.60	0.38	0.78	0.46
22	0.32	2.05	0.87	1.80	1.10
35-36	0.32	2.08	0.85	1.75	1.07
37	0.32	1.55	1.31	2.71	1.66
34	0.25	3.12	0.71	1.46	0.97
32	0.19	3.58	1.04	2.16	1.56
23	0.17	2.59	0.38	0.79	0.59
			Avg.	1.57	0.98
Oil					
Pd21	0.45	1.87	1.19	2.44	1.29
38	0.41	1.40	0.44	0.90	0.50
39	0.34	1.46	0.89	1.84	1.09
35	0.32	3.61	0.90	1.86	1.14
36	0.32	2.33	1.06	2.19	1.34
37	0.32	1.73	1.07	2.21	1.35
23	0.17	3.63	0.66	1.37	1.02
			Avg.	1.83	1.10
SF6					
Pd38	0.41	(1.45)	0.61	1.26	0.69
39	0.34	(1.49)	0.64	1.32	0.78
35	0.32	(2.88)	0.82	1.70	1.04
36	0.32	(2.88)	0.77	1.60	0.98
37	0.32	(2.22)	1.14	2.36	1.44
			Avg.	1.64	0.98
$Chaff + Oil$,	Z_s=	0.32-0.45	Avg.	1.76	1.04
$Chaff + Oil$,	Z_s=	0.17-0.25	Avg.	1.45	1.03

and contrary to the conventional Gaussian plume model, with its $Z = 0$ mode, all eight plumes lofted. The median maximum height of lofting was $Z \approx 0.65$, just under the $Z \approx 0.77$ observed in the tank.

The elevated-release plumes showed even more variability. Much of the variability correlated with $\bar{w}$, which ranged from about -0.3 to + 0.3 m/s near the height of releases, and with the mode of vertical velocity, $\hat{w}$, which ranged from -1.0 to +0.6 m/s. Yet, the median $\hat{w}/U$, which determines the initial slope of the plume centerline, was -0.3; in convective scaling coordinates, the slope is $\hat{w}/w^*$, which had a median of -0.32. Fully consistent with this, when I took a time average of the five periods at the most well -represented height (Z_s=0.32 to 0.34), the oil fog C_y showed a downslope on the $X - Z$ plot very close to -0.32. This is qualitatively similar to the results of simulations, although WD's (1978, 1987) and Lamb's (1978, 1979) plots show slopes closer to -0.5.

The primary interest in elevated plume behavior, from a regulatory and design point of view, is the magnitude and location of the maximum ground concentrations. Table III is provided here to summarize the CONDORS results for each case in which the tracer mappings appear adequate for approximation of C_{ymax} and X_{max}. Some extrapolation to the surface has been made, from $\bar{z} \approx 70$ m for the chaff and 50m for oil fog. Only PPC -corrected chaff is given; the uncorrected chaff C_{ymax} averaged 20% larger. The inclusion of SF_6 values in this table may seem curious, as it was measured at only one arc distance, about 1200 m. By unexpected good luck, it happened that this distance was near the X_{max} for the oil fog for these five periods, with no more than a factor of 1.27 difference. The SF_6 C_y averaged only 10% less than the oil fog C_{ymax}, so it seems worthwhile including the gas sample results. Table III also includes results from laboratory tank and LES numerical model simulation.

As suggested in Briggs (1985), on the basis of WD's and Lamb's simulation results and a few preliminary (1982) CONDORS data points, about 2/3 of the oil and (PPC) chaff maximum impacts were near $x = 2z_sU/w^*$, ranging $\pm30\%$. This is equivalent to a $-0.5w^*$ plume centerline effective descent rate. The four points with the lowest Z_s, 0.17, 0.19, and 0.25, were the primary exceptions, with $x \approx 3.2\, z_sU/w^*$, ranging $\pm20\%$. However, I would not credit this latter result with much statistical significance.

Table III compares the observed C_{ymax} with the Gaussian model prediction and the empirical fit suggested in Briggs (1985):

$$C_{ymax}Z_s = 0.484(1 + 2Z_s) \tag{5}$$

Ignoring reflection from z_i, all Gaussian models give a maximum $\int \chi dy$ when $\sigma_z = z_s$, resulting in $Uz_s \int \chi dy/Q = C_yZ_s = [2/(\pi e)]^{1/2}$=0.484. Including reflection from z_i results in a transcendental equation whose solution can be closely approximated, within 0.6%, by $C_{ymax}Z_s = 0.484[1 + (2Z_s)^{20}]^{0.05}$. Thus, it makes an extremely sharp transition from 0.484 at $Z_s < 0.5$ to 0.484 $(2Z_s)$ at $Z_s > 0.5$. This approximation has been used in the table. A scan of the $C_{ymax} \div$ Gaussian column reveals that the Gaussian prediction substantially underpredicts the great bulk of CONDORS observations. The averages for the simulations, chaff oil, and SF_o all agree that C_{ymax} is 60 to 80% larger than the Gaussian prediction, overall. However, the simulations show a trend towards increasing underprediction with increasing Z_s. The field observations encompass a smaller range of Z_s and have much larger scatter, which tends to obscure any trend; however, one is discernable if chaff and oil observations are combined and then divided into two ranges of Z_s, as is done at the bottom of Table III.

When C_{ymax} is divided by the prediction of Eq.5, the trend with Z_s disappears. The simulations, chaff, and SF_6 all give average ratios within 1% of 0.97, while the oil fog gives 1.10. The oil fog ratios show less scatter, but with a small data set like this, inclusion or exclusion of just done datum can substantially change some statistics. I would prefer to look at the oil, chaff, and SF_6 results as a whole, to say that they substantially agree with all of the elevated-source simulations on C_{ymax}; they show no more than a few percent bias when compared to Eq.5, but show a large scatter about its predictions, with a standard deviation of about ±35% and a range somewhat exceeding ±50%.

This paper has been reviewed in accordance with the U.S. Environment Protection Agency's peer and administrative review policies and approved for presentation and publication. Mention of trade names or commercial products does not constitute endorsement or recommendation for use.

References

Barad, M.L. (ed), 1958. *Project Prairie Grass; a field program in diffusion.* Geophys. Res. Paper No. 59, Vols. I and II, AFCRF-TR-58-235, Air Force Cambridge Research Center, Bedford, MA.

Briggs, G.A., 1985. *Analytical parameterizations of diffusion: The convective boundary layer.* J. Climate Appl. Meteor. **24**, 1167-1186.

Briggs, G.A., 1988. *Analysis of diffusion experiments:* Chapter 2 in Lectures on Air Pollution Modeling, A. Venkatram and J.C. Wyngaard, Eds. Amer. Meteor. Soc., Boston, 63-117.

Briggs, G.A., 1989. *Field measurements of vertical diffusion in convective conditions.* Preprints Sixth Joint Conference on Applications of Air Pollution Meteorology, Anaheim, CA, American Meteorological Society, 167-170.

Briggs, G.A., 1992. *Plume dispersion in the convective boundary layer. Part II: analyses of the CONDORS field experiment data.* Submitted to J. Appl. Meteor.

Briggs, G.A., W.L. Eberhard, J.E. Gaynor, W.R. Moninger and T. Uttal, 1986. *Convective diffusion measurements compared with laboratory and numerical experiments.* Preprints Fifth Joint Conf. on Applications of Air Pollution Meteorology, Chapel Hill, NC, Amer. Meteor. Soc., 340-343.

Briggs, G.A., and K.R. McDonald, 1978. *Prairie Grass revisited: optimum indicators of vertical spread.* Proc. Ninth Int. Tech. Meeting on Air Pollution Modeling and its Application, No. 103, NATO, 209-220. [Available as contribution 78/8, Atmospheric Turbulence and Diffusion Division, Oak Ridge, TN 37831].

Deardorff, J.W., 1970. *Convective velocity and temperature scales for the unstable boundary layer for Rayleigh convection.* J. Atmos. Sci., **27**, 1211-1213.

Deardorff, J.M., and G.E. Willis, 1982. *Ground-level concentrations due to fumigation into an entraining mixed layer.* Atmos. Environ., **16**. 1159-1170.

Deardoff, J.M., and G.E. Willis, 1984. *Ground-level concentration fluctuations from a buoyant and a non-buoyant source within a laboratory convectively mixed layer*. Atmos. Environ. **18**, 1297-1309.

Eberhard, W.L., W.R. Moninger, T. Uttal, S.W. Troxel, J.E. Gaynor and G.A. Briggs, 1985. *Field measurements in three dimensions of plume dispersion in the highly convective boundary layer*. Preprints Seventh Symp. on Turbulence and Diffusion, Boulder, Amer. Meteor. Soc., 115-118.

Eberhard, W.L., W.R. Moninger, and G.A. Briggs, 1988. *Plume dispersion in the convective boundary layer. Part I: CONDORS field experiment and example measurements*. J. Appl. Meteor. **27**, 599-616.

Hanna, S.R., 1986. *Lateral dispersion from tall stacks*. J. Appl. Meteor. **25**, 1426-1433.

Izumi, Y., and J.S. Caughey, 1976. *Minnesota 1973 atmospheric boundary layer experiment data report*. Air Force Cambridge Research Laboratory Environmental Research Papers, No. 547, AFCRL-TR-0038, 28 pp.

Kaimal, J.C., W.L. Eberhard, W.M. Moninger, J.E. Gaynor, S.W. Troxel, T. Uttal, G.A. Briggs, and G.E. Start, 1986. *Project CONDORS: Convective diffusion observed by remote sensors*. Boulder Atmos. Observatory Rep.#7, NOAA Wave Propagation Lab., Boulder, Colorado, [NTIS #PB86-222 221/AS], 305 pp.

Lamb, R.G., 1978. *A numerical simulation of dispersion from an elevated point source in the convective planetary boundary layer*. Atmos. Environ. **12**, 1297-1304.

Lamb, R.G., 1979. *The effects of release height on material dispersion in the convective planetary boundary layer*. Preprints Fourth Symposium on Turbulence, Diffusion, and Air Pollution. Reno, Amer. Meteor. Soc., 27-33.

Moninger, W.R., W.L. Eberhard, G.A. Briggs, R.A. Kropfli, and J.C. Kaimal, 1983. *Simultaneous radar and lidar observations of plumes from continuous point sources*. 21st Radar Meteorology Conference. Amer. Meteor. Soc., Boston, MA, 246-250.

Nieuwstadt, F.T.M., 1980. *Application of mixed-layer similarity to the observed dispersion from a ground-level source*. J. Appl. Meteor. **19**, 157-162.

Nieuwstadt, F.T.M., 1992 *A large-eddy simulation of a line source in a convective atmospheric boundary layer. -I. Dispersion characteristics* Atmos. Environ. **26A**, 485-495.

Nieuwstadt, F.T.M., and J.P.J.M.M. De Valk, 1987. *A large eddy simulation of buoyant and non-buoyant plume dispersion in the atmospheric boundary layer*. Atmos. Environ. **21**, 2573-2587.

Willis, G.E., and J.W. Deardorff, 1976. *A laboratory model of diffusion into the convective planetary boundary layer*. Quart. J. Roy. Meteor. Soc. **102**, 427-445.

Willis, G.E., and J.W. Deardorff, 1978. *A laboratory study of dispersion from an elevated source within a modeled convective planetary boundary layer*. Atmos. Environ. **12**, 1305-1311.

Willis, G.E., and J.W. Deardorff, 1981. *A laboratory study of dispersion from a source in the middle of the convectively mixed layer*. Atmos. Environ. **15**, 109-117.

Willis, G.E., and J.W. Deardorff, 1983. *On plume rise within a convective boundary layer* Atmos. Environ. **17**, 2435-2447.

Willis, G.E., and J.W. Deardorff, 1987. *Buoyant plume dispersion and inversion entrapment in and above a laboratory mixed layer*. Atmos. Environ. **21**, 1725-1735.

Willis, G.E., and N. Hukari, 1984. *Laboratory modeling of buoyant stack emissions in the convective boundary layer*. Preprints, Fourth Joint Conference on Applications of Air Pollution Meteorology, Portland, OR, American Meteorology Society, 24-26.

Yaglom A.M., 1972. *Turbulent diffusion in the surface layer of the atmosphere*. Izv. Atm. Oceanic. Phys. 8, 333-340.

A ONE DIMENSIONAL WIND PROFILE MODEL WITH STABILITY AND EDDY VISCOSITY ESTIMATIONS FROM SODAR DATA

HANS DAHLQUIST
Military Weather Service, Air force H.Q., S-107 84, Stockholm, Sweden.

(Received October 1991)

Abstract. In a one-dimensional wind profile model, methods for eddy viscosity and stability estimations from sodar data have been evaluated with soundings. For eddy viscosity parameterization the ageostrophic method and mixing-length theory have been investigated. Three methods for estimating the static stability have been evaluated; a wind profile adjustment method, gravity wave analysis of sodar backscatter and flux profile functions for windspeed and standard deviation of vertical windspeed. The wind profile model with variable momentum flux (VF) with height shows better results than an earlier constant flux model (Bergström, 1986). The VF model can be used for extending the sodar profile up to 1500 m.

1. Introduction

1.1. General Background

In Scandinavia almost 20 Doppler acoustic sounders (sodars) have been installed for meteorological research, environmental applications and wind energy mapping. At some airports the instrument is also used for windshear warnings and fog prediction (Dahlquist, 1986). In all these applications a knowledge of the static stability (N) and the eddy viscosity (K_M) of the Planetary Boundary Layer (PBL) is of key importance. For most operational users of sodar, only the directly measured wind profile and the time evolution of the backscatter are used. Stability and K_M are normally derived by other methods, such as regular soundings, tethered balloons or meteorological towers. Those methods have limitations in time and space resolution. It is therefore important to develop methods by which the instrument also can serve as a sensor to provide these additional parameters. Here methods for stability and K_M profile estimations from sodar data will be tested in a one-dimensional wind profile model.

For the commercial sodars available today, the average maximum range is normally around 300-600 m in an urban and noisy environment. In several applications wind profiles up to 1500-2000 m are desirable.

1.2. Objective

The main objective of this paper is to see to what extent information about atmospheric dynamic processes above the sodar range can be extracted from the sodar data in order to estimate a wind profile up to the top of the PBL.

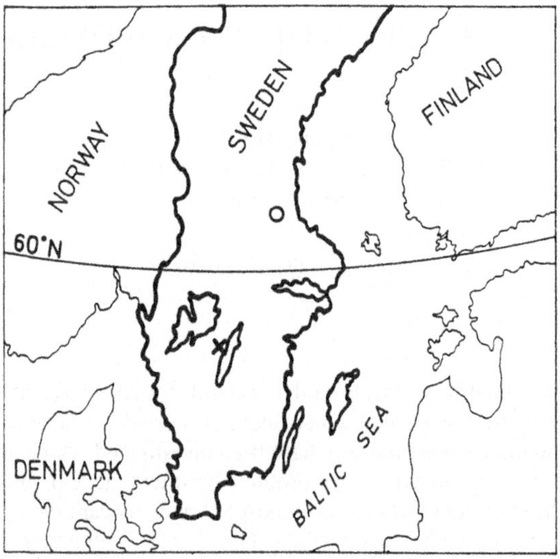

Fig. 1. Map of Scandinavia with Karlsborg at the cross and Jädraäs at the ring.

2. The One-Dimensional Wind Profile Model

2.1. Basic Model Equations

The horizontal components of the Navier-Stokes equation, after a Reynolds decomposition and omission of the horizontal derivatives of the shear stresses, can be written (Holton, 1979);

$$\frac{du}{dt} = f(v - v_g) - \frac{\partial}{\partial z}(\overline{u'w'}) \tag{1}$$

$$\frac{dv}{dt} = -f(u - u_g) - \frac{\partial}{\partial z}(\overline{v'w'})$$

where v_g and u_g are the geostrophic wind components and f is the Coriolis parameter. Assuming the eddy viscosity hypothesis we may write;

$$-(\overline{u'w'}) = K_M \frac{\partial u}{\partial z}$$

$$-(\overline{v'w'}) = K_M \frac{\partial v}{\partial z}. \tag{2}$$

By inserting (2) into (1) and assuming stationarity we obtain two differential equations for u and v;

$$K_M \frac{d^2 u}{dz^2} + \frac{du}{dz}\frac{dK_M}{dz} + f(v - v_g) = 0 \tag{3a}$$

$$K_M \frac{d^2 v}{dz^2} + \frac{dv}{dz}\frac{dK_M}{dz} - f(u - u_g) = 0. \tag{3b}$$

This set of equations can be solved analytically if K_M is constant with height (the Ekman spiral). However, with a constant eddy viscosity profile it is not possible to make a realistic simulation of rapid changes in wind with height. Therefore K_M is allowed to vary with height. This introduces new parameters to be estimated such as static stability, Ekman Layer thickness and boundary conditions for the wind. The geostrophic wind (G) is found by assuming a constant K_M but the detailed structure of the wind profile, between the lower measured and the upper estimated boundary, is derived with a variable momentum flux. For convenience we introduce:

$$W = u - u_g + i(v - v_g).$$

Multiplying (3b) with the imaginary unit, i, and adding (3a) to (3b) we have;

$$K_M \frac{d^2W}{dz^2} + \frac{dW}{dz}\frac{dK_M}{dz} - ifW = 0. \tag{4}$$

Since K_M is allowed to vary freely with height, the ordinary differential equation (4) can only be solved numerically. The numerical scheme used here is a two-point boundary value shooting method (Press et al., 1987). This is an iterative procedure where trial integrations that satisfy the lower boundary conditions are made. The discrepancies from the desired upper boundary conditions for the u and v components are used to adjust the lower boundary conditions, until boundary conditions at both endpoints are satisfied to a desired accuracy. In order to solve (4) we need boundary conditions at the lower boundary, ($z = Z_t$) and at the top ($z = H_S$), see Fig. 2:

(i) The measured sodar wind at Z_t

(ii) The geostrophic wind (G) at H_S.

The Ekman Layer thickness ($H_S - Z_S$) is proportional to the square root of the average eddy viscosity $\bar{K}_M$. Too thick a layer in relation to $\bar{K}_M$ will give rise to a shear, due to too large a flux of momentum, so the numerical solution will have difficulties in satisfying the boundary conditions at H_S. This means that the boundary value G, will be reached below the upper endpoint and the only possibility for a numerical solution is an unrealistic wiggling profile above that point with supergeostrophic winds, etc. As a first estimate of $H_S - Z_S$ we use:

$$\delta_E = [\frac{\bar{K}_M}{f/2}]^{1/2} \tag{5}$$

where δ_E is the Ekman layer thickness (Pedlosky, 1986), based on a K_M which must be determined from the sodar profile. If too high a K_M value has been used for the shooting, a new value of H_S is picked at the lowest point where the ageostrophic components are zero. How much the average K_M at the top of the sodar profile must be reduced, in order to obtain solutions at H_S or below for most of the profiles, has been tested numerically.

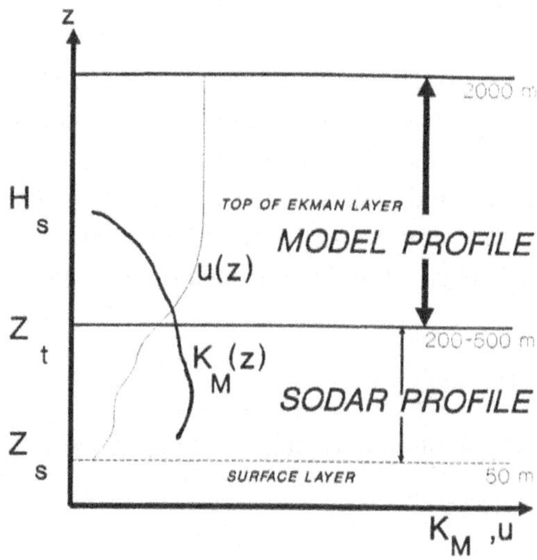

Fig. 2. Model structure and parameters.

In addition we need the eddy viscosity profile ($K_M(z)$) from the top of the sodar profile (Z_t) to H_S. The final K_M profile is found through an iterative process where the shear of the model wind profile and the static stability are used to calculate the K_M profile. The initial shear for this iterative procedure is calculated from a model profile with a constant K_M. From that shear a first non-constant K_M profile is calculated with the same N. The next iteration step will give another K_M profile from the new shear and so on until the difference between two successive model profiles is less than the desired accuracy.

3. Eddy Viscosity Parameterization

A rough estimate of K_M at the top of the sodar profile will be used both to estimate G and H_S. Furthermore a K_M profile above the sodar range must also be parameterized to obtain the final solution. Here two methods will be discussed, the ageostrophic method and mixing-length theory.

An example of two windprofiles with different given eddy viscosity profiles are shown in Figs. 3 and 4. Note the slow veering with height when the eddy viscosity is assumed constant (12 m^2/s, Fig. 3) and the rapid adjustment both in speed and direction when the eddy viscosity increases from 1 to 12 m^2/s in the lowest 200 m (Fig. 4).

If there is a strong drop in the eddy viscosity profile at some level, as there is in an inversion, a veering of the wind takes place here (Lapworth, 1987). No vertical flux of momentum through the inversion means that the forcing from above cannot be seen below. The profile in Fig. 4 is an example of this rapid adjustment to the geostrophic wind just above an inversion.

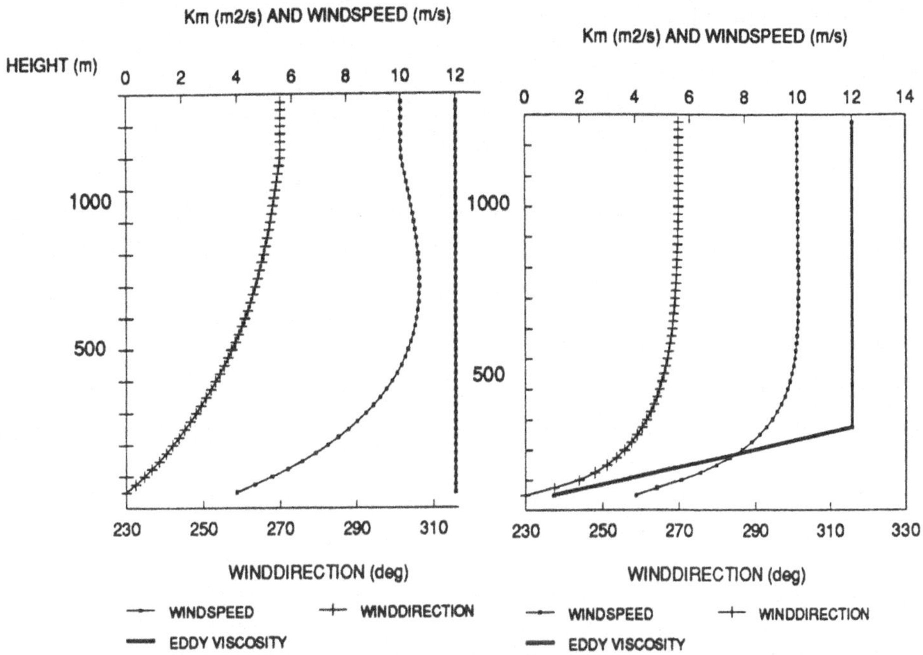

Fig. 3. (left) and 4 (right). Wind profiles derived from equation (4) with different eddy viscosity profiles (solid straight lines) under 200 m. The given geostrophic wind is 270/10 and the wind at the lower boundary is 230/4 (degrees and m/sec).

3.1. The Ageostrophic Method

One method that has been used for K_M profiles in the PBL is the ageostrophic method (Schmitz-Peiffer et al.,1987).The method can be applied to wind profiles in a geostrophically balanced PBL if the conditions are stationary, homogeneous and barotropic. Since these conditions almost never are satisfied, the results turn out to be rather unsatisfactory.

It is worth noting that the PBL is never in complete geostrophic balance since the timescales of turbulence given as $1/N$ or $1/S$, where N is the *Brunt Väisälä* frequency and S is the vertical shear, are much smaller than the geostrophic timescale $(1/f)$. The difference in timescale is between a factor 10 to 100 when the Richardson Number (Ri) is about unity. All attempts to find geostrophic parameters with turbulent closure techniques will therefore be difficult but not impossible.

To solve the equations for the stress, it is not necessary to derive the static stability explicitly. Equations (3a) and (3b) can be integrated from the top of the surface layer (Z_S) to a height Z and we have:

$$K_{MZ}\frac{\partial u}{\partial z}|_Z - K_{MZ_s}\frac{\partial u}{\partial z}|_{Z_s} = f\int_{Z_s}^{Z}(v_g - v)dz \tag{6}$$

$$K_{MZ}\frac{\partial v}{\partial z}|_Z - K_{MZ_s}\frac{\partial v}{\partial z}|_{Z_s} = f\int_{Z_s}^{Z}(u_g - u)dz. \tag{7}$$

Solving (6) and (7) for K_{MZ} gives:

$$K_{MZ} = \frac{f[\frac{\partial v}{\partial z}|_{Z_s} \int_{Z_s}^{Z} (v_g - v)dz + \frac{\partial u}{\partial z}|_{Z_s} \int_{Z_s}^{Z} (u_g - u)dz]}{\frac{\partial u}{\partial z}|_{Z} \frac{\partial v}{\partial z}|_{Z_s} - \frac{\partial v}{\partial z}|_{Z} \frac{\partial u}{\partial z}|_{Z_s}}. \tag{7}$$

The high sensitivity to input data is due to the mathematical structure of eq. (8). Both the numerator and the denominator involve small differences between large, almost equal numbers.

3.2. Mixing-Length Theory

Mixing-length theory is one way to obtain a K_m profile in and above the sodar measurement range assuming that a representative stability can be derived from the sodar data.

The eddy diffusivity increases with the length scale of the turbulence ($l_{k,n}$) and the magnitude of the vertical wind shear (S) (Stull, 1988);

$$K_M = l_k^2 S. \tag{8}$$

The mixing length can be related to the Richardson Number by (Estournel and Guedalia, 1987);

$$\frac{l_k}{l_n} = 1 - 5Ri \qquad 0 \leq Ri < 0.164 \tag{9}$$

and

$$\frac{l_k}{l_n} = (\frac{1}{1 + 41Ri})^{0.84} \qquad Ri \geq 0.164$$

where

$$Ri = \frac{N^2}{S^2} \quad and \quad S^2 = (\frac{\partial u}{\partial z})^2 + (\frac{\partial v}{\partial z})^2. \tag{10}$$

The neutral mixing length (l_n) is given by:

$$l_n = \frac{1}{1/kz + 1/\lambda} \tag{11}$$

where $\lambda = 0.0004|G|/f$ and k the von Kármán constant ($\cong 0.35$)

4. Estimation of Geostrophic Wind

The upper endpoint boundary value is the geostrophic wind. This wind can be given by an analyzed surface or a 850 mb chart, assuming that there is no thermal wind. If G is not available, an estimation must be made from the sodar profile data. The method used here to find the geostrophic wind is to adjust an approximate solution to (4) with constant $\bar{K}_M$ to the measured sodar profile and then solve for the geostrophic wind conponents u_g and v_g.

A solution to (4) assuming a constant $\bar{K}_M$ with height is given by

$$\begin{aligned} u_t &= u_g + e^{-\alpha}[(u_s - u_g)\cos\alpha + (v_s - v_g)\sin\alpha] \\ v_t &= v_g + e^{-\alpha}[-(u_s - u_g)\sin\alpha + (v_s - v_g)\cos\alpha] \end{aligned} \tag{12}$$

where $\alpha = (Z_t - Z_s)(f/2\bar{K}_m)^{1/2}$ and Z_s, v_s and u_s denote values from the bottom (50 m) of the measured wind profile. Correspondingly Z_t, v_t and u_t are the height and speed components at the top of the sodar profile, usually between 200 - 500 m.

Solving for u_g and v_g gives;

$$u_g = \frac{(v_t - v_s)e^{-\alpha}\sin\alpha - u_t(e^{-\alpha}\cos\alpha - 1) + u_s(e^{-2\alpha} - e^{\alpha}\cos\alpha)}{1 - 2e^{-\alpha}\cos\alpha + e^{-2\alpha}} \tag{13}$$

$$v_g = \frac{(u_s - u_t)e^{-\alpha}\sin\alpha - v_t(e^{-\alpha}\cos\alpha - 1) + v_s(e^{-2\alpha} - e^{\alpha}\cos\alpha)}{1 - 2e^{-\alpha}\cos\alpha + e^{-2\alpha}} \tag{14}$$

The $\bar{K}_M$ is here the estimated average eddy viscosity at the top of the sodar profile, derived with mixing-length theory (section 3.2). Through the numerical trial and error procedure described earlier, $\bar{K}_M$ can be reduced by up to 75% for very stably stratified situations. The eddy viscosity decreases with height for stable stratification and is about constant when the stratification is neutral.

5. Methods for Stability Estimations

For the mixing-length method described in section 3.2 a Brunt-Väisälä frequency N must be estimated from the top of the sodar data that is representative for the rest of Ekman layer above, H_S - Z_t in Fig. 2. Three methods for estimating N will be discussed.

5.1. WIND PROFILE METHOD

One approach to the stability problem is to minimize the RMS error between a measured sodar profile and Ekman profiles with different K_M given by varying the stability (Thaning, 1985).

The inverted Monin-Obukhov Length ($1/L$), is varied in steps from stable (0.1) to unstable (-0.1) for the Ekman profiles. The stability is then given by the value of $1/L$ for which the RMS error of the model u and v components with respect to measurements has a minimum. See Fig. 5.

The problem associated with this method is that surface-layer theory is applied to measurements that are mostly in the Ekman layer. The lowest sodar wind is at 50 m and quite often that is above the surface-layer. Another problem is that the method can not always find a minimum in RMS error during unstable or convective

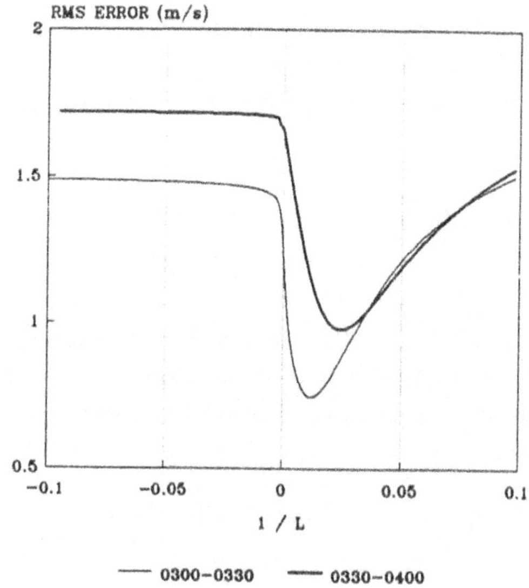

Fig. 5. RMS error in u and v components between model and measurements versus $1/L$ for two sodar profiles with 30 min average.

situations. This is clearly shown in Fig. 6. The PBL is slightly stable before 0830. After 0830 no minimum is found and $1/L$ goes to either + 0.1 or - 0.1. Even the other surface-layer parameter data, Z_s and friction velocity (u_*), are of no use after 0830.

In this wind profile model, this method will not be used because it does not usually converge when the stratification is unstable.

5.2. Flux Profile Method

By combining the non-dimensional flux profile functions f_u and f_w for horizontal windspeed (U) and standard deviation of vertical windspeed (σ_w) respectively (Businger et al., 1971; Dyer, 1974; and Pasquill and Smith, 1983) one obtains;

$$\sigma_w = u_* f_w \tag{16}$$

$$U = u_* f_U \tag{17}$$

$$\sigma_w = U \frac{f_w}{f_U} \tag{18}$$

where u_* is the friction velocity.

Since f_w/f_U is a function of L, Z, boundary-layer height (H) and roughness length (Z_o), L can be estimated at a certain height for a given Z_o from (18) if also U and σ_w are known.

All sodars have a typical minimum level or noise in σ_w, usually between 0.1 and 0.3 m/s. Both Thaning (1985) and Thomas (1986) have shown that, by using

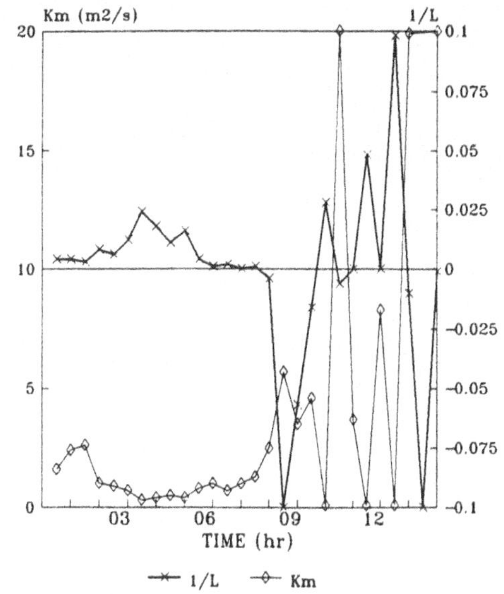

Fig. 6. Stability ($1/L$) and eddy viscosity in the surface layer from sodar data with a windprofile adjustment method. From measurements in Jädraäs 25 July 1989.

only the windspeed and σ_w, it is possible to distinguish between stable, neutral and unstable stratifications. The method can almost determine the six Pasquill categories, an often referenced scheme for static stability suggested by Pasquill (1961). Surface wind speed in relation to insolation and cloudiness gives the stability class. For the most stable Pasquill classes (F and E), σ_w is close to the noise level of the instrument. The variances can be written as

$$\sigma_w^2 = \sigma_{wt}^2 + \sigma_{w\,noise}^2 \tag{19}$$

where σ_{wt} is the real variance of vertical windspeed.

Finally by combining (18) and (19);

$$\sigma_w^2 = U^2(\frac{f_w}{f_U})^2 + \sigma_{w\,noise}^2. \tag{20}$$

An error in f_w/f_U from the sodar data will result in a larger error in $1/L$ when the stratification is stable than it will during unstable conditions, see Fig. 7.

In Fig. 8, the unmarked area in stability can not be resolved for a sodar with a noise level in σ_w equal to 0.3 m/s. For windspeeds under 1 m/s, only neutral to slightly stable and unstably stratified conditions are possible to estimate.

High frequency cutoff due to low pulse repetition frequency of the instrument (Kristensen et al., 1986) and difficulties in finding the Doppler peaks cause this noise.

The mixing-length parameterization needs a Brunt-Väisälä (N) instead of an L as a stability parameter. For a stable PBL a buoyancy length scale for turbulence

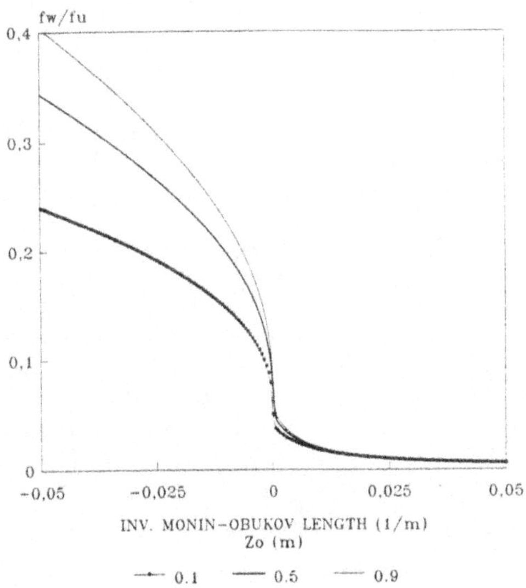

Fig. 7. Flux profile functions (f_w/f_U) versus stability $(1/L)$ at 200 m for a 300 m PBL height at three different roughness lengths (Z_o) from (18).

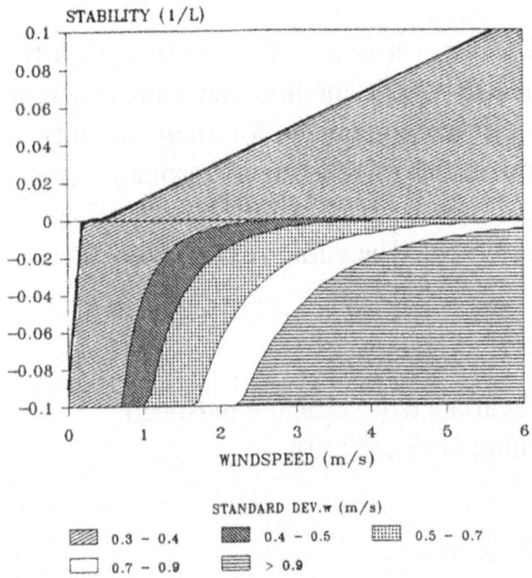

Fig. 8. Stability as a function of windspeed for different σ_w. Height is 125 m and Z_o is 0.5 m. From equation (20).

(l_b) is given by;

$$l_b = \frac{\sigma_{wt}}{N}.$$

The buoyancy length scale indicates the supression of vertical motions by the static stability. For neutral stratification (N=0) the buoyancy length scale goes to infinity. For stable N, l_b varies from a few meters to several hundred meters for normally measured σ_w (0.2-0.8 m/s). This is the same range as for the other turbulent length scale L. Since the wind profile model uses a first-order closure in an eddy viscosity parameterization with mixing-length theory, the unstable cases will be regarded as neutral ($L > 0$).

If we therefore assume that l_b is proportional to L, we get;

$$N = \frac{C_N \sigma_{wt}}{L} \tag{21}$$

where C_N is a proportionality constant that has to be determined from the data set, see section 6.3.

5.3. Spectral Analysis of Gravity Waves

The stability in the stable case can be estimated from a spectral analysis of gravity waves if no other method works. If an inversion layer is present, there will often be detectable oscillations registered with the sodar. The method is similar to the one described by Carlsson (1985) with the main difference that we are here looking at a time series of integrated backscatter instead of integrated water vapor content from a ground based radiometer.

The Brunt-Väisälä frequency of a gravity wave is proportional to the static stability of the inversion layer (Holton, 1972, pp 168);

$$N = \sqrt{\frac{g}{\bar{\theta}} \frac{d\theta}{dz}}. \tag{22}$$

In equation (22) g is acceleration due to the gravity and $\bar{\theta}$ is the average potential temperature of the layer. Since the sodar profile also gives the vertical windshear, a local R_i can be estimated from (11).

In Fig. 9, the stability is plotted as a function of wave period. The so-called Nyquist period in the diagram indicates how often one must sample in order to resolve a frequency corresponding to a certain stability. An isothermal layer has a wave period of 340 sec or N=0.0185 Hz.

6. Data and Results

The data set from Karlsborg has been used to find a method to estimate the stability for the parameterization of the eddy viscosity. The wind profile model has been compared and evaluated with the same data set.

6.1. Sodar Data

From the early seventies with analogue non-Doppler sodars to the digital instrument that we use today, the technicial evolution has been tremendous. During the last 15 years several international intercomparisons have taken place which show the

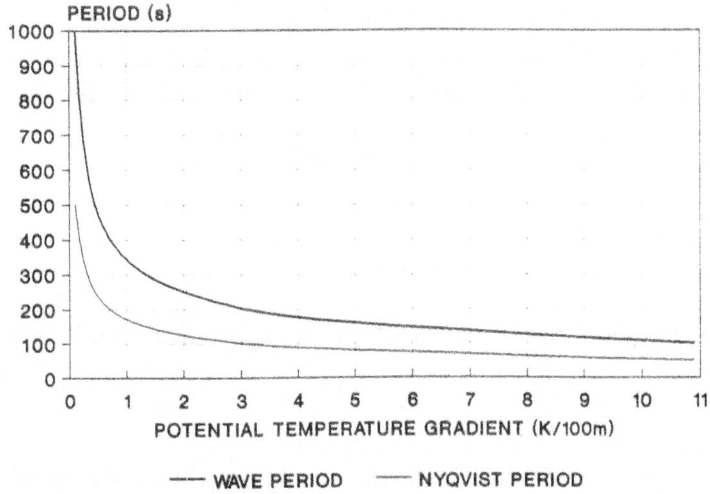

Fig. 9. Wave and Nyquist period for gravity waves as a function of vertical potential temperature lapse rate when average $\bar{\theta}$=273K.

capability of the sodar to measure wind vectors (Asimakopoulos et al. (1983), Gaynor et al. (1983) and Mastrantonio et al. (1982)).

Sodars have a limited measurement range due to strong attenuation in the optimum acoustic frequencies that are used (Beran et al., 1972). Usually the range varies from a few hundred meters up to over 1 km depending on:

* Atmospheric conditions (inversions, convection, windspeed)
* Noise level and spectra (cars, aircraft, trees)
* Transmitted power and frequency (50-1000 W, 1 - 3 KHz)
* Antenna size
* Integration time ((5-60 min)

The output from the instrument is usually a time-height plot of the backscatter intensity, and for Doppler sodars as in the present case, also windprofiles with their standard deviations, see Fig. 10.

6.2. Radar Wind Measurements and Other Meteorological Data

Soundings and radar wind raw data at 0400 or 0500 and 1300 hours from mid February 1987 to mid December 1988 at Karlsborg have been collected. Additional data are surface chart geostrophic winds, 850 mb winds and comments from meteorologists on site. The total number of complete soundings, radar and sodar profiles is 403.

Karlsborg is also a regular sounding station but the measurements are of almost no use in this application since the standard software (CORA, Väisälä) does not take into account small changes in the wind below approx 900 m. Regardless of the measured wind below 900 m, only a linear profile is given. From approximately

Mon Jan 23 14:20:15 19

INT TIME: 30 OF 30 MIN. SOND:107 107 106 107 LIMIT: 0% VERT COMP: off MMODE:

height	hor wind			vert velocity			hor wind U			hor wind V			echo
m	deg	m/s	sigd	m/s	std	Q	m/s	std	Q	m/s	std	Q	magn
450	235	16.9		-0.3	0.41	2	-13.9	1.03	4	9.6	0.00	2	
425	231	14.4					-11.3	6.12	7	9.0	0.80	3	
400	230	16.9		-0.0	0.31	2	-13.0	1.33	7	10.7	2.03	6	
375	231	15.8	3.0	0.1	0.68	12	-12.3	1.53	13	9.9	1.37	10	
350	227	15.5	2.3	0.2	0.35	9	-11.5	1.66	13	10.4	1.79	15	148
325	224	14.3	3.5	0.3	0.61	21	-10.0	1.99	21	10.2	1.35	18	111
300	222	14.5	2.5	0.2	0.71	22	-9.9	1.51	30	10.6	1.13	27	136
275	223	13.2	5.5	0.4	0.69	32	-9.1	1.96	37	9.6	1.24	31	131
250	219	12.0	6.8	0.4	0.71	40	-7.7	2.57	43	9.2	1.47	41	176
225	217	11.0	4.4	0.2	0.72	52	-6.7	1.85	57	8.7	1.44	50	152
200	216	9.9	7.8	0.2	0.58	62	-5.9	2.07	62	7.9	1.90	48	160
175	215	9.1	8.0	0.1	0.69	68	-5.3	2.19	65	7.4	1.68	45	142
150	214	6.8	8.6	0.2	0.72	73	-3.9	2.04	68	5.6	2.26	54	150
125	209	4.1	17.2	-0.0	0.75	83	-2.0	2.10	86	3.6	2.62	65	133
100	203	2.3	24.0	0.1	0.45	100	-0.9	1.74	95	2.1	2.03	91	150
75	204	1.5	38.1	0.1	0.52	99	-0.6	1.64	101	1.4	1.89	95	116
50	242	1.2	61.1	-0.2	0.28	90	-1.1	1.94	78	0.6	1.00	100	173

Fig. 10. Sodar printout with 30 min integration time. The columns are from the left; height to the center of measured volume, wind direction, windspeed and standard deviation of wind direction (σ_d). The next nine columns are data from the three antennas, 'magn' is a relative magnitude of backscatter and 'Q' is the number of received soundings (pulses) out of emitted 107 or 106 ones during the integration time (Sensitron Sodar, Sweden).

500 m, the data are filtered with a 5 min running average. With normal ascent rate, this corresponds to a running average over 1000 m.

6.3. Stability Estimations with Flux Profile Method

With the flux profile method in section 5.2., stability has been estimated from sodar data. At 100, 125 and 150 m, an average $1/L$ was calculated from (20). The boundary layer height is assumed constant (400 m) because of the small effect in $1/L$ resulting from a varying H.

For the site, $Z_o = 0.5m$ has been used. The roughness inhomogeneities are significant at the site close to the shoreline of the lake. The regular soundings within an hour from the time of the sodar data have been used to estimate N from (22). Since the integration time of the sodar was 15 min, sometimes up to five sodar records have been used and compared with one sounding.

Fig. 11 shows a plot of the result with N from stably stratified soundings and N from sodar with C_N chosen to 12 in order to maximize the number of datapoints between the dotted lines. Here the raw data from the sounding were not available; only the vertical profile as recordered by the operator is used to calculate N. The criterion for a significant point in the sounding, according to the sounding instructions, allows for an error in N of about 0.02 Hz when the stratification is neutral. For a stable sounding (N=0.025), the uncertanty is less, namely $\pm$ 0.004 Hz. The data are therefore very scattered but containing a signal.

Both the resolution in the soundings and roughness inhomogeneities cause this significant scatter. The proportionality constant C_N in eq. (21) can now be determined by minimizing the difference between the sodar-derived N and the N

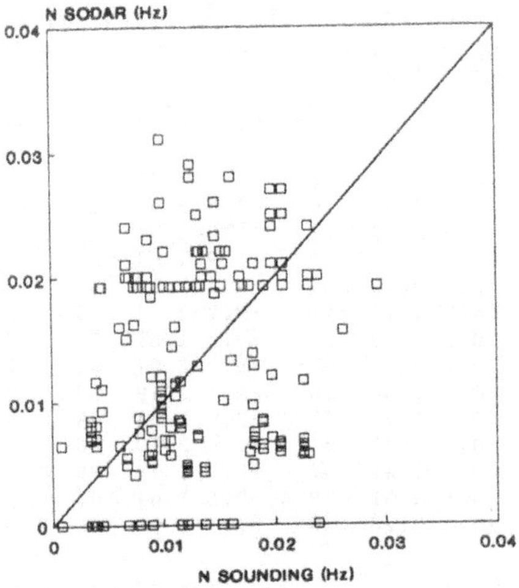

Fig. 11. Brunt Väisälä values from stable soundings versus estimated N from sodar data, 19 Jan to 20 Dec 1988, 211 cases. Dotted lines mark the uncertainty range when the temperature gradient is calculated over a 20 mb thick layer according to WMO (World Meteorological Organization) sounding instructions.

calculated from the sounding. The value of C_N can roughly lie between 5 and 20 to keep the majority of points between the dotted lines in Fig. 11. Further measurements with high resolution low-level soundings during a year at a homogeneous site would be needed to determine C_N more accurately. The scatter due to uncertainty in the flux profiles and known problems with the sodar to measure a correct σ_w, especially in stable stratification, would in any case remain (Gaynor et al., 1983). The estimated N will however give an indication if the profiles are unstable, neutral, stable or very stably stratified. Before 19 Jan, 1988, the σ_w data were not saved and the method can only be evaluated from about 40% of the total data set.

6.4. Stability from Gravity Waves

In order to analyze the gravity waves, raw data from the sodar have been recorded. This means that each pulse from the three antennas is sampled at every height interval. Here the time series of vertically amplified and integrated backscatter over 75 m is analyzed.

The backscatter magnitude from the emitted acoustic pulse normally decreases with height. If an inversion layer is present, the backscatter magnitude increases with height and has a maximum near the top of the inversion. The reason is that this region generates small-scale turbulence due to shear or destabilization by cooling when clouds are present. Turbulent inhomogeneities with a scale equal to half the wave length of the emitted pulse (7 cm at 2400 Hz) cause the most effective acoustic backscatter. If the amplitude of the gravity waves is about 50-100 m, as in

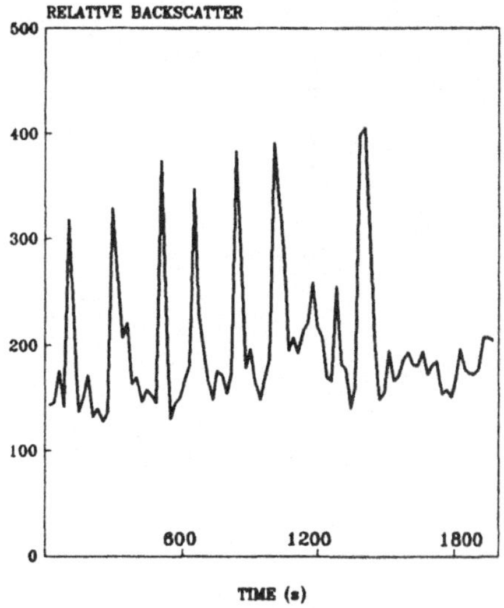

Fig. 12. Vertically integrated backscatter as a function of time, 0415-0446, 8 September 1988.

Fig. 12, they can be detected with this method.

The backscatter mode has a pulse every 21 sec so that periodic oscillations with a minimum period (P) of the sampling period (42 sec) can be resolved. According to (22) and with $N = 2\pi/P$ this period corresponds to a potential temperature increase of over 6 K/100 m. The pulse repetition frequency (PRF) is therefore enough for a spectral analysis of the most stable gravity wave with the backscatter. The power spectra for 8 September, 1988 (Fig. 13) has the strongest peak at 0.00595 Hz, corresponding to a wave period of 168 s.

Applying (22) to the case in figs. 12 and 13 with $P \approx 168$ s yields;

$$N = \frac{2\pi}{P} = 0.0374Hz$$

or an increase in $\bar{\theta}$ of almost 4 K/100 m, which corresponds to a temperature increase of about 3 K/100 m. The sounding at that time shows a ground-based inversion up to 170 m with a temperature increase totalling 5.6 K, which corresponds to a gradient of 3.3 K per 100 m. For this particular case the method is thus useful. It is however difficult in general to detect gravity waves with this method, especially at this site close to a runway with frequent afterburner takeoffs of military jets.

Fig. 14 shows an analysis of a time-height matrix of sodar backscatter raw data. The strongest echoes (black) are from an oscillating inversion on top of a dense fog. The amplitude of the oscillations is here between 50 and 60 m. At the bottom, below approx. 40 m, no data are available due to mechanical noise from the outgoing pulse.

The timescale for the gravity wave is about an hour. For the stably stratified PBL the natural variations in radiation, geostrophic wind or background noise on

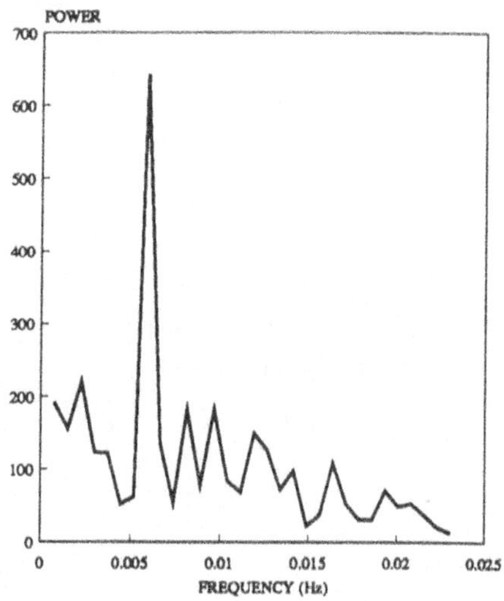

Fig. 13. Power spectra 8 September, 1988.

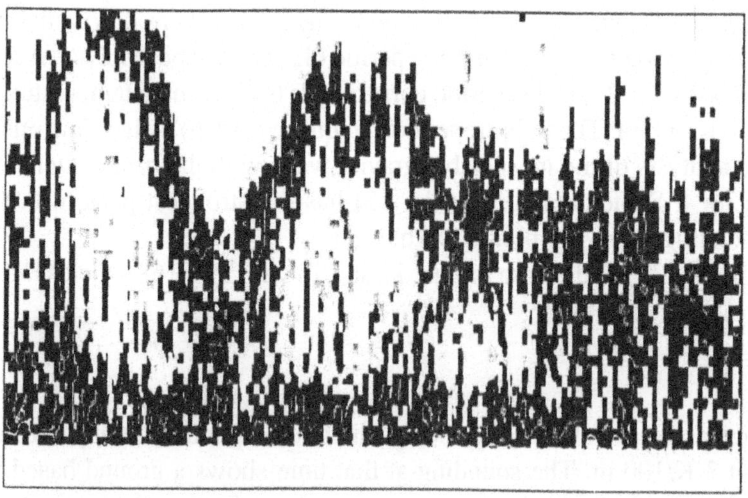

Fig. 14. Analysis of a time-height matrix of backscatter raw data. One sounding every 21 sec and 5 m spacing vertically. The weakest backscatter is grey and the strongest is black. Height range is 40-350 m. From Karlsborg 1040-1132, 5 September 1988.

that timescale are often substantial. The example from 8 September (Fig. 12) was an exception. An improvement in the method would be to add the signal from the vertical Doppler raw data and make a vertical coherence spectra or a correlation matrix of the wind variations.

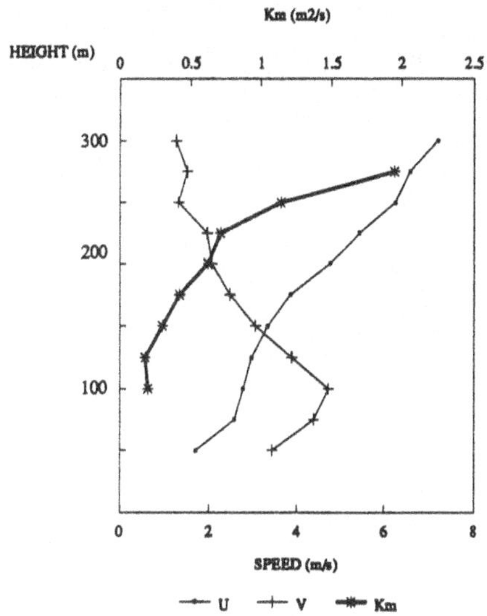

Fig. 15. u, v and resulting K_M profiles from Karlsborg sodar 0500, 17 Oct 1987.

6.5. Eddy Viscosity Estimations with the Ageostrophic Method

Most sodar profiles show one or more negative K_M values after applying (8). This can be an indication of either a geostrophically unbalanced profile, non-stationarity, thermal wind or convection. The last phenomenon is a well known problem in K-parameterization because of counter-gradient transport of momentum. Numerical problems are also important here. Only 90 out of 286 sodar profiles (31.5%) had no negative K_M values at all.

This might be an indication of how often we have a PBL in almost geostrophic balance, and how numerically sensitive the method is to a small denominator in (8). The denominator is a difference between two relatively large numbers. In the numerator we have an integration of the ageostrophic wind components multiplied by the shear in the surface layer (Z_s). The wind at the bottom is more sensitive to local effects and technical maintenance of the instrument than farther up. Therefore a small error in surface-layer shear will cause large errors in the estimated K_M profile.

During the daytime, convective and unstably stratified periods dominate. The number of profiles with no negative eddy viscosity is therefore almost twice as large during nighttime (65.5%) compared to daytime (34.5%).

Balloon soundings that have been used earlier are probably in most cases not in balance with the geostrophic wind due to the short timescale they represent. A 15 or 30 min time average from an acoustic sounder is probably a more representative windprofile of a PBL in near geostrophic balance.

Fig. 15 is an example of a profile during near-neutral conditions. The wind profile has kinks at the bottom but the amplitudes are too small to be seen in the resulting K_M profile.

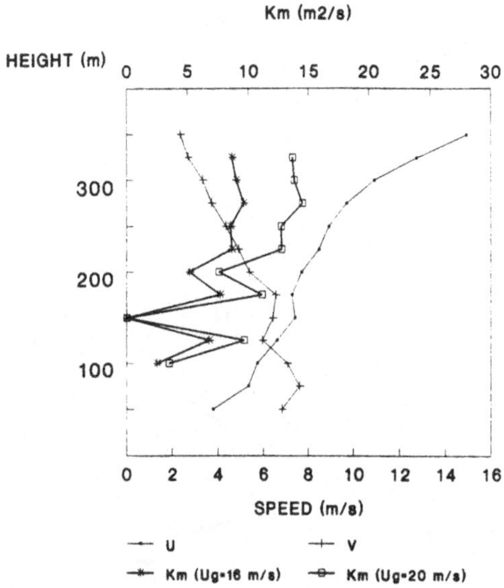

Fig. 16. u, v and K_M profiles from Karlsborg sodar data 0500, 17 May 1988 for two different geostrophic windspeeds.

Fig. 16 shows another case with a weakly stable layer between 125 and 175 m. Note the kink in the v profile there and the relatively smooth K_M profile above, indicating a very well mixed and balanced profile in the strong wind above (15.1 m/s at the top).

The absolute value of K_M depends upon the given or estimated G. The shape of the eddy viscosity profiles does not vary with G.

The eddy viscosity at the top of the profile (8 and 13 m^2/s) can be used as an aid as a first guess of δ_E if the stratification is close to neutral and conditions are stationary.

Another way to show that the method is consistent with theory is to plot the number of negative K_M in each of the 194 sodar profiles and the average K_M versus the estimated geostrophic windspeed. In Fig. 17 the number of negative profiles is about equal for all windspeed classes, indicating that non-geostrophic balance is independent of windspeed. The average eddy viscosity increases with increasing windspeed as expected.

Equation (8) has been applied to analytical Ekman profiles and numerical solutions of equation (4) to check the sensitivity of the method. Results from analytical profiles are trivial. In Fig. 18 numerically computed profiles of u and v, with a given K_M profile, are shown. The variation of the estimated K_M from (8) around the given values is due to the numerical integration technique (fifth-order Runge-Kutta with adjustable stepsize, Press et al., 1987). The sensivity of the method given in section 3.1 is clearly shown.

The sodar profiles from Karlsborg with a 15 or 30 min integration time sometimes gave very smooth K_M profiles with (8). The vertical resolution of the sodar

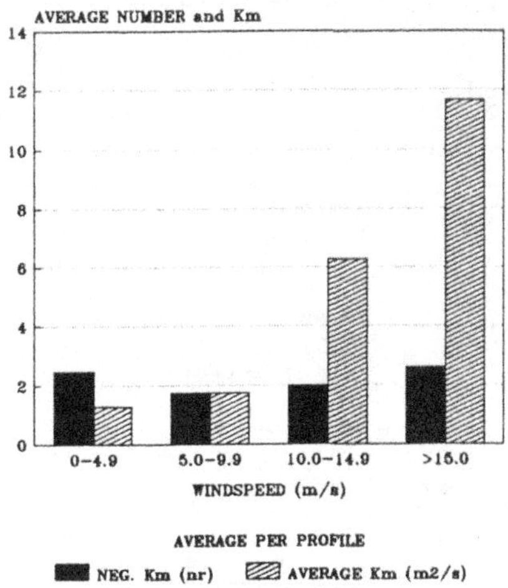

Fig. 17. Number of negative K_M and average K_M versus estimated G for four different windspeed classes. 327 profiles Jan.-Nov. 1988.

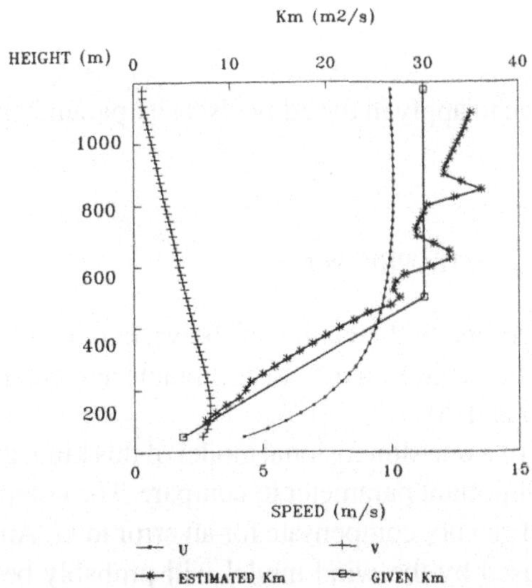

Fig. 18. Model profiles of u and v from equation (4) and K_M from eq. (8).

data is in this case 25 m. The best results from (8) are obtained with differences taken over 100 m layers. A similar result was found by Schmitz-Peiffer (1987). At least 4 wind values, at 25 and 75 m for the shear at the top of the surface layer and at 50 and 150m for the lowest shear in the Ekman layer, are needed to calculate K_M at one level (100m). For a profile up to 500 m there will be 15 eddy viscosity values.

The method is almost useless when the stratification is unstable. Then the

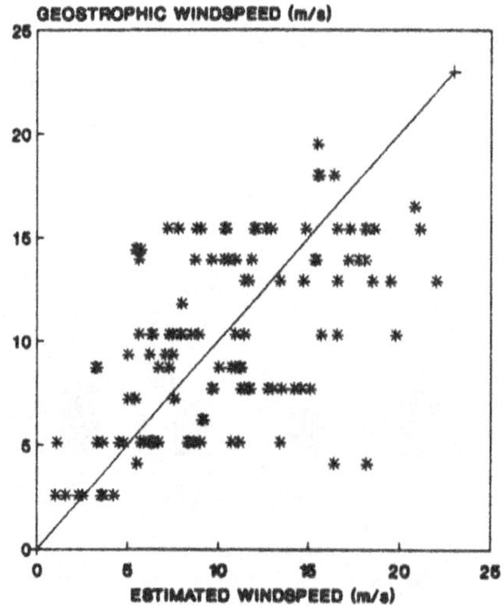

Fig. 19. Estimated geostrophic windspeed from model compared with measured wind gradient from surface synoptic charts.

method is too sensitive to apply in the eddy viscosity parameterization for the wind profile model.

6.6. ESTIMATION OF GEOSTROPHIC WIND

The Ekman adjustment method (section 4) shows, compared with manually estimated G from synoptic surface charts, an acceptable agreement both in speed and direction, see figs. 19 and 20.

For an evaluation of a one-dimensional model of this kind, the geostrophic wind is perhaps the most important parameter to compare. The computed profile is fixed at the boundaries and can not compensate for an error in G. Another consideration is that the errors caused by the wind model will probably be smaller than those introduced by (for example) the thermal wind (Thaning et al., 1990).

Fig. 19 is a plot of estimated windspeed compared with manually measured geostrophic wind from surface charts (130 cases). The model underestimates high geostrophic windspeeds. One reason is that the sodar normally measures too low a windspeed when the 10 m windspeed is over 10-15 m/s. The wind-generated noise increases with windspeed. High wind gusts will therefore not be detected, causing too low an average windspeed.

The sensitivity in eddy viscosity magnitude for the method is shown in Fig. 21 for an analytical profile with G given as 270 deg, 10 m/s and constant $\bar{K}_M = 5m^2/s$. Note that the given geostrophic wind is found just at 5 m^2/s.

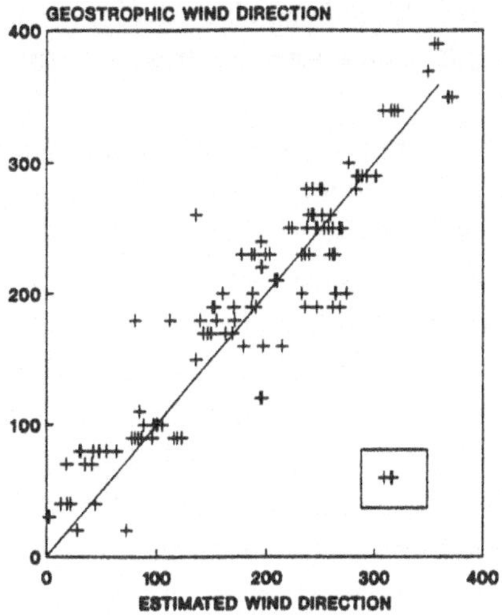

Fig. 20. Measured geostrophic wind direction from surface charts versus sodar estimations with an Ekman adjustment method. From Karlsborg 19 Jan to 20 Dec, 1988 (130 profiles). The direction is corrected for the gap between 360° and 0°. Situations with an easterly sea breeze are marked with a rectangle in the lower right corner.

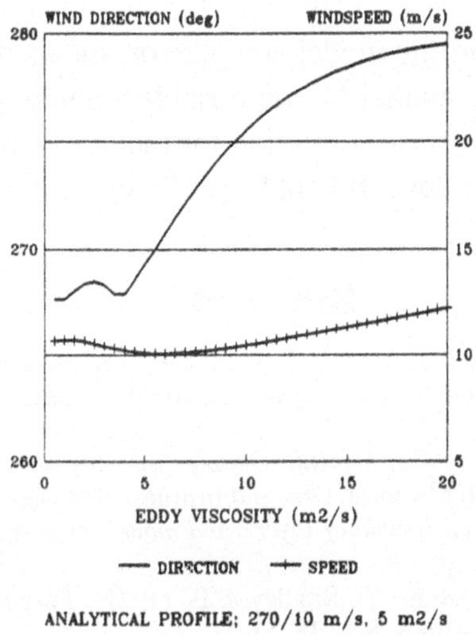

Fig. 21. Estimated geostrophic windspeed and wind direction versus eddy viscosity for the Ekman adjustment method.

7. Discussions and Conclusions

Three methods for stability estimations with sodar data, for a one-dimensional wind profile model, have been discussed. The windprofile method only works under neutral or stable stratification, and in those cases the method has limited usefulness. The gravity wave method can also only be used when the stratification is stable and when detectable oscillations are present during stationary conditions for about an hour or more.

Here the flux profile method has been chosen for the windprofile model in combination with mixing-length theory for the eddy viscosity parameterization. By assuming a proportionality between a buoyancy length scale and a turbulent length scale, a rough value of N is estimated from sodar and compared with N given from soundings. The flux profile method works both when the stratification is stable and unstable but all unstably stratified cases will be regarded as neutrally stratified ($N = 0$) for the K_M parameterization. In most cases the layers above 100 m are stably stratified. For the Karlsborg one year data set 90.2% of the soundings were stably stratified in all reported layers below 1500 m.

The ageostrophic method agrees well with given K_M when analytically and numerically generated profiles are used. When the method is applied to sodar data during stationary conditions:
(i) The shape of the K_M profiles often indicate when stable layers are present.
(ii) Estimated values K_M increase with increasing windspeed while non-geostrophic balance is independent of windspeed.
(iii) The method can not be used during convective or unstable conditions except to indicate that the wind profile is not stationary, barotropic or in geostrophic balance.

The result of the windprofile model comparison shows that the average errors and standard deviations are smaller for the variable momentum flux model than for an earlier constant flux model. A correction for thermal wind would help, as well as a correction in roughness for different wind directions.

References

Asimakopoulos, D. N., Mousley, T. J., Helmis, C. G., Lalas, D.R. and Gaynor, J. (1983): *Quantitative Low-Level Acoustic Sounding and Comparison with direct Measurements*. Boundary-Layer Meteorol. **27**, 1-26.

Beran, D.W., and Clifford, S.F. (1972):*Acoustic Doppler measurements of the total wind vector.* Preprints AMS 2nd Symp. on Meteorol. Obs. and Instrum., San Diego, Calif., pp. 412-417.

Bergström, H., (1986):*A simplified boundary layer wind model for practical application*. J. Appl. Met, **25**, No. 6, 813-824.

Businger, J.A., Wyngaard, J.C., Izumi, Y., Bradley, E.F., (1971): *Flux profile relationships in the atmospheric surface layer.* J Atmos. Sci., **18**, 51-73.

Dahlquist, H., (1984):*An operational forecasting system for stratus and fog with sodar.* Proc. Nowcasting-II Symp., Norrköping, Sweden, 3-7 Sept. (ES SP-208), 347-351.

Dyer, A.J., (1974):*A review of flux-profile relations*. Boundary-Layer Meteor.,**1**. 363-372.

Carlsson, I., (1985):*Theory and observations of gravity waves in an inversion layer.* Report DM-**46**. MISU, Stockholm, Sweden, pp 48.

Estournel, C. and Guedalia, D., (1987):*A new parameterization of eddy diffusivitities for nocturnal boundary-layer modeling*. Boundary-Layer Meteor., **39**, 191-203.

Holton, J. R., (1972): *An Introduction to Dynamic Meteorology*. Academic Press. New York, pp 318.

Gaynor, J. E., Kaimal, J. C., et al, (1983):*Evaluation of Wind Parameters Measured by four Doppler Sodars.*, NOAA/ERL, Boulder, CO. AMS Fifth symposium. Meteorological observations and Instrumentation, 488-491.

Kristensen, L. and Gaynor, J. E., (1986):*Errors in Second moments Estimated from Monostatic*

Doppler Sodar Winds. Part I: Theoretical Description. Part II: Application to Field Measurements., Journal of Atmospheric and Oceanic Technology., Sept 1986, 523-534.

Lapworth, A. J., (1987):*Wind profiles through boundary-layer capping inversions.*, Boundary-Layer Meteor., **39**, 333-378.

Mastrantonio, G. and Fiocco, G., (1982):*Accuracy of wind velocity determinations with Doppler sodars.*, J. Appl. Meteor., **21**, 823-830.

Pasquill, F., (1961):*The estimation of the dispersion of windborne material.* Met. Mag., 90, **33**.

Pasquill, F and Smith, F. B., (1983): *Atmospheric diffusion.* Ellis Horwood Ltd, England.

Pedlosky, J., (1986): *Geophysical Fluid Dynamics.*, Springer-Verlag., 2nd edition. pp 710.

Press, W. H. et al. (1987):*Numerical Recipes.*, Cambridge University Press., pp 818.

Schmitz-Peiffer, A. et al. (1987):*The ageostrophic method-an update.* Boundary-Layer Meteorol. **39**, 269-281.

Stull, Roland B., (1988): *An introduction to boundary layer meteorology.* Kluwer Academic Publishers, pp 666.

Thaning, L. (1985):*En metod att uppskatta atmosfärens stabilitet från sodarmätningar.* FOA rapport **E 40020**, pp 44.

Thaning, L., Arnoldsson, G., Dahlquist, H. Grandin, G. and Peterson, E. (1990): Vindprofilmodell, *utvärdering.* FOA rapport **E 40045**-4.5., pp 42.

Thomas, P., (1986):*Stability classification by acoustic remote sensing.* Atmos. Res., **20**: 165-172.

INTERCOMPARISON OF TURBULENCE DATA MEASURED BY SODAR AND SONIC ANEMOMETERS

P. THOMAS and S. VOGT
Institut fur Meteorologie und Klimaforschung Kernforschungszentrum/Universitat Karlsruhe, Germany.

(Received October 1991)

Abstract. The capability of SODAR to measure the mean wind field in the lower boundary layer is well known and documented. Therefore, mean wind data are easily obtainable by means of the SODAR-technique, and are used to simulate the transport of pollutants after their release into the atmosphere. But when calculating the diffusion of pollutants, information about atmospheric turbulence is needed, too.

In principle, a SODAR can measure turbulence data like the standard deviation of the vertical wind speed or horizontal wind direction. But when measuring turbulence data with a SODAR, one is beset by a host of limitations like volume sampling, spatial and temporal separation of sampling volume, attenuation of the acoustic waves and the slow speed of sound. Therefore, successful turbulence measurements with SODAR are not numerous and little is known about the quality of these data.

In this context an intercomparison between a REMTECH-SODAR and a sonic anemometer mounted at the 100 m level of our meteorological tower was performed in summer 1990 at the Kernforschungszentrum Karlsruhe. The intercomparison is in two parts:

(1) Half hour mean values of the standard deviation of the vertical wind speed are intercompared by scatter plots and by a linear regression and correlation analysis.

(2) During 7 periods, 2 hours each, and covering atmospheric stabilities from unstable to slightly stable, the instantaneous vertical wind speeds were measured by both instruments and spectra were calculated.

The intercomparison demonstrates that DOPPLER-SODAR sounding is a reliable technique to determine besides the mean field, also athmospheric turbulence data like Sigma(w).

1. Instruments and Measurements

The SODAR was the monostatic three-dimensional instrument A0 manufactured by the French REMTECH enterprise. The SODAR emits a double frequency pulse of 150 ms duration with a centered frequency of 1600 Hz.Analysis of the backscattered echo using FFT is automatically done by the SODAR software. The rejection of echos by software is mainly due to fixed echos and ambient noise. Furthemore, calibrations are performed at regular and irregular time intervals to take care of temperature variations.

The sonic anemometer is a KAIJO DENKI model DAT 300 mounted at the 100 m level of a 200 m high meteorological tower. The tower is an open lattice-type structure of a square cross section with a side length of 1.5 m. The boom bearing the sonic anemometer extends 4 m to the north.

The SODAR is located about 200 m north of the tower in a clearing (60 m x 100 m). The clearing is surrounded by a forest with pines of about 20 m in height.

From June 1 to July 31, 1990, both instruments measured simultaneously 30-min mean values of the standard deviation σ_w of the vertical wind speed w at 100

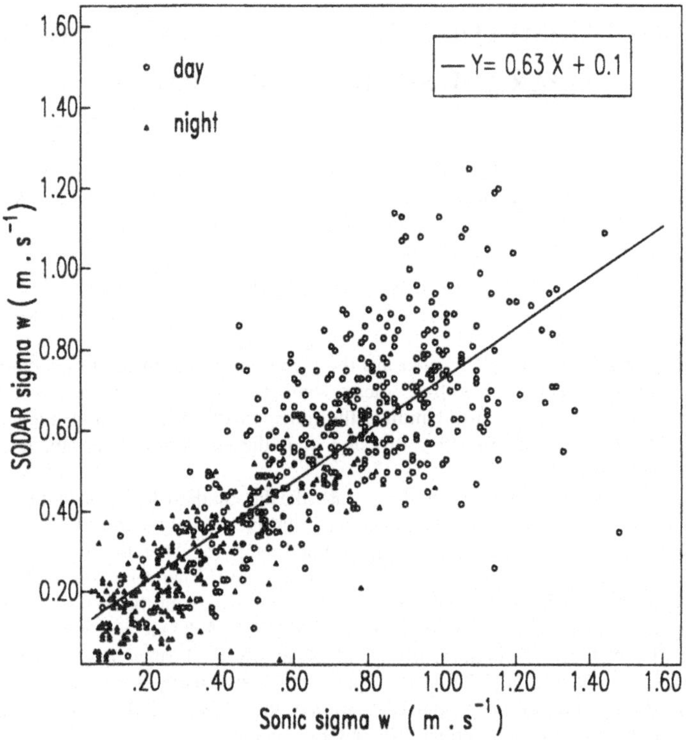

Fig. 1. Scatter plot of the standard deviation σ_w measured at 100 m AGL.

m above ground level (AGL). In a second campaign in August and September 1990 the instantaneous vertical wind speed w was measured at 100 m AGL during seven test runs lasting more than two hours each. The sampling rates were 1.06 s and about 5 s for the sonic anemometer and the SODAR, respectively. The sampling rate of the SODAR is a mean value. In the SODAR time series, periodically and stochastically distributed gaps exist caused by the calibration and rejection of data performed by the SODAR software. To yield a higher data rate during the instantaneous measurements, the SODAR was operating only with the vertically pointing antenna at a relatively high emission rate.

2. Results of the Long Term Intercomparison

Figure 1 shows a scatter plot of all data pairs of σ_w measured simultaneously. The calculated regression line and its equation are indicated. In Fig. 2 the relative differences between σ_w-values are plotted versus the horizontal wind speed measured simultaneously by the sonic anemometer at 100 m AGL. In Tab. I the results of the regression and correlation analysis are compiled. The bias (B) is the difference of the mean values of both time series, it is the systematic difference. The word 'precision' means the root mean square difference (RMSD), when before calculating RMSD the time series of one of the instruments is corrected for the bias.

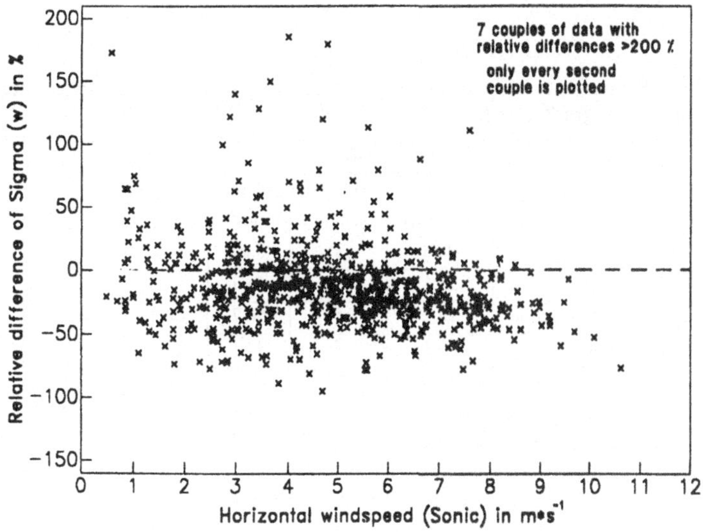

Fig. 2. Relative difference $((\sigma_w^{SODAR} - \sigma_w^{SONIC})/\sigma_w^{SONIC}$ versus sonic horizontal wind speed measurements at 100 m AGL.

TABLE I

Comparison of 30-min mean values of σ_w measured by sonic anemometer and SODAR 100 m AGL

	ALL	DAY	NIGHT
Number of data pairs	1509	857	536
Mean value sonic X (cm s^{-1})	58	75	36
Mean value SODAR Y (cm s^{-1})	47	59	29
Bias: Y - X (cm s^{-1})	-12	-16	-7
Root mean square difference (cm s^{-1})	22	26	14
Precision $(RMSD^2 - B^2)^{0.5}$1(cm s^{-1})	18	20	12
Correlation coefficient	0.82	0.65	0.84
Central European Time		7:00 a.m.	9.00 p.m.
		7:30 p.m.	5:30 a.m.

From Figs. 1 and 2 and from Tab. I the following conclusions can be drawn:

- The SODAR generally underestimates σ_w. The main reasons for this underestimation is the inability of the SODAR to detect small-scale fluctuations due to its large sampling volume and its low sampling rate. So high frequency wind fluctuations are suppressed in the σ_w calculation.
- The absolute differences between σ_w measured by both instruments are small at low σ_w-values during nighttime and during low horizontal wind speed situations.
- The relative differences between σ_w measured by both instruments decrease with horizontal wind speed and show no diurnal variations.

TABLE II
General data of time series of instantaneous vertical wind speed and significant mean meteorological parameters for 7 tests runs.

			SODAR					
Run No.	Day/ Month	Time (CET)	% of observed data	Mov.av. length (s)	% of transf. data	Temp. °C	Speed U (ms^{-1})	σ_θ (degrees)
1	01/08	13h09-15h12	11.1	7	46.3	29.3	5.7	15.0
2	10/08	10h41-12h36	16.2	5	50.7	24.1	4.6	13.9
3	15/08	09h50-11h40	13.1	5	44.1	21.2	6.4	10.1
4	16/08	17h20-18h40	16.9	3	40.1	16.5	7.9	7.0
5	04/09	14h19-16h30	21.7	5	65.1	17.7	6.5	6.2
6	05/09	10h58-1315	20.6	5	61.3	14.7	3.9	13.7
7	11/09	08h22-10h39	25.7	5	70.4	11.7	1.6	16.0

CET: Central European Time
U: Horizontal wind speed

- The correlation between both time series is better during nighttime, when it is calm and stable than during daytime, when it is windy and unstable.
- Similar results have already been published in Kaimal et al. (1984),Chintawongvanich et al. (1989) and Thomas and Vogt (1990).

3. Comparison of Spectra

General data concerning the seven test runs during which the vertical wind speed has been measured instantaneously are compiled in Tab. II. In the last column of Tab. II the mean standard deviation σ_θ of the horizontal wind direction is indicated. It corresponds to the whole time of a test run and has been measured by a vector vane,(Von Holleuffer-Kypke et al.,1984). From σ_θ and u it can be seen that during all episodes unstable conditions prevailed except for No. 4 and 5, when slightly stable conditions occured with high wind speed.

As already stated in Section 2 there are gaps in the SODAR time series resulting in poor data coverage of only 11.1% up to 25.7% (see column 4 in Tab. II) as compared to equally spaced 1 s sampling. Low pass filtering with an interpolation using Cressman weight functions (Jones, 1972), was applied to the time series of both instruments,

- to produce time series with equal time steps of 1 s,
- to get more SODAR-data (see column 6 in Tab. II),
- to reduce aliasing (backfolding of power above the Nyquist frequency).

The moving averaging length (time window) to interpolate the SODAR data is indicated in column 5 of Tab. II. Finally spectra were computed by Discret

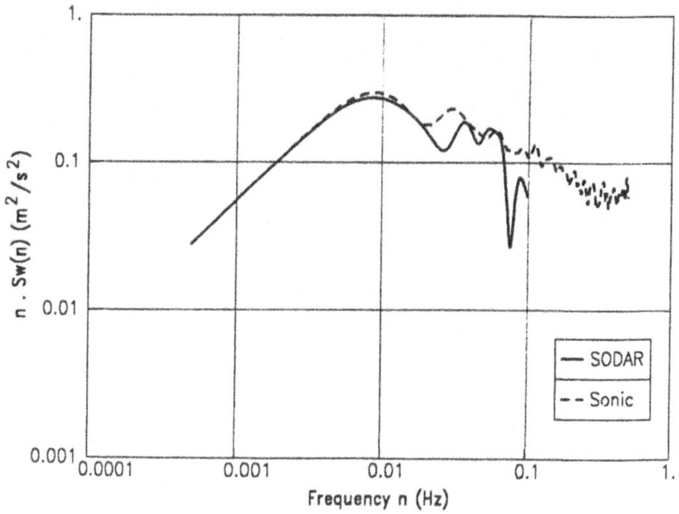

Fig. 3. Spectra of vertical wind speed, run No. 2 during unstable stratification.

Fourier Transformation of the autocovariances of the vertical wind speed using a Tukey-Hanning spectral window (Jenkins et al., 1968 and Priestley, 1981).

By way of example, logarithmic power spectra of the vertical wind speed are plotted in Figs. 3 and 4 corresponding to runs No. 2 and 5, respectively. These plots show the following general features of the spectra and demonstrate the characteristics of both instruments:

- The upper limit of the frequency n of the spectra are:
  . The Nyquist frequency 0.5 Hz in the case of the sonic anemometer.
  . a frequency less than about 0.15 Hz in the case of the SODAR due to its lower sampling rate and a poor confidence level at higher frequencies.
- The SODAR spectrum is attenuated at frequencies $n \geq u/4d$ due to its large sampling volume (u and d are the horizontal wind speed and the horizontal diameter of the sampling volume, respectively).
- The SODAR spectra have sharp minima that are caused by the sampling process. The frequencies of these minima correspond to an approximately periodic gap of data in the time series.
- In contrast to spectra measured during unstable conditions, those belonging to stable stratification and high wind speed situations
  . have lower power,
  . have peaks shifted to higher frequencies that are more difficult to estimate,
  . are underestimated by the SODAR especially at high frequencies, as this instrument cannot detect small scale eddies that are more frequent during stable stratification.

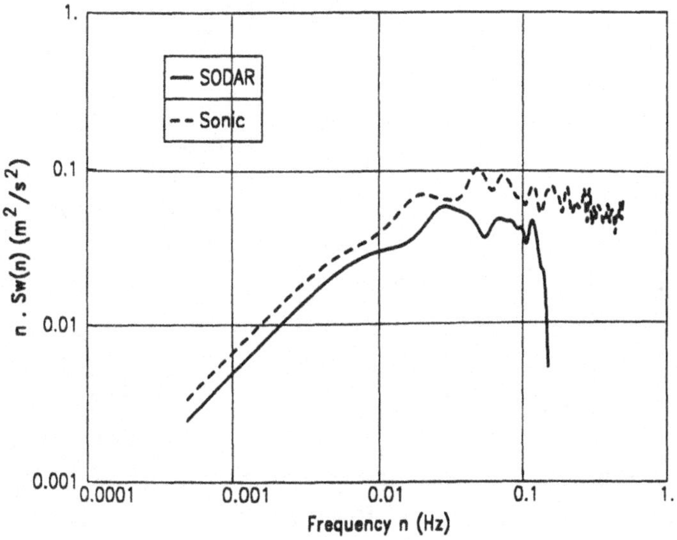

Fig. 4. Spectra of vertical wind speed, run No. 5 during slightly stable stratification with high wind speed.

4. Summary and Conclusion

During a long term intercomparison lasting two months 30-min mean values of the standard deviation σ_w of the vertical wind speed w measured at 100 m AGL with a sonic anemometer and a SODAR agreed well. But there was a general underestimation by the SODAR. During seven test runs lasting more than two hours and covering different meteorological situations, the vertical wind speeds were measured by both instruments instantaneously at 100 m AGL, from which spectra were calculated. These spectra are underestimated by the SODAR at high frequencies and during stable stratification. This underestimation is also responsible for the underestimation of σ_w, as the square of σ_w is equal to the integral over the spectra.

The peaks of the spectra measured by both instruments coincide well. The spectra can be used to estimate the outer turbulent scale wavelength and the integral scale of turbulence. Both parameters agree very well when calculated from sonic and SODAR measurements, respectively.

The whole intercomparison demonstrated that the Doppler-SODAR is a reliable remote sensing instrument to determine not only mean wind field data but atmospheric turbulence data as well.

References

Chintawongvanich P., Olsen R. and Biltoft C.A., 1989. *Intercomparison of Wind Measurements from two Acoustic Doppler SODARs, a Laser Doppler Lidar, and In Situ Sensors*. J. Atmos. Oceanic Technol., Vol. 6, 785-797.

Jenkins G.M. and Watts D.G., 1968. *Spectral Analysis and Its Applications*. San Francisco, Holden-Day.

Jones R.H., 1972. *Aliasing with Unequally Spaced Observations*. J. Appl. Met., Vol. 11, No. 2,245-254.

Kaimal J.C., Gaynor J.E., Finkelstein P.L., Graves M.E. and Lockart T.J., 1984. *An evaluation of Wind Measurements by Four Doppler SODAR*. *BAO* Rep. No. 5, Wave Propagation Laboratory, NOAA/ERL, Boulder, CO, USA 110 pp.

Priestley M.B., 1981. *Spectral Analysis and Time series*. Volumes 1 and 2, Academic Press, New-York.

Thomas P. and Vogt S., 1990. *Measurement of Wind Data by Doppler SODAR and tower Instruments: An Intercomparison*. Meteorol. Rdsch., Vol. 42, 161-165.

Von Holleuffer-Kypke R., Hübschmann W.G., SüB F., Thomas P., 1984. *Meteorologisches Informationssystem des Kernforschungszentrums Karlsruhe*. Atomenergie, Kerntechnik, Vol. 44, No. 4,300-304.

LIDAR MEASUREMENTS OF PLUME STATISTICS

HANS E. JØRGENSEN and TORBEN MIKKELSEN
Department of Meteorology and Wind Energy, Risø National Laboratory, DK-4000 Roskilde, Denmark

(Received October 1991)

Abstract. Surface-layer aerosol diffusion experiments have been conducted using artificial smoke plume releases at ground level over flat and homogeneously vegetated terrain at the Meppen proving grounds in the Federal Republic of Germany (1989). At fixed downwind locations in the range out to 800 m from the source, instantaneous crosswind plume profiles were detected repetitively at high spatial (1.5 m) and temporal (3 sec) intervals by use of a mini LIDAR system. The experiments were accompanied by measurement of the surface-layer mean wind and turbulence quantities by sonic anemometers. On the basis of measured crosswind concentration profiles, the following statistics were obtained: 1) Mean profile, 2) Root mean square profile, 3) Fluctuation intensities,and 4)Intermittency factors. Furthermore, some experimentally determined probability density functions (pdf's) of the fluctuations are presented. All the measured statistics are referred to a fixed and a 'moving' frame of reference, the latter being defined as a frame of reference from which the (low frequency) plume meander is removed. Finally, the measured statistics are compared with statistics on concentration fluctuations obtained with a simple puff diffusion model (RIMPUFF) developed at Risø.

1. Introduction

A full-scale experimental study on the natural variability and fluctuations in dispersing aerosol plumes has been conducted in the atmospheric surface layer, using LIDAR for remote sensing of instantaneous concentration profiles in cross-sections from a surface released plume.

Measurements of the near-ground, cross-wind concentration profiles were obtained 2 m above the ground at ranges between 100 and 800 m from the source with high spatial (1.5 m) and temporal (3 sec) resolution.

The experiments are supported by detailed measurements of wind and turbulence by use of sonic anemometers, meteorological towers, and frequent radiosonde balloons.

Experiments presented in this paper were obtained from the BOREX '89 campaign described later. The three presented experiments covered the meteorological situations of neutral, unstable, and stable with concentration measurements 160 m downwind from the release point.

The main objective of our study has been to establish a data base, consisting of series of detailed instantaneous crosswind concentration profiles, from which fundamental plume statistics relating to uncertainties in dispersion modelling can be studied.

Boundary-Layer Meteorology **62:** 361–378, 1993.

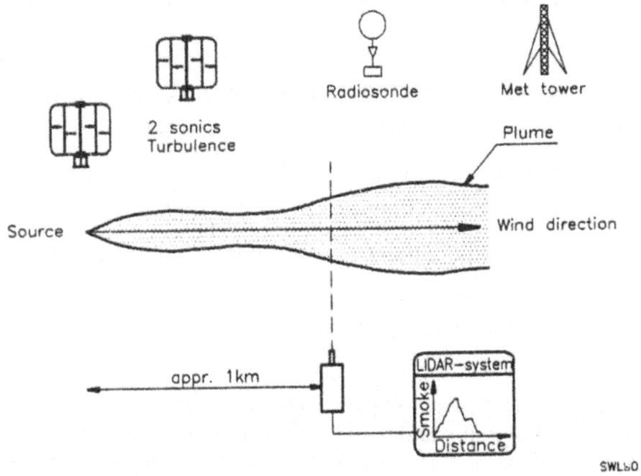

Fig. 1. Birdseye view of experimental setup for the diffusion experiments, showing (schematically): Source position, smoke plume, sonic anemometers and the mini LIDAR system.

2. Experiments

The BOREX '89 dispersion experiment took place over flat and homogeneously vegetated terrain, situated in Meppen, Germany, inside a military proving area. This site is ideal for micrometeorological dispersion experiments due to its flatness and homogeneity. In August the vegetation consists of naturally growing grasses and heather plants, which produce a homogeneously distributed roughness of the order of 1 cm. The general outline and setup for the experiments is shown in Fig. 1.

2.1. Meteorological Instrumentation

Meteorological instrumentation was established to support the plume dispersion experiments. Mean wind and turbulence, including atmospheric stability, were monitored by two 3-axis sonic anemometer/thermometers (Kaijo Denki type DAT 310) on 10 m tall meteorological masts, located immediately upwind of the release point.

2.2. Generation and Detection of Aerosol Plumes

2.2.1. Source of aerosols.

A powerful and sturdy aerosol generator, designed and constructed by Dynamit-Nobel, Germany, was used to produce continuous surface releases of submicron aerosol plumes consisting of conglomerated SiO_2 and NH_4Cl, which could easily be detected by the LIDAR system. The aerosols were formed by mixing the liquids $SiCl_4$ and a 25% water solution of NH_4OH in the neutralizing stoichiometric ratio (1:3.2) into a strong wind venturi. The flow rate of the aerosol was fixed at a constant rate. Depending on the atmospheric stability and the flow rate of the

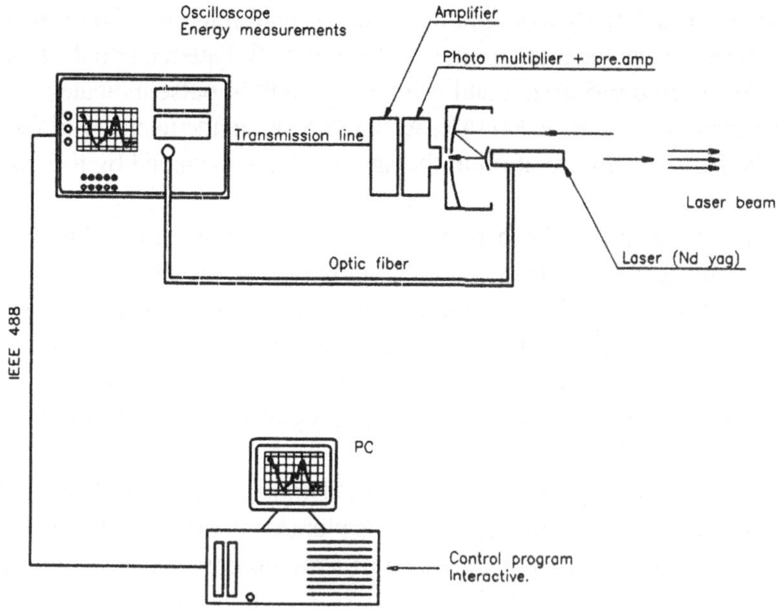

Fig. 2. Schematic view of the lidar system that shows the path between the controlling PC, digitizer receiver system and the laser.

chemicals, the plumes were visible several kilometers downwind, and detectable even farther downwind.

2.2.2. Aerosol measurements.

During the BOREX campaigns, the particle size distributions were measured at several downwind distances, both inside and in the margins of the meandering plume, by means of an optical particle counter (type Polytec HC-15). This instrument covered a particle radius size ranging from 0.16 um to 5.4 um, with 126 sizing bins. The distributions were found to be remarkably consistent and reproducible in terms of size interval and shape.

2.2.3. Lidar.

The instrument for detection of instantaneous crosswind profiles of plume concentrations is a Mini-LIDAR system, originally designed by the German Aerospace Research Establishment, Department of Opto-electronics, and built under license for our purpose by IBS GmbH, Grafrath (Germany). It consists of a pulsed laser (Nd:YAG 1064 nm), a 15 cm telescope with photodetector and preamplifiers, a transient recorder, and a PC-controlled data storage and monitoring facility. The design of the LIDAR system is illustrated in Fig. 2. The principle of the LIDAR is similar to the well known principle of remote detection using RADAR, only the LIDAR uses coherent light (from a laser) instead of microwaves: Pulsed laser radiation of very short duration (with a pulse length less than 1 m) is transmitted horizontally through the aerosol plume, where a small fraction proportional to the number of aerosols in the measurement volume is backscattered into the telescope. The time between transmission and reception of a light pulse indicates the range between the particles and the LIDAR. At the same time, the intensity of the signal reflects the particle concentration in the small volume occupied by the travelling light pulse.

Including the length of the laser pulse, the response time of the photo-detector, and of the pre and transient recorder amplifiers, the overall impulse response time of the LIDAR system was limited to 10 nanoseconds, which corresponds to an effective distance constant of the instrument equal to 1.5 m. The laser beam divergence was 3 milliradian, corresponding to a 0.60 m pulse diameter at 200 m range. The plumes were scanned with a frequency of 1/5 Hz at various downwind distances perpendicular to the plume axis.

The photo detector signals were sampled and digitized at 250 MHz, corresponding to a fixed point measurement through the plume every 0.6 m. With an effective distance constant of the instrument equal to 1.5 m, this corresponds to an sampling rate of 2.5.

3. Data Registration and Processing

3.1. Wind and Turbulence

Signals from the sonic anemometers were processed in real time during the experiments via an on line, fast scanning data acquisition system. This system provided consecutive 10 min averages of the mean wind speed and direction, in addition to the covariance matrix of the fluctuating wind components (u', v', w') and temperature (T'). In this way the key parameters for classification of the meteorological conditions and atmospheric stability, including the turbulence intensities, surface-layer shear stress (u_*), and the surface-layer heat flux (H_o), were monitored during the experiments. In addition, 10 Hz logging of the raw sonic anemometer signals was performed for subsequent spectral analysis and for use in the dispersion simulation.

3.2. LIDAR - Signals

Data processing of the measured backscatter profiles have been undertaken in order to correct the LIDAR measurements for range dependence and for extinction.

The LIDAR equation (see Measures, 1984)) was subsequently solved numerically to yield volumetric backscatter coefficients as a function of range by various techniques. To compensate for the extinction, an estimation of the extinction-to-backscatter ratio (25.0) was used in the LIDAR equation, appropriate for the artificial aerosol plume being measured.

A background visibility of 20 km, corresponding to an extinction coefficient at 1064 nm of 2E-4 [m^{-1}] was used as an initial estimate in the iteration procedure used to solve the LIDAR equation. A threshold value applied for clipping the background noise was set to 1.5 times this value. By inclusion of the system constant for the mini LIDAR in the iteration procedure, the photo detector signals could be transformed into cross-section profiles of volumetric backscatter coefficients for the plume. The backscatter coefficients in turn, are assumed to be proportional to the particle number concentration per unit volume, i.e., to the concentration. This assumption is justified when the shape of the particle size distribution, and the optical properties of the particles, do not change during an experiment (Evans, 1988). However the constant of proportionality must be calibrated from one experiment to another with aerosol concentration measurements taken from inside the plume.

4. Results and Discussion

4.1. Weather during Experiments

Thursday, August 17, 1989, was a clear and sunny day with few, scattered cumulus clouds riding on top of a 1200 m inversion cap. Light winds came from the west with speeds ranging from 4 - 6 m/s during the day. A transition from unstable to stable flow was encountered around sunset at 19:00.

Fig 3 shows the measured time series of the 10 min averaged quantities of wind speed and direction, temperature, mechanical shear stress and surface heat flux as measured by the sonic anemometer. The shading marks the time interval of the unstable period between 16:29 and 17:45.

Fig 4 shows the turbulent energy spectra of the three velocity components (u', v', w'), calculated from the 10 Hz timeseries measured during the unstable period between hrs. 16:40 to 17:35.

4.2. The Crosswind Plume Structure

In the time interval between 16:29 and 17:45 (the unstable period), a series of 872 crosswind plume profiles were taken near the ground with the LIDAR system positioned 160 m downwind of the source.

Fig. 5 shows a sequence of these data, containing 80 measured profiles between 17:03 and 17:09. The profiles are presented in two different frames of reference, - a 'Fixed frame', and a 'Moving frame'.

The 'Fixed frame' (Fig. 5 left) presents the profiles as measured from the fixed position of the LIDAR.

The 'Moving frame' (Fig. 5 right) presents the same data set, but with the low frequency plume meander component removed. In the moving frame the profiles have been aligned in the horizontal crosswind direction in such a way that their individual center of mass positions coincide. This is equal to removing the larger scale turbulent eddies (greater than or equal to the plume size), responsible for plume meandering. Measurements and calculations in this frame are consequently representative for 'in-plume' statistics, and for the relative diffusion process governing puff dispersion.

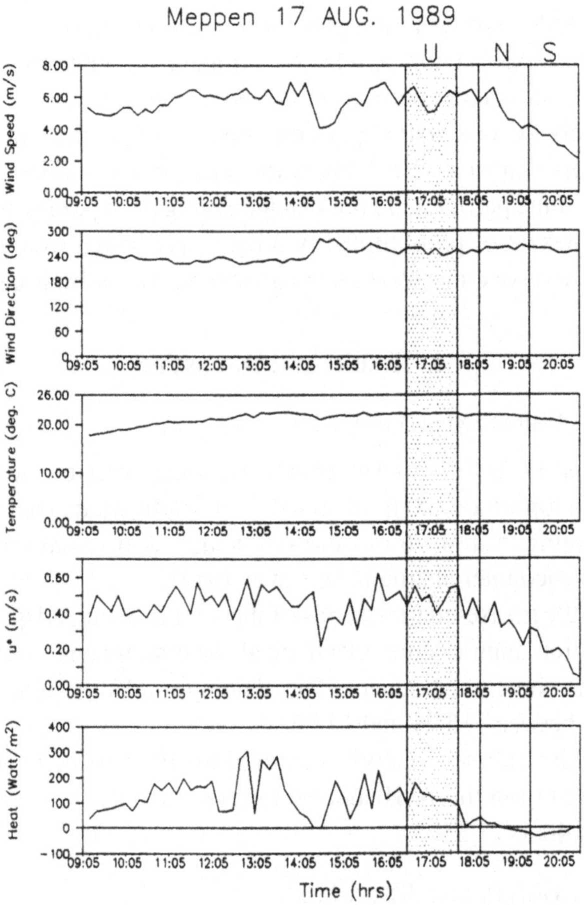

Fig. 3. 10 min averaged meteorological quantities, measured on August 17, BOREX 89. The shaded period marks an unstable (U) section, selected for detailed analysis in this paper. The period 18:15–19:25 marks a neutral (N) run, and the period 19:23 to 20:33 marks a stable (S) run. Quantities shown are: Wind speed, Direction, Temperature, Mechanically induced shear stress turbulence, and turbulence transported heat flux.

Statistics of the crosswind plume structure, based on the entire subset of 872 profiles, are presented in Figure 6. Also here the data are presented in a 'fixed frame' (left) and a 'moving frame' (right). Starting from the top the figure shows:

1) Mean concentration profiles
2) Fluctuation concentration profiles (root mean square)
3) Intensity profiles (defined as the ratio of 2 over 1)
4) Intermittency profiles (defined as the fraction of time with nonzero concentration values).

Significant differences can be seen to exist between statistics calculated in the

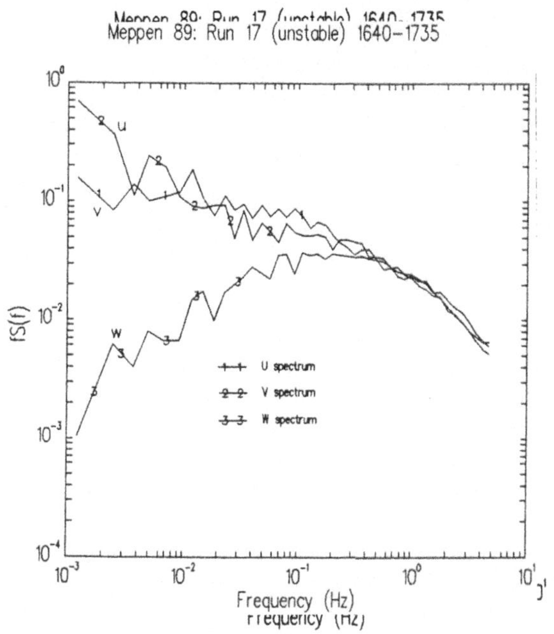

Fig. 4. Ensemble average of 5 velocity spectra of the wind components u', v', w' in the unstable case.

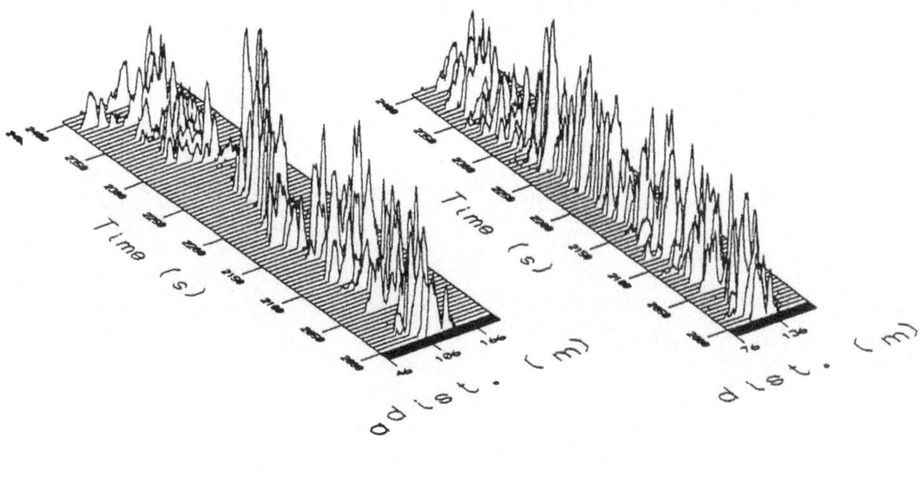

Fig. 5. Fixed frame (left) and Moving frame (right) profiles.

two different frames of references.

Where the mean and fluctuation (RMS) profiles in the fixed frame exhibit significant irregularities, the moving frame profiles become almost Gaussian in form. Also the centerline peak concentration is seen to be about 5 times higher in

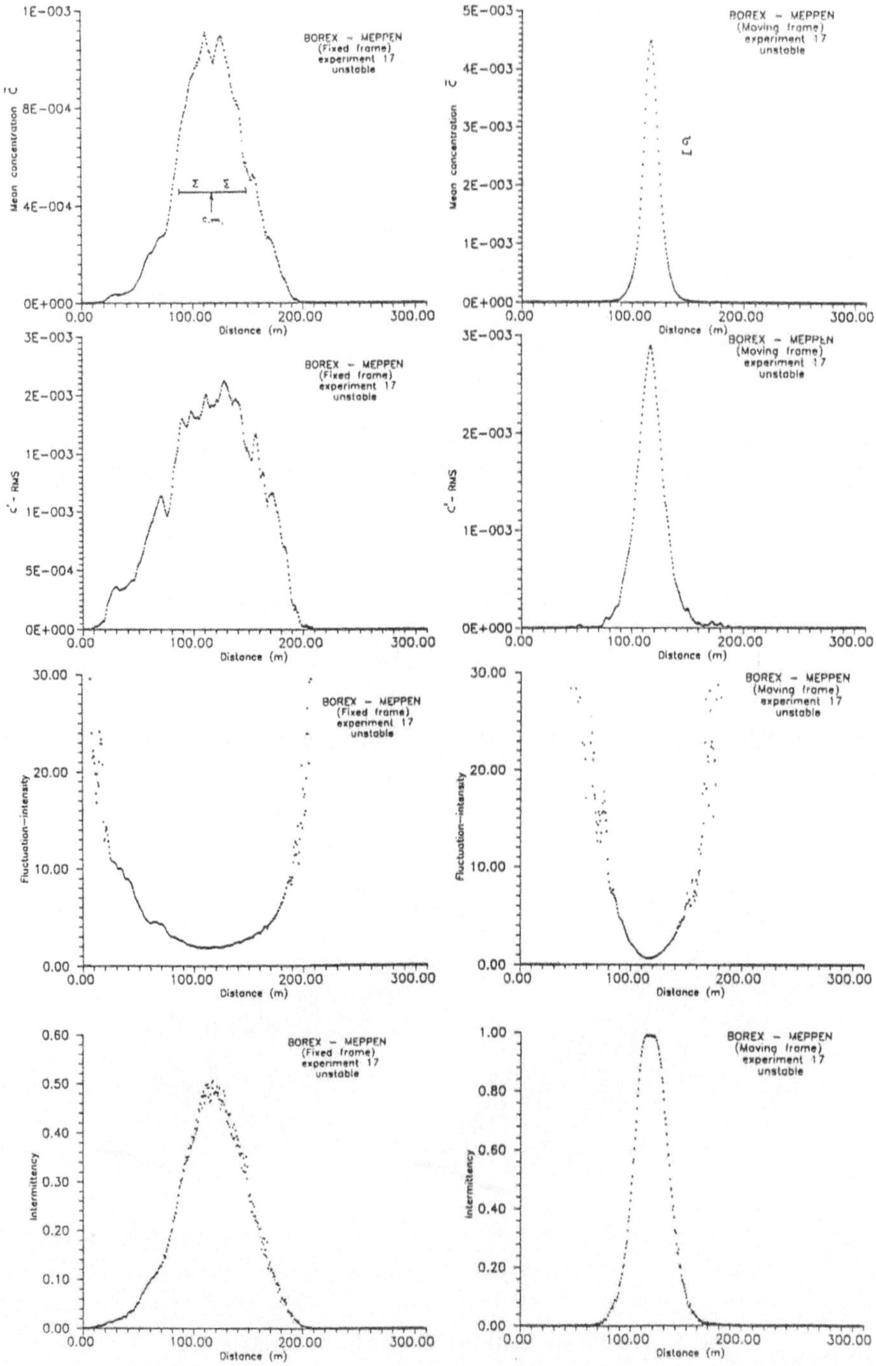

Fig. 6. Crosswind structure from unstable case. Shown for fixed (left) and moving (right) frames, reference are: 1) Mean profiles, 2) Fluctuation profiles, 3) Intensity profiles, and 4) Intermittency profiles. Abscissae shows distance from the LIDAR. Concentrations are expressed in values above threshold (2E-4m^{-1}). Plume standard deviations are: 29.6 m in fixed and 8.9 m in moving frame.

the moving frame, and the plume width (sigma value) is much smaller. Furthermore, the RMS profile in the centerline of the plume is reduced by some 30% during the transformation from fixed to moving frame. This shows that at this downwind distance, the horizontal meandering represents 30% of the RMS in the centerline.

The fixed frame intensity profile exhibits the anticipated, characteristic U-shape often found (Hanna (1984); Weber *et al.* 1988)), but in this study with amazing symmetry and height also for the 'moving frame' case. The fluctuation intensity in the centerline is also reduced going from fixed to moving frame. This is due to the relatively high increase of the corresponding mean concentration at the centerline.

Another effect demonstrating the difference between fixed frame and in-plume statistics (moving frame) where the horizontal meander has been removed is the profound difference in peak intermittency, exceeding the 99% level in the moving frame case while barely reaching the 50% level in the fixed frame case. The intermittency factors reported may be somewhat arbitrary due to their dependence on the subjectively introduced threshold value for removal of background noise, but the profound difference between fixed frame and moving frame statistics is clear. Finally, it should be noted that there is a great deal of similarity between the defined 'moving frame' statistics and the 'conditional plume statistics' investigated by Sawford (1987), Wilson *et al.* (1985).

The traditional definition of 'conditional statistics' of plume concentrations only discriminates in-plume concentrations from zero readings. Our definition of in-plume statistics (moving frame) discriminates the statistics with respect to the position within the ensemble averaged instantaneous plume. i.e., we discriminate statistics measured near the centerline of the instantaneous plume from the statistics measured near the edges of the plume. This is possible only due to the high spatial resolution obtained with the LIDAR system.

4.3. THE CONCENTRATION PDF'S

For every crosswind position measured by the LIDAR system, the probability density function (pdf) could be calculated. Some pdf's have been made on the basis of the 872 sequential concentration profile measurements obtained in the unstable case. These are shown in Fig. 7. Fig. 7 distinguishes again these measured pdf's between a fixed (left) and a moving frame (right) as before, but it also shows the centerline pdf (top), and plume margin pdf (bottom). The most important scaling parameter for the pdf's is the intermittency factor. The shape of the (99% intermittent) moving frame centerline pdf differs from the three other pdf's of lower intermittency by having a nonzero modal value. This is of interest for puff modeling.

When compared with other experimentally determined pdf's, it is important to keep in mind that our pdf's have been measured with an effective distance constant (i.e., line averaging) of the LIDAR system equal to 1.5 m., which is responsible for a filtering (a low pass averaging) of the smallest scale fluctuations. In our case concentration fluctuations of a scale less than 1.5 m have been filtered out (See

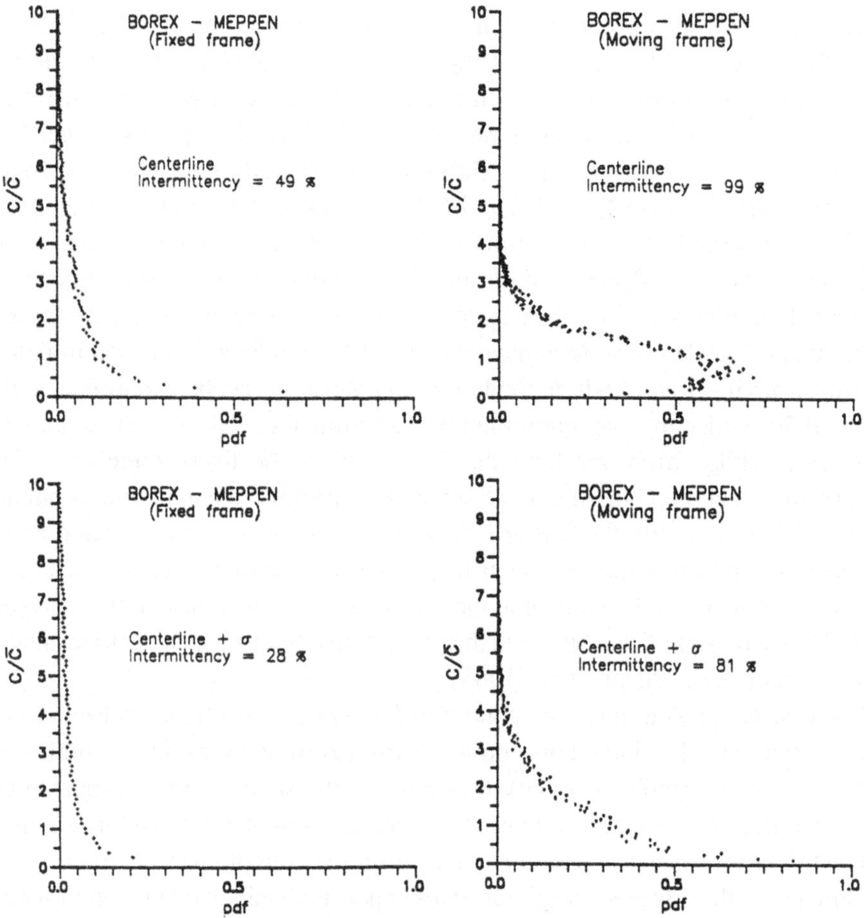

Fig. 7. Probability distribution functions from unstable case. Shown are fixed (left) and moving frame (right) pdf's at centerline (top) and at plume margin (bottom).The pdf's are shown with reversed axis.

Mikkelsen *et al.*, 1990).

4.4. LIDAR MEASUREMENTS VERSUS RIMPUFF

On the basis of the measured concentration fluctuation statistics, a 'data vs. model' comparison study has been made with the puff dispersion model RIMPUFF (Mikkelsen *et al.* 1984) . Three different stabilities (unstable/neutral/stable) have been modelled (see Fig 3).

4.4.1. RIMPUFF model

RIMPUFF is a Gaussian mesoscale puff dispersion model designed for real time simulations of hazard releases to the atmosphere. The model applies to stationary meteorology and in conjunction with a flow model, also to non uniform terrain with moderate topography on a horizontal scale from 0 to 100 km, Mikkelsen *et al.*(1984), Thykier-Nielsen *et al.* (1990).

Input for the model can be either modelled or measured winds, with corresponding turbulence intensities or stability classes.

Gaussian puffs are advected in the flow field at a user specified time interval. The model uses two different methods to calculate the growth of the puffs, either by use of simultaneously measured values of the turbulence intensities (for instantaneous puff growth), or through specification of Pasquill stability classes combined with a power law description of the (10-min averaged) plume diffusion (See Mikkelsen *et al.* (1984) for details).

4.4.2. Input to RIMPUFF

For the present study, the wind data for driving RIMPUFF have been obtained by high speed sampling of wind speed and direction from a tower located upwind and near the release point.

Turbulence intensities, both lateral and vertical, used for calculating the puff growth, were produced from a 2 min running mean of the 10 Hz sampled sonic data. A spatially uniform but time varying wind field was specified through calculated 10 sec mean values of the instantaneously measured wind speeds and directions.

During the simulation experiments, the source strength in the model was kept constant, and a puff was released every 5 sec. The model receptor points for monitoring timeseries of concentrations were as in the experimental setup, which corresponds to 512 receptors located along a line each 0.6 m in the crosswind direction, 160 m downwind from the release point. The instantaneous concentrations modelled in this way were sampled at the many receptor points at 1/5 Hz.

4.4.3. Comparison

The strength of the aerosol generator producing the visible smoke plume was kept constant (within a few per cent) although its strength not known in absolute terms. This however has no importance because the LIDAR is able to measure concentrations only in relative terms. Consequently, in order to compare LIDAR measured and RIMPUFF modeled statistics, the measured mean concentration profile has been normalized with the area of modelled mean profile. This is the only adjustment used for the comparison, except for a scaling of the intermittency profiles which will be explained latter.

Results of the measured and modelled mean profiles are shown in Fig. 8. It is seen that RIMPUFF is able to predict the mean crosswind concentration profile for the unstable and near neutral case. However, in the stable case the measured mean concentration profile exhibits a very narrow peak (where the meander has died), which the puff model is apparently not able to resolve. Results of the RMS profiles are compared in Fig. 9. The result also shows good agreement between the measured and modelled RMS. But also here the modelled RMS differs significantly from the measured RMS in the stable case. This is due to the absence of larger scale meandering of the plume. The in-plume fluctuations then become relatively more dominating, and the larger difference occurs because RIMPUFF is not able

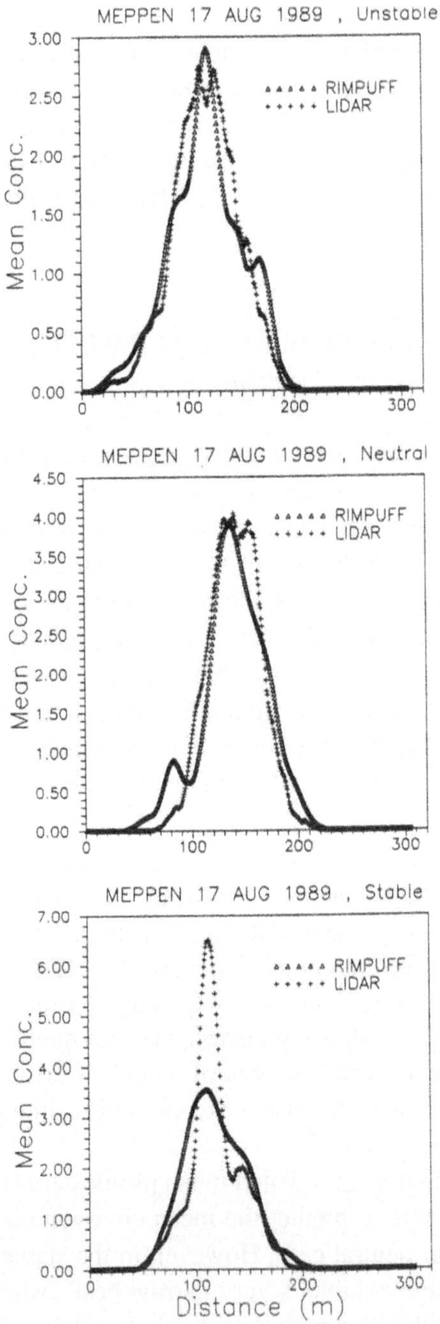

Fig. 8. Modelled and measured mean concentration profile.

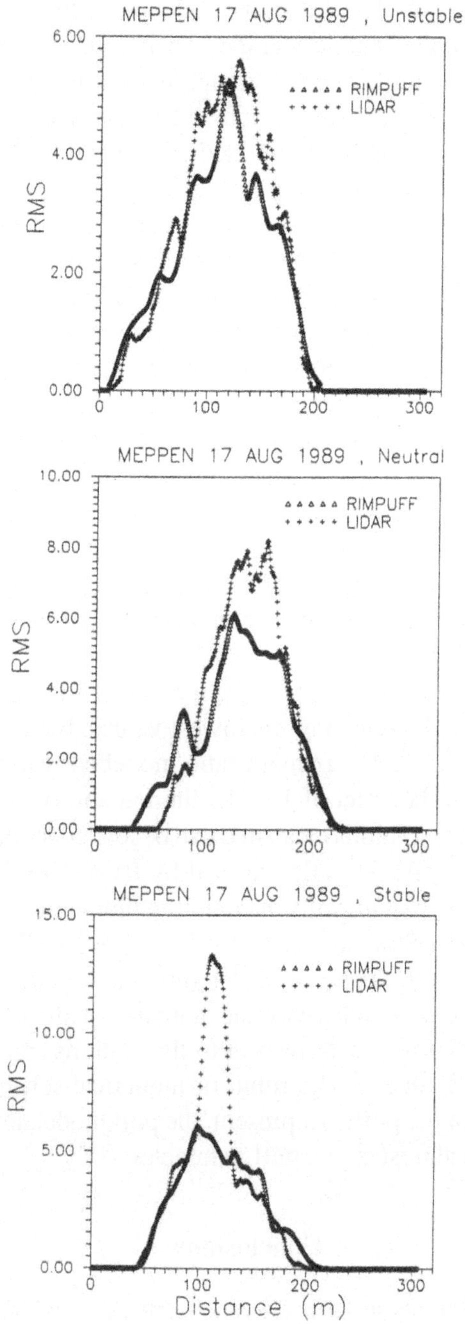

Fig. 9. Modelled and measured crosswind RMS.

to produce in-plume fluctuations due to the individual puff's homogeneity and uniform concentration distribution (Gaussian). Results of the fluctuation intensity are shown in Fig. 10. Both the measured and the modelled data show the characteristic U-shape. But one should also notice that the uncertainties increase on the plume edges because of high intermittency. A comparison of intermittency factors are shown in Fig. 11 exhibits excellent agreement in the neutral and unstable case. The intermittency produced by RIMPUFF is based on a specified threshold cut-off of concentrations. The threshold value used is calculated on the basis of the measured data. In each case (unstable/neutral/stable) the fraction is calculated between the mean concentration in the centerline and the noise cut-off level. This fraction is used to produce a threshold in the modelled case on the basis of the modelled centerline mean concentration. In general the model shows good agreement between observed and modelled concentration fluctuation data, especially when meandering is the dominating source of fluctuations as it is in the unstable and neutral case. The variance produced by the model is mainly related to the meandering of the plume. A small amount of the variance is due to fluctuations in C_0 (max concentration in the puffs) which is created by fluctuations in the dispersion parameters (sigma values), which in turn are due to measured input of turbulence intensities determining puff sizes. This is insufficient to produce fast internal fluctuations in the mean plume and was originally included to account for non-stationary dispersion characteristics on a longer time scale (5-10 min's).

The pdf in the crosswind centerline for the neutral case is shown in Fig 12. The modelled pdf shows fair agreement at the lowest values, but the agreement stops at $C_{0,max}/C_m = 8$ which is the maximum ratio modelled, whereas the measured pdf continues to ratios of the order of 15. The limited ability of the puff model to simulate the highest concentrations measured can be related to the lack of in-plume fluctuations in the puffs. The pdf's determined by RIMPUFF has a very similar nature to the one obtained by Gifford's meandering plume model (Gifford, 1959), and the extension given by Sawford *et al.* (1985). A common feature for the pdf's mentioned is that they fail in the sense of determining the tails, and therefore are more or less useless in determining extreme statistics of the concentrations. An improvement of the puff model with respect to fluctuations and extreme statistics would naturally be to develop an algorithm or numerical scheme accounting for the internal fluctuation of the puffs. At present, the puff model deals with ensemble averaged, and therefore almost static, puff parameters only.

5. Conclusions

We have produced continuous surface releases (lasting several hours) of submicron aerosol plumes which were detected by our LIDAR system. The LIDAR was oriented to measure backscatter profiles inside the aerosol plume at right angles to the plume axis, and at various downwind locations. The experiments analyzed in this paper are concentration profiles taken 160 m downwind from the source.

Since the concentration fluctuations at this range are primarily dominated by plume meandering, and not so much by in-plume fluctuations, a simple puff model

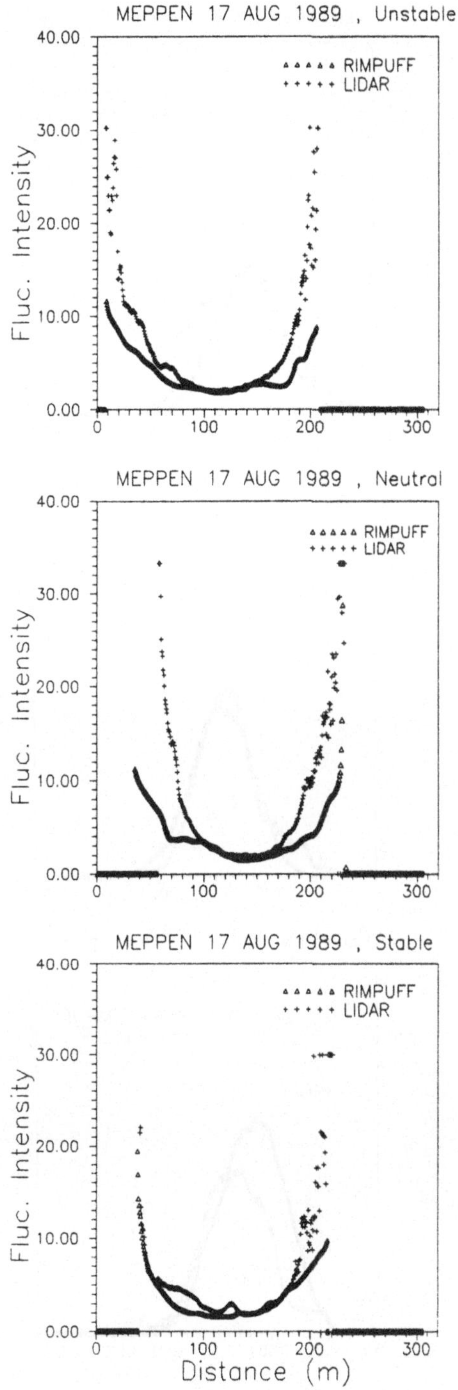

Fig. 10. Modelled and measured fluctuation intensity.

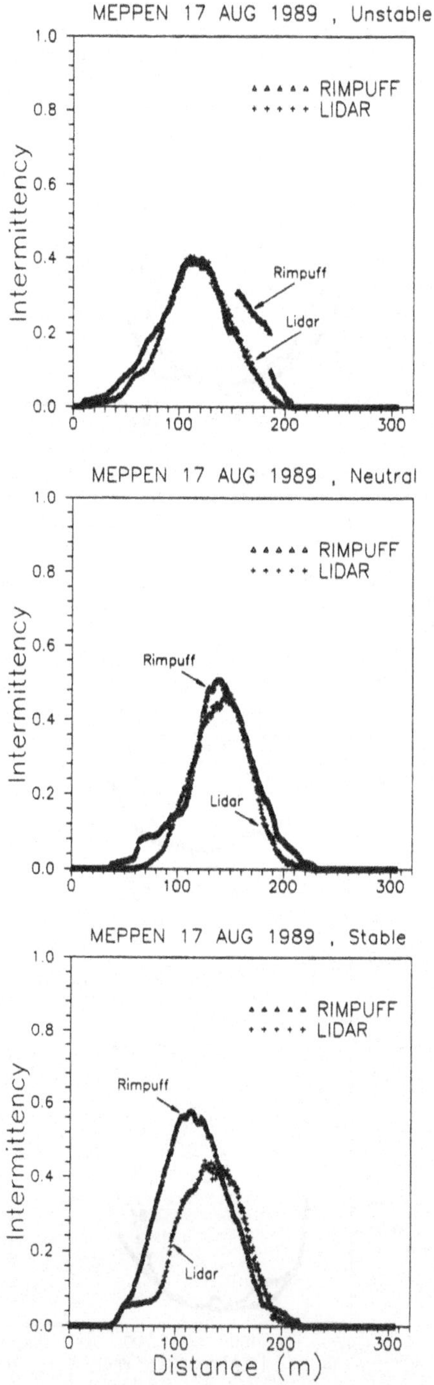

Fig. 11. Modelled and measured intermittency.

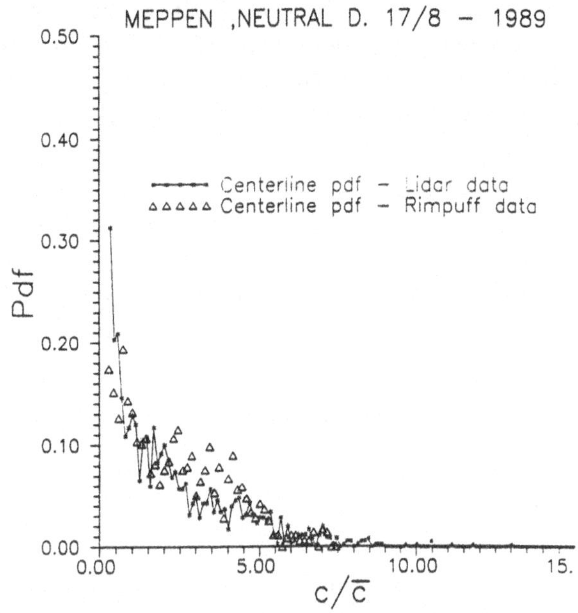

Fig. 12. Modelled and measured Pdf from the crosswind centerline in the neutral case.

driven by measured winds, is able to reproduce most of the measured fluctuation statistics. The ability of the model is worse under stable conditions, but still acceptable for real time applications.

The Lidar system has in addition provided us with new and important information on the in-plume statistics of smoke plumes. Although the individual instantaneous crosswind profiles do not at all seem to be gaussian, the mean instantaneous crosswind profile can be regarded as gaussian, as the moving frame measurement shows. This is also the idea behind using 'gaussian' puffs for the ensemble averaged clouds.

The BOREX data base, from which these experiments were taken, provides the real time modeller with a series of reference dispersion experiments in which the level of naturally occurring fluctuations in plume concentrations have been investigated. In addition, the data set establishes a reference data set for comparison and evaluation of near source atmospheric dispersion models.

Acknowledgements

This work has been sponsored by the Commission of the European Communities (Nuclear Safety Research, Radiation Protection Program, DG XII.D.) under contract No. BI7-0017 for the assessment and uncertainty evaluation in real time decision support systems.

We gratefully acknowledge the following research groups who actively collaborated in this project: Harald Weber and Welfhart aufm Kampe, Amt für Wehrgeophysik, German Military Geophysical Office (GMGO), Traben-Trarbach, Germany. Christian Werner, Hartmut Herrmann and Jürgen Streicher, German Aerospace Research Establishment (DLR), Institute of Opto-electronics, Oberpfaffenhofen, Germany and Stephan Borrmann, Institute of Meteorology, University of Mainz, Germany. In addition we want to thank our skilled technicians, who made these experiments possible: Arent Hansen and Søren W. Lund (Risø), Udo, Bernt and Klaus (GMGO). And finally thanks to the people at the 'Erprobungsstellung 91', Meppen, in particular to Mr. Lother Lüer, for his willing collaboration and assistance.

References

Evan, B.T.N., (1988): *Sensitivity of the Back scatter/Extinction ratio to changes in Aerosol properties: implication for LIDAR*. In: Applied Optics, vol 27 pp. 3299 (Aug. 1988).

Gifford, F.A. (1959): *Statistical Properties of a Fluctuation Plume Dispersion model*. Adv. Geophys., 6 pp 117-138.

Hanna, S.R. (1984): *The Exponential Probability Density Function and Concentration Fluctuations in Smoke Plumes*. In: Boundary Layer Meteorology vol 29, pp. 361–375, 1984.

Measures, R.,M., (1984): *Laser Remote Sensing – Fundamentals and Applications*. John Wiley & sons, New York, 1983.

Mikkelsen, T., S.E Larsen, and S. Thykier-Nielsen (1984): *Description of the Risø Puff Diffusion Model*. In Nuclear Technology, Vol 67, oct. 1984, pp. 56–65

Mikkelsen, T., H. E. Jørgensen, W. aufm Kampe, H. Weber and S. Borrmann (1990): *The Effect of Finite Sampling Volumes on Measured Concentration Probability Density Functions*. In: Ninth Symposium on Turbulence and Diffusion (Published by the American Meteorological Society, Boston, MA, USA).

Sawford B.L., and H. Stapountzis (1986): *Concentration Fluctuations According to Fluctuating Plume in One or Two Dimensions*. Boundary Layer Meteorology, vol 37 (1986), pp 89–106.

Sawford B.L., (1987): *Conditional Concentration Statistics in the Atmospheric Boundary Layer*. Boundary Layer Meteorology, vol 38 (1987), pp 209–223.

Thykier-Nielsen, S., T. Mikkelsen, R. Kamada, S.A. Drake (1990): *Wind Flow Model Evaluation Study for Complex Terrain*. In: Ninth Symposium on Turbulence and Diffusion (Published by the American Meteorological Society, Boston, MA, USA).

Weber, H., W. aufm Kampe and T. Mikkelsen (1988): *Concentration Fluctuations Measured in the Atmospheric Surface Layer*. In: Eighth Symposium on Turbulence and Diffusion (Published by the American Meteorological Society, Boston, MA, USA).

Wilson, D.J., Fackrell, J.E., and Robins, A.G., (1985): *Intermittency and Conditionally Averaged Concentration Fluctuation Statistics in Plumes*. Atmos. Eviron., vol 19, pp. 1053-1064 (1985)

SPATIAL STRUCTURE OF A JET FLOW AT A RIVER MOUTH

B. SHTEINMAN, E. MECHREZ and A. GUTMAN
Kinneret Limnological Laboratory, Israel Oceanographic & Limnological Research, P.O.B. 345, Tiberias, Israel 14102

(Received October 1991)

Abstract. The present work concentrates on the latest data measured in the Jordan river flow in lake Kinneret. Spectral characteristics of fluctuating velocity components have been obtained and processed. The three-dimensional structure of turbulence along the flow has been described. The main features of the jet flow turbulence in the river mouth are: a) The supply of turbulent energy changes due to different mechanisms along the flow. b) The structure of turbulence formed in the river decays rapidly along the flow, and c) In the sand area and beyond it, a significant generation of turbulent energy occurs. Quantitative estimations of the above effects were carried out.

1. Introduction

In this paper we emphasize the latest data measured in the Jordan river flow into Lake Kinneret. The purpose of this on-site study is to measure turbulence in specific flow conditions existing in river mouths, including: a) A free jet flow with the turbulence formed within the river and transformed into lake turbulence; b) A specific bottom relief, including ripples, a sand bar and a sharp deepening after the bar crest.

This study is one of several on-site investigations of jet flow turbulence in river mouths that we intend to perform. The complete investigation includes two other stages:

- Channel mouth laboratory study that was carried out at the Transcaucasia Regional Hydrometeorological Institute, (TRHI) at Baku (see Shteinman *et al.*, 1992)
- Development of a conceptual model for turbulence at a river mouth.

2. Experimental Set up and Data Processing

The experiments were performed during May-June 1991. The profile of the Jordan River mouth was measured with the echosounder along the center line. The recorded profile is presented in Figure 1. Large accumulations on the bed and a sand bar, seen in Figure 1, are explained by significant sediment transport in the Jordan during high water seasons. The flow velocity fluctuations were recorded every 25 m downstream from the exit-cross section up to 300 m, at up to four depths, wherever possible. The measuring devices are velocity fluctuation meters, developed at TRHI. Within each 15 min record, the 3-D velocity fluctuations of the flow have been measured every 0.1 sec. Only stationary processes are treated in the present

Boundary-Layer Meteorology **62:** 379–383, 1993.

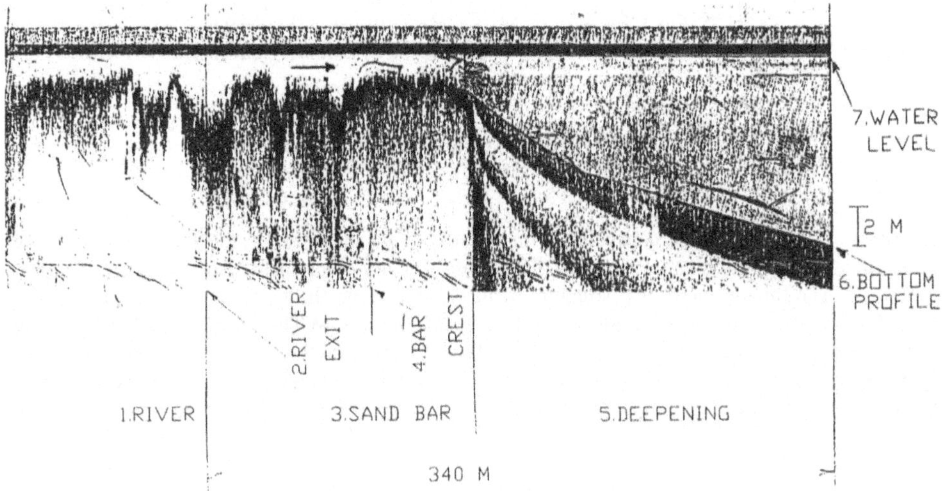

Fig. 1. Bottom relief of the river Jordan mouth: 1. River section; 2. River exit; 3. Mouth sand bar; 4. Bar crest; 5. Depth's drop; 6. Bottom profile; 7. Water level.

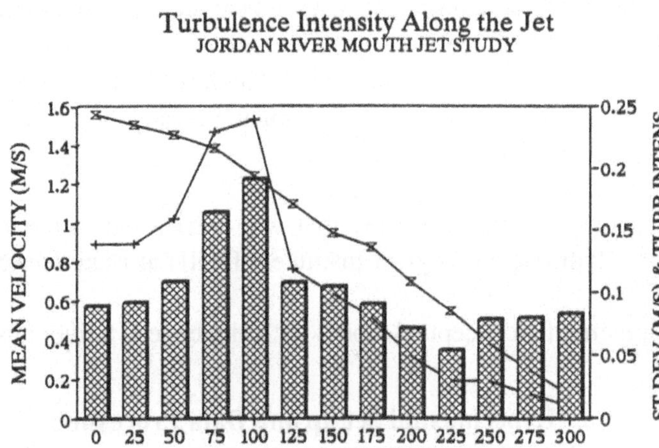

Fig. 2. Turbulence intensity along the surface jet flow in the Jordan river mouth.

study, therefore raw data containing errors or trends were sorted out at the editing stage of the data processing. Spectral characteristics of the three components of velocity fluctuations have been computed. The resulting turbulence parameters are presented in the following.

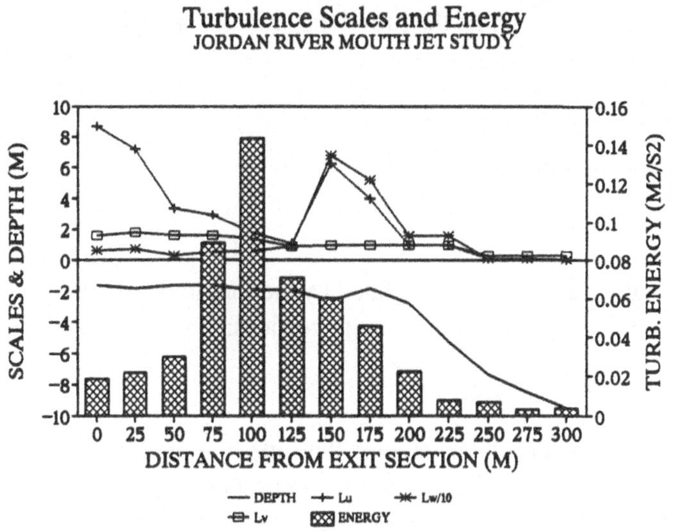

Fig. 3. Turbulence scales and energy along the surface jet flow in the Jordan river mouth.

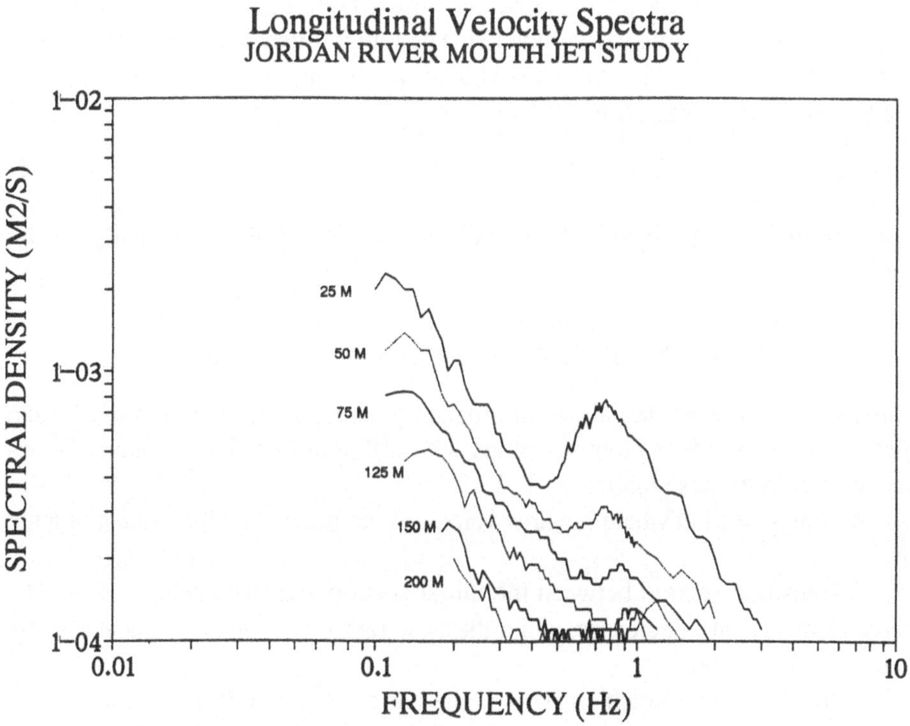

Fig. 4. Spectral density of longitudinal velocity at six cross-sections of the Jordan river mouth.

3. Turbulence Parameters of the Jet Flow

The obtained data on velocity components (u, v, w) along the flow provide parameters of turbulence at the river mouth. The standard deviations of velocity components $(\sigma_u, \sigma_v, \sigma_w)$ define the turbulence kinetic energy E:

$$E = \frac{1}{2}(\sigma_u^2 + \sigma_v^2 + \sigma_w^2) \tag{1}$$

and the turbulence intensity K:

$$K = \sigma_u/U \tag{2}$$

where U is the average velocity calculated by the longitudinal velocity component u, and the ratio between the scales of large vortices:

$$L_u : L_v : L_w = \sigma_u^2 : \sigma_v^2 : \sigma_w^2. \tag{3}$$

It is accepted in river turbulence studies that one may take the vertical scale L_w equal to the river depth H. We follow this assumption up to the point where the jet flow separates from the river bed after passing the bar crest. After this point the spectral characteristics of the velocity autocorrelation function show a well defined horizontal interface between the jet and the near bottom layers. In this case the vertical dimensions of the jet are chosen to approximate L_w. Two others scales L_u and L_v are obtained using expression (3). The longitudinal scale of large vortices L_u was also obtained in another independent way, using the mean frequency ω of longitudinal velocity pulsations:

$$L_u = U/\omega. \tag{4}$$

Both definitions of L_u have led generally to similar results, thus justifying the chosen approach.

4. Distribution of Turbulence Parameters along the Flow

Figures 2 and 3 present the above turbulence parameters measured in the Jordan river mouth. Three river mouth sections with different turbulence characteristics may be seen from these figures.

1. An initial section where the flow velocities are practically the same along the flow.

2. A transition section between the initial section and the sand bar crest. The longitudinal velocity decreases, while the flow rate is practically the same along the flow in this section.

3. A free flow section which starts at the bar crest and ends at the zone of full velocity fading. The water surface longitudinal slope drops to zero through this section, the flow becomes pure inertial and the flow jet separates from the bed (in most cases).

The first two of the above sections (first 100 m in our case) show the following jet flow turbulence characteristics:

- Turbulence intensity and energy rise to their maximum values;

- longitudinal vorticity scale decreases while the lateral scale increases and starts to prevail.

The last free flow section is characterized by the following features of its turbulence:

- Turbulence energy drops to very small values;
- longitudinal intensity drops to some value and stays constant;
- the lateral scale of vorticities is an order of magnitude bigger than that of the longitudinal;
- all three components of vorticity scales drop to the size of common lake vorticities.

5. Turbulent Energy Generation along the Flow

The above features point to different mechanisms of turbulent energy generation, which are switched on as the jet flow goes through the river mouth. The first two sections of the river mouth are characterized by two zones of energy supply, each observed as peaks on the spectral density curve (see Figure 4), like in the case of river turbulence. These zones cause two groups of vortices: a large one with the vertical scale equal to the depth, and another one with size equal to about the distance between the bed ripples' crests. Significant generation of turbulent energy at the second section, as observed in Figure 3, may be explained by the jet-stagnant water contact zone, at which the whirls build up. The third section is characterized by the flow separation from the bed and the whirls build up in the jet-lower layers interface. This affects turbulent energy generation. As can be seen from Figure 4, the inertial '-5/3' interval on the spectral density curve is no longer observed at this section, which might stem from a strong anisotropy generated in the bar crest area. A complete description of spectral characteristics and the mechanisms of energy generation and dissipation in river mouths will be presented elsewhere.

References

Shteinman B, Gutman A., Mechrez E. (1992): *Laboratory Study of the Turbulent structure of a Channel Jet Flowing into an Open Basin. Boundary-Layer Meteorology* **62**, 411–416, 1993 (this volume).

LIMITATION OF USING MULTI-TYPE SENSORS IN COMBINATION FOR VERIFYING ATMOSPHERIC PHENOMENA

S. EGERT, E. EYAL, J. SIVAN and R. REICHMAN
Israel Institute For Biological Research, P. O. B. 19, Ness-Ziona, Israel

(Received October 1991)

Abstract. The feasibility of combining two sensors for monitoring gas concentrations over a large area, is demonstrated via controlled simulation experiments. The network considered was a lidar combined with a point detector – the MIRAN spectrophotometer equipped with a flow-through gas cell. The first has a spatial scanning capacity and a very good time resolution, whereas the second has an inherent temporal memory effect. Algorithms for operating the two sensors coincidentally, taking their physical characteristics into account, were constructed. A good fit was found between measured concentrations in the MIRAN and reconstructed concentrations, using the lidar data for the same environment. Thus complementary operation can be achieved. The MIRAN temporal memory effect can be overcome by correlating the concentration calculated over different time intervals.

1. Introduction

Two types of detectors are usually combined in order to: a) test new systems. (The information from one detector serves to verify results of to the other.) b) gain information that otherwise could not be achieved, because the two detector types are based on different physical characteristics. The second issue is important especially when each detector type complements the information gained by the other, otherwise the system information is incomplete.

The best way to monitor the distribution of gas concentrations over a wide area, is to utilize a single system, which is a lidar (Measures, 1984). This system is capable of covering large atmospheric sections, mapping them either in planar or volume coverage, during short time periods. However, the main limitation is the need of line of sight along the direction to be monitored. This can sometimes be achieved only partially because some locations are not on a line of sight from the lidar. The combination between this system and point detectors located at these specific sites may help to overcome the problem.

Monitored gas information is usually analyzed on a temporal basis, making it possible to follow cloud spread. The example below shows the potential difficulties in the using two sensors and the method making possible this temporal analysis when two different detector types are operated together.

2. Instrumentation and Method

The goal of the multi-detector experiment was to operate two different detectors in the same environment and outline a method to extract the necessary information from the combination. Therefore, the two detector types were operated in a con-

Boundary-Layer Meteorology **62:** 385–391, 1993.

trolled environment where the detected gas was the atmospheric tracer SF_6 instead of a naturally occurring gas. The appropriate monitoring system is the CO_2 lidar (Uthe, 1986).

The lidar traced this gas by applying the Differential Absorption Method (DIAL) (Measures, 1988). This method is based on spectral discrimination of the gas from its background. The power of two spectral lidar signals - one in the absorption peak of the gas and the other, a reference spectral line which is not absorbed, are monitored continuously. The signal in the absorption peak will decrease as the gas concentration increases, while the power of the reference line will not change. According to the lidar equation (Measures, 1988) the logarithm of the signal ratio is proportional to the gas concentration.

The samplers which are usually used in conjunction with the lidar systems are point detectors. Their time constant is very long relative to the nearly real-time lidar system, even when they are operated at successive short time intervals.

The relatively fast MIRAN (MIRAN - IA) gas monitor was chosen as the second detector. This instrument is a spectrophotometer connected to a gas cell. The monitored gas flows through the cell at a constant rate. A light source is viewed through the gas by the appropriate optical detector. The light is spectrally filtered to the known absorption band of the monitored gas, so that the gas concentration in the cell can be measured by the degree of signal attenuation.

In order to assure that both instruments function in the same environment, the gas was restricted to a closed cell situated in the open atmosphere (Fig. 1). The lidar beam entered the cell through a window at its front, and reflected back from a target at its rear end. The gas sampled into the MIRAN was aspirated from the cell through ten equivalent nozzles, located on a line parallel to the lidar line of sight. The gas was injected into the cell in known amounts and mixed uniformly within its volume using a slow fan. The gas level control in the cell, lidar and MIRAN were operated using the same computer system. Thus gas levels, time between gas injections and intervals between lidar and MIRAN operation could be controlled.

The cell was operated both as a closed cavity, and as a half open one. During the second mode of operation, the cell was filled to a known concentration, then the front window was opened. The gas slowly diffused out of the cell, allowing a gradual decrease in its concentration inside the cell. This allowed operation of the two systems in slowly changing conditions.

The lidar system was operated at 10Hz where every ten signals were integrated to increases system stability. The MIRAN volume is 5.6 l and its flow rate cannot exceed 10 l/min. Therefore, MIRAN measurements were taken every five second in order to allow a change of more than 10% of the enclosed gas volume between consecutive measurements.

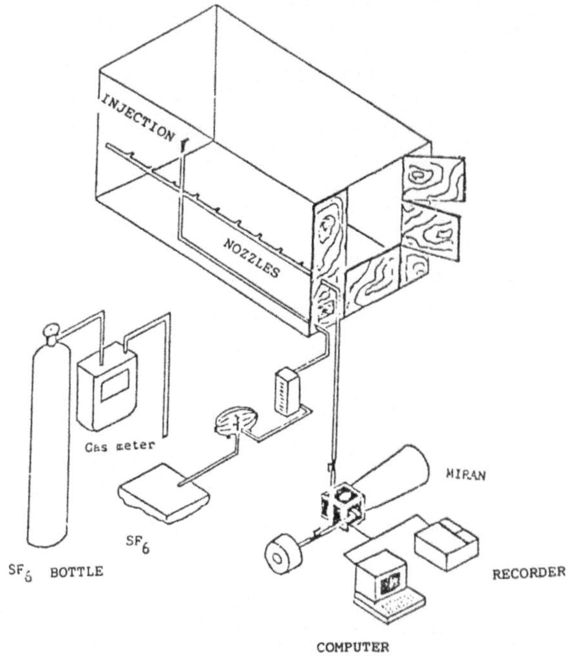

Fig. 1. Schematic layout of the gas monitoring cell. SF_6 injection mechanism, monitoring nozzles and MIRAN unit are shown.

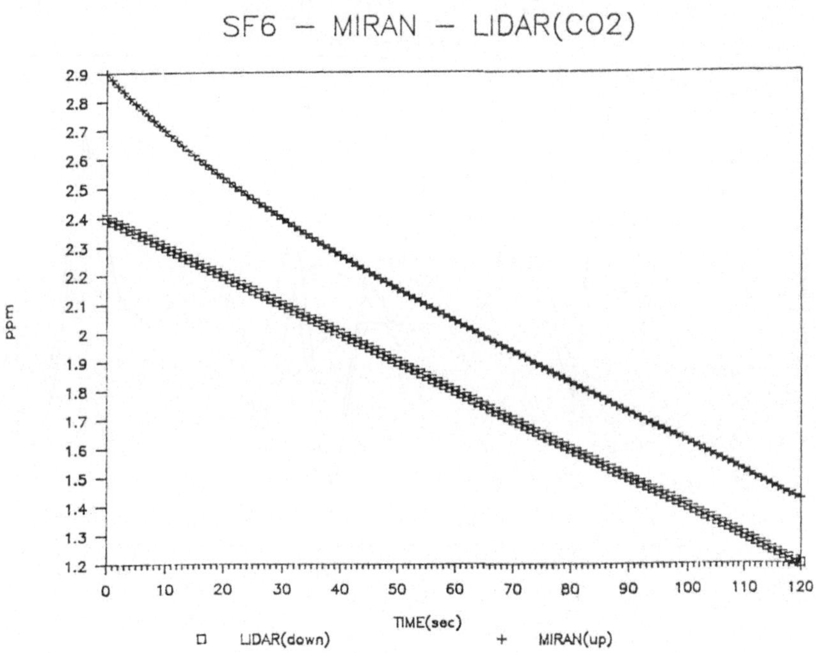

Fig. 2. Theoretical simulation of gradually decreasing concentration as a function of time. Lower curve describes the field concentration, whereas the upper one is the MIRAN concentration.

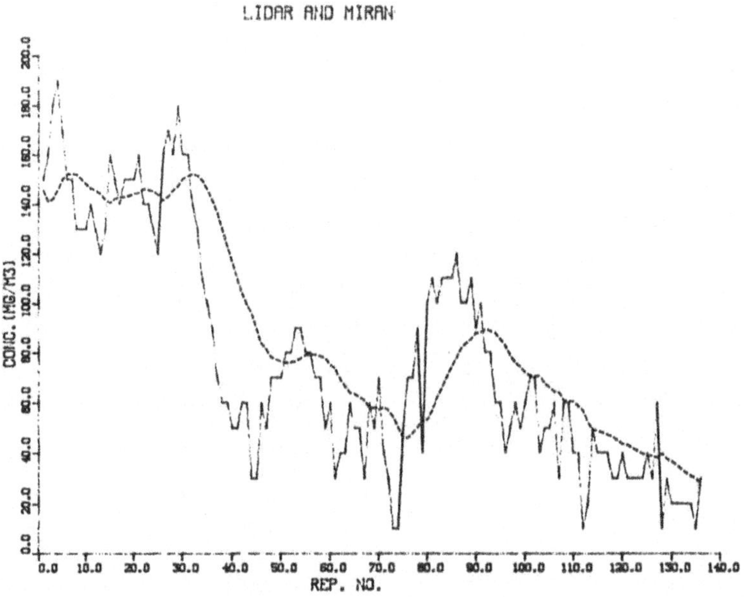

Fig. 3. Theoretical simulation of concentration changes as a function of time, monitored by lidar and MIRAN coincidentally. Continuous trace is the lidar measurement affected by system noise. Dashed line gives the MIRAN values showing both the averaging effect and time lag.

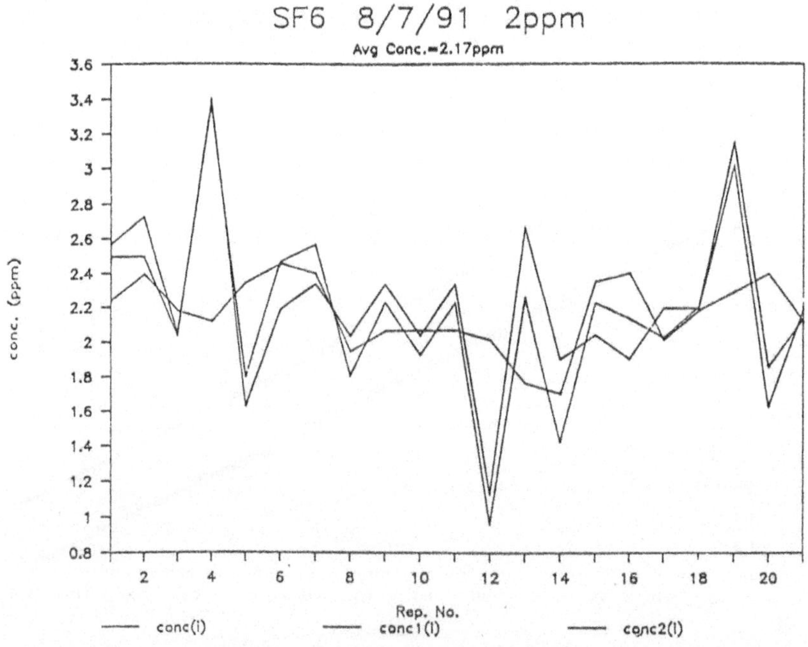

Fig. 4. Constant concentration of 2 PPM SF_6 monitored by lidar and MIRAN coincidentally. The MIRAN reading is 2 PPM (error of 5%). Three successive lidar readings give the lidar degree of variation. Average lidar reading is 2.2 PPM.

3. Operation Algorithm for the Two Systems in Conjunction

In order to construct the operation algorithm for the two systems in conjunction, their physical characteristics must be studied in detail. The lidar signals give the gas concentration information almost in real-time. This results from the short time of flight of the transmitted and received beam to and from the cell, which is of order of micro-seconds.

The MIRAN behaviour is totally different. Since the maximum gas flow rate is relatively slow, the MIRAN operation can be described by a recursion process as follows.

In every period of time, a small portion of the MIRAN volume is replaced. This replacement causes a mixing of the incoming gas with the existing gas in the cell. Thus a small portion of the 'new' information is mixed with the 'old'. At successive time intervals, a newer amount, with possibly different concentration, is mixed with the previous mixture and the process continues. The mathematical description of this process is given in the set of equations (1).

$$\begin{aligned} &DC_i(0) = (V_{in}/V_m) * (C_O - C_i(0)) \\ &C_i(1) = C_i(0) + DC_i(0) \\ &* \qquad\qquad\qquad\qquad\qquad\qquad\qquad\qquad\qquad\qquad\qquad\qquad (1) \\ &* \\ &* \\ &C_i(n) = C_i(n-1) + DC_i(n-1), \end{aligned}$$

where DC_i is the change in monitored MIRAN concentration, V_{in} is the volume change during one cycle of MIRAN operation, V_m is the MIRAN volume and C_i(n) is the true concentration in the cell at time t_n.

This mode of operation results both in averaging and time lag relative to the lidar information. The following examples describe this.

The simple case of gradually decreasing concentration is given in Fig. 2 – lower curve. The points appearing on this curve can be theoretically monitored and presented by the lidar system in real time. Therefore the line structure is unchanged. At the starting point t_o, the MIRAN still refers to the 'past' of this process, and therefore exhibits a higher concentration. When the data points of the upper curve are entered into MIRAN eqs. (1), gradually, the upper curve results. It exhibits a decrease along a different slope with higher values. This behaviour is a result both from the time lag and the averaging nature of the instrument.

Fig. 3 shows another simulation of concentration changes. This simulation refers to lidar system noise and show decreasing oscillations due to atmospheric effects. The dashed trace gives the MIRAN behaviour, when the two instruments start at

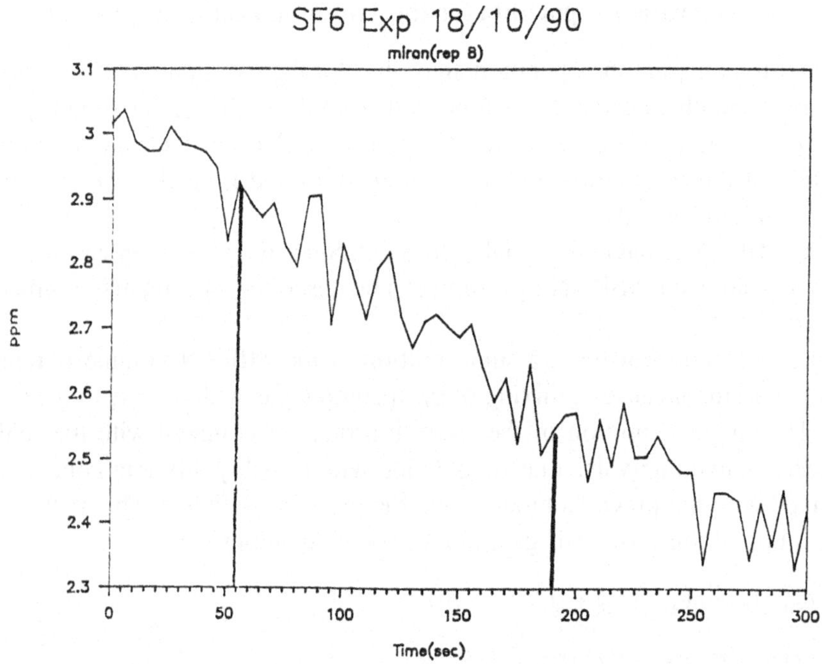

Fig. 5. MIRAN monitoring of gradual decrease in concentration as a function of time. Lidar was operated during the time period marked by the two lines.

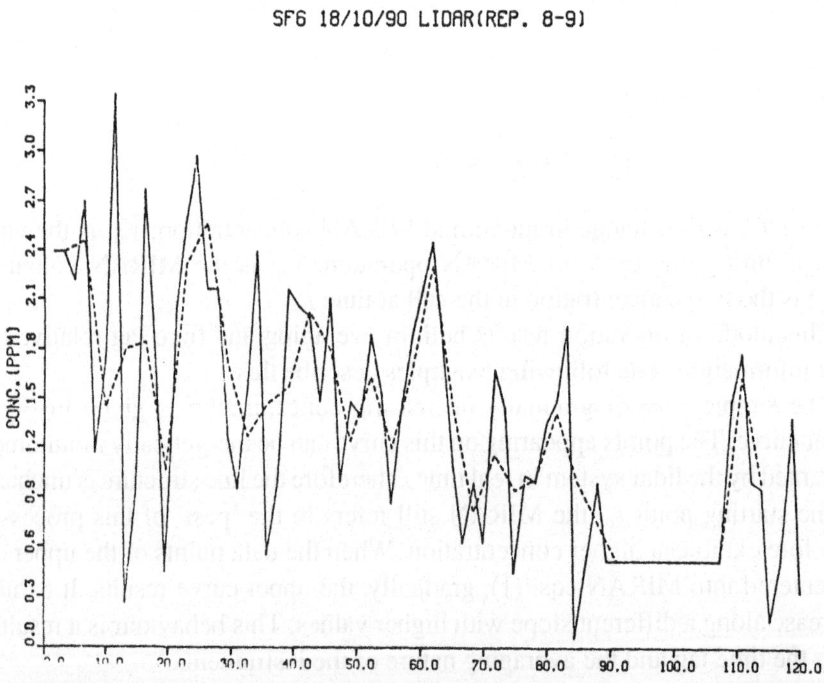

Fig. 6. Lidar measurements during the time period given in Fig. 5. The appropriate difference between MIRAN and lidar in both starting value and slope are evident.

the same concentration. The difference in readings of the two instruments is very clear, and so are the averaging and time lag contributions.

Experimental verification of the phenomena described above was performed in two stages: constant and varying concentrations. The constant concentration monitoring event is shown in Fig. 4. This is the only case where the results from both systems must coincide, and show the same average concentration. This figure shows a test where a concentration of 2 ppm of SF_6 was maintained in the closed cell and monitored constantly by the MIRAN. The curves show three lidar measurements that give the average value of 2.17 ppm. Figs. 5 and 6 show the MIRAN and lidar data along a decreasing slope, where the lidar data are taken over a shorter time interval within that of the MIRAN. The difference of the two slopes is evident, as shown in the theoretical simulations (Fig. 2).

When the lidar and MIRAN are used in the monitoring of different environmental cells, the MIRAN true concentration may be predicted with inverse recursion according to equation (1). This process defines the concentration that is the measured MIRAN value, in the volume that was replaced during each monitoring interval. The calculated value and not the monitored value is the one that can be used to complement the lidar measurement in order to provide a reliable multi-type sensor system.

4. Conclusions

The method of operation and limitations of multi-type sensors in combination for verification of atmospheric phenomena were demonstrated for a specific example. This example exhibits the detector characteristics which are responsible for the mismatch in the monitored concentration values. Though the example is specific, some general conclusions can be drawn.

1. Prior to network operation, the necessary spatial and temporal resolution must be predicted. This will determine the slowest usable component and system complexity.
2. Detectors having different temporal characteristics cannot usually begin operation coincidentally. (There must be a time delay between them.)
3. Slow sensors having a 'time memory' can be operated together with fast detectors in the same network, with the proper operation algorithm.

References

Measures, R.M.,1984: 'Laser Remote Sensing' Wiley-Interscience Publications, New York.

Measures, R.M.,1988: 'Laser Remote Chemical Analysis' John Wiley, New York.

Uthe, E., 1986: 'Airborne CO_2 Dial Measurements of Atmospheric Tracer Gas Concentration Distributions.' Appl. Opt. Vol 25, 2492.

MIRAN-IA: General Purpose Gas Analyzer, Foxboro Analytical, 140 Water St Box 449, South Norwalk, CT 06856.

OBSERVATIONS OF THE DIURNAL OSCILLATION OF THE INVERSION OVER THE ISRAELI COAST

JOSEPH BARKAN AND YIZHAK FELIKS
Israel Institute For Biological Research P. O. B. 19 Ness-Ziona, Israel

(Received October 1991)

Abstract. Vertical profiles of temperature and humidity were measured over the sea in two series of 48 hours each, during the summer near the Israeli coast of the Mediterranean. A prominent inversion was observed in the temperature profiles. In the first series the average height of the inversion base was 350m and in the second, 600m. In the inversion a very sharp decrease of the absolute humidity was found. Below the inversion down to the sea surface the atmosphere was well mixed. A significant diurnal oscillation was observed at the height of the inversion base. During the day the inversion moved up and during the night it moved down. This movement was 250m in the first series and 450m in the second. The movement of the inversion base was almost adiabatic. It is suggested that the fluctuation in the height of the inversion base is mainly due to the developing breeze.

1. Introduction

In their survey on dispersion of pollutants in the coastal strip of Israel, Manes et al. (1976) found that the frequency of inversion occurrences is highest in July (84%). August (80.6%) and September (82%), are close behind. The frequency is considerably lower in June, only 61.6%. But the base of the inversion is the lowest in this month. In 50% of the inversion cases, the base height was less than 500m. In August, on the other hand, only 25% of the cases were below 500m.

Similar findings were reported by Rindsberger et al. (1969, 1972, 1974, 1978) in several papers on the subject of air pollution in the coastal area.

Shaia and Jaffe (1976), in their work about the inversions in Israel, based on radiosonde data from Beit-Dagan (some 7 km inland), found that a subsidence inversion exists during the summer months over Israel, approximately 80% of the time.

The upper air synoptic system which governs the Eastern Mediterranean during the summer months is the subtropic anticyclone (Fig. 1). This system, which is a mean feature of the general circulation, has an annual oscillation. In the winter months its center moves south of 30°N, whereas in the summer months it moves northward and covers the region. It causes large-scale subsidence over the area, and is responsible for the almost continuous existence of the inversion in this season. Occasional penetration of cold air into the Eastern Mediterranean basin from the north, causes temporary withdrawal of the subtropic anticyclone and, consequently, weakening of the subsidence inversion, lifting its base, or even causing its complete disappearence.

Near the surface, the only synoptic system present in the summer months is a warm trough, which is an extension of the Indian monsoon low. Its axis begins at

Boundary-Layer Meteorology **62:** 393–409, 1993.

Fig. 1. Mean subtropic anticyclone in the summer months at 500mb.

the Persian Gulf and terminates in western Anatolia (Fig. 3a). It causes a permanent westerly flow over Israel. The behavior of this trough depends, almost entirely, on the upper air systems. When the subtropic anticyclone is strong, the Persian trough is weak. Consequently, the westerly flow weakens, less moist and cool marine air penetrates into the region, the inversion base descends and the temperature rises. The weakening of the anticyclone brings the opposite results. The prominent mesoscale system during the summer is the sea and land breeze. The activity level of this system depends on the former two. A weak upper air anticyclone and a deep Persian trough cause a vigorous sea-land breeze circulation and vice versa.

Two series of ship-based atmospheric measurements were carried out in the summer of 1987 near Ashdod Harbour, some 40 km south of Tel-Aviv (Fig. 2), using a tethered balloon system. Simultaneous measurements of the wind, temperature and humidity were made on the pier of Ashdod Harbour. As far as we know, these marine based measurements in the Eastern Mediterranean basin are the first of their kind.

2. Description of the Measurements

The measurements were made west of Ashdod Harbour (Fig. 2), along a straight line. Two series of measurements were conducted from aboard the ship. Each one

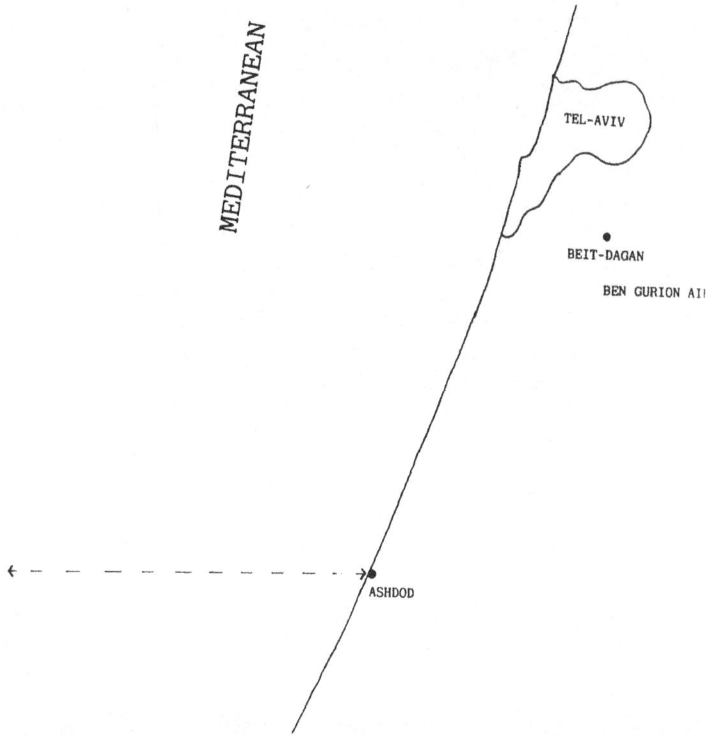

Fig. 2. Map of the coastal region of Israel. The dashed line beginning in Ashdod is the ship track.

lasted approximately 48 hours. In each series, vertical profiles of wind, temperature and absolute humidity were measured. The profiles extended up through the top of the inversion. The measurements were made when the ship was not moving. For details of the measurements see Tables I and II.

The first series were conducted between June 9, 1987, 10:51 and June 11, 1987, 05:34; the second, between August 2 1987, 14:45 and August 4, 1987, 04:51. All hours are in local summer time.

The instruments were placed on a ship, with a large and comparatively empty deck. From 20m up we used a tethered balloon system, made by A. I. R., Boulder, Colorado. The system consists of a zeppelin-shaped balloon with a radiosonde attached below. The balloon is tied to a powerful electronic winch by a very strong and light kevlar line. The sonde measures temperature, absolute humidity, wind speed, wind direction and atmospheric pressure, once every 10 seconds. The data were transmitted to a small receiving station, connected to a PC computer. The data were stored on its magnetic memory.

Simultaneous measurements were made on the outermost pier of Ashdod Harbour. Wind direction and speed were measured at 10m height. Temperature and absolute humidity at the height of 2m.

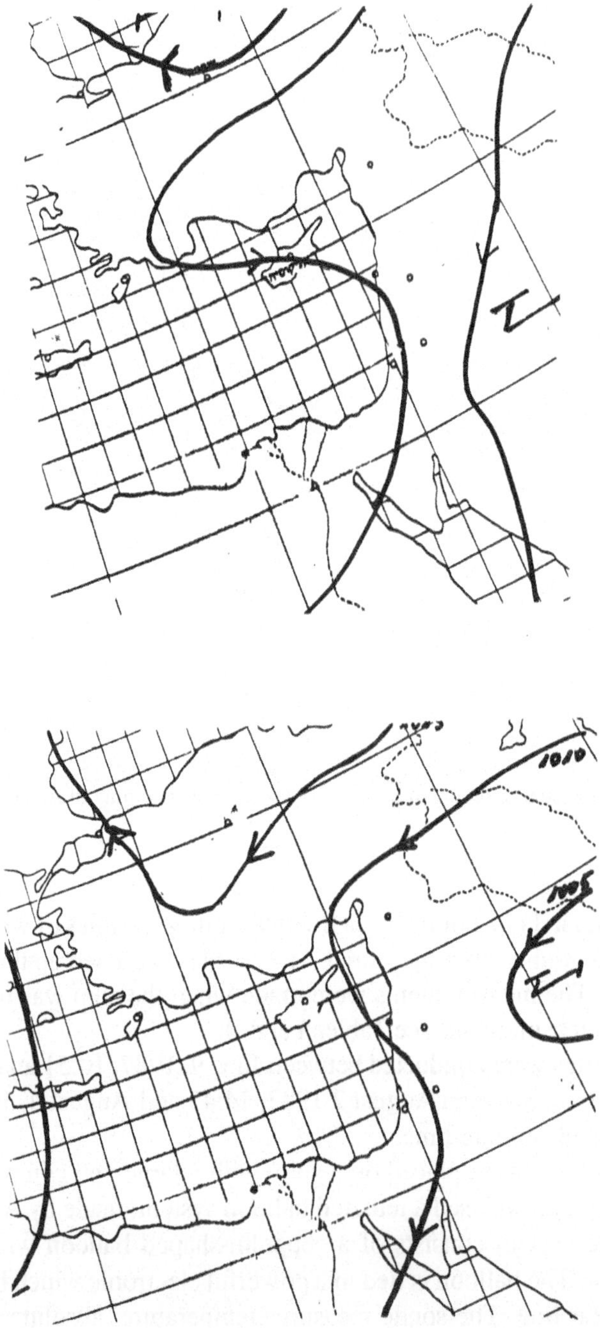

Fig. 3. Surface isobars at 5mb intervales for 18:00 GMT: (a) June 9, 1987. (b) June 10, 1987. The Persian trough is located over the northestern part of the Eastern Mediterranean.

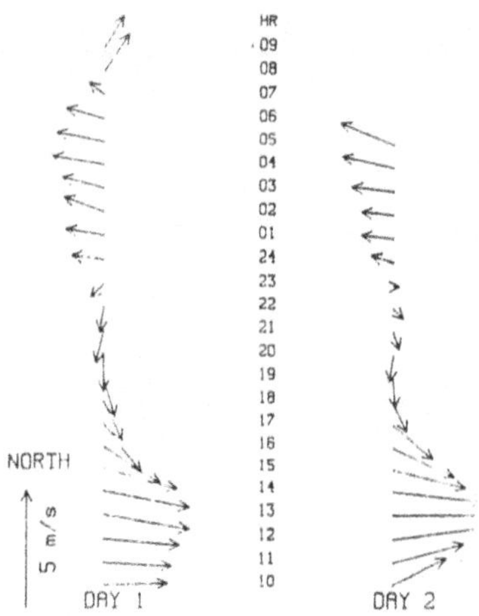

Fig. 4. The surface wind at the shoreline of Ashdod as a function of time. Day 1 begins on June 9, 10:00 LST till June 10 , 09:00. Day 2 begins on June 10, 10:00 till June 11, 05:00.

TABLE I
Balloon measurements June (Series 1)

Serial no.	Date	Time of measurement (LST)	Distance from Shore (km)
1	6.9	10:51-12:20	4
2	6.9	13:51-16:08	7
3	6.9	17:14-18:29	11
4	6.9	21:21-22:38	7
5	6.10	01:05-02:32	7
6	6.10	06:28-07:51	15
7	6.10	10:00-11:02	2
8	6.10	13:01-14:36	11
9	6.10	19:33-20:41	16
10	6.11	02:50-04:03	7

2.1. The First Series (June 9–11)

In the beginning of the first series, the seasonal trough was near its average in position and strength. Later it weakened somewhat, but not much (Fig. 3).

The wind on shore (Fig. 4) was typical of a sea-land breeze system. During the day it blew from west to north-west. Gradually rotating clockwise, it reached the north in the evening. At night it continued to veer and became an easterly to south-easterly land breeze. The land breeze was much lighter than the daytime sea breeze.

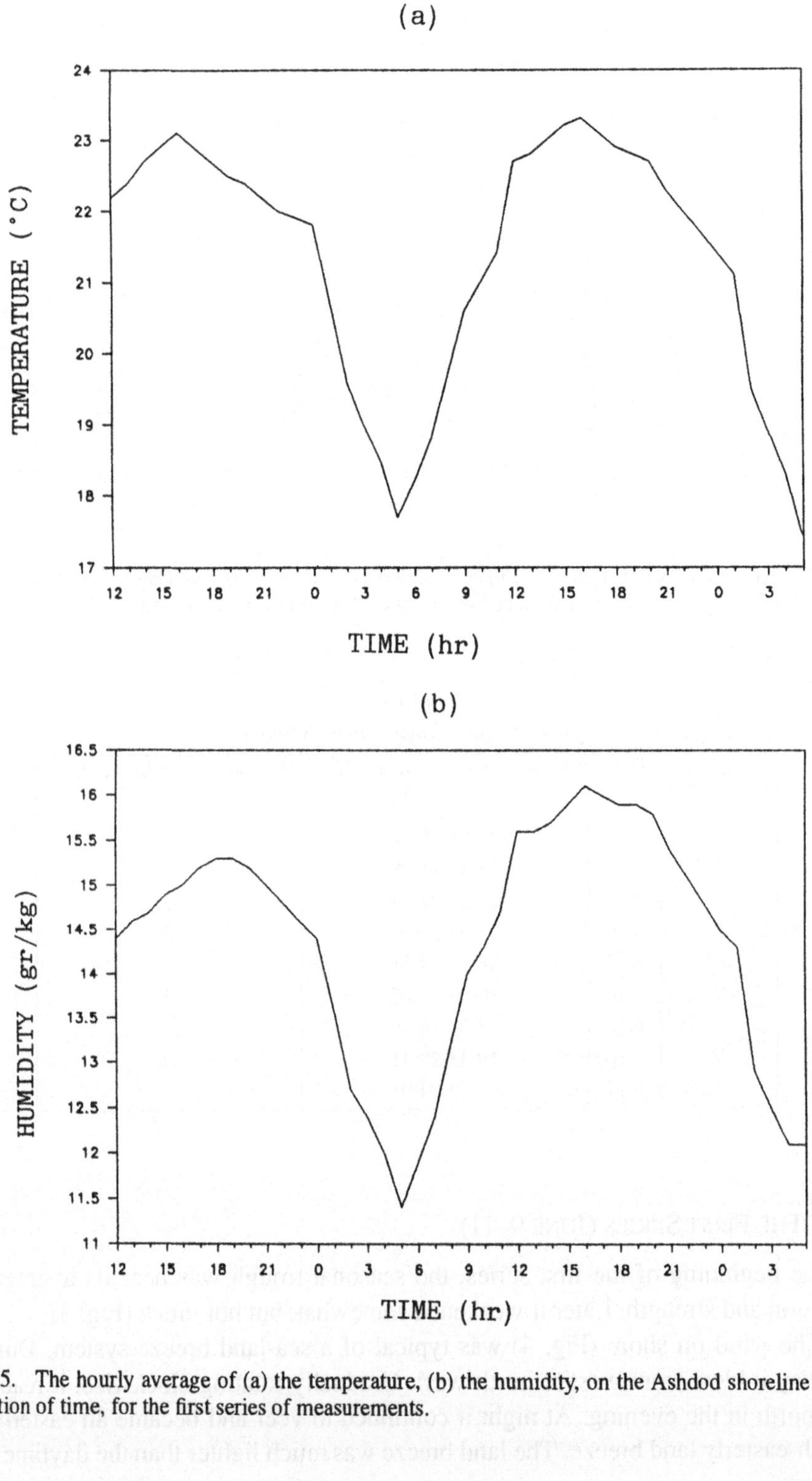

Fig. 5. The hourly average of (a) the temperature, (b) the humidity, on the Ashdod shoreline as a function of time, for the first series of measurements.

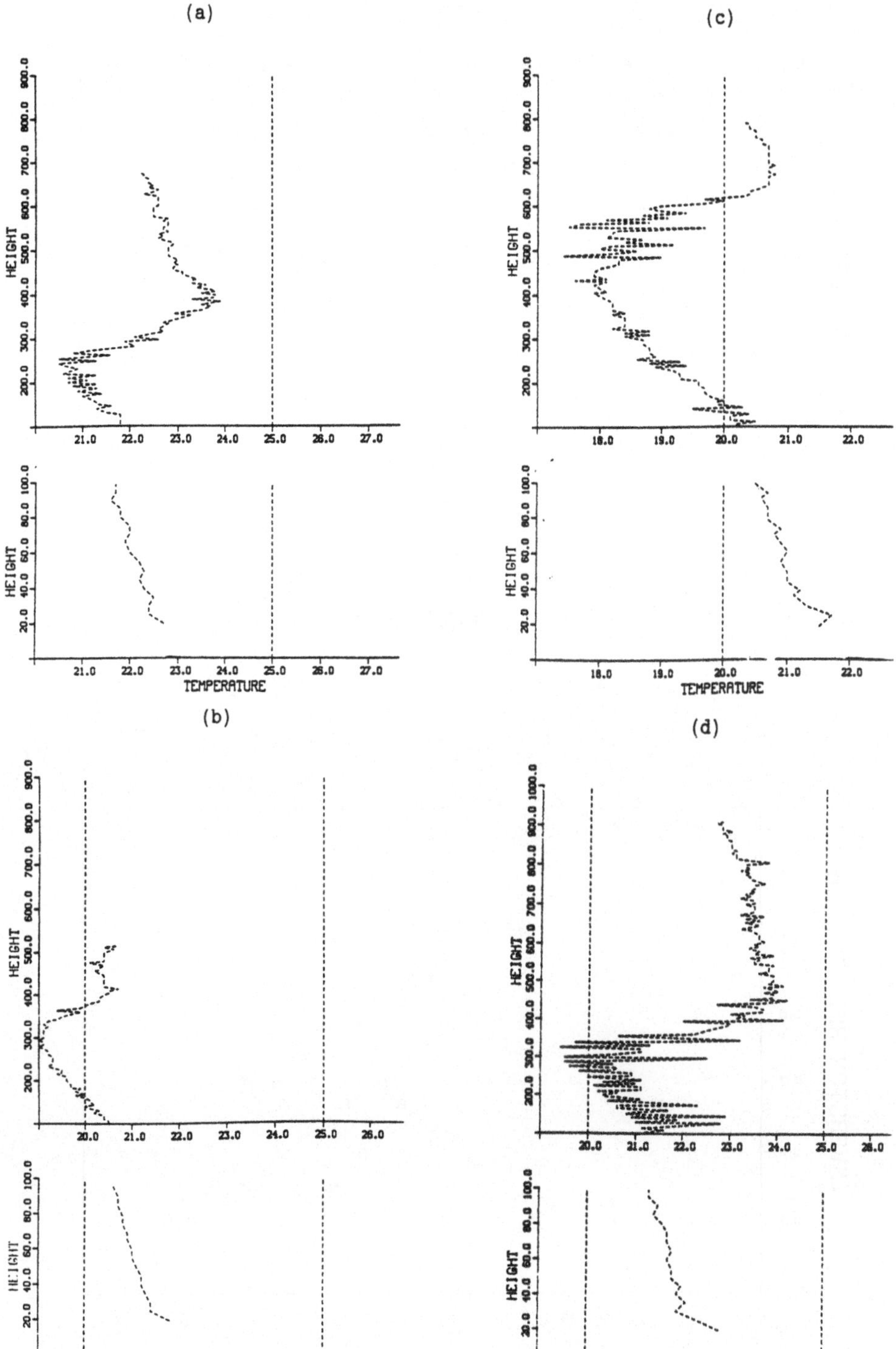

Fig. 6. Vertical temperature profiles (see Table I). (a) Serial no. 2. (b) Serial no. 4. (c) Serial no. 5. (d) Serial no. 8.

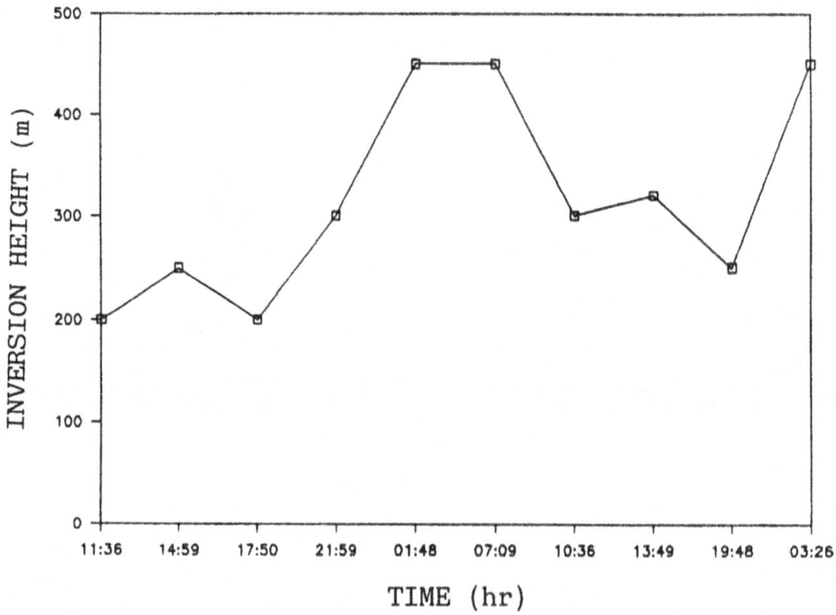

Fig. 7. The height of the inversion base as a function of time for the June series.

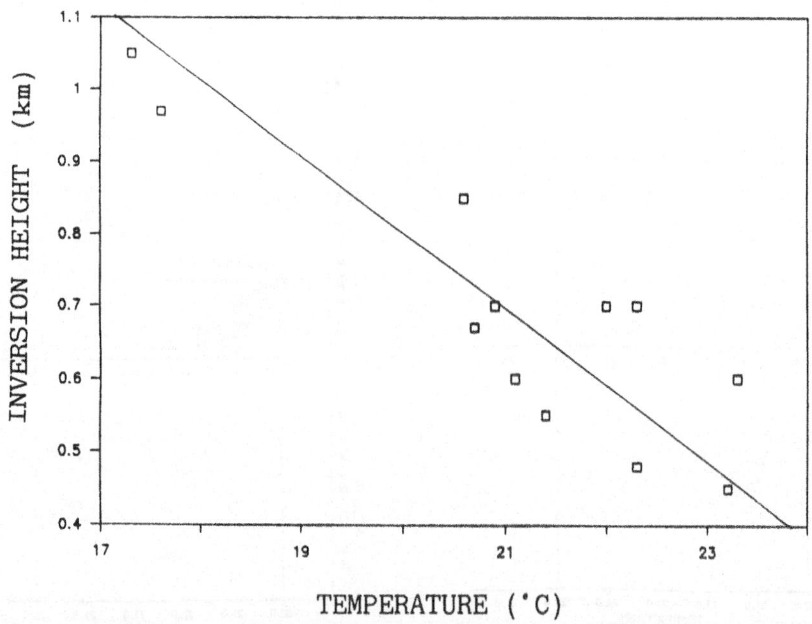

Fig. 8. The height of the inversion base versus the temperature at the inversion base, and the regression line for June series.

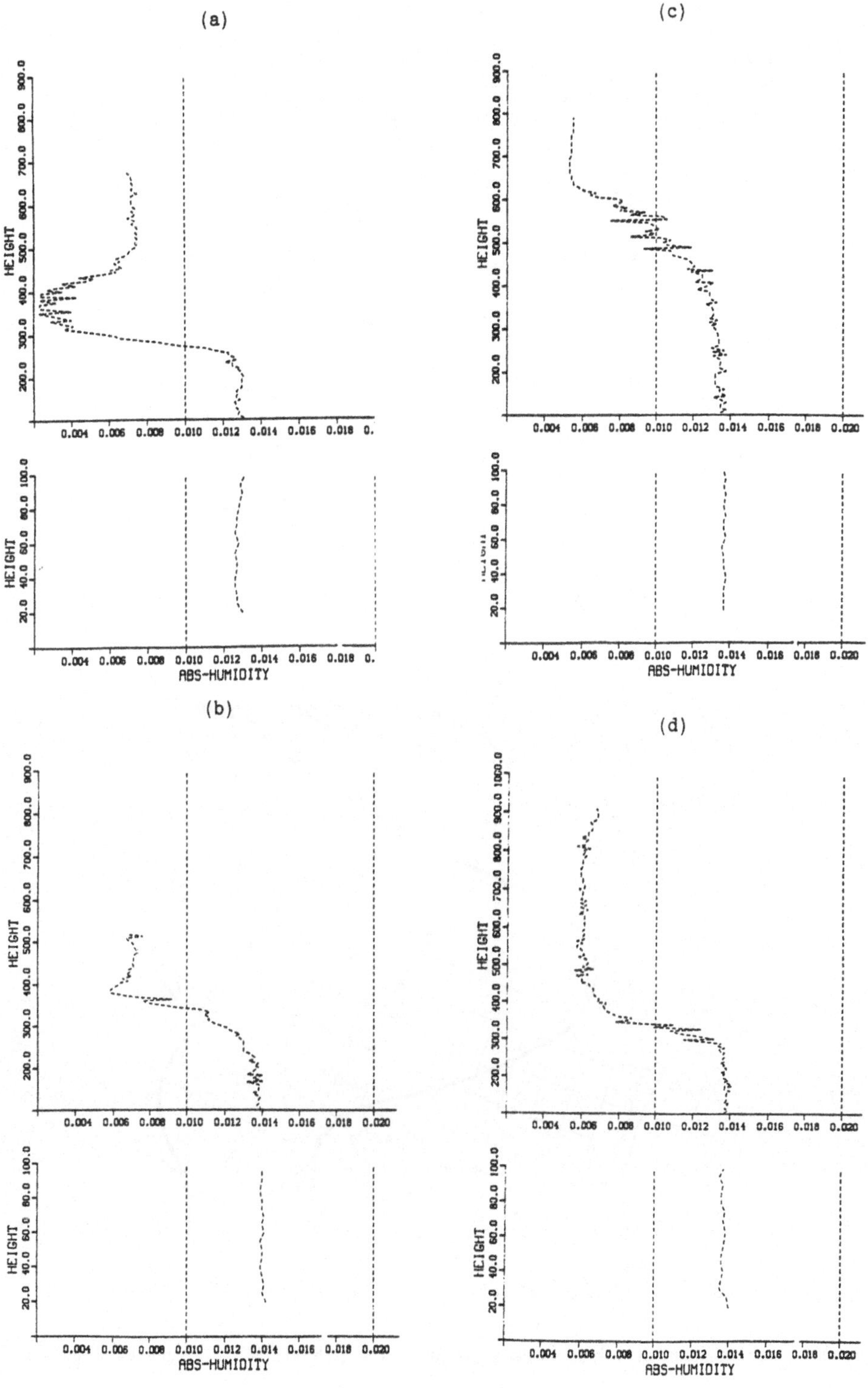

Fig. 9. Same as Fig. 6, but for the humidity.

Fig. 10. Surface isobars at 5mb intervals (a) August 2, 1987, 18:00 GMT. (b) August 4, 1987, 00:00 GMT.

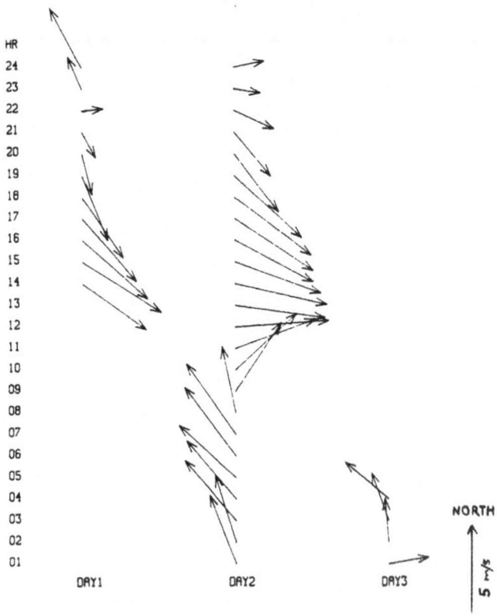

Fig. 11. The horizontal wind at Ashdod shoreline as a function of time. Day 1 begins on August 2, 14:00 LST. Day 3 ends on August 4, 04:00 LST.

TABLE II
Balloon measurements August (Series 2)

Serial no.	Date	Time of measurement (LST)	Distance from Shore (km)
1	8.2	16:51-17:47	7.0
2	8.2	22:35-23:17	17.5
3	8.3	00:38-01:13	7.5
4	8.3	02:32-04:31	2.5
5	8.3	07:06-07:40	12.5
6	8.3	10:08-10:48	12.5
7	8.3	12:09-12:57	17.5
8	8.3	14:19-15:02	12.5
9	8.3	16:29-17:15	22.5
10	8.3	22:37-23:12	12.5
11	8.4	00:46-01:27	17.5
12	8.4	04:04-04:41	7.5

The diurnal variation of temperature and absolute humidity (Fig. 5) was prominent also. Both the temperature and humidity maxima occured in the afternoon, coinciding with the peak of the sea breeze. The minima, on the other hand, occured at dawn, together with the land breeze peak.

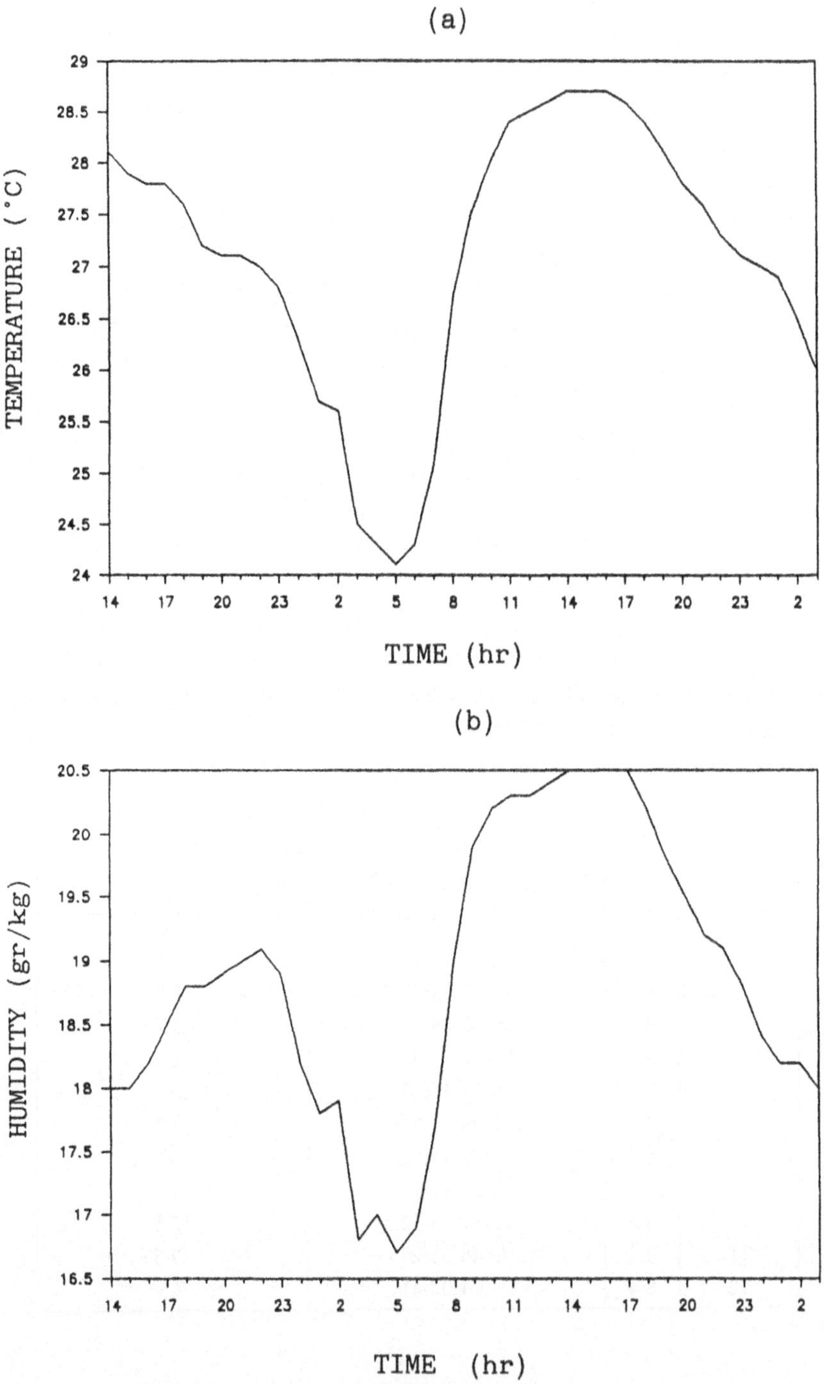

Fig. 12. The hourly average of (a) the temperature, (b) the humidity, on Ashdod shoreline as a function of time for the second series of measurements.

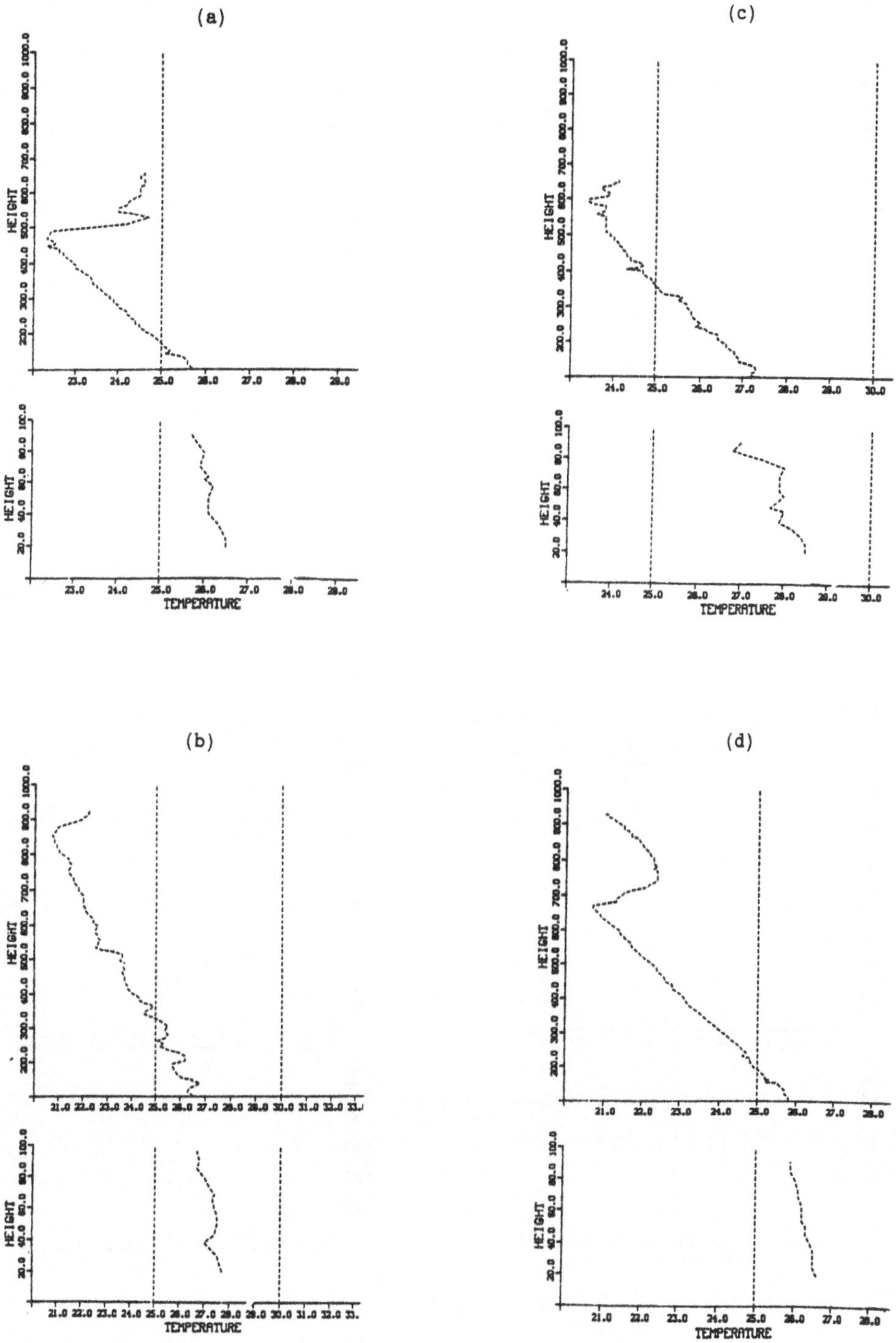

Fig. 13. Vertical profiles of the temperature. (a) Serial no. 2 (b) Serial no. 6. (c) Serial no. 8. (d) Serial no. 10.

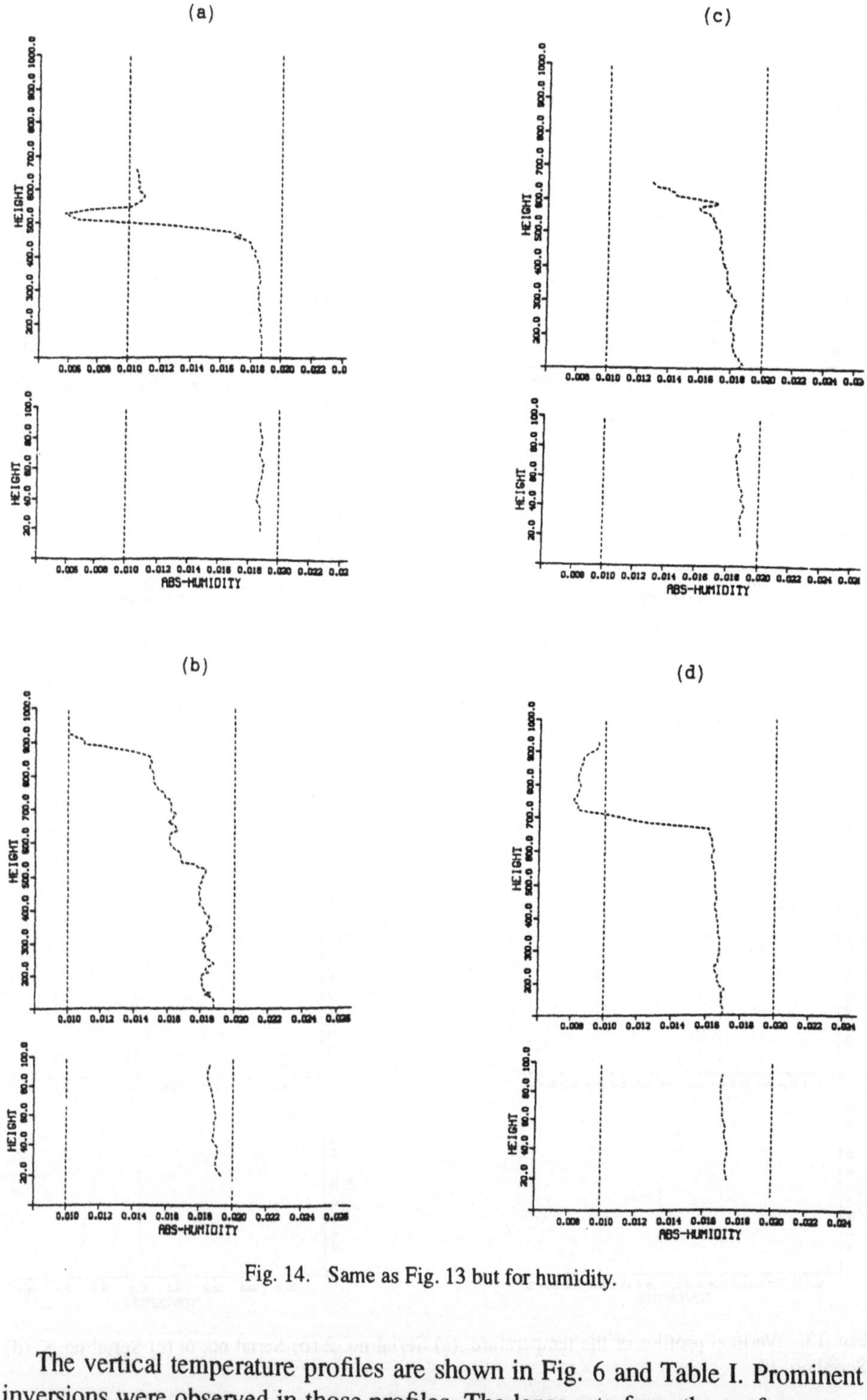

Fig. 14. Same as Fig. 13 but for humidity.

The vertical temperature profiles are shown in Fig. 6 and Table I. Prominent inversions were observed in those profiles. The lapse rate from the surface up to the inversion base, appeared to be close to the adiabatic, i.e. 1°C per 100m. Inside

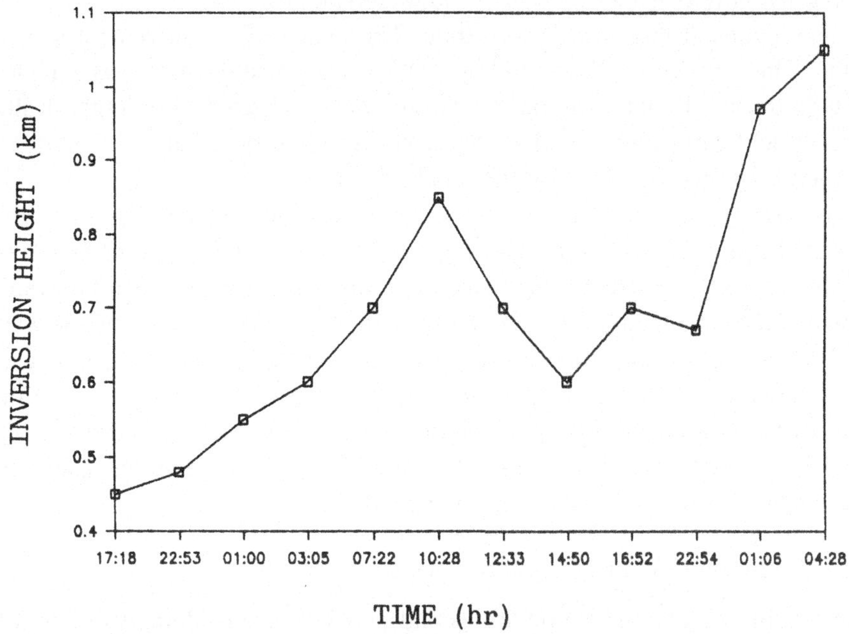

Fig. 15. The height of the inversion base as a function of time for the August series.

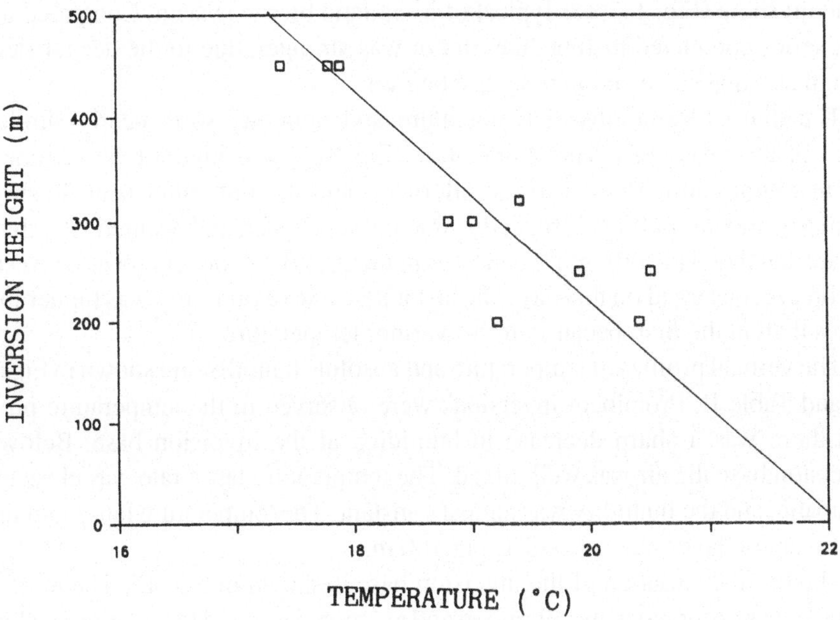

Fig. 16. Same as Fig. 8 but for the August series

the inversion layer the lapse rate was about −3°C−−4°C per 100 m. The cause of the noise observed near the inversion base is not clear to us. It might be the result of internal gravity waves reflected from the inversion layer.

The inversion base varied with time. The total vertical movement was about 250m. This variation is shown in Fig. 7. We can see that the base was at its lowest altitude during the day and early evening hours and reached its highest altitude at night and early morning. The very low inversion base between 250-500m is frequently observed in June (Manes et al., 1976).

In Fig. 8 we have plotted the temperature versus the height of the inversion base and the linear regression line. The slope of this line is 9.7°C per 1000 m, i.e. the adiabatic lapse rate. Thus we conclude that the inversion base move upward and downward adiabatically. The correlation coefficient of the regression line was 0.9.

The vertical profiles of absolute humidity are shown in Fig. 9. Between the sea surface and the base of the inversion the humidity is almost constant due to intensive mixing. In the inversion a sharp decrease in absolute humidity of 10 gr/kg per 100 m was observed. The diurnal oscillation in the height of the sharp change in the humidity follows that of the inversion base.

2.2. The Second Series (August 2–4)

The synoptic conditions during the second series were considerably different from those of the June series. At the beginning, the Persian trough was quite weak but it deepened quickly and became deeper than average (Fig. 10). The behavior of the wind on shore (Fig. 11) was typical of a sea-land breeze system. Compared to the first series conducted in June, the breeze was stronger, due to the deeper Persian trough and higher inversion base (see below).

The diurnal variations of temperature and humidity were nearly similar to those of the first series. The temperature (Fig. 12a) was higher than in June, but the maximum still occured in the afternoon and the minimum near dawn. The humidity maximum (Fig. 12b) on the first day was lower, and occured later than on the second day. This difference can be explained by the stronger and more westerly sea breeze, observed on this day. The absolute value of humidity was higher in this series than in the first, because of the warmer temperature.

The vertical profiles of temperature and absolute humidity are shown in Figs. 13, 14 and Table II. Prominent inversions were observed in the temperature profiles and there was a sharp decrease in humidity, at the inversion base. Below the inversion base the air was well mixed. The temperature lapse rate was close to the adiabatic, and the humidity was almost constant. The temperature lapse rate inside the inversion layer was −2−−3°C per 100 m.

The diurnal variation of the inversion base is shown in Fig. 15. The minimum altitude was observed in the afternoon and evening hours and the maximum altitude after midnight and the early morning hours. The total vertical movement was about 450 m. In Fig. 16 the temperature versus the height of the inversion base and the linear regression line are shown. The slope of this line is 9.4°C per 1000 m, i. e. an adiabatic lapse rate. Thus, the movement of the inversion base was adiabatic, as in the first series.

The height of the inversion base in August was much higher than in June, due to the deeper Persian trough. The total vertical movement in August was twice that in June. This fact can probably be attributed to the stronger sea breeze.

3. Summary

Two series of measurements in June and August 1987 were conducted over the sea near the Israeli coast of the Mediterranean. Each series lasted 48 hours. Vertical profiles of temperature and humidity were measured every four hours at different distances from the coast. On the shoreline, continuous measurements of wind, temperature and humidity were taken.

In both series it was found that: (a) The wind at the shoreline was dominated by the sea and land breeze system. (b) Prominent inversions were observed in the vertical temperature profiles. Below the inversion down to the sea surface, the air was well mixed and the temperature profile was almost adiabatic. A sharp decrease in humidity at the inversion base was observed. Below the inversion down to the sea surface the humidity was almost constant. (c) A prominent diurnal oscillation in the inversion base was found. During daytime the inversion base moved downward reaching its minimum height in the afternoon and evening. During the night it moved upward reaching its maximum height in the late night and early morning. (d) The movement of the inversion base is almost adiabatic. We assume that this diurnal oscillation is mainly due to sea and land breezes as shown by Feliks (1992).

The main differences between the measurements of June and August were as follows: In August the Persian trough was deeper. The sea and land breezes were stronger. The average height of the inversion base was greater by about 250m, and the lapse rate in the inversion was weaker. The amplitude of the diurnal oscillation in August was twice as large as in June. We assume that the deeper Persian trough and the higher and weaker inversion lead to a stronger breeze. This breeze causes the larger amplitude of the diurnal oscillation of the inversion base.

References

Feliks, Y., 1992: A numerical model for the estimation of the diurnal fluctuation of the inversion due to the sea breeze. *Boundary-Layer Meteor.* **62**, 151–161, 1993 (this volume).

Manes, A., J. Tzur, and M. Rindsberger, 1976:Estimate of the environmental hazards from uncontrolled dispersion of poisonous material. Isr. Meteor. Serv.. Beit Dagan.

Rindsberger, M., 1969: Meteorological aspects of air pollution potential in the area of greater Tel-Aviv. Isr. Meteor. Serv.. Series E, no. 16, Beit-Dagan.

Rindsberger, M., 1974: Analysis of the mixing depth over Tel-Aviv. Isr. J. Earth Sci., 23, 13-17.

Rindsberger, M., M. Segal, and A. Manes, 1978: Environmental guidelines for the 'L' project in Tel-Aviv. Isr. Meteor. Serv.. Series E, no. 34, Beit-Dagan.

Shaia, J. S. and S. Jaffe, 1976: Midday inversions over Beit-Dagan. Isr. Meteor. Serv.. Series A, no. 33, Beit-Dagan.

LABORATORY STUDY OF THE TURBULENT STRUCTURE OF A CHANNEL JET FLOWING INTO AN OPEN BASIN

B. SHTEINMAN, A. GUTMAN and E. MECHREZ

Kinneret Limnological Laboratory, Israel Oceanographic & Limnological Research, P.O.B. 345, Tiberias, Israel 14102

(Received October 1991)

Abstract. The turbulent structure of a river jet flowing into a lake was simulated in the laboratory. The main purpose of the study was to look for the turbulent coherent structures at a river mouth up to the point where the river defined turbulence ceases to exist. Experiments were conducted at the Caucasia Regional Hydrometeorological Institute at Baku, and final analysis of the data was performed at the Kinneret Limnological Laboratory in Israel. To model the river outgoing jet properly, a sand bottom was applied, which formed the ripples and the bar as in natural conditions. Several days after the beginning of the laboratory experiment, when the bar and the ripples had reached a quasi steady state, the turbulent jet measurements began. Experimental data were gathered by simultaneous recording of longitudinal velocities at different distances between the channel outlet and the sand bar. The records were processed and related turbulent characteristics of the jet flow were obtained. Values of mean velocities, water depths along the jet axis, spectrum of joint velocities at chosen frequencies, and maximal and average size of the coherent structures, were calculated. The patterns of the above parameters enable the description of the decay process of a free turbulent jet flowing out a chanel.

1. Introduction

The purpose of this laboratory study is to model turbulence at specific flow conditions existing in river mouths, including:

a) a free jet flow with turbulence formed within the river and transformed into lake turbulence;

b) a specific bottom relief, including ripples, sand bar and sharp deepening after the bar crest.

This laboratory study is the first stage of an investigation of turbulence in river mouths flow. Two other stages consist of:

– River mouth on-site studies, including the Jordan River flow into Lake Kinneret (Sea of Galilee).

– Development of a conceptual model for turbulence at a river mouth.

2. Experimental Set Up and Data Processing

The experimental set up was constructed at the Transcaucasia Regional Hydrometeorological Institute (TRHI), Baku. It includes a glass channel of 2 m width, 40 m length, with a receiving reservoir 15 m wide and a bottom slope of 20 deg. Sand was applied to form ripples and a sand bar under stationary flow conditions, as found in nature. The average depth within the channel was about 0.5 m.

Boundary-Layer Meteorology **62:** 411–416, 1992.

The measuring devices used were velocity fluctuation meters, developed at TRHI. Simultaneous records of the longitudinal velocity fluctuations of the surface flow have been processed for pairs of closely spaced points at several distances along the central line.

Only stationary processes are presently investigated, therefore raw data containing errors or trends were sorted out at the editing stage of the data processing. Spectral characteristics of velocity fluctuations have been computed. The resulting turbulence parameters are presented below.

3. Coherent Turbulent Structures

We are looking for the coherent turbulence structures in river mouths. We use spectral characteristics of turbulent flows, as defined in random processes theory.

Let us first consider the time autocovariance $R_u(\tau)$ and the power spectral density $S_u(\Omega)$ (see Landahl, 1986). We define also the joint covariances $R_{uv}(\tau)$ over the longitudinal velocity fluctuations $u(t) = u(x,t)$ and $v(t) = u(x + \Delta x, t)$ at two points spaced a small distance apart. Then the cross-spectral density $S_{uv}(\Omega)$ is:

$$S_{uv}(\Omega) = (1/2\pi) \int e^{i\Omega\tau} R_{uv}(\tau) d\tau = Co_{uv}(\Omega) + iQu_{uv}(\Omega), \tag{1}$$

and the coherence $Coh_{uv}(\Omega)$ will be:

$$Coh^2_{uv}(\Omega) = \frac{Co^2_{uv} + Qu^2_{uv}}{S_u S_v}. \tag{2}$$

We start the analysis by looking at the shape of the time autocovariance $R_u(\tau)$. All measured functions $R_u(\tau)$ cross the τ-axis; the time period τ_o of the first intersection $R_u(\tau_o) = 0$ defines the scale of the largest most energetic coherent structures L^τ_{max}, according to the value of the mean longitudinal velocity U at the point of measurement:

$$L^\tau_{max} = U\tau_o. \tag{3}$$

A Fourier transform of the function $R_u(\tau)$ yields the power spectral density $S_u(\Omega)$. It is a decreasing function with several peaks. The highest peak at low frequency Ω_o reflects the largest coherent structures. The time period $1/\Omega_o$ leads to another independent estimation of the length scale, L_{max}, of the largest turbulence coherent structures:

$$L^\Omega_{max} = U/\Omega_o r. \tag{4}$$

The peaks in spectral density characterize a hierarchy of coherent structures at the point where measurements were made. By looking for the distribution of

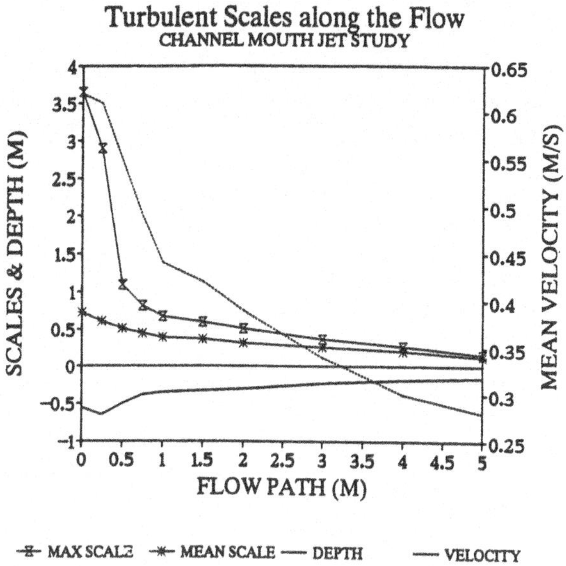

Fig. 1. The largest and mean scales of turbulent coherent structures of jet flow in the channel mouth.

the heights and positions of the peaks at different points along the flow, the flow turbulence at the river mouth may be analyzed. As usual, we define the mean frequency Ω_{mean} of all significant peaks in $S_u(\Omega)$ and the corresponding mean scale L_{mean} of coherent structures in the turbulent flow.

The measured coherence functions $Coh_{uv}(\Omega)$ also have several significant peaks. This shows that different points along the flow are related by shared coherent structures appearing at a number of frequencies. The larger the distance between the points, the higher should be the contribution of largest coherent structures in the coherent functions. We are using the weights of the peaks at different frequency ranges as 'tracers' for relative quantities of coherent structures of different sizes along the flow.

Let Σ be the total area under all significant peaks in $Coh_{uv}(\Omega)$. Then Σ_{low}, Σ_{high}, and Σ_{int} stand for relative contributions of the peaks at low, high and intermediate frequency ranges, according to some chosen limits (1 and 2 Hz in our case). The weights of the peaks are defined as follows:

$$F_{low} = \Sigma_{low}/\Sigma$$

$$F_{high} = \Sigma_{high}/\Sigma \qquad (5)$$

$$F_{int} = 1 - F_{low} - F_{high}.$$

All the above defined parameters are presented in Table I. Two additional parameters are the values of the coherence at frequencies Ω_o and Ω_{mean}.

TABLE I

Turbulent parameters along the surface of the jet flow in the channel mouth.

L (m)	ΔL Steps (m)	U Velocity (m/sec)	H Depth (m)	$Coh.$ at Max. Energy Freq.	Freq. Max. Energy (Herz)	$Coh.$ at Avg. Energy Freq.	Freq. Avg. Energy (Herz)	Low Freq. Weights	High Freq. Weights	Max. Scale (m)	Mean Scale (m)
0	0.03	0.62	0.55	0.68	0.17	0.32	0.86	0.41	0.25	3.65	0.72
	0.05	0.62	0.55	0.69	0.17	0.32	0.86	0.43	0.24	3.65	0.72
	0.1	0.62	0.59	0.74	0.17	0.32	0.86	0.49	0.21	3.65	0.72
	0.2	0.62	0.65	0.75	0.17	0.35	0.86	0.5	0.24	3.65	0.72
0.25	0.03	0.61	0.65	0.62	0.21	0.37	1	0.38	0.26	2.9	0.61
	0.05	0.6	0.63	0.65	0.22	0.39	1	0.41	0.23	2.73	0.6
	0.1	0.59	0.53	0.73	0.22	0.41	1	0.46	0.12	2.68	0.59
	0.2	0.57	0.52	0.58	0.23	0.29	1.07	0.36	0.39	2.48	0.53
0.5	0.03	0.55	0.5	0.55	0.51	0.36	1.08	0.22	0.43	1.08	0.51
	0.05	0.54	0.5	0.55	0.51	0.37	1.08	0.29	0.36	1.06	0.5
	0.1	0.53	0.5	0.54	0.56	0.4	1.08	0.29	0.29	0.95	0.49
	0.2	0.51	0.48	0.51	0.6	0.48	1.08	0.2	0.3	0.85	0.47
0.75	0.03	0.49	0.38	0.4	0.6	0.38	1.09	0.18	0.44	0.82	0.45
	0.05	0.48	0.38	0.42	0.61	0.4	1.09	0.24	0.36	0.79	0.44
	0.1	0.47	0.38	0.42	0.61	0.52	1.09	0.24	0.3	0.77	0.43
	0.2	0.45	0.36	0.4	0.63	0.59	1.1	0.18	0.31	0.71	0.41
1	0.03	0.44	0.35	0.37	0.65	0.33	1.12	0.16	0.53	0.68	0.39
	0.05	0.44	0.35	0.4	0.65	0.35	1.12	0.22	0.47	0.68	0.39
	0.1	0.43	0.35	0.42	0.65	0.39	1.13	0.22	0.4	0.66	0.38
	0.2	0.43	0.33	0.4	0.7	0.63	1.13	0.2	0.32	0.61	0.38
1.5	0.03	0.42	0.32	0.39	0.71	0.34	1.14	0.16	0.52	0.59	0.37
	0.05	0.42	0.32	0.45	0.71	0.3	1.14	0.22	0.52	0.59	0.37
	0.1	0.41	0.3	0.46	0.76	0.3	1.21	0.22	0.53	0.54	0.34
	0.2	0.41	0.3	0.4	0.76	0.37	1.21	0.2	0.45	0.54	0.34
2	0.03	0.39	0.3	0.38	0.76	0.3	1.22	0.16	0.56	0.51	0.32
	0.05	0.39	0.3	0.38	0.76	0.35	1.22	0.16	0.54	0.51	0.32
	0.1	0.38	0.27	0.38	0.84	0.35	1.26	0.16	0.54	0.45	0.3
	0.2	0.38	0.27	0.4	0.84	0.37	1.31	0.2	0.34	0.45	0.29
3	0.03	0.34	0.23	0.37	0.92	0.3	1.31	0.15	0.7	0.37	0.26
	0.05	0.34	0.23	0.4	0.92	0.36	1.31	0.16	0.54	0.37	0.26
	0.1	0.34	0.22	0.4	0.97	0.36	1.31	0.18	0.5	0.35	0.26
	0.2	0.33	0.22	0.4	0.97	0.37	1.32	0.18	0.5	0.34	0.25
4	0.03	0.3	0.18	0.36	1.11	0.2	1.43	0.12	0.74	0.27	0.21
	0.05	0.29	0.18	0.37	1.12	0.2	1.45	0.15	0.74	0.26	0.2
	0.1	0.28	0.18	0.39	1.17	0.2	1.5	0.15	0.72	0.24	0.19
	0.2	0.28	0.18	0.39	1.24	0.2	1.55	0.18	0.68	0.23	0.18
5	0.03	0.28	0.15	0.3	1.75	0.14	2.15	0.04	0.87	0.16	0.13
	0.05	0.28	0.15	0.3	1.75	0.14	2.15	0.06	0.84	0.16	0.13
	0.1	0.26	0.15	0.32	2.17	0.14	2.6	0.11	0.79	0.12	0.1
	0.2	0.25	0.15	0.32	3.12	0.21	3.57	0.11	0.61	0.08	0.07

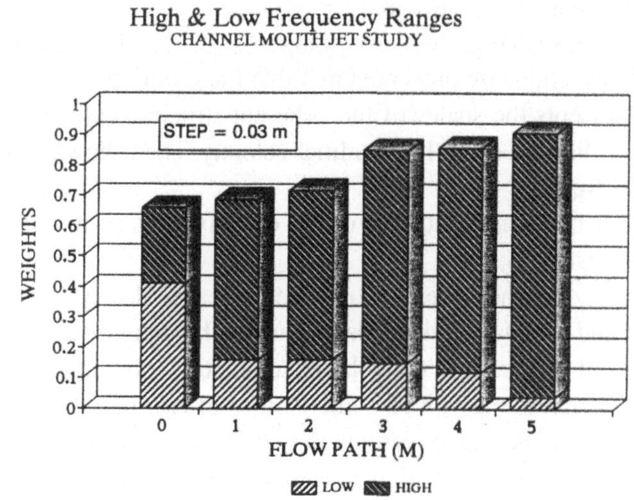

Fig. 2. The weights of high and low frequency ranges in the coherence peaks between two points separated by 0.03m.

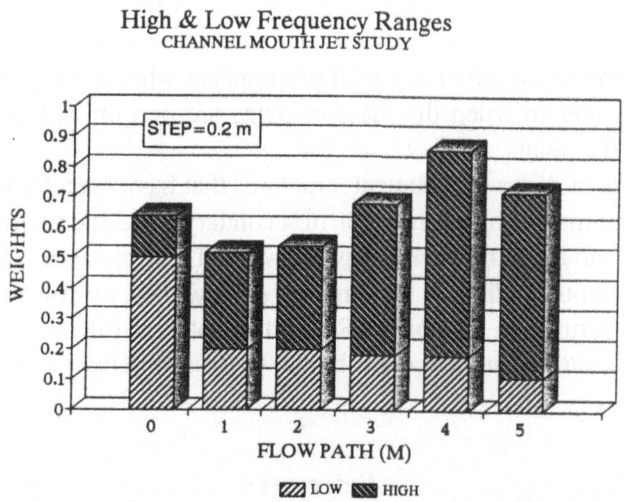

Fig. 3. The weights of high and low frequency ranges in the coherence peaks between two points seperated by 0.2m.

4. Results and Discussion

The main object of this laboratory study is the turbulent coherent structures passing through the channel mouth. The following four parameters are chosen to represent the coherent structures behaviour:

- maximal length scale L^{τ}_{max}
- mean length scale L_{mean}

- weight of the peaks in the low frequency range F_{low}
- weight of the peaks in high frequency range F_{high}. The above parameters, defined in the previous section, are presented in Table I and in Figures 1-3.

Figure 1 presents the scales of the coherent structures along the flow, as well as the depth profile and mean longitudinal velocity. The following features may be seen from this figure:

a) the largest coherent structures decay quickly, before the mean longitudinal velocity decreases;
b) the mean size of the coherent structure decreases slowly.

Figures 2 and 3 show the interplay between the weights of high and low frequency ranges of the peaks in the coherence function, obtained by simultaneously recording the longitudinal velocity fluctuations at two points at a small distance apart. The intermediate range weights may be obtained from the figures by summing the weights of low and high ranges and comparing to one. The following features of the coherent structures may be mentioned:

a) large coherent structures decay quickly, and their relative weights decrease to several percent when measured with a small step size between the points (0.03 m in Figure 2), and to ten percent at a larger step (0.2m in Figure 3);
b) small size coherent structures prevail at large distances down the flow, especially when measured with a small step size between the points;
c) intermediate sized structures tend to disappear when measured with a small step, and remain when the measurements are performed with a 0.2m step between the points.

The process of decay of turbulent structures that have been formed within the channel is accompanied by a generation of secondary turbulence in the contact zone of the channel mouth jet flow with stagnant water. These processes are best studied with measurements of all three components of flow velocity. Such measurements have been performed in the Jordan River flow into Lake Kinneret. Preliminary results are presented in another contribution to this Conference (Shteinman *et al.*, 1992).

References

Landahl, M.(1986): *Turbulence and random processes in fluid mechanics*. Cambridge University Press

Shteinman B, Mechrez E., Gutman A.,(1992): *Spatial Structure of the Jet Flow in the River Mouth*. *Boundary-Layer Meteorology* **62**, 379–383, 1993 (this volume).

ACTINIC FLUXES: THE ROLE OF CLOUDS IN PHOTODISSOCIATION

PETER G. DUYNKERKE and MICHIEL VAN WEELE
University of Utrecht, Utrecht, The Netherlands.

(Received October 1991)

Abstract. The role of clouds in photodissociation is examined by both modeling and observations. It is emphasized that the photodissociation rate is proportional to the actinic flux rather than to the irradiance. (The actinic flux concerns the energy that is incident on a molecule, irrespective of the direction of incidence. The irradiance concerns the energy that is incident on a plane.) A 3-layer model is used to calculate the actinic flux above and below a cloud, relative to the incident flux, in terms of cloud albedo, zenith angle and the albedo of the underlying and overlying atmosphere. Cloud albedo is mainly determined by cloud optical thickness. An expression for the in-cloud actinic flux is given as a function of in-cloud optical thickness. The 3-layer model seems to be an useful model for estimation of photodissociation rates in dispersion models. Further, a multi-layer delta-Eddington model is used to calculate irradiances, actinic fluxes and photodissociation rates of nitrogen dioxide $J(NO_2)$ as a function of height in inhomogeneous atmospheres. For the considered wavelength interval [290-420 nm], Rayleigh scattering, ozone absorption and Mie scattering and absorption by cloud drops and aerosols should be taken into account. It is stressed that both models are one-dimensional and as such are unable to deal with partial cloudiness. It is shown that if no clouds are present, the actinic flux depends primarily on the solar zenith angle. The actinic flux usually increases with height. For cloudy atmospheres, another important parameter with respect to the actinic flux is added: cloud optical thickness, which determines cloud albedo. It can be shown that in-cloud characteristics and cloud height are less important in describing the effect of a cloud on the actinic flux (outside the cloud). The in-cloud values of the actinic flux can exceed the values outside the cloud. Finally, using the photostationary state relationship, good agreement is found between model results and aircraft measurements.

1. Introduction

The chemistry of the atmosphere is driven by solar radiation, which dissociates certain molecules into reactive atoms or free radicals. This process is called photodissociation. Models which describe the chemistry of the atmosphere must therefore include an accurate description of the radiation processes which initiate the photodissociation process. In the troposphere the wavelength region of interest is approximately from 290 to 730 nm. Below 290 nm stratospheric O_2 and O_3 block out most of the ultraviolet radiation and beyond 730 nm no photochemistry of interest takes place.

The photodissociation rate is determined by the product of the absorption cross-section, photodissociation quantum yield and the actinic flux. The actinic flux is a radiation quantity which will be considered in somewhat more detail in this paper. The calculation of solar radiation begins with prescribing the incident flux at the top of the atmosphere and includes the scattering and absorption of light in the atmosphere and at the surface. Most of the research on radiation is concerned with the calculation of the irradiance, which expresses the flow of energy through a (horizontal) surface. As mentioned above, for photodissociation the important

Boundary-Layer Meteorology **62:** 417–432, 1993.

quantity is not the irradiance but the actinic flux. The latter is the flux of energy through a spherical surface. Physically the difference between the irradiance and the actinic flux arises because the irradiance describes the flow of radiant energy through the atmosphere while the actinic flux concerns the available energy at a particular point in the atmosphere that determines the probability of an encounter between a photon and a molecule.

The major variables that control the atmospheric radiation, and thus affect the atmospheric photochemical reactions, are the time of the day, time of year, latitude, atmospheric state (gases, aerosols and clouds) and surface albedo. Under cloud-free conditions, the solar radiation is mainly affected by gas absorption (i.e. H_2O, O_3), Rayleigh scattering and extinction (scattering and absorption) due to aerosols. The parameterization of these processes is relatively well known and can be modelled quite easily (Finlayson-Pitts and Pitts, 1986). Compared with the cloudless atmosphere, relatively little theoretical work has been done on the influence of the presence of clouds on photodissociation. Madronich (1987) studied the effect of clouds on the actinic flux by using the delta-Eddington radiative transfer model (Shettle and Weinman, 1970; Joseph *et al.*, 1976). The main input parameters for this kind of model are the optical depth (τ), single scattering albedo (ω) and assymmetry factor (g). Madronich (1987) showed model results for different values of τ, ω and g. Some characteristic features of the effect of clouds on the actinic flux were demonstrated by Madronich (1987) using a simple one-dimensional homogeneous cloud model without absorption, based on geometric considerations only. This approach also forms the base of the model presented by Thompson (1984).

2. Observations

2.1. SYNOPTIC SITUATION

On 15 September 1989 the weather in the Netherlands was determined by a low pressure system with a central pressure of about 970 hPa. At 1200 UTC the low pressure system was located off the west coast of Scotland and moving northwards towards Iceland. At this time the warm front associated with the frontal system was located right over the Netherlands.

During the whole day the sky was completely covered with clouds, mainly in the form of layered cloud consisting of stratocumulus, stratus and some cumulus. The cloud top was around 1500 m (Figure 1) and cloud base lifted from about 200m in the night to about 800 m around noon, and descended again to around 400m during the afternoon.

In Figure 1 we have plotted the data from the radiosonde ascent in De Bilt (52°06'N and 5°11'E, close to station 627 in Figure 2) at 1200 UTC. The wind was blowing from the west throughout the whole boundary layer. The wind speed increased from about 2 m/s at the surface to about 15 m/s at cloud top, above which it remained nearly constant with height. The relative humidity increased from about

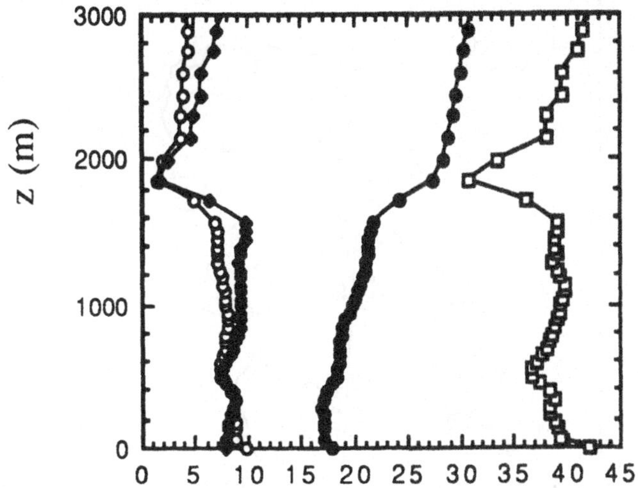

Fig. 1. The radiosonde ascent at 1200 UTC on 15 September 1989 in De Bilt: (open circles) specific humidity q [g/kg], (full circles) potential temperature θ [C], (open squares) equivalent potential temperature θ_e [C] and (diamonds) relative humidity [%] divided by 10 as a function of height.

78% to near 100% at cloud top (Figure 1). Above the boundary layer the air was quite dry with a specific humidity of about 4 g/kg. The potential temperature shows a dry adiabatic layer up to about 400 m. From about 400 m up to cloud top we have a wet-adiabatic lapse rate which can be seen from the constant value of the equivalent potential temperature up to cloud top. Above the boundary layer we have a stable lapse rate of about 2.5 K/km.

2.2. Aircraft Observations

On 15 September 1989 GEOSENS made aircraft observations of chemical components in and outside a stratocumulus deck over the Netherlands, Belgium and the North Sea (Van Broekhuizen and Van Kuijk, 1990). Because on this day the wind was blowing from the west, measurements were made along two horizontal tracks running almost from north to south: ABCD and EFGH in Figure 2. Moreover, an ascending and a descending spiral flight were made both at point B (0725 UTC) and F (1222 UTC).

The aircraft observations were made by a twin-engined Piper Chieftain (PA 31-350) with a typical cruise speed of about 70 m/s. The observations of interest for this study are the concentrations of the gas phase components of NO, NO_x and O_3, the measurements of which are all based on the chemiluminescence method. The NO_x and NO concentrations are measured by a modified 8840 instrument of Monitor Labs and the O_3 concentration by a modified 8002 instrument of Combustion Engineering. The gas concentrations are measured every second and are on-line averaged over 5 s. These 5-s averaged values were used in this study.

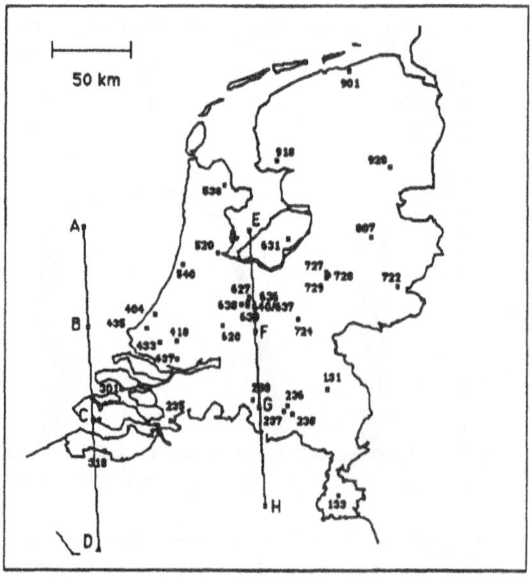

Fig. 2. The location of the 34 RIVM (National Institute of Public Health and Environmental Protection) surface stations within the Netherlands and the horizontal aircraft tracks ABCD and EFGH over the Netherlands, Belgium and the North sea.

3. Model

The effect of clouds on the actinic flux was studied by Madronich (1987) by using the delta-Eddington radiative transfer model (Shettle and Weinman, 1970; Joseph *et al.*, 1976). This model forms the basis of the model presented in this section. The output of the model consists of actinic fluxes, up- and downward (diffuse and direct) irradiances and photodissociation constants (in this paper only of NO_2) at all specified levels. As such it is possible to analyse the height dependence of the actinic flux etc. in cloudy atmospheres.

For the calculation of radiative transfer through the atmosphere, the different extinction processes have to be parameterized. As we are interested in photodissociation, only the radiation with a wavelength in the near-UV region has to be considered, e.g. in the case of nitrogen dioxide the range from 290-420 nm in the troposphere. In the atmosphere radiation of these wavelengths is attenuated by Rayleigh scattering, ozone absorption, (Mie-)scattering by water droplets and (Mie-)scattering and absorption by aerosol particles. The optical properties of (the combination of) these processes - the optical thickness τ, single scattering albedo ω and asymmetry factor g, will be calculated for each layer of the model.

3.1. Radiative Processes

In the model Rayleigh scattering is parameterized in the manner of Slingo and Schrecker (1982). However, in our case calculations are made monochromatically, instead of using spectral bands. With the use of the hydrosatic balance and the gas law the Rayleigh optical thickness for a layer of thickness $\Delta p(> 0)$ can be expressed by

$$\tau_R = 0.9793 \frac{R_d}{g_o} \frac{(n_s^2 - 1)^2}{\lambda^4} \Delta p \quad (1)$$

with the pressure p in hPa, the wavelength λ in microns; R_d the gas constant for dry air, g_o the gravity constant and n_s the refractive index of air at standard temperature and pressure. The values of the refractive index are taken from Edlen's (1953) formula. The single scattering albedo for this process is equal to one, the asymmetry factor is zero.

The optical thickness by ozone absorption can be defined as the product of the absorption cross-section (ACS) of ozone and the ozone column of the layer considered. The ACS-data were taken from Finlayson-Pitts and Pitts (1986). The vertical ozone distribution is the standard profile of Lacis and Hansen (1974) with the total ozone amount equal to 348 D.U. (Dobson units) (0.348 cm STP), a typical (year averaged) value for mid-latitudes. This profile does not include eventually increased ozone levels in the lower troposphere (see section 5). The single scattering albedo for the ozone absorption process is zero, the asymmetry factor is not relevant.

For the case of Mie scattering the general equation defining the optical thickness for a layer (thickness Δz) containing particles with a size distribution n(r) is considered (e.g. Stephens, 1978):

$$\tau = \int_o^{\Delta z} \int_o^{\infty} n(r) \pi r^2 Q_{ext}(r) dr dz \quad (2)$$

where Q_{ext} is the extinction efficiency (van de Hulst, 1981). The multiple scattering by cloud drops can be parameterized in terms of the liquid water path W (i.e. in a layer of thickness Δz) and the effective radius of the size distribution as proposed by Stephens (1978) and Slingo and Schrecker (1982):

$$\tau_d = \frac{3}{2} \frac{W}{\rho_l r_{eff}} \quad (3)$$

with:
ρ_l= density of liquid water (in kg/m^3)
r_{eff}= effective radius of droplet size distribution (in m)
W = liquid water path (in kg/m^2):

$$W = \int_o^{\Delta z} \rho_0 q_l dz \quad (4)$$

where ρ_0 is the density of air (in kg/m^3) and q_l is the liquid water content in (kg/kg). In this parameterization it is assumed that the extinction efficiency is equal to two. This corresponds with the assumption of large drops with respect to the wavelength.

The single scattering albedo can be derived from the theory of anomalous diffraction (van de Hulst, 1981). The single scattering albedo can be defined in terms of the effective efficiency coefficients for absorption and extinction (van de Hulst, 1981)

$$\omega = 1 - \frac{Q_{abs,eff}}{Q_{ext,eff}}. \tag{5}$$

We assume again that the (effective) extinction efficiency is equal to two. The effective absorption efficiency for a layer that contains a mixture of particles can be calculated if the size distribution and the refractive index as a function of wavelength are known. Data for the refractive index of water were taken from Hale and Querry (1973). The size distribution is assumed to be independent of height. We use the modified gamma distribution of Hansen and Travis (1974). This distribution is characterized by two parameters: the effective radius r_{eff} and the effective variance v_{eff} (see for definitions Hansen and Travis, 1974; van de Hulst, 1981). The single scattering albedo as a function of wavelength can be expressed in terms of these parameters as

$$\omega_d = 1 - \frac{1}{3}Ar_{eff} + \frac{1}{8}A^2r_{eff}^2(1 + v_{eff}) \tag{6}$$

with

$$A = \frac{8\pi m_i}{\lambda}$$

where λ is the wavelength and m_i is the (wavelength-dependent) imaginary part of the refractive index. A typical value for the effective radius of stratocumulus clouds is 10 μm. Eventually, this value can be varied for different cloud layers. The effective variance is typically 0.2. We can read from equation (6) that the width of the size distribution only appears in the second-order term.

The asymmetry factor g depends on the real part of the refractive index and the single scattering albedo. Following van de Hulst (1981) we find:

$$g = \frac{1 + \omega\gamma}{1 + \omega} \tag{7}$$

where γ is determined by the real part of the refractive index. For clean water with a refractive index of around 1.334, van de Hulst (1981) gives γ =0.74. Thus, because ω_d is nearly one we find typically g_d=0.87, i.e. the forward scattered fraction of the incident radiation is equal to $g_d^2 \approx 0.76$. Again it is assumed in equation (7) that the drops are large with respect to the wavelength.

Aerosol optical properties are basically approached in the same way as cloud drops. However, the assumption of large particles with respect to the wavelength is no longer valid. Fortunately aerosol extinction is, especially compared to the multiple scattering by cloud drops, a relatively minor effect. Therefore, the optical parameters for aerosol extinction are calculated numerically with the theory of anomalous diffraction. No distinction is made between different types of aerosol. It is assumed that the extinction by the particles is partly due to absorption. Following Finlayson-Pitts and Pitts (1986) we assume that the index of refraction of the aerosol n_p is independent of height or wavelength: n_p=1.50-0.02i. In the model we use the so called haze-L modified gamma distribution of Deirmendjian (1969) as suggested by Demerjian *et al.* (1980). The effective particle radius of this distribution is 0.5 μm. The effective variance is 0.42. The height distribution is taken from Demerjian *et al.* (1980).

Under these assumptions we can calculate (numerically) the optical thickness τ_p for aerosol extinction with equation (2) and the single scattering albedo ω_p with equation (5). It is difficult to establish a realistic value for the asymmetry factor (g_p) for aerosol scattering. We apply equation (7) although this equation is strictly only valid for large aerosol particles with respect to the wavelength. For the assumed real part of the refractive index of 1.50, van de Hulst gives $\gamma = 0.64$. For a decreasing size of the aerosol particles it can be expected that the asymmetry factor will also decrease.

3.2. SOLVING METHOD

Joseph *et al.*(1976) showed that combining a Dirac-delta function and the usual two-term approximation of the Eddington method of the delta-Eddington phase function results in a similar transfer equation as for the Eddington method as obtained by Shettle and Weinman (1970), only with transformed parameters (τ, ω and g). This yields two coupled equations fo the isotropic (I_0) and the anisotropic (I_1) parts of the radiance:

$$\frac{\partial I_1}{\partial \tau'} = -3(1-\omega')I_o + \frac{3\omega'}{4\pi}S_o e^{-\tau'/\mu_0} \tag{8}$$

$$\frac{\partial I_0}{\partial \tau'} = -(1-\omega' g')I_1 + \frac{3\omega'}{4\pi}g'\mu_0 S_0 e^{-\tau'/\mu_0} \tag{9}$$

$$\tau' = (1-\omega g^2)\tau \tag{10}$$

$$\omega' = \frac{(1-g^2)}{1-\omega g^2)}\omega \tag{11}$$

$$g' = \frac{g}{1+g} \tag{12}$$

where S_0 is the incoming solar flux at the top of the atmosphere and μ_0 is the cosine of the solar zenith angle. These equations are solved by the method described by Shettle and Weinman (1970). The similarity between the original parameters and their transformed counterparts was observed earlier by van de Hulst (1974).

One of the advantages of the transformation is that the asymmetry factor is effectively reduced, because the results of the Eddington approximation improve for more isotropic scattering. The delta-Eddington approximation is superior for the description of radiative transfer in cloudy atmospheres as this method is very accurate with strong (asymmetric) multiple scattering and small absorption (King and Harshvardhan, 1986).

The model equations can be solved analytically for a homogeneous atmosphere (i.e. constant single scattering albedo ω and constant asymmetry factor g). This solution will be of use in the next section. However, in order to solve the equations for a vertically inhomogeneous atmosphere, the atmosphere is divided into a number of layers which are each characterized by constant values of ω and g. These parameters in turn are determined by the extinction processes treated in section 3.1: Rayleigh scattering (index R), ozone absorption (index O_3), Mie scattering by water drops (index d) and scattering and absorption by aerosol particles (index p).

The optical thickness τ of a layer is calculated as the sum of the optical thickness of each process for that layer. The single scattering albedo and the asymmetry factor of a layer are calculated by weighting the contribution of the different processes. With known values for the optical thickness and $\omega_d . g_d$, ω_p and g_p (section 3.1) the following characteristics for each layer are obtained (Slingo and Schrecker, 1982):

$$\tau = \tau_{O3} + \tau_R + \tau_d + \tau_p \tag{13}$$

$$\omega = \frac{\tau_R + \omega_d \tau_d + \omega_p \tau_p}{\tau} \tag{14}$$

$$g = \frac{g_d \omega_d \tau_d + g_p \omega_p \tau_p}{\omega \tau} . \tag{15}$$

Now the radiative transfer equation can be solved by the method of Shettle and Weinman (1970) and Joseph *et al.* (1976). Imposing that the downard diffuse irradiance at the top of the atmosphere is zero and prescribing the albedo of the ground results in a set of 2N linear equations (N is the number of layers). This is easily solved by rewriting the matrix in tri-diagonal form which can be solved very efficiently (e.g. Press *et al.*, 1986).

4. Analytic 3-Layer Model

Apart from the numerical model described in the preceding section it is useful to have also a simplified approach to get better insight in the effect of clouds on the actinic flux. It was shown by Madronich (1987) that for a direct beam that is incident under an angle θ_0 with the normal of a surface (e.g. a cloud) with albedo A, the actinic flux is enhanced by a factor (1+2 A cos θ_0). Thus, for direct radiation the actinic flux above a reflecting surface can be enhanced theoretically by a factor three. For diffuse radiation we obtain a factor (1+A).

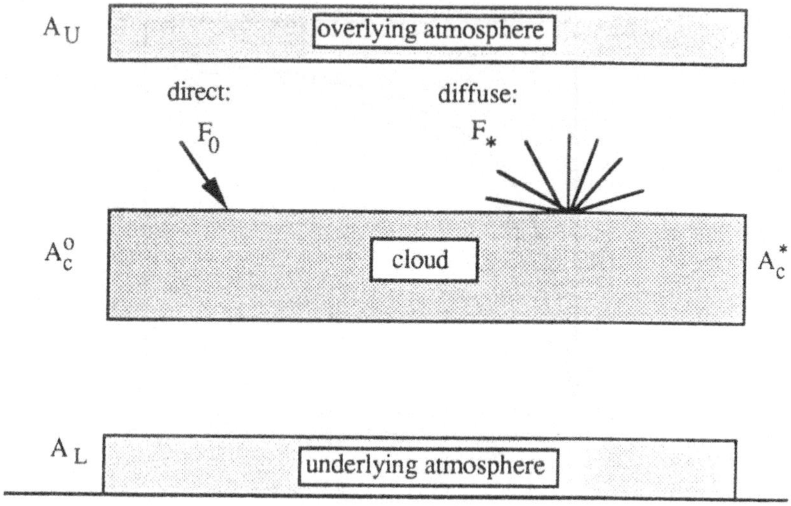

Fig. 3. A layer cloud with albedo A_C^0 and A_C^* for direct and diffuse radiation, respectively. The radiative properties of the overlying and the underlying atmosphere are prescribed with albedo A_U and A_L, respectively.

Inspired by the simple cloud model presented by Madronich (1987), we consider a one-dimensional three-layer atmosphere (Figure 3), i.e. a cloud layer is considered with a layer above that represents the overlying atmosphere and a layer below that represents the underlying atmosphere (including the Earth's surface).

We consider a direct flux (F_0 incident at an angle θ_0 (=$\cos^{-1}\mu_0$) and a diffuse downward flux ($F*$) at cloud top. Absorption in the cloud layer is neglected, because it is observed that absorption in clouds is very small for visible and UV light. The single scattering albedo for these wavelengths is typically larger than 0.999. Further, it is assumed that all radiation that is transmitted through the cloud is fully diffuse (i.e. a cloud optical thickness of at least 4, such that the transmission is at most $e^{-4} < 2\%$). All surfaces are Lambertian, that is, all reflected light is diffuse.

Under these assumptions, we can write down the following equations for the actinic flux at cloud top (F_A) and at cloud base (F_B) in terms of F_0, μ_0 and $F*$ – i.e. the downward fluxes that can be expected at cloud top if no cloud is present –, A_L, A_U and the cloud albedo for diffuse light (A_C^*) and for direct light (A_C^0):

$$F_{A\downarrow} = F_0 + F_* + A_U F_{A\uparrow} \tag{16}$$

$$F_{A\uparrow} = 2\mu_0 A_C^0 F_0 + A_C^* F_* + (1 - A_C^*) F_{B\uparrow} + A_C^* A_U F_{A\uparrow} \tag{17}$$

$$F_B = (1 - A_C^*) F_{A\downarrow} + A_C^* F_{B\uparrow} \tag{18}$$

$$F_{B\uparrow} = A_L F_{B\downarrow}. \tag{19}$$

Combining the upward and downward contributions:

$$F_B = F_{B\downarrow} + F_{B\uparrow} \tag{20}$$

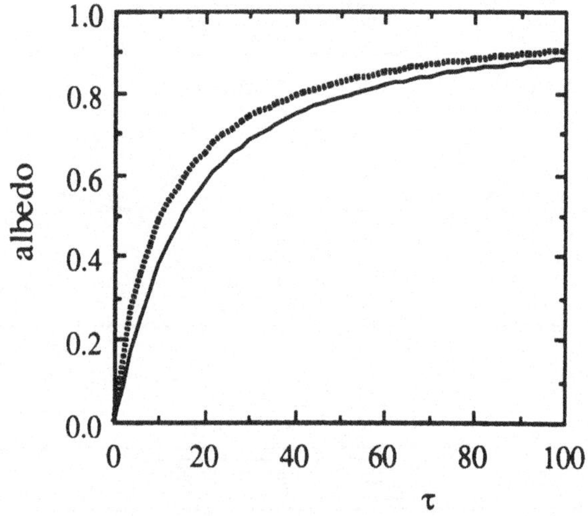

Fig. 4. The albedo for diffuse (dotted line) and direct radiation (full line) as a function of the cloud optical thickness (τ).

$$F_A = F_{A\downarrow} + F_{A\uparrow}. \tag{21}$$

These equations can easily be rewritten in terms of the incoming fluxes F_0 in $F*$.

The parameters A_L and A_U express the influence of reflections from the surrounding atmosphere. Knowledge of the absorption in the surrounding atmosphere is not required, only the reflected light. It can be expected that A_L is for a large part determined by the albedo of the Earth's surface. Variations in both A_L and A_U can be expected as a result of cloud height, aerosol loading, ozone concentrations etc. Estimations for A_L and A_U can be obtained with the multi-layer model (section 3).

For the determination of cloud albedo, a distinction is made between a cloud albedo for direct light and an albedo for diffuse light, because it is observed that cloud albedo is not only a function of cloud optical thickness but also of the angle of incidence. An analytic relationship can be obtained by solving model equations (8) and (9) for a single non-absorbing cloud layer (Shettle and Weinman, 1970). For the boundary conditions, it is assumed that a direct beam is incident at cloud top, while no diffuse radiation enters the cloud at cloud top, nor at cloud base. The expression for the cloud albedo of a non-absorbing cloud as a function of cloud optical thickness τ_C for a beam incident at an angle θ_0 (represented by $\mu_0 = cos\theta_0$) in the delta-Eddington approximation is given by:

$$A_C^0(\tau_C, \mu_0) = 1 - \frac{(2 + 3\mu_0) + (2 - 3\mu_0)e^{-\tau_C(1-g^2)/\mu_0}}{\{4 + 3(1 - g)\tau_C\}} \tag{22}$$

It is assumed that the asymmetry factor g is constant for the whole cloud. This is reasonable in the limit of large drops with respect to the wavelength. A typical value for clouds is $g = 0.87$ (equation (7)).

The albedo for diffuse radiation can be approximated in a similar way. The model equations are solved for diffuse downward flux at cloud top. This yields

$$A_C^*(\tau_C) = 1 - \frac{4}{\{4 + 3(1-g)\tau_C\}}. \tag{23}$$

The cloud albedos given by Madronich (1987) in Table 1 of his paper satisfy equations (22) and (23). In Figure 4 the diffuse albedo and the direct albedo for $\theta_0 = 50^o$ are shown as a function of cloud optical thickness ($g = 0.87$).

As stated in the preceeding section the model equations can be solved for a single homogeneous cloud layer. This yields an expression for the in-cloud actinic flux as a function of in-cloud optical thickness. Boundary conditions at cloud top and at cloud base are provided by the equations for F_A and F_B given above. Replacing S_0 by the direct part of the incoming actinic flux F_0 at cloud top, gives the actinic flux in the cloud as a function of the in-cloud optical thickness in the delta-Eddington approximation in terms of F_A, F_B and F_0:

$$F_C(\tau, \mu_0) = 4\pi I_0(\tau) + F_0 e^{-\tau(1-g^2)/\mu_o} \tag{24}$$

with

$$I_0(\tau) = B_1 - \frac{3}{4\pi}\mu_0^2 F_0 e^{-\tau(1-g^2)/\mu_0} - B_2(1-g)\tau,$$

$$B_1 = \frac{1}{4\pi}[(3\mu_0^2 - 1)F_0 + F_A],$$

$$B_2 = \frac{F_A - F_B + F_0(3\mu_0^2 - 1)(1 - e^{-\tau_C(1-g^2)/\mu_0})}{4\pi(1-g)\tau_C}.$$

Finally, it is desirable to relate in-cloud optical thickness to in-cloud height. If we assume that the liquid water content and therefore the volume of the drops (proportional to r^3) increases linearly with height, the optical thickness is then determined by the surface of the drops (proportional to r^2). Thus, we can express the in-cloud optical thickness in the relative in-cloud height z' by

$$\tau(z') = \tau_C[1 - z'^{5/3}] \tag{25}$$

with

$$z' = \frac{z - z_b}{z_t - z_b}$$

where τ_C is the total optical thickness of the cloud, z_b is cloud base and z_t is cloud top. However, clouds will have generally a more complicated vertical variation of optical thickness.

5. Results and Discussion

5.1. THE MULTI-LAYER MODEL

In this paper we shall mention only some of the most characteristic results of the multi-layer model, especially with respect to clouds. Detailed model calculations of the actinic flux (and photodissociation rates) as a function of the most important model input parameters were performed by Demerjian *et al.* (1980).

It is stressed that the actinic flux in cloudless atmospheres is dominantly determined by solar zenith angle. Due to the increase of the contribution of upward radiation, the actinic flux increases with height, while the increase is stronger for larger zenith angle.

Further, it was found that ozone concentrations have only a small influence on the photodissociation rate of nitrogen dioxide. This can be understood, because ozone absorption is only important for wavelengths smaller than around 330 nm, while the phodissociation of NO_2 occurs for wavelengths up to 420 nm.

In the lowest parts of the troposphere the effect of aerosol or haze can be very strong, again especially for large zenith angles. For high aerosol loading in combination with high sun, radiation can be trapped in an aerosol layer. In such cases the actinic flux can be a maximum near the top of this aerosol layer.

This trapping of radiation can be more pronounced in clouds. Near cloud top the actinic flux in the cloud can be enhanced relative to the actinic flux above the cloud. If the albedo for direct light is smaller than the albedo for diffuse light, i.e. when direct light is transmitted more efficiently than diffuse radiation, the incoming direct radiation is trapped. This effect occurs for $\mu_0 > 2/3$, that is for zenith angles of around 48° or smaller (see equations (22) and (23)).

The actinic flux above and below clouds depends primarily on cloud optical thickness and solar zenith angle. In-cloud characteristics are important insofar as these affect cloud optical thickness. The effect of cloud height on the change of the actinic flux is small.

5.2. THE 3-LAYER MODEL

In order to apply the three-layer model presented in section 4, it is required to find appropriate values for the Albedo's A_L and A_U of the underlying atmosphere (including the Earth's surface) and the overlying atmosphere, respectively. These albedos represent the second order (and higher) contributions from the surrounding atmosphere that contribute to the change of the actinic flux below and above the cloud.

Typical values for these albedos were estimated from calculations with the multi-layer model. For a typical boundary-layer cloud with cloud base around 500 m above ground level, we find typically A_L=0.16 and A_U=0.30. The higher the cloud base, the larger the albedo of the underlying atmosphere.

We can calculate easily from equations (16)-(21) the actinic flux for the case of

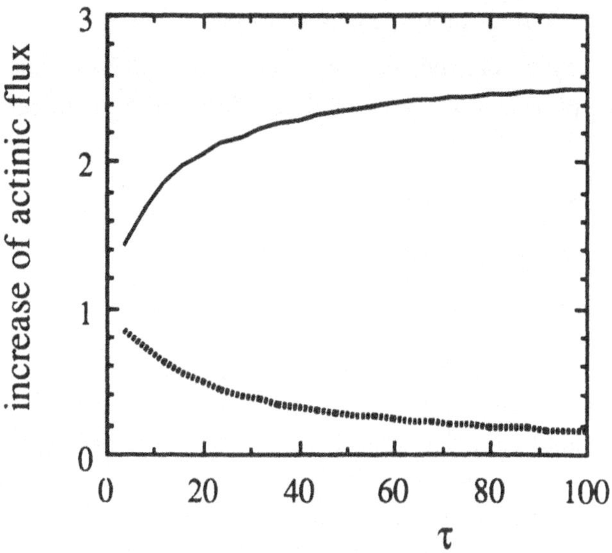

Fig. 5. The relative increase in the actinic flux at cloud top (full line) and at cloud base (dotted line) due to the presence of the cloud as a function of the cloud optical thickness (τ) for direct radiation with a zenith angle of 50°.

no cloud ($A_C^* = A_C^0 = 0$). This yields:

$$F_{clr} = F_A = F_B = \frac{1 + A_L}{1 - A_L A_U}(F_0 + F*). \tag{26}$$

Further, we can analyse the maximum actinic flux enhancement above the cloud ($A_C^* = A_C^0 = 1$). For only direct radiation incident at cloud top, we find for μ_0=1

$$\{\frac{F_A}{F_0}\}_{max} = 1 + 2\frac{1 + A_U}{1 - A_U}. \tag{27}$$

For only diffuse incident radiation we find

$$\{\frac{F_A}{F*}\}_{max} = 1 + \frac{1 + A_U}{1 - A_U}. \tag{28}$$

Combining equations (26) and (28) we find with $A_L = 0.16$ and $A_U = 0.30$ for diffuse radiation a maximum enhancement of around 2.3 and for direct radiation (overhead sun) around 3.9 compared to clear sky actinic flux. In the atmosphere a combination of direct and diffuse radiation is always incident at cloud top. This yields maximum values somewhere in between.

In figure 5 the increase of actinic flux above and below the cloud relative to the clear sky value is shown as a function of cloud optical thickness for direct radiation with a zenith angle of 50°. It was assumed that no direct radiation is transmitted through the cloud. Therefore, the model can not be applied for $\tau < 4$.

5.3. COMPARISON WITH AIRCRAFT OBSERVATIONS

In order to compare the calculated photodissociation rate of NO_2 with observations, we can apply the photostationary state relationship. Three chemical reactions are involved:

$$NO_2 + h\nu \quad J_{NO_2} \rightarrow NO + O \tag{29}$$

$$O + O_2 + M \quad k_2 \rightarrow O_3 + M \tag{30}$$

$$O_3 + NO \quad k \rightarrow NO_2 + O_2. \tag{31}$$

The first reaction describes the photodissociation of nitrogen dioxide. The photodissociation rate is denoted by J_{NO_2}. The radiation incident on a molecule is denoted by $h\nu$. The O-radical produced is very reactive, mainly with O_2 to ozone. The molecule M (e.g. N_2 or O_2) absorbs the excess energy and stabilizes the O_3 molecule formed. The ozone formed reacts with nitric oxide (NO). Reaction (30) is very fast. Therefore, combining (29) and (30) shows that NO_2 is dissociated into NO and O_3, whereas (31) gives a production of NO_2 due to the reaction of NO with O_3. If production and destruction of NO_2 are in equilibrium, the photostationary state holds

$$\frac{NO\,O_3}{NO_2} = \frac{J_{NO_2}}{k}. \tag{32}$$

The reaction rate k is a well known function of temperature. Therefore, we can apply equation (32) in order to relate measured concentrations of NO, NO_2 and O_3 with J_{NO_2} values.

At 1222 UTC near point F in Figure 2 both an ascending and a descending spiral were made. The ascending spiral was started at about 900 m up to 1800 m height, whereas the descending spiral was made from 1800 m down to about 600 m. As an input for the radiation model, we take a cloud base at 650 m and cloud top at 1600 m. For the liquid water density, we take a linear profile with a maximum of about 0.14 g/m^3 at cloud top. This yields a liquid water path of 133 g/m^2. The optical thickness of the cloud with $r_{eff} = 10^{-5}$m, thus becomes τ_d= 10 (equation (3)).

In Figure 6 the photostationary state ratio $NO\,O_3/NO_2$ as a function of height is shown for both spiral flights, together with the ratio J_{NO2}/k as calculated by the multi-layer model for a cloud with an optical thickness of 10. For comparison the cloud-free photodissociation rate as a function of height is shown, too. The large increase in the photodissociation rate above the cloud and in the upper part of the cloud can be seen clearly. The photodissociation rate decreases with height just above the cloud. Higher in the troposphere an increase with height is found as a result of scattering of downward radiation.

Finally, it is remarked that the comparison between the photodissociation rate of NO_2 and measured concentrations is restricted by the assumption that the photostationary state relationship holds. This requires high time and space resolution

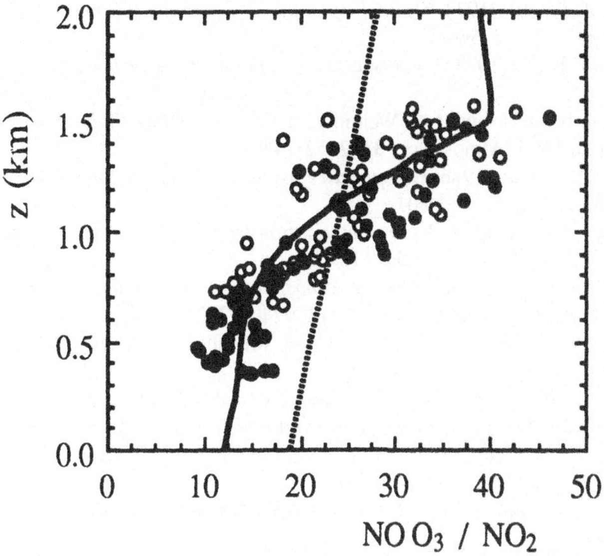

Fig. 6. The photostationary state ratio NO $O_3/NO2$ as a function of height for the upward (open circles) and downward (full dots) spiral made near point F (Figure 2) at 1222 UTC. The lines represent J/k as a function of height calculated by the multi-layer model for a cloud with a cloud optical thickness of 10 (solid line) and for a clear sky (dotted line) on 15 September.

of the data. The 5-s averages of the spiral flights seem to satisfy this requirement satisfactorily. Even if deviations from the photostationary state relationship were important, the relative variation with height would probably be rather constant. It is recommended that measurements of actinic fluxes and J_{NO2} be performed in and around clouds in combination with concentration measurements, in order to verify the model results presented in this paper more directly. Finally it is recommended that the effect of partial cloudiness be incorporate into the models.

Acknowledgements

The aircraft observations used in this study were made by GEOSENS b.v. within the VROM project 64.19.07.01: 'An investigation into the chemical composition of cloud water.'

References

Deirmendjian, D., 1969. *Electromagnetic scattering on spherical polydispersions*, Elsevier Publ. Co., New York, USA.

Demerjian, K.L., K.L. Schere, and J.T. Peterson, 1980. *Theoretical estimates of actinic (spherically integrated) flux and photolytic rate constants of atmospheric species in the lower troposphere.* Adv. Environ. Sci. Technolo., **10**, 369-459.

Edlen, B., 1953. *the dispersion of standard air*. J. Opt. Soc. Am., **43**, 339-344.

Finlayson-Pitts, B.J., and J.N. Pitts jr., 1986. *Atmospheric Chemistry*. John Wiley and Sons, New York, USA.

Hale, G.M. and M.R. Querry, 1973. *Optical constants of water in 200 nm to 200 μm wavelength region*. Appl. Opt., **12**, 555-563.

Hansen, J.E. and L.D. Travis, 1974. *Light scattering in planetary atmospheres*. Space Sci. Rev.,, **16**, 527-610.

Joseph, J.H., Wiscombe, W.J. and J.A. Weinman, 1976. *The delta- Eddington approximation for radiative flux trasnfer*. J. Atmos. Sci., **33**, 2452-2459.

King, M.D. and Harshvardhan, 1986. *Comparative accuracy of selected multiple scattering approximations*. J. Atmos. Sci., **43**, 784-801.

Lacis, A.A., and J.E. Hansen, 1974. *A parameterization for the absorption of solar radiation in the Earth's atmosphere*. J. Atmos. Sci., **31**, 118-133.

Madronich, S., 1987. *Photodissociation in the Atmosphere: 1. Actinic flux and the effects of ground reflections and clouds*. J. Geophys. Res. D., **92**, 9740-9752.

Press, W.H., Flannery, B.P., Teukolsky, S.A., and W.T. Vetterling, 1986. *Numerical Recipes*, Chap. 2, p.40-41, Cambridge Univ. Press, Cambridge.

Shettle, E.P., and J.A. Weinman, 1970. *The transfer of solar irradiance through inhomogeneous atmospheres evaluated by Eddington's approximation*. J. Amos. Sci., **27**, 1048-1055.

Slingo, A., and H.M. Schrecker, 1982. *On the shortwave radiative properties of stratiform water clouds*. Quart. J.R. Met. Soc., **108**, 407-426.

Stephens, G.L., 1978. *Radiative properties of extended water clouds, Part II*. J. Atmos. Sci., **35**, 2123-2132.

Thompson, A.M., 1984. *The effect of clouds on photolytic rates and ozone formation in the unpolluted troposphere*. J. Geoph. Res. D, **89**, 1341-1349.

Van Broekhuizen, H.J. and A. van Kuijk, 1990. *The chemical composition of cloud and rain water, 1st measuring campaign, 15 September 1989*. GEOSENS b.v., PO Box 12067, 3004 GB Rotterdam, The Netherlands (in Dutch).

van de Hulst, H.C., 1974. *The spherical albedo of a planet covered with a homogeneous cloud layer*. Astron. Astrophys., **35**, 209-214.

van de Hulst, H.C., 1981. *Light scattering by small particles* Dover Publ., New York, USA.

PART IV

ROUND TABLE DISCUSSION: INTERACTIONS AND FEEDBACK BETWEEN THEORY AND EXPERIMENT

THE ROUND TABLE DISCUSSION: INTERACTIONS AND FEEDBACK BETWEEN THEORY AND EXPERIMENT

EDITED BY S. R. HANNA

Panel Members: Philip Chatwin, Han van Dop, Steven Hanna (Chairman), Michael Poreh, Brian Sawford, Roland Stull

(Received October 1991)

S. HANNA: "Interactions and feedback between theory and experiment" is a very general topic that covers everything that we do. It is interesting to begin with the basic scientific method which we were taught when we were about 10 years old in our first science class. The scientific method consists of several steps, beginning with the statement of an objective or a hypothesis. I feel that in some research projects, this step is not taken, and people are going about their business without any clear objective in mind. In the case of the research discussed this week, there are two types of modeling objectives. One is the development of a comprehensive model, such as a model for the ozone distribution over Israel. Other examples of comprehensive models are meteorological models which handle the sea breeze and the depth of the marine boundary layer, models which simulate the mesoscale flows in the Jordan Valley, and models explaining ozone and SO_2 distributions. These research projects have a very comprehensive objective. The second type of objective is more specific and is generally oriented towards one particular aspect of the problem. An example of this type of project is the Thorney Island field study, which took place in England. The scientific objective was to determine what happens to a column of dense gas when it is instantaneously released into the atmosphere at ground level. Although there are not any real-world industrial sources that are exactly like this, the field data allow us to investigate a very narrow aspect of the problem.

Another issue is whether wind tunnel data are just as good as field data. I have often looked at the Fackrell and Robins wind tunnel data, which have been used by several people on this panel and in the audience to try and validate models which predict the concentration fluctuations in a plume. However, I see many things happening in the Fackrell and Robins wind tunnel study that I have not seen happening in the field. There seems to be a double hump in the cross-wind distribution of concentration fluctuation intensities observed in the wind tunnel. Many of the models attempt to simulate this double hump, which has not been observed in the field. Also, mesoscale eddies are important in the real atmosphere but cannot be simulated in a wind tunnel. These mesoscale eddies determine whether the concentration intensity $\sigma_c/\bar{C}$ approaches 0 or 1 or some other constant. In the real atmosphere, for as long a sampling time as you have, there are always mesoscale eddies. These eddies will cause fluctuations and $\sigma_c/\bar{C}$ could approach any non-zero value, depending on the particular characteristics of the mesoscale eddy field.

The final point that I wanted to make is that one main use of experimental data is to determine which parameters are really important for the scientific phenomenon that you are studying. These parameters can be identified by applying factor analysis and multivariate analysis to field data in order to determine if there are fundamental functional relations that should be simulated by the theory. Or, before deriving any theory, you can simply plot your data in a similarity form and calculate the correlations. This will tell you which of the parameters are strongly related to the other ones, what types of similarity analysis should be done, and which factors are unimportant. If there is a parameter that is not correlated with anything else, then perhaps it is not important enough to be included in the theory.

R. STULL: Steve suggested that I give a specific example, namely the relationship between experiments, theory, and applications for transilient matrix theory.

We finally figured out how to measure the transilient mixing coefficient in the atmosphere. Picture the following scenario: Assume you have a tall tower and at some height on this tower you release a series of small smoke puffs, and you track the movement of those puffs – say by lidar – over a finite time interval. Each puff might end up in a different location. After a large number of puffs have been tracked you have a distribution of puff destinations. This is in an absolute dispersion sense; you are not looking at the spread of individual puffs, but the movement of puff centers. Thus you have, more or less, a distribution of puff destinations for one source height. That distribution, if you describe it as a probability distribution that has an integral of unity, gives you one column of the transilient matrix, by definition. Assume that you continue to track those smoke puffs over longer times and of course they move to different places. You will have a different distribution for longer times and it will give you a different transilient matrix. Remember that the matrices depend on the time interval as well as on the turbulent mixing.

In this way you can measure one column of the transilient matrix for different times. Of course if different puffs are released at different source heights, then you can get the full transilient matrix. You know what all the air is doing, where it is coming from, and where it is ending up. Now step back from this field experiment and consider what you have – smoke puffs released from particular heights are like smoke emissions from a smoke stack. That means that if you know the transilient matrix which describes the characteristics of the flow and the turbulent mixing, you can immediately pick out any column of your transilient matrix to determine how this smoke disperses from a smoke stack at any height – say, tall smoke stacks in one column of the matrix, shorter stacks in a different column.

For example, for any one particular type of smoke stack you could look at the transilient matrices for different time intervals and see how the distribution will change effectively with distance downwind of this stack. These transilient matrices can be measured in the real atmosphere or simulated with large eddy simulation models. What we have seen so far, is that these transilient matrices obey similarity theory; namely, for the convective mixed layer the matrix has a certain pattern. With a measured matrix for some particular state, like free convection or a neutral boundary layer or a combination of the two, one can scale the matrix with respect to the depth of the mixed layer and with respect to appropriate scaling velocities like u_* or w_*. These empirically measured matrices can then be applied in Israel, United States, Germany or wherever, once you "unscale" the matrix using the similarity variable. That is a way to go from field experiments using the theory of transilient turbulence to an application for different smoke stacks.

The theory has potential for site studies and planning for future power plants or industries. You can ask yourself what happens if you increase the height of a smoke stack; you can look right into this transilient matrix and pick out different columns and see what the smoke will do. If you can measure the mixing from all possible heights and destinations, then you have all the information for any height stack that you want. It is already there in the matrix. This potential application has not been used much yet. Hopefully in the next year or two my colleague, Eloranta (who has a lidar at Wisconsin) and I will be measuring mixing with smoke puffs in the real atmosphere. This will be our first attempt to measure the thing in real life, and it is exciting for us. We think that we can capture with transilient matrices not only random-like situations, such as dispersion, but also the deterministic behavior, such as that speculated by Deardorff, simulated by Lamb, and measured by Briggs.

M. POREH: Alternatively, all you have to do is use the prediction schemes to give you exactly the value of the transilient matrixes, the probabilities, the similarities, and so on. You do not even have to go to real experiments.

R. STULL: You are right, if our parameterizations were perfect, then we would not have to use experiments. However, our parameterizations are not perfect. So I think that at least for various special cases it is really important to measure transilient matrices for the real atmosphere. Once we confirm that the measured matrices match the simulated or predicted mixing from theory, then perhaps – if the scientific community is willing, we can continue to use theories to calculate the matrices without having to do expensive field experiments.

B. SAWFORD: I think that discussing the interaction between theory and experiment is a difficult thing to do, as Steve pointed out in the beginning. Most of us are very much used to working with the whole range of scientific tools, from theory through experiment, field work and so on, so here we are tending to state the obvious.

I would like to try and put a philosophical point of view on some of these ideas. I think the first point to make is obvious, which is that the whole is greater than the sum of the parts. In fluid mechanics in general and atmospheric science in particular, we need to have at our finger tips, all three tools of theory, laboratory experiments, and field experiments. They are very much complementary; no one of them will give the whole story to most problems. It is really quite dangerous to be just a theoretician or an experimentalist and not give due regard to the other activities. Perhaps the most dangerous thing to be is just a modeler and not be properly aware of the experimental aspects of the work. Steve also mentioned the need to have a clear objective in whatever you are doing. This is a very important point and I suspect that sometimes experimentalists tend to charge in and make a range of measurements without perhaps thinking how those measurements might be used, and this is sometimes a failing in the data. We come across that problem as theoreticians when we try to test our theories using data. We find that the data are inadequate because something that we need for our theory has not been measured. That is a very good example of an instance where there needs to be interaction between the theoreticians and the experimentalists. Perhaps on a slightly controversial note, I would like to suggest that fluid mechanics is basically an empirical and experimental science. I know that there are lots of exceptions and I have managed to think of some. I think that for instance the Kolmogorov similarity theory is one example of something that came out theoretically and then was confirmed experimentally. At about the time that it was confirmed Kolmogorov stood up and said that the theory was not quite right, that we have some adjustments to the theory, and away they went again with intermittency corrections. That is one example where theory came first. By and large, fluid mechanics is an experimental science. A lot of the theories are descriptive and they involve unknown constants, "fudge factors", whatever you like, which at the end of the day require the experimental data to determine. One other point that I might raise here is a question really; where do the various sorts of large numerical calculations fit into the scheme of things, are they experiments or are they theories? I tend to think of complete turbulence calculations, direct numerical simulations of turbulence, for instance, as experiments. Basically they are solving Navier-Stokes equations exactly for particular resolution. It is rather like doing a laboratory experiment for that resolution. But you obviously have a whole range of numerical calculations, going through large eddy simulations, where you are doing a little bit of modeling and still doing a lot of almost exact calculations. Those sort of things sit on the borderline between experiment and theory. Perhaps carrying on this theme of the importance of experimental work, I can close by saying that today as more and more computing power becomes available, there seem to be more and more people taking up computers and becoming "computer jockies".

Therefore, I would like to make an appeal for people to bear in mind that the ultimate confirmation or ultimate truth comes back to making experiments. I think that the computer people are very much in need of good experimental data so that when they are doing large eddy simulation, for instance, where there is still some modeling involved, they can keep their work on track.

D. RIDE: I think as scientists we have yet to make a proper distinction between descriptive and predictive models. As scientists we tend to include as much science as we can, although we have had two papers at this conference which clearly demonstrate that there is a limit to the usefulness of putting more physics into our models. But many predictive models require a far greater simplicity. For instance, if you have a descriptive model which needs the first, second, and third moments of your concentrations, you may very well get a good descriptive fit. When you come to prediction, most of us would agree that you can predict a mean concentration, say, with a fair degree of confidence. You can also perhaps predict the variance, but few of us would predict the third moment, apart from saying that in many situations, probably the pdf is strongly positively skewed. As a user of models, I come constantly up against the problem that I am dealing with elegant models developed by scientists who, in a way, want to demonstrate that they have done good science. However, the elegant models are not always good predictive models.

B. SAWFORD: To some extent, when I talked about predictive and descriptive models, I was perhaps thinking a little more generally than you obviously are, in the sense of predictive theories being those which actually come up with a new result which is then demonstrated experimentally. Most of the models we deal with I would regard as descriptive in the sense that we have some sort of range of phenomena that we observe, whether it be dispersion in smoke plumes or whatever it is, and our models are really just empirically describing that phenomenon. They make prediction of detail, of course, of concentrations and things like that, but I was thinking perhaps more generally and philosophically about the terms "predictive" and "descriptive".

H. KAPLAN: So far, most of the models are designed to calculate average concentrations, and many experiments are designed to fit parameters to those models. We know from theoretical works and from experimental measurements, that fluctuations around the average are of the same order of magnitude, or even an order of magnitude greater. Do you think that the next generation of models and experimental fitting to these models – and I am speaking about applied models, not theoretical ones – will be to design models to predict probabilities and not average concentration?

B. SAWFORD: I think that it is difficult to know where we are going in terms of being able to properly build models of higher order properties of plumes and dispersion, especially in a practical sense. My feeling is that we will eventually fall back on the sorts of things that Steve showed the other day, which are fairly simple models which have lots of problems with them. But because the problems we are dealing with are so noisy, so hard to measure properly, in a practical sense those simple models are the best models. Making a model more complicated doesn't necessarily improve things, especially from a regulatory point of view. When you necessarily improve things, especially from a regulatory point of view. When you are asking fairly simple questions, even when you ask about variation of concentrations and so on, you are still going to be talking in fairly broad terms. Whenever you try and make specific comparisons in the atmosphere, you run into the enormous variability, aside from variability of the turbulent part itself, of all the other factors that Steve discussed in his paper.

R. STULL: I think that part of the work that researchers do is based on the regulatory environment of what is required. Is a 1/2 hour or a 1 hour average concentration required? Do we not care about fluctuations in the regulatory sense? If that's the case, then a lot of our effort, spent in finding mean concentration values, has been driven by certain needs – by regulatory needs, among other things. If these regulatory needs change, then our research goals might change.

P. CHATWIN: It seems to me that we know something that most regulatory authorities don't know – quite often what regulatory authorities want is nonsense, scientifically. I actually feel that given the need to chase dollars, there is a conflict between the need to get the dollars for research and what we think is proper research. I am constantly faced with this conflict. I feel that we as a community, and I'm talking as scientists and engineers, not as regulators, ought to oppose this and say that what is really wanted is models of good scientific sense, based on sound physics. Perhaps I will say more about that later.

S. HANNA: I think that, unfortunately, we are dealing with a complex question, involving scientific, social, political, economic, and legal issues. In the United States, lawyers usually make the regulations. We have suggested to the EPA that concentration fluctuations are important and that they ought to account for PDF's and so on. However, whenever EPA scientists bring this up with the lawyers who make the rules on air quality, they just don't want to deal with the uncertainties that are implied. Lawyers like absolute truths, and don't want to hear that the mean concentration is 100 $\mu g/m^3$ with a 95% confidence range from 50 to 200 $\mu g/m^3$.

P. CHATWIN: That's exactly the point.

M. GRABER: I am speaking as a regulator. Laws are being written by lawyers but they know nothing about lognormal distributions, etc. They put into the law whatever the scientists tell them, and if it is a 1/2 hour average or a 1 hour average, or this type of distribution, or another, it should come from the scientific community, who have to convince the lawyers to write those laws according to what science dictates. I think when it comes to regulations it is always a compromise based on the law, and input from lawyers, and what the scientific discipline has to offer. That is my experience.

P. CHATWIN: I think this is exactly right. There is obviously a keen scientific interest in fluctuations and in the statistical nature of gas dispersion which is not reflected in the laws and regulations. Governments use science which is usually at least 10 or 15 years old. I think that it is difficult to get the lawyers to accept proper advice, given the present scientific controversy.

M. GRABER: They have such difficulty in understanding us. You need lots of patience to tell a lawyer, even if he is intelligent (and most of them are intelligent), to understand what you mean by those numbers and formulas. It can be done, but it takes a lot of effort and patience.

S. HANNA: This discussion points out that there is a great need for communicators who can translate the findings of the scientists into something that can be understood by lawyers and law-makers.

M. POREH: I want to make a confession. I don't study theoretical diffusion problems because I want to provide prediction tools for regulators. Sometimes we don't tell them all we know, so they will continue to support our study. For example, we don't know whether concentration fluctuations are really the decisive factor, in fact, on human beings concerning, for example SO_2.

I think we can agree on one thing – that turbulent diffusion in general, and in the atmosphere in particular, is a difficult problem. We must be very careful about using methods that don't have a theoretical basis. On the other hand we realize that such methods limit perhaps our ability to provide theoretical methods to analyze such non-linear problems. The overall capability of numerical work is increasing, when for $10000, you can put a powerful computer or two on your desk. Most of the students feel, and I think they are right, that they want to work on theoretical modeling. They would rather not worry about experimental work, since it is much more difficult, and you need more funding. Universities provide computers, they don't provide equipment for large experiments. We should realize that most of the numerical schemes are based on semi-empirical closures, and therefore we do need data. Field studies are a must; after all, you want to study the real atmosphere. However we must realize the inherent limitations of these studies. For example, because of the diurnal cycle, we don't get to steady state. Sometimes we don't get good averages, if averaging time is not sufficient, particularly for events that do not occur very often, such as relatively high concentrations. When I looked at some laboratory data, I found that once in 60000 points I got a value of instantaneous fluctuations that was 15 times the average. However, when I calculate the probability that this will happen in the atmosphere, it is clear that this fluctuation is very difficult to find at a particular location.

With respect to small-scale physical modeling, as an experimentalist I celebrate the success of the most simple physical model – the water tank, which is not correct from a Reynolds number point of view, but which does provide a basic understanding of the complicated convective boundary layer. I was happy to hear the results of the CONDORS experiment; actually the main conclusion was that the Deardorff water tank experiments were correct. I know they are not perfect, and we must realize that physical models are models. They do not and will never provide a full, accurate description of the atmosphere. The Reynolds number is not correct. However water tanks provide good data, controlled data, with which we can test our models, by changing the Reynolds number and seeing how the model can predict something that we know very well. This is not the case in the atmosphere. I think we have a lot of work to do together.

When I look back, a lot has been accomplished in the last twenty years, and I hope that we can say so 10 years from now.

G. KALLOS: This discussion seems like the discussions we have in our department once or twice a year about experimental, modeling, or theoretical work. I don't think it is true that there is a gap between theory, modeling, and observations. This is something we have to accept. If something is not in our field, we have to look carefully and maybe think about it – someone may come and present a new idea. We cannot reject it immediately. Both methods, experiment and theory, are tools. First of all we have to define some basic concepts of the problem we are studying and then we look at the theories. Then we look at the tools available, and then we choose the right tools. For example, if we have a series of screwdrivers and we want to repair a watch, we need a specific screwdriver. If we want to repair a car we need another screwdriver. It is the same for the experimental work as it is for the modeling work. For an experimental work I would ask some basic questions: what is the time scale or the space scale of the phenomenon we are studying? How representative are the selected monitoring locations? The same applies for the models. Is the right physics included in the equations? What are the scales and how are they described by a model? What are the assumptions we make? If we cannot control these, the results (and consequently our conclusions) have nothing to do with reality. What is the behavior of the natural phenomena? We are trying to describe nature on a piece of paper. What is needed is a careful design first, according to the key parameters, the development of a conceptual model, and finally the selection of the most appropriate tools. This selection must be done according to the status of knowledge. It does not matter if it is experimental or modeling. I think that the combination of both approaches always gives better results. Therefore, there is need for cooperation between experimental and modeling groups.

M. POREH: I fully agree except that we have to realize that we are not free in reality to choose always the right technology, because we would have to be experts in the technology. The example of the large eddy simulation is very striking. It appears that most scientists cannot afford to take this approach. When you look, you find that this technology has been used particularly in large research centers.

G. KALLOS: I think this is the art of science, to choose the most appropriate tools for the specific problem. We cannot rely only on experimental measurements or only on modeling results. In experimental work, we can make the same mistakes. Let's say that we choose the wrong way of sampling, or the wrong sensor or whatever, and the whole thing collapses. It is not only the modeling work which suffers from such mistakes. I strongly disagree with people who say that they evaluate model results with some experimental results. I will never accept this argument, or this methodology. For me, it is something in between. I can say that this is a comparison between experimental work and modeling work, and not evaluation. Experimental results suffer a lot from time and space representativeness. I agree with Dr. Poreh on the cost of the technology we want to use. This problem is more severe in experimental studies.

G. BRIGGS: If I can add to that: social and regulatory needs are always far beyond the solidly established science, so we are always forced to extrapolate models further than data would permit. An awful example is the extrapolation of Pasquill's original curves 100 times beyond the distance of most of the data – all stable and unstable curves were based on the Prairie Grass experiment, which only extended to 800 m from the source. But when the regulatory need was for distances of hundreds of km, someone was not afraid on logarithmic coordinates to extend Pasquill's lines rather drastically. I agree that experiments can never give us all the information that we need. My own experiment, CONDORS, as an example, tests only the passive case, while most sources in the real world are buoyant. But the intent of this game is to build confidence in both theoretical and laboratory models and in numerical simulations. Laboratory tanks can be used for far more varieties of experiments that I can undertake in the field, for 1/4 million dollars or so. This is true likewise for large eddy simulations. But only through field experiments can we find the weaknesses and limitations in these tools.

H. VAN DOP: I would like to look at some current problems. We all know that K-theory fails in many cases. We are looking for alternatives to that theory in order still to be able to obtain a reasonable description of the transport processes in the atmospheric boundary layer. An example is for instance the transilient turbulent theory which is a way to get a better description of transport. Some say that Lagrangian models are an alternative to describe transport and that they have features which are not included in the K-theory. I think that modern theories have certain advantages and whether you want to apply specific theories or not depends on the practical situation where you want to apply them. To come to my second point: In my view we really need many more observations and much more interaction with experimentalists, involved in cloud physics and transport processes. Here we have a case that is a combination of processes. It is turbulence, buoyant transport and condensation where various phases are involved. The work is all based on well established theory: the turbulent dynamic and thermodynamic equations. So basically the theory is there, but it is the synthesis that is lacking, which is the main point that I want to make. Atmospheric science, contrary to other sciences, is a science where we have to live from synthesis, from putting things together and it is unlike many other fields where a problem is approached by breaking it into tiny pieces. We look at these tiny pieces in detail keeping everything else constant and well defined. That is analytical science. In the atmosphere we are doing "synthetical" science. We have to deal with many processes at the same time: turbulent transport, thermodynamics, chemistry, and radiation. These processes influence each other and that is where the problem arises. What is going on in the atmosphere is a combination of thermodynamic, chemical and turbulent processes in which each, on its own, is quite well understood, but put together, the interacting system is largely not understood. For instance, take the ocean and atmospheric circulation in climate science; we are not able to predict the future state of the atmosphere, not only with regard to physical parameters like temperature or cloud cover, but also with regard to chemical composition. We know roughly what the chemical composition of the atmosphere is now, but nobody dares to say what it will be in 50 years, not because we do not understand the science, but because we have a problem of putting it together. That is where we need better links with experimentalists. We use insufficient observations and that is why we, the modelers, have so much freedom; there are no self-correcting observations. That is my main remark but I have an additional point. Because of the complexity of models, verification is often very difficult. In a simple analytic experiment you can say: "this is my observation, this is my theory," compare them and you can say whether the theory is wrong or right. In complex atmospheric experiments there are often so many parameters involved that it is hard to tell whether you have made a satisfactory model. So, we are often accepting models by consensus. That means that a group of peers agrees on this or that model, or on specific techniques and parameterizations, which are considered to be up to the common standard of knowledge. Therefore, we accept a model which contains pieces of these "agreed" elements as an acceptable model. That is a way of deciding upon model performance which you will not encounter in other sciences.

W. SADEH: I would like to make a comment on an aspect of turbulent diffusion that bothers me. If I go back to conferences of 20 years ago and, unfortunately, I do not have here a tape recording of these conferences, I feel that we had exactly the same discussions. I remember the first time I met Steve Hanna more than 15 or 20 years ago and then we had the same type of discussions: we need one more experiment (!), we need one more model (!), what is our objective? I think we are faced here with a problem which is much more basic. We have been measuring the same things for many years. The differences between the results of measurements of the plumes that I have seen at this conference is that we have better instrumentation and better computers to analyze the data. The question is: **Do we have better understanding and or knowledge?** We have models, we try to develop one model after another, and we make measurements to death! It is much easier to develop a model if we do an experiment and collect some data and to publish a paper instead of sitting down and asking a couple of basic and simple questions: **What do we know and what we do not know?** and, **what we are missing in our understanding of the problem?** And, I would like to give an example. To some extent, all of us are in the business of a concept called "boundary layer" introduced at the beginning of the century by Prandtl. We have a concept, a good concept. And, we try to see how this concept can be applied to turbulent diffusion in the atmosphere, and we are trying hard to apply this concept to solve a host of specific problems. The next question is: Do we have an objective to try to gain a basic understanding of turbulent diffusion? When my students ask me this question I have a simple answer: **We are fortunate that we are in this field because we shall be in business for life!** The chances of a general understanding or a general solution are so minimal that you have to look at each individual problem and solve it. If we were to present the turbulent diffusion problem in those terms to the regulators, the lawyers, and the politicians, we would have a much better chance of success. I do not see in the foreseeable future any way to achieve a full understanding of turbulence or of turbulent diffusion that will enable us to come up with an equation, like some basic fluid mechanics equations such as the Bernoulli equation, that will give us a reasonable operational answer. Let's face the facts. The problem of turbulent diffusion is so complex that we have to do more basic thinking and less measurements and modeling. The issue is that the problem is so complex and of such a nature that in all likelihood a general approach is unfeasible, and not realistic. If we look at each individual problem and if we try to solve each individual problem through a synergism of thinking, analysis, computer simulation, well designed and planned measurements and modeling we will be much better off. But, the thinking about the problem prior to computer simulation, modeling and measurements is what is first needed.

H. VAN DOP: Of course I agree with this statement. Apart from a few details you are right. All the laws and all our knowledge will not improve the situation very much because these laws are already known. It is the synthesis which is the big problem but I do not agree with you that there will not be any progress, because 20 years ago the weather prediction was for one day ahead but now it is for six days ahead with equivalent skill. So there is progress. And you can say that physical insight has not changed since that time, but progress has come as an improved understanding of a complex system.

W. SADEH: For years I was a weather forecaster, and I remember using synoptic maps, drawing isobars and making forecast for pilots. When I look at the forecasting system today, it is very sophisticated. To what extent is it better than the simple forecasts that I gave to pilots 25 years ago?

P. ALPERT: I want to comment on that. I think that we, like all people, like to talk about forecasts. And I think what Han van Dop just said is true. We do have 5 to 6 day forecasts for the general flow of the atmosphere that have the same accuracy we had 20 years ago for one day forecasts.

P. CHATWIN: I think that you will not be surprised that many of the points I wanted to make have already been made. Like Michael, my comments are basically driven by my prime interest, to understand better the behavior of the concentration field in turbulent flow, and particularly properties other than the ensemble mean. I believe that from the scientific point of view, a statistical description is essential. As we have already discussed, I do believe that, eventually, it will be recognized, not only by scientists, engineers, meteorologists and oceanographers, but also by lawyers and politicians, that scientific descriptions, even of weather forecasts, ought to be statistical. The probability that it will rain tomorrow in Eilat is very low, although the probability that it will rain in London is 50 percent (or 75 percent). The recognition of that as a correct sort of description seems to me to be progress. Like everyone else on the panel, I stress that experiments are essential. I feel like Professor Kallos, nevertheless, that there is a need to recognize that experiments are not gospel. We must ask certain questions about them, as well as models. I just want to mention three points that were in my paper and I do think they are important. Steven Hanna, Brian Sawford, and one or two others already referred to the needs for clear objectives in investigations, and I think it is important in doing experiments involving turbulence to make sure we define the underlying ensemble. I also think that it is important to investigate, ask questions about, and check that we know the relationship between the measured concentrations and the real concentrations. I also believe that we have not perhaps been as critical or as careful in investigating phenomena like instrument smoothing and the treatment of noise. I want to emphasize again that molecular diffusion is a key process. If I could just single out one sort of experiment that would interest me (and, I think, many others), it would be that we pay more attention to the effect of the source, and particularly source size, on the structure of the concentration fluctuations.

M. GRABER: Could you explain your last comment on the size of the source?

P. CHATWIN: The concentration distribution you observe when you release contaminants in turbulent flow comes from a source; it may be an explosion, or may be a continuous release like from a smoke stack. I think that it is well understood theoretically, that the bigger the source, roughly speaking, the smaller the concentration fluctuations. I can give you references to that. As far as I know, what is not known is how long these effects persist, and what parameters are most important. I think that this is a general point that has not yet been investigated to an adequate extent. There have been experiments, but not very many.

P. ALPERT: I have a question regarding the transilient matrix method. In a sense it is not different than trying to solve the turbulence problem in second, third, and higher order closure. I see the mechanistic way of thinking in the transilient method, but I don't see how the principles of physics. In contrast, I do see how the principles of physics are used in the second, third, or higher order closure.

R. STULL: The transilient matrix is just a statistic. It allows you to retain some measure of non-local mixing, which you don't have in most local methods. When you are looking for a turbulent closure method in the transilient matrix, you can try to close this non-local mixing. In the transilient method, you can describe this effect, whereas in the second-order closure method, everything is local. Of course, with second-order closure, people hope that some of the non-local mixing is brought in through the higher order terms, but in reality, the closure is entirely local. You're taking only the local gradients or the mean values, or it could be the second or third moments. As we have seen by the CONDORS experiment, there are very many non-local processes, and a lot of people are trying to find different ways to capture the large eddy, or the coherent structure of the non-local process. Whether it is a CONDORS experiment, where we see the deterministic results, or a stochastic simulation with the Langevin formula, or the transilient approach, there are many different ways to capture the coherent structure and its effects, which are not captured in most of the K- theory models.

M. POREH: I'd like to make a comment regarding the pessimism that has been voiced. The type of questions we are asking today are much more advanced and sophisticated than the ones we asked 20 years ago. Then, we did not understand the convective boundary layer, and used σ_z and σ_y curves. We are embarrassed today that we used them. We didn't know how to predict well the stability of the atmosphere because we used discrete stability classes. The contributions of Jeff Weil showed clearly that we were mistaken. We interpreted many stable cases as neutral. Our basic understanding is on a much higher level than 20 years ago.

S. HANNA: I would like to get back to what Dr. Kallos said. Some model developers have pointed out that the mesoscale numerical model predicts an average over a grid, whereas the observation is only at a point. Therefore, you really shouldn't be attempting to evaluate the predictions of the mesoscale model with a point measurement. It is sometimes claimed that the model produces better results than the point measurements. However, I have always felt that the point data must be worth something in the evaluations. How should observations from a mesoscale wind network be used to evaluate the predictions of a mesoscale model?

G. KALLOS: If I want to check the performance of my model (if it is good or bad, close to reality or not) I need a set of data which must have a certain scale. When we put in a monitoring station, we have two options: to put it on the roof of the building or next to it; of course the results are completely different. What I would like to see sometimes is a better way, a method of filtering the data, for example, to obtain a data set which will be representative in a certain space and time scale comparable to that of the model used. Mesoscale and several other scales of models provide in general a better picture of a phenomenon because of the time and space resolution which cannot be obtained even with a very dense monitoring network. It is difficult to tell whether a model always gives better results. I never said that, and I don't believe in that. My disagreement is in assuming that experimental results are always true because they are measured. Evaluation of experimental measurements is equally important as model evaluation.

P. CHATWIN: Such a model is usually deterministic, but what you are looking at is one realization of a statistical process. Experimenters (and regulators) should ask modelers to provide estimates of the statistical uncertainties associated with their model predictions.

S. HANNA: I think we have reached the end of our time and I would like to thank you for your participation in this session.

AUTHOR INDEX

The manufacturer's authorised representative in the EU is Springer
Nature Customer Service Centre GmbH, Europaplatz 3, 69115 Heidelberg,
Germany. If you have any concerns regarding our products, please
contact ProductSafety@springernature.com

Printed and bound by CPI Group (UK) Ltd, Croydon, CR0 4YY
30/07/2026
02182028-0007